国外优秀数学著作
原版系列

U0063167

微分方程与包含的拓扑方法（英文）

Topological Methods for Differential Equations and Inclusions

［美］约翰·R.格雷夫 (John R. Graef)
［美］约翰尼·亨德森 (Johnny Henderson)
［阿］阿卜杜勒加尼·奥哈比 (Abdelghani Ouahab)
著

哈尔滨工业大学出版社
HARBIN INSTITUTE OF TECHNOLOGY PRESS

黑版贸登字 08-2021-061 号

图书在版编目(CIP)数据

微分方程与包含的拓扑方法 = Topological Methods for Differential Equations and Inclusions:英文/(美)约翰·R. 格雷夫(John R. Graef),(美)约翰尼·亨德森(Johnny Henderson),(阿)阿卜杜勒加尼·奥哈比(Abdelghani Ouahab)著. —哈尔滨:哈尔滨工业大学出版社,2024.1

ISBN 978-7-5767-0940-7

Ⅰ.①微…　Ⅱ.①约…②约…③阿…　Ⅲ.①微分方程-英文②拓扑-英文　Ⅳ.①O175②O189

中国国家版本馆 CIP 数据核字(2023)第 154427 号

WEIFEN FANGCHENG YU BAOHAN DE TUOPU FANGFA

策划编辑　刘培杰　杜莹雪
责任编辑　刘立娟　李兰静
封面设计　孙茵艾
出版发行　哈尔滨工业大学出版社
社　　址　哈尔滨市南岗区复华四道街 10 号　邮编 150006
传　　真　0451-86414749
网　　址　http://hitpress.hit.edu.cn
印　　刷　哈尔滨市石桥印务有限公司
开　　本　787 mm×1 092 mm　1/16　印张 24.5　字数 345 千字
版　　次　2024 年 1 月第 1 版　2024 年 1 月第 1 次印刷
书　　号　ISBN 978-7-5767-0940-7
定　　价　98.00 元

(如因印装质量问题影响阅读,我社负责调换)

Dedications

John Graef dedicates this work to his wife Frances and the memory of his parents.

Johnny Henderson dedicates this work to his wife Darlene.

Abdelghani Ouahab dedicates this work to his parents, wife Zohra, his chidren Hemza, Fatima-Zohra, and Abdelhka, his sisters, brothers, and all the members of his family (Ouahab, Hidaoui, Baliki). He also wishes to express his gratitude to Professors Toufika Mossaoui and Smaïl Djebel for inviting him to the Department of Mathematics at E.N.S. Kouba-Algeria during 2015-16 to give a course in multivalued analysis and fixed point theory for post graduate students. He also thanks Professors Yousfate Abderrahmane and Mouffak Benchohra for their encouragement.

The authors thank the team at Taylor and Francis Group/CRC Press and Editor Sarfraz Khan for their interest in this project.

Contents

Introduction

There are a great variety of motivations that lead mathematicians to study dynamical systems having velocities uniquely determined by the state of the system and loosely related to it. Broadly speaking, the motivation for a part of this study involves replacing differential equations such as

$$y' = f(y),$$

where f is a single-valued function, by differential inclusions

$$y' \in F(y), \tag{0.1}$$

where F is a multivalued function. A system of differential inequalities

$$y'^i \le f^i(x_1, \ldots, x_n), \ i = 1, \ldots, n,$$

can also be considered as a differential inclusion. If an implicit differential equation

$$f(y, y') = 0$$

is given, then we can put $F(y) = \{v : f(y, v) = 0\}$ to reduce it to a differential inclusion. Differential inclusions are used to study ordinary differential equations with a relatively unknown right-hand side.

Example 0.1. *Consider the equation with a discontinuous right-hand side*

$$y' = 1 - 2\operatorname{sgn}(y),$$

where

$$\operatorname{sgn} y = \begin{cases} +1, & \text{if } y > 0, \\ -1, & \text{if } y < 0. \end{cases}$$

The classical solution of above problem is given by

$$y(t) = \begin{cases} 3t + c_1, & \text{if } y < 0, \\ -t + c_2, & \text{if } y > 0. \end{cases}$$

As t increases, the classical solution tends to the line $y = 0$, but it cannot be continued along this line, since the map $y(t) = 0$ so obtained does not satisfy the equation in the usual sense (for $y'(t) = 0$ and the right-hand side has the value $1 - 2\operatorname{sgn} 0 = 1$). Hence there are no classical solutions of the initial value problem starting with $y(0) = 0$.

Therefore a generalization of the concept of a solution is required. To get around discontinuous right-hand side nonlinearities, the problem may be regarded as a differential inclusion

$$y'(t) \in F(y(t)), \quad \text{a.e } t \in [0, \infty), \quad y(0) = y_0, \tag{0.2}$$

where $F : \mathbb{R}^n \to \mathcal{P}(\mathbb{R}^n)$ is a vector set-valued map into the set of all subsets of $\mathbb{R}^n$, which can be defined in several ways. The simplest convex definition of F is obtained by the so-called Filippov regularization [117]:

$$F(y) = \bigcap_{\varepsilon > 0} \overline{co\,(f(\{y \in \mathbb{R}^n |\, \|y\| \le \varepsilon\} \backslash M))},$$

where $conv\, F(y)$ is the convex hull of f, M is a null set (i.e $\mu(M) = 0$, where μ denotes the Lebesgue measure in $\mathbb{R}^n$), and ε is the radius of the ball centered at y.

In the 1930s, differential inclusions were initiated by the Polish and French mathematicians Zaremba and Marchaud [209,292,293]. They were mostly interested in existence results and some qualitative properties. Zaremba and Marchaud studied respectively the so-called paratingent equations and contingent equations. In the 1960s, Ważewski [284–286] proved that one can use more "classical" solutions than contingent or paratingent ones since the solutions are understood in the Carathéodory sense as absolutely continuous functions satisfying (0.1) almost everywhere. One of the most important examples of differential inclusions comes from control theory. Filippov and Ważewski have considered the control system

$$y'(t) = f(y, u), \ u \in U,$$

where u is a control parameter. It appears that the control system and the differential inclusions

$$y' \in F(y, U) = \bigcup_{u \in U} f(y, u)$$

have the same trajectories. If the set of controls depends on y, that is $U = U(y)$, then we obtain the differential inclusion

$$y' \in F(y, U(y)).$$

The equivalence between a control system and the corresponding differential inclusion is the central idea used to prove existence theorems in optimal control theory. Since the dynamics of economical, social, and biological systems are multi-valued, differential inclusions serve as natural models in macro systems with hysteresis. A differential inclusion is a notion which generalizes an ordinary differential equation. Therefore all problems considered for differential equations such as existence of solutions, continuation of solutions, dependence on initial conditions and parameters, are present in the theory of differential inclusions. Since a differential inclusion usually has many solutions starting at a given point, new issues appear, such as investigation of topological properties of the set of solutions, selection of solutions with given properties, evaluation of the reachability sets (viability theory), etc. In order to solve the above problems, special mathematical techniques were developed. Differential inclusions have been the subject of an intensive study by many researchers in recent decades; see for example [24,33,34,138,173,180,275]. For further readings and details, we refer to the books by Andres and Górniewicz [24], Aubin and Celina [33], Aubin and Frankowska [34], Deimling [97], Górniewicz [138], Hu and Papageorgiou [173], Kamenskii [180], and Tolstonogov [275].

The contents of this book are laid out as follows. In Chapter 1 we present the essential background in multivalued analysis that will be utilized in the remainder of the book. In addition to introducing basic concepts such as upper and lower semicontinuity of multivalued functions, weak continuity, open, closed, and compact maps, and sums and products of semicontinuous maps, much of the the notation to be used is also explained. A section on examples of multivalued maps is included here. Chapter 2 begins with a description of the Hausdorff-Pompeiu metric topology. Relationships to the Vietoris topology are discussed, and some fixed point theorems for maps on complete metric spaces are also given.

The measurability of multivalued maps is the main focus of Chapter 3. The existence of measurable selection functions plays a major role in this chapter as well. In this regard, the Lusin property and the Scorza-Dragoni property play important roles here too. Chapter 4 is devoted to the study of continuous selection theorems with special attention paid to Michael's selection theorem. The important topic of linear multivalued operators is the subject of Chapter 5.

In Chapter 6 a variety of fixed point theorems are presented including those of Kakutani, Fan, Leray and Schauder, Tychonoff, and Krasnosel'skii, as well as a number of variations of Krasnosel'skii's theorem including some in Fréchet spaces. In this chapter, measures of noncompactness are introduced into the discussion. Fixed point theorems for the sum of two multivalued operators and for operators in topological vector spaces are also examined. Generalized metric and Banach spaces are studied in Chapter 7. It is also in this chapter where the notion of convergent matrices is introduced.

Fixed point theorems in vector metric spaces are examined in Chapter 8. This includes versions of Schaefer's fixed point theorem, the nonlinear alternative of Leary-Schauder type, Perov's theorem, and the Covitz and Nadler fixed point theorem.

In Chapter 9 we study systems of impulsive differential equations on the half-line.

Chapter 10 is devoted to random fixed point theorems especially those of Krasnosel'skii and Perov types.

In Chapter 11 we introduce the study of semigroups and in Chapter 12 linear and semilinear systems of differential inclusions are studied. Inclusions with impulse effects are included here as well.

Random systems of differential equations are considered in Chapter 13. In particular we examine the Cauchy problem and boundary value problems for coupled systems of the second order. In Chapter 14, random fractional differential equations involving the Hadamard fractional derivative are studied.

In Chapter 15 we turn our attention to difference equations. Existence of solutions to initial value problems is considered as are questions of the compactness of the solution space.

Existence of solutions and the compactness of the solution space for discrete inclusions are considered in Chapter 16. Semilinear systems of discrete equations are considered in Chapter 17. Boundary value problems for systems of second order difference equations are examined in Chapter 18.

There is a bibliography of more than 300 relevant references.

Chapter 1

Background in Multi-valued Analysis

1.1   Some notions and definitions

Let E be topological space (metric space, normed space or locally convex spaces). Denote by $\mathcal{P}(E) = \{Y \subset E : Y \neq \emptyset\}$, $\mathcal{P}_{cl}(E) = \{Y \in \mathcal{P}(E) : Y \text{ is closed}\}$, $\mathcal{P}_b(E) = \{Y \in \mathcal{P}(E) : Y \text{ is bounded}\}$, $\mathcal{P}_{cv}(E) = \{Y \in \mathcal{P}(E) : Y \text{ is convex}\}$, and $\mathcal{P}_{cp}(E) = \{Y \in \mathcal{P}(E) : Y \text{ is compact}\}$. Let X and Y be two spaces and assume that for every point $x \in X$ a nonempty closed subset $\varphi(x)$ of Y is given (sometime we will assume only that $\varphi(x) \neq \emptyset$); in this case, we say that φ is a *multivalued mapping* from X to Y and write $\varphi \colon X \to \mathcal{P}(Y)$. More precisely a multivalued map $\varphi \colon X \to \mathcal{P}(Y)$ can be defined as a subset $\varphi \subset X \times Y$ such that the following condition is satisfied:

$$\text{for all } x \in X \text{ there exists } y \in Y \text{ such that } (x, y) \in \varphi.$$

In what follows, the symbol $\varphi \colon X \to Y$ is reserved for single-valued mappings, i.e. $\varphi(x)$ is a point of Y. Let $\varphi \colon X \to \mathcal{P}(Y)$ be a multivalued map. We associate with φ the *graph* (interchangeably, we use the notations $\Gamma(\varphi)$ or Γ_φ) of φ by putting:

$$\Gamma(\varphi) = \{(x, y) \in X \times Y \mid y \in \varphi(x)\}$$

as well as two natural projections $p_\varphi \colon \Gamma_\varphi \to X$, $q_\varphi \colon \Gamma(\varphi) \to Y$ defined as follows: $p_\varphi(x, y) = x$ and $q_\varphi(x, y) = y$, for every $(x, y) \in \Gamma(\varphi)$.

The point-to-set mapping $\varphi \colon X \to \mathcal{P}(Y)$ extends to a set-to-set mapping by putting:

$$\varphi(A) = \bigcup_{x \in A} \varphi(x), \quad \text{for } A \subset X;$$

then $\varphi(A)$ is called the *image* of A under φ. If $\varphi \colon X \to \mathcal{P}(Y)$ and $\psi \colon Y \to \mathcal{P}(Z)$ are two maps, then the *composition* $\psi \circ \varphi \colon X \to \mathcal{P}(Z)$ of φ and ψ is defined by:

$$(\psi \circ \varphi)(x) = \bigcup \{\psi(y) \mid y \in \varphi(x)\}, \quad \text{for every } x \in X.$$

If $X \subset Y$ and $\varphi \colon X \to \mathcal{P}(Y)$, then a point $x \in X$ is called a *fixed point* of φ provided $x \in \varphi(x)$. Let

$$Fix(\varphi) = \{x \in X \mid x \in \varphi(x)\}.$$

For $\varphi \colon X \to \mathcal{P}(Y)$ and any subset $B \subset Y$, we define the small counter image $\varphi_-^{-1}(B)$ and the large counter image $\varphi_+^{-1}(B)$ of B under φ as follows:

$$\varphi_-^{-1}(B) = \{x \in X \mid \varphi(x) \cap B \neq \emptyset\}, \quad \varphi_+^{-1}(B) = \{x \in X \mid \varphi(x) \subseteq B\}.$$

It will be convenient to use the above definitions for arbitrary sets B, that is, we do not necessarily require that $B \subset Y$. For $\varphi \colon X \to \mathcal{P}(Y)$ we define $\varphi^{-1} \colon Y \to \mathcal{P}(X)$ by

$$\varphi^{-1}(y) := \varphi_+^{-1}(\{y\}) = \{x \in X : y \in \varphi(x)\}.$$

In particular, if $f : X \to Y$ is single-valued, then by our tacit identification of single-valued and multivalued maps, we have $f^{-1} : Y \to \mathcal{P}(X)$, in this case

$$f^{-1}(y) := \{x \in X : y = f(x)\} = f_+^{-1}(\{y\}) = f_-^{-1}(\{y\}).$$

Proposition 1.1. *For $\varphi : X \to \mathcal{P}(X)$ we have*

$$\Gamma(\varphi^{-1}) := \{(y, x) \in Y \times X : (x, y) \in \Gamma_\varphi\}.$$

In particular, $(\varphi^{-1})^{-1} = \varphi$, $dom(\varphi^{-1}) = \varphi(X)$, $dom(\varphi) = \varphi^{-1}(Y)$.

Proof. For fixed $y \in Y$ the set $\varphi^{-1}(y)$ consists of all $x \in X$ with $y \in \varphi(x)$, that is, with $(x, y) \in \Gamma(\varphi)$. $\qquad\square$

If $\varphi : X \to \mathcal{P}(Y)$ and $A \subset X$, then by $\varphi|_A : A \to \mathcal{P}(Y)$ we will denote the restriction of φ to A. If, moreover, $\varphi(A) \subset B$, then the map $\widetilde{\varphi} : A \to \mathcal{P}(B)$, $\widetilde{\varphi}(x) = \varphi(x)$ for every $x \in A$ is the contraction of φ to the pair (A, B).

Below, we will summarize properties of images and counter-images.

Proposition 1.2. *Let $\varphi : X \to \mathcal{P}(Y)$ be a multivalued map, $A \subset X$, $B \subset Y$, and $B_j \subset Y$ for $j \in J$; then we have:*

$$\varphi_+^{-1}\big(\varphi(A)\big) \supseteq A,$$
$$\varphi\big(\varphi_+^{-1}(B)\big) \subseteq B,$$
$$X \setminus \varphi_-^{-1}(B) = \varphi_+^{-1}(Y \setminus B),$$
$$X \setminus \varphi_+^{-1}(B) = \varphi_-^{-1}(Y \setminus B),$$
$$\varphi_+^{-1}\left(\bigcup_{j \in J} B_j\right) \supset \bigcup_{j \in J} \varphi_+^{-1}(B_j),$$
$$\varphi_+^{-1}\left(\bigcap_{j \in J} B_j\right) = \bigcap_{j \in J} \varphi_+^{-1}(B_j),$$
$$\varphi_-^{-1}\big(\varphi(A)\big) \supset A,$$
$$\varphi\big(\varphi_-^{-1}(B)\big) \supset B \cap \varphi(X),$$
$$X \setminus \varphi_-^{-1}(B) = \varphi^{-1}(Y \setminus B),$$
$$\varphi_-^{-1}\left(\bigcup_{j \in J} B_j\right) = \bigcup_{j \in J} \varphi_-^{-1}(B_j),$$
$$\varphi_-^{-1}\left(\bigcap_{j \in J} B_j\right) \subset \bigcap_{j \in J} \varphi_-^{-1}(B_j).$$

The proof of Proposition 1.2 is straightforward and we leave it to the reader.

For given maps $\varphi, \psi : X \to \mathcal{P}(Y)$, we define $\varphi \cup \psi : X \to \mathcal{P}(Y)$ and $\varphi \cap \psi : X \to \mathcal{P}(Y)$ as follows:

$$(\varphi \cup \psi)(x) = \varphi(x) \cup \psi(x) \quad \text{and} \quad (\varphi \cap \psi)(x) = \varphi(x) \cap \psi(x),$$

for every $x \in X$. Of course the map $\varphi \cap \psi$ is defined provided that $\varphi(x) \cap \psi(x) \neq \emptyset$, for every $x \in X$. As an easy observation, we obtain:

Proposition 1.3. *Let $\varphi, \psi : X \to \mathcal{P}(Y)$ be such that $\varphi \cap \psi$ is defined and let $B \subset Y$; then we have:*

$$(\varphi \cup \psi)_+^{-1}(B) = \varphi_+^{-1}(B) \cap \psi_+^{-1}(B),$$
$$(\varphi \cap \psi)_+^{-1}(B) \supset \varphi_+^{-1}(B) \cup \psi_+^{-1}(B),$$
$$(\varphi \cup \psi)_-^{-1}(B) = \varphi_-^{-1}(B) \cup \psi_-^{-1}(B),$$
$$(\varphi \cap \psi)_-^{-1}(B) \subset \varphi_-^{-1}(B) \cap \psi_-^{-1}(B).$$

If we have two maps $\varphi \colon X \to \mathcal{P}(Y)$ and $\psi \colon Y \to \mathcal{P}(Z)$, then for any $B \subset Z$ we obtain:

Proposition 1.4.

$$
\begin{aligned}
(\psi \circ \varphi)_+^{-1}(B) &= \varphi_+^{-1}\big(\psi_+^{-1}(B)\big), \\
(\psi \circ \varphi)_-^{-1}(B) &= \varphi_-^{-1}\big(\psi_-^{-1}(B)\big).
\end{aligned}
$$

Finally, let us consider two maps $\varphi \colon X \to \mathcal{P}(Y)$ and $\psi \colon X \to \mathcal{P}(Z)$. Then we define the Cartesian product $\varphi \times \psi \colon X \to \mathcal{P}(Y) \times \mathcal{P}(Z)$ of φ and ψ by putting:

$$(\varphi \times \psi)(x) = \varphi(x) \times \psi(x) \quad \text{for every } x \in X.$$

As an easy observation, we get:

Proposition 1.5. *Let $B \subset Y$ and $D \subset Z$; then we have:*

$$
\begin{aligned}
(\varphi \times \psi)_+^{-1}(B \times D) &= \varphi_+^{-1}(B) \cap \psi_+^{-1}(D), \\
(\varphi \times \psi)_-^{-1}(B \times D) &= \varphi_-^{-1}(B) \cap \psi_-^{-1}(D).
\end{aligned}
$$

1.2 Examples of multivalued mappings

In this section we give some general examples of multivalued maps.

Example 1.6. (Inverse function) *Let $f : X \to Y$ be a single valued map from X to Y. Then its inverse can be considered as a multivalued map $F_f : Y \to \mathcal{P}(X) \cup \{\emptyset\}$ defined by*

$$F_f(y) = f^{-1}(y) = \{x \in X : f(x) = y\}, \ y \in Y.$$

Example 1.7. (Implicit function) *Let $f : X \times Y \to Z$ and $g : X \to Z$ be single valued maps such that for every $x \in X$ there exists $y \in Y$ such that*

$$f(x, y) = g(x).$$

The implicit function defined by f and g is a multivalued map $F : X \to \mathcal{P}(Z)$ given by

$$F(x) = \{y \in Y : \ f(x, y) = g(x)\}.$$

Example 1.8. (Multivalued dynamical systems) *Dynamical systems determined by autonomous ordinary differential equations without the uniqueness property are multivalued maps.*

Example 1.9. *Let $f : X \times Y \to \mathbb{R}$ be a single valued map. Assume there is $r > 0$ such that for every $x \in X$ there exists $y \in Y$ such that $f(x, y) \le r$. Then we can define the multivalued map $F_r : X \to Y$ by*

$$F_r(x) = \{y \in Y : f(x, y) \le r\}.$$

Example 1.10. (Metric projection) *Let A be a compact subset of a metric space (X, d). Then, for every $x \in X$, there exists $a \in A$ such that*

$$d(a, x) = d(x, A).$$

We define the metric projection $P : X \to \mathcal{P}(A)$ by

$$P(x) = \{a \in A : d(a, x) = d(x, A)\}, \ x \in X.$$

Example 1.11. (Bouligand cone) *Let C be a closed subset of $\mathbb{R}^n$. The multivalued map $T_C : \mathbb{R}^n \to \mathcal{P}(\mathbb{R}^n)$ defined by*

$$T_C(x) = \left\{ v \in \mathbb{R}^n : \frac{\lim\limits_{h \to 0^+} \inf d(x + hv, C)}{h} = 0 \right\}$$

is called the Bouligand contingent cone to C at $x \in C$.

Example 1.12. (Implicit differential equations) *Let $f : [a, b] \times \mathbb{R}^n \times \mathbb{R}^n \to \mathbb{R}^n$ be a continuous function. Consider the implicit differential equation*

$$x'(t) = f(t, x(t), x'(t)), \quad t \in [a, b], \tag{1.1}$$

where the solution is understood in the sense of almost everywhere in $[a, b]$. Consider the multivalued map $F : [a, b] \times \mathbb{R}^n \to \mathcal{P}(\mathbb{R}^n)$ defined by

$$F(t, x) = Fixf(t, x, \cdot) = \{y \in \mathbb{R}^n : f(t, x, y) = y\}$$

and the differential inclusion

$$x'(t) \in F(t, x(t)), \quad t \in [0, 1]. \tag{1.2}$$

Clearly, problem (1.1) is equivalent to (1.2).

Example 1.13. (Control problem) *Consider the control problem*

$$x'(t) = f(t, x(t), u(t)), \quad t \in [a, b], \quad x(a) = x_0 \in \mathbb{R}^n, \tag{1.3}$$

where $f : [a, b] \times \mathbb{R}^n \times \mathbb{R}^m \to \mathbb{R}^n$ is a continuous function and $u \in U \subseteq \mathbb{R}^m$. To solve (1.3), we define the multivalued map $F : [0, 1] \times \mathbb{R}^n \to \mathcal{P}(\mathbb{R}^n)$ by

$$F(t, x) = \{f(t, x, u) : u \in U\}.$$

Then solutions of (1.3) are those of the differential inclusion

$$x'(t) \in F(t, x(t)), \quad t \in [0, 1], \quad x(a) = x_0. \tag{1.4}$$

Problem (1.1) is equivalent to (1.2).

Example 1.14. (Duality map) *Let E be a normed space and E^* its dual space. The multivalued mapping $F : E \to \mathcal{P}(E^*)$ defined by*

$$F(x) = \{f \in E^* : \|f\| = \|x\| \text{ and } f(x) = \|x\|^2\}$$

is called the dual map from E into E^*.

1.3   Vietoris topology

Throughout this section, (X, τ) is a Hausdorff topological space (that is, τ denotes the Haudorff topology on X). Given $A \in \mathcal{P}(X)$, we define

$$A^- = \{B \in \mathcal{P}(X) : A \cap B \neq \emptyset\} \text{ (those sets in } X \text{ that "hit" } A)$$

and

$$A^+ = \{B \in \mathcal{P}(X) : B \subseteq A\} \text{ (those sets in } X \text{ that "miss" } A^c).$$

Definition 1.15. • *The "upper Vietoris topology" (denoted by $\widehat{\tau}_{UV}$) is generated by the base*

$$\mathcal{L}_{UV} = \{U^+ : U \in \tau\}.$$

• *The "lower Vietoris topology" (denoted by $\widehat{\tau}_{LV}$) is generated by the subbase*

$$\mathcal{L}_{LV} = \{U^- : U \in \tau\}.$$

• *The "Vietoris topology" (denoted by $\widehat{\tau}_V$) is generated by the subbase $\mathcal{L}_{UV} \cup \mathcal{L}_{LV}$.*

Remark 1.16. *It follows from the above definition, that a basic element for the Vietoris topology $\widehat{\tau}_V$ is given by*

$$B(U, V_1, \ldots, V_n) = \{A \in \mathcal{P}(X) : A \subseteq U, A \cap V_k \neq \emptyset, k = 1, \ldots, n\},$$

where $U, V_1, \ldots, V_n \in \tau$.

The Vietoris topology is "natural" in the following sense.

Lemma 1.17. *If $I : X \to \mathcal{P}(X)$ is the injection map defined by $I(x) = \{x\}$, then $I(\cdot)$ is continuous when $\mathcal{P}(X)$ is equipped with the $\widehat{\tau}_V$-topology.*

Proof. Let $U \in \tau$. Then we have

$$I^{-1}(U^+) = \{x \in X : \{x\} \subseteq U\} = U \in \tau.$$

Similarly, if $V_1, \ldots, V_n \in \tau$, then

$$I^{-1}(\cap_{k=1}^n V_k^-) = \{x \in X : \{x\} \cap V_k \neq \emptyset, k = 1, \ldots, n\} = \cap_{k=1}^n V_k^- \in \tau.$$

Therefore $I(\cdot)$ is continuous into $\mathcal{P}(X)$ with the Vietoris topology $\widehat{\tau}_V$. $\square$

Example 1.18. *The Vietoris topology $\widehat{\tau}_V$ is not the finest topology on $\mathcal{P}(X)$ for which $I(\cdot)$ is continuous.*

To see this, let X be an infinite set equipped with the cofinite topology τ_c

$$\tau_c = \{U : U \backslash X, \text{is finite set}\} \cup \{\emptyset, X\}.$$

Then the closed subsets of X are $\emptyset, X$, and finite subsets of X. Let $\mathcal{F}$ denote the family of nonempty, finite subsets of X. Then $I^{-1}(\mathcal{F})$ is an open set in $(P(X), \widehat{\tau}_V)$ and contains some infinite sets. So $\mathcal{F} \notin \widehat{\tau}_V$ and thus $I(\cdot)$ remains continuous if on $\mathcal{P}(X)$, we consider the stronger topology obtained by $\mathcal{F}$ to the original subset $\mathcal{L}_{UV} \cup \mathcal{L}_{LV}$.

As in the above example, let $\mathcal{F}$ denote the family of nonempty and finite subsets of X.

Proposition 1.19. *The family $\mathcal{F}$ is dense in $(\mathcal{P}(X), \widehat{\tau}_V)$.*

Proof. If $U \in \tau$ is nonempty, then U contains a finite subset and so $U^+ \cap \mathcal{F} \neq \emptyset$. Similarly, if $V_1, \ldots, V_n \in \tau$ are nonempty, let $x_k \in V_k$, $k = 1, \ldots, n$. Then $\{x_k\}_{k=1}^n \in (\cap_{k=1}^n V_k^-) \cap \mathcal{F}$. Thus, $\mathcal{F}$ intersects every element in the base of $\widehat{\tau}_V$ and so $\mathcal{F}$ is dense as claimed. $\square$

An immediate interesting consequence of the above proposition is the following lemma.

Lemma 1.20. *([173]) If (X, τ) is a separable Haudorff space, then $(\mathcal{P}(X), \widehat{\tau}_V)$ is a separable topological space.*

For an analyst, the most interesting topological spaces are those which are at least Hausdorff (i.e., a T_2-space). The next proposition tells us that under some additional, reasonable conditions on X, the topological space $(\mathcal{P}_{cl}(X), \widehat{\tau}_V)$ has nice separation properties.

Lemma 1.21. *If (X, τ) is a regular topological space, then $(\mathcal{P}_{cl}(X), \widehat{\tau}_V)$ is a Hausdorff topological space.*

Proof. Let $A, B \in \mathcal{P}_{cl}(X)$ and assume that $A \neq B$. Then $A \cap X \backslash B$ or $B \cap X \backslash A$ in nonempty. Suppose $A \cap X \backslash B \neq \emptyset$ and let $a \in A \cap X \backslash B$. Then since by hypothesis X is regular, we can find U_1, $U_2 \in \tau$ such that $a \in U_1, B \subseteq U_2$ and $U_1 \cap U_2 = \emptyset$. Note the U_1^- and U_2^+ are disjoint elements in $\widehat{\tau}_V$ and $A \in U_1^-$, while $B \in U_2^+$. So indeed $\widehat{\tau}_V$ is a Hausdorff topology. $\qquad\square$

Lemma 1.22. *If (X, τ) is a Hausdorff topological space, then (X, τ) is compact if and only if $(\mathcal{P}_{cl}(X), \widehat{\tau}_V)$ is compact.*

1.4   Continuity concepts

The three Vietoris topologies introduced in Section 1.3 lead to corresponding continuity concepts for multifunctions.

Definition 1.23. *Let $F : X \to \mathcal{P}(Y)$ be a multifunction (set-valued map).*

- *If $F : X \to (\mathcal{P}(Y), \tau_{UV})$ is continuous, then $F(\cdot)$ is said to be upper semicontinuous (briefly, u.s.c.).*

- *If $F : X \to (\mathcal{P}(Y), \tau_{LV})$ is continuous, then $F(\cdot)$ is said to be lower semicontinuous (briefly, l.s.c.).*

- *If $F : X \to (\mathcal{P}(Y), \tau_V)$ is continuous, then $F(\cdot)$ is said to be continuous (or Vietoris continuous).*

We present a local version of the above definition.

Definition 1.24. *Let $F : X \to \mathcal{P}(Y)$ be a multifunction (set-valued map).*

- *F is said to be upper semicontinuous at $x_0 \in X$ if and only if for each open subset U of Y with $F(x_0) \subseteq U$, there exists an open V of x_0 such that for all $x \in V$, we have $F(x) \subseteq U$.*

- *F is said to be lower semicontinuous at $x_0 \in X$ if the set $\{x \in X : F(x) \cap U \neq \emptyset\}$ is open, for any open set U in Y.*

Using the definition of the three Vietoris topologies, we immediately deduce the following results. We recall that a set $\mathcal{M}$ with a preorder $\succeq$ is directed, if every finite subset has an upper bound. A generalized sequence is a map

$$\mu \in \mathcal{M} \mapsto x_\mu \in X,$$

where (X, τ) is an topological space. An element $x \in X$ is the limit of $(x_\mu)_{\mu \in \mathcal{M}}$ if, for every neighborhood $\mathcal{V}$ of x, there exists $\mu_0 \in \mathcal{M}$ such that x_μ belong to $\mathcal{V}$, for all $\mu \succeq \mu_0$.

Proposition 1.25. *For a multifunction* $F : X \to \mathcal{P}(Y)$, *the following are equivalent:*

a) *F is upper semi continuous (u.s.c.)*

b) *$F_+^{-1}(V)$ is open in X for every $V \subseteq Y$ open.*

c) *For every closed $C \subseteq Y$, $F_-^{-1}(C)$ is closed in X.*

d) *$\overline{F_-^{-1}(D)} \subseteq F_-^{-1}(\overline{D})$.*

e) *For any $x \in X$, if $\{x_\alpha\}_{\alpha \in J}$ is a generalized sequence, $x_\alpha \to x$, and V is an open subset of Y such that $F(x) \subseteq V$, then there exists $\alpha_0 \in J$ such that, for all $\alpha \in J$ with $\alpha \geq \alpha_0$, we have $F(x_\alpha) \subseteq V$.*

Proof. We proceed along the pattern $a) \Rightarrow b) \Rightarrow c) \Rightarrow d) \Rightarrow e) \Rightarrow a)$.

- $a) \Rightarrow b)$. Let W be open in Y, then

$$F_+^{-1}(W) = \{x \in X : F(x) \subset W\}.$$

We will now show that $F_+^{-1}(W)$ is an open set in X. Let $x \in F_+^{-1}(W)$; then $F(x) \subset W$. Since F u.s.c, there exists $V(x) \in \mathcal{N}(x)$ such that

$$F(V(x)) \subset W \Rightarrow V(x) \subset F_+^{-1}(W).$$

Hence, $F_+^{-1}(W)$ is open in X.

- $b) \Rightarrow c)$. Let Q be a closed set in Y, then

$$F_-^{-1}(Q) = \{x \in X : F(x) \cap Q \neq \emptyset\}$$

and

$$X \backslash F_-^{-1}(Q) = \{x \in X : F(x) \subset X \backslash Q\} = F_+^{-1}(X \backslash Q).$$

Since Q is a closed set in Y, $X \backslash Q$ is an open set in Y. From $b)$, we have $F_+^{-1}(X \backslash Q)$ is open in X. Thus $F_-^{-1}(Q)$ is closed in X.

- $c) \Longrightarrow d)$. Let D be a subset of Y. Then

$$D \subset \overline{D} \Longrightarrow F_-^{-1}(D) \subset F_-^{-1}(\overline{D}) \Longrightarrow \overline{F^{-1} - (D)} \subset \overline{F_-^{-1}(\overline{D})}.$$

Since $F_-^{-1}(\overline{D})$ closed, $F_-^{-1}(\overline{D}) = \overline{F_-^{-1}(\overline{D})}$. Thus,

$$\overline{F_-^{-1}(D)} \subset F_-^{-1}(\overline{D}).$$

- $d) \Longrightarrow e)$. Let $\{x_\alpha\}_{\alpha \in J}$ be a generalized sequence, $x \in X$, $x_\alpha \to x$ and let V be an open set in Y such that $F(x) \subset V$. We will show that there exists $\alpha_0 \in J$ such that for all $\alpha \geq \alpha_0$, we have $F(x_\alpha) \subset V$. Assume that this is not the case. Then for all $\alpha \in J$, there exists $\beta \in J$ such that $\beta \geq \alpha$ and $F(x_\beta) \not\subset V$. This implies that $x_\beta \in F_-^{-1}(Y \backslash V)$, and thus $x_\beta \in \overline{F_-^{-1}(Y \backslash V)}$. Since $x_\alpha \to x$, we can easily show that $x_\beta \to x \in \overline{F_-^{-1}(Y \backslash V)}$. From $d)$, we have $x \in F_-^{-1}(Y \backslash V)$, which is in contradiction with $F(x) \subset V$.

- $e) \Longrightarrow a)$. Let $x \in X$ and V be an open set in Y such that $F(x) \subseteq V$. Suppose that for all $V \in \mathcal{N}(x)$, we have $x_v \in V$, such that $F(x_v) \cap Y \backslash V \neq \emptyset$. Let

$$R = \{[x_v, V] \in V \times \mathcal{N}(x) : x_v \in F_-^{-1}(Y \backslash V)\}.$$

We introduce a partial ordering on R, by declaring that $[x_v, V] \leq [x_{v'}, V']$ if and only if $V' \subset V$. Our claim is that R with this partial ordering becomes a directed set. Indeed, let $[x_v, V]$, $[x_{v'}, V'] \in R$. Since $V \cap V' \in \mathcal{N}(x)$, there exists $x_{v \cap v'} \in V \cap V'$ such that $x_{v \cap v'} \in F_-^{-1}(Y \backslash V \cap V')$. We consider $[x_{v \cap v'}, V \cap V'] \in R$. It is clear that $[x_v, V] \leq [x_{v \cap v'}, V \cap V']$ and $[x_{v'}, V'] \leq [x_{v \cap v'}, V \cap V']$. Define $\phi : R \to \mathcal{N}(x)$ by $[x_v, V] \to \phi([x_v, V]) = V$. Clearly, $\phi(R)$ is cofinal in $\mathcal{N}(x)$. For any $[x_v, V]$, let $x_{\phi[x_v, V]} = x_v$. We will show that $x_v \to x$. Let $V' \in \mathcal{N}(x)$; then there exist $x_{v'} \in V'$ such that $x_{v'} \in F_-^{-1}(Y \backslash V)$. So for any $[x_v, V] \geq [x_{v'}, V']$, we have $x_v \in V \subset V'$. Hence, $x_v \to x$. Since $F(x) \subset V$, by e), there exists $[x_v, V] \in R$ such that $[x_{v'}, V'] \geq [x_v, V]$ implies $F(x_{v'}) \subset V$. Thus, $x_{v'} \notin F_-^{-1}(Y \backslash V')$ which is a contradiction.

$\square$

The corresponding result for lower semicontinuity reads as follows.

Proposition 1.26. *For a multifunction $F : X \to \mathcal{P}(Y)$, the following are equivalent:*

a) *F is lower semicontinuous (l.s.c.)*

b) *For every $V \subseteq Y$ open, $F_-^{-1}(V)$ is open in X.*

c) *For every closed $C \subseteq Y$, $F_+^{-1}(C)$ is closed in X.*

d) *$\overline{F_+^{-1}(D)} \subseteq F_+^{-1}(\overline{D})$.*

e) *$F(\overline{A}) \subseteq \overline{F(A)}$, for every set $A \subseteq X$.*

g) *For any $x \in X$, if $\{x_\alpha\}_{\alpha \in J}$ is a generalized sequence, $x_\alpha \to x$, then for every $y \in F(x)$ there exists a generalized sequence $\{y_\alpha\}_{\alpha \in J} \subset Y$, $y_\alpha \in F(x_\alpha)$, $y_\alpha \to y$.*

Proof. Again, our pattern is $a) \Rightarrow b) \Rightarrow c) \Rightarrow d) \Rightarrow e) \Rightarrow f) \Longrightarrow g) \Longrightarrow a)$.

- $a) \Rightarrow b)$. Let V be open in Y; then

$$F_-^{-1}(V) = \{x \in X : F(x) \cap V \neq \emptyset\}.$$

We will now show that $F_-^{-1}(V)$ is an open set in X. Let $x \in F_-^{-1}(V)$, then $F(x) \cap V \neq \emptyset$. Since F is l.s.c., there exists $U(x) \in \mathcal{N}(x)$ such that

$$F(z) \cap V \neq \emptyset \text{ for all } z \in U(x), \text{ so } U(x) \subseteq F_-^{-1}(V).$$

Hence, $F_-^{-1}(V)$ is open in X.

- $b) \Rightarrow c)$. Let Q be a closed set in Y; then

$$F_+^{-1}(Q) = \{x \in X : F(x) \subseteq Q\}$$

and

$$X \backslash F_+^{-1}(Q) = \{x \in X : F(x) \cap X \backslash Q \neq \emptyset\} = F_-^{-1}(X \backslash Q).$$

Since Q is a closed set in Y, $X \backslash Q$ is an open set in Y. From b), we have $F_-^{-1}(X \backslash Q)$ is open in X. Thus, $F_+^{-1}(Q)$ is closed in X.

- $c) \Longrightarrow d)$. Let D be a set in Y; then we have

$$F_+^{-1}(D) \subseteq F_+^{-1}(\overline{D}) \Longrightarrow \overline{F_+^{-1}(D)} \subseteq \overline{F_+^{-1}(\overline{D})}.$$

From $c)$, we obtain

$$\overline{F_+^{-1}(D)} \subseteq F_+^{-1}(\overline{D}).$$

- $d) \Longrightarrow e)$. Let A be a subset in X. We will show that $F(\overline{A}) \subseteq \overline{F(A)}$. Assume that $F(\overline{A}) \nsubseteq \overline{F(A)}$. Then there exists $y \in F(\overline{A})$, such that $y \notin \overline{F(A)}$, and thus there exists $V(y) \in \mathcal{N}(y)$, with $V(y) \cap F(A) = \emptyset$. This implies that

$$A \subseteq F_+^{-1}(Y \backslash V(y)).$$

From $d)$, we have

$$A \subseteq \overline{F_+^{-1}(Y \backslash V(y))} = F_+^{-1}(Y \backslash V(y)).$$

Since $y \in F(\overline{A})$, there exists $x \in \overline{A}$, such that $y \in F(x)$. Since $x \in \overline{A}$, we have a generalized sequence $\{x_\alpha\}_{\alpha \in J}$ in A, $x_\alpha \to x$. Hence, $x \in F_+^{-1}(Y \backslash V(y))$, and by definition of F_+^{-1} we obtain $F(x) \cap V(y) = \emptyset$, which is a contradiction to $y \in F(x)$.

- $e) \Longrightarrow g)$. Let $\{x_\alpha\}_{\alpha \in J}$ be a generalized sequence, $x \in X$, $x_\alpha \to x$, and $y \in F(x)$. Set $A = \{x_\alpha : \alpha \in J\}$, where J is a directed set. By $e)$, we have

$$F(\overline{A}) = F(A \cup \{x\}) \subseteq \overline{F(A)}.$$

Let

$$R = \{[x_\alpha, V] \in V \times \mathcal{N}(y) : x_v \in F_-^{-1}(V)\}.$$

Since $y \in F(\overline{A})$, this implies that $y \in \overline{F(\{x_\alpha : \alpha \in J\})}$. Then $R \neq \emptyset$. We introduce a partial ordering on R, by declaring that $[\alpha, V] \leq [\alpha', V']$ if and only if $V' \subset V$ and $\alpha' \leq \alpha$. Our claim is that R with this partial ordering becomes a directed set. Indeed, let $[\alpha, V]$, $[\alpha', V'] \in R$. Then since J is directed, there exists $\beta \in J$ such that $\alpha \leq \beta$ and $\alpha' \leq \beta$. Also because $y \in \cap_{\alpha \in J} \cup_{\beta \geq \alpha} F(x_\beta)$ and $V \cap V' \in \mathcal{N}(y)$, we can find $\gamma \in J$, $\gamma \geq \alpha$, such that $x_\gamma \in F_-^{-1}(V \cap V')$. Then $[\alpha, V] \leq [\gamma, V \cap V']$ and $[\beta, V'] \leq [\gamma, V \cap V']$. So R is directed. Define $\phi : R \to \mathcal{N}(y)$ by $[\alpha, V] \to \phi([\alpha, V]) = \alpha$. Clearly, $\phi(R)$ is cofinal in J. For any $[\alpha, V] \in R$, let $y_{\phi([\alpha, V])} \in F(x_\alpha) \cap V$. Also, $x_{\phi([\alpha, V])} = x_\alpha$. Since $\phi(R)$ is cofinal in J, we have $x_{\phi([\alpha, V])} \to x$. We will show that $y_{\phi([\alpha, V])} \to y$. Let $V' \in \mathcal{N}(y)$. Then there exists $\alpha' \in J$ such that $x_{\alpha'} \in F_-^{-1}(V')$. So for any $[\alpha, V] \geq [\alpha', V']$, we have $y_{\phi([\alpha, V])} \in V \subseteq V'$, which implies that $y_{\phi([\alpha, V])} \to y$.

- $g) \Longrightarrow a)$. Let $x \in X$ and W be an open set in Y such that $F(x) \cap W \neq \emptyset$. We will show that there exists $V \in \mathcal{N}(x)$, such that $F(z) \cap V \neq \emptyset$, for all $z \in W$. Assume that is not the case. Then for all $V \in \mathcal{N}(x)$, we have $x_v \in V$, such that $F(x_v) \cap V = \emptyset$. Let

$$R = \{[x_v, V] \in V \times \mathcal{N}(x) : x_v \in F_-^{-1}(V)\},$$

and

$$\phi : R \to \mathcal{N}(x) \text{ by } [x_v, V] \to \phi([x_v, V]) = V.$$

As in Proposition 1.25, we can prove that R is directed and $\phi(R)$ is cofinal in $\mathcal{N}(x)$. For any $[x_v, V]$, let $x_{\phi[x_v, V]} = x_v$. We will show that $x_v \to x$. Let $V' \in \mathcal{N}(x)$; then there exists $x_{v'} \in V'$ such that $x_{v'} \in F_-^{-1}(V)$. So for any $[x_v, V] \geq [x_{v'}, V']$, we have $x_v \in V \subset V'$. Hence, $x_v \to x$. Since $F(x) \cap W \neq \emptyset$, then there exists

$y \in F(x) \cap W$. By g), there exists $y_{\phi([x_v, V])} \in F(x_v)$ such that $y_{\phi([\alpha, V])} \to y$. But $y_{\phi([\alpha, V])} \in F(x_v) \subseteq Y \backslash W$. Thus, $y \in Y \backslash W$, which is a contradiction.

Now $[x_{v'}, V'] \geq [x_v, V]$, implies $F(x_{v'}) \subset V$. Thus, $x_{v'} \notin F_-^{-1}(V')$, which is a contradiction.

$\square$

Remark 1.27. *In the case where X and Y are topological spaces with countable bases, we may take usual sequences instead of generalized ones in conditions e) and g) of Propositions 1.25 and 1.26, respectively.*

In the single-valued case, both notions are equivalent to continuity:

Proposition 1.28. *For a multifunction $f : X \to Y$, the following are equivalent:*

a) f is continuous at $x_0 \in X$.

b) $f : X \to \mathcal{P}(Y)$ is upper semicontinuous at $x_0 \in X$.

c) $f : X \to \mathcal{P}(Y)$ is lower semicontinuous at $x_0 \in X$.

Proof. For every open set V in Y we have

$$f^{-1}(V) = f_+^{-1}(V) = f_-^{-1}(V).$$

$\square$

Example 1.29. *The following set-valued mappings are upper semicontinuous:*

1) $F : \mathbb{R} \to \mathcal{P}(\mathbb{R})$ defined by

$$F(x) = \begin{cases} \{1\}, & x > 0, \\ \{-1, 1\} & x = 0, \\ \{-1\} & x < 0. \end{cases}$$

2) $F : \mathbb{R} \to \mathcal{P}(\mathbb{R})$ defined by

$$F(x) = \begin{cases} \{x + 1\}, & x > 0, \\ [-1, 1] & x = 0, \\ \{x - 1\} & x < 0. \end{cases}$$

3) $F : \mathbb{R} \to \mathcal{P}(\mathbb{R})$ defined by $F(x) = [f(x), g(x)]$, where $f, g : \mathbb{R} \to \mathbb{R}$ are l.s.c. and u.s.c. functions, respectively.

Example 1.30. *The following set-valued mappings are lower semicontinuous:*

1) $F : \mathbb{R} \to \mathcal{P}(\mathbb{R})$ defined by

$$F(x) = \begin{cases} [a, b], & x \neq 0, \\ \{\alpha\}, & x = 0, \end{cases}$$

where $\alpha \in [a, b]$.

2) $F : \mathbb{R} \to \mathcal{P}(\mathbb{R})$ defined by

$$F(x) = \begin{cases} [0, |x| + 1], & x \neq 0, \\ \{1\}, & x = 0. \end{cases}$$

3) $F : \mathbb{R} \to \mathcal{P}(\mathbb{R})$ defined by $F(x) = [f(x), g(x)]$, where $f, g : \mathbb{R} \to \mathbb{R}$ are u.s.c. and l.s.c. functions, respectively.

4) Let $X = Y = [0, 1]$. Define

$$F(x) = \begin{cases} [0, 1], & x \neq \frac{1}{2}, \\ [0, \frac{1}{2}], & x = \frac{1}{2}. \end{cases}$$

In general, the concepts of upper semicontinuity and lower semicontinuity are distinct. The following standard example illustrates this.

Example 1.31. *Let $X = Y = \mathbb{R}$. Define*

$$F_1(x) = \begin{cases} \{1\}, & x \neq 0, \\ [0, 1], & x = 0, \end{cases} \quad and \quad F_2(x) = \begin{cases} \{0\}, & x = 0, \\ [0, 1], & x \neq 0. \end{cases}$$

We can easily show that F_1 is u.s.c. but not l.s.c., while F_2 is l.s.c. but not u.s.c.

1.5 Upper semicontinuity and closed graphs

Another useful continuity notion related to the previous ones, can be defined using the graph of a multifunction.

Definition 1.32. *A multifunction is said to be closed if its graph Γ_F is a closed subset of the space $X \times Y$.*

Here are some equivalent formulations.

Theorem 1.33. *The following conditions are equivalent:*

a) The multifunction F is closed.

b) For every $(x, y) \in X \times Y$ such that $y \notin F(x)$, there exist neighborhoods $V(x)$ of x and $W(y)$ of y such that $F(V(x)) \cap W(y) = \emptyset$.

c) For generalized sequences $\{x_\alpha\}_{\alpha \in J} \subset X$ and $\{y_\alpha\}_{\alpha \in J} \subset Y$, if $x_\alpha \to x$, and $y_\alpha \in F(x_\alpha)$ with $y_\alpha \to y$, then $y \in F(x)$.

Proof. Our pattern follows $a) \Rightarrow b) \Rightarrow c) \Rightarrow a)$.

- $a) \Rightarrow b)$.

  Let $(x, y) \in X \times Y$ be such that $y \notin F(y)$. Then $(x, y) \notin \Gamma_F$, and this implies that $(x, y) \in X \times Y \backslash \Gamma_F$. Since Γ_F is closed, there exists $(V(x), W(y)) \in \mathcal{N}(x) \times \mathcal{N}(y)$, such that $V(x) \times W(y) \cap \Gamma_F = \emptyset$. We will show that $F(V(x)) \cap W(y) = \emptyset$. Suppose that there exists $z \in F(V(x)) \cap W(y)$. Then there exists $r \in V(x)$, such that $z \in F(r)$, and this implies that $(r, z) \in \Gamma_F$, which is a contradiction.

- b) $\Rightarrow$ c).
  Let $\{x_\alpha\}_{\alpha \in J}$ be a generalized sequence such that

$$x_\alpha \to x, \ y_\alpha \in F(x_\alpha), \ y_\alpha \to y.$$

Assume that $y \notin F(x)$. Then there exist $(V(x), W(y)) \in \mathcal{N}(x) \times \mathcal{N}(y)$, such that $F(V(x)) \cap W(y) = \emptyset$. Now

$$x_\alpha \to x \Longrightarrow \exists \alpha_0 \in J \text{ such that } \forall \ \alpha \geq \alpha_0; \text{ we have } x_\alpha \in V(x),$$

and

$$y_\alpha \to y \Longrightarrow \exists \alpha_1 \in J \text{ such that } \forall \ \alpha \geq \alpha_1; \text{ we have } y_\alpha \in W(y).$$

Since J is is directed, then there exists $\beta \in J$ such that $\alpha_0, \alpha_1 \leq \beta$, and hence for all $\alpha \geq \beta$, we have $x_\alpha \in V(x)$ and $y_\alpha \in W(y)$, with $y_\alpha \in F(x_\alpha)$. Then $F(V(x)) \cap W(y) \neq \emptyset$ which is a contradiction.

- c) $\Rightarrow$ a).
  Let $(x_\alpha, y_\alpha) \in \Gamma_F$, $\alpha \in J$, $x_\alpha \to x, y_\alpha \to y$ and $y_\alpha \in F(x_\alpha)$. From c), we obtain that $y \in F(x)$. Hence Γ_F is closed.

$\square$

Example 1.34. *Let $f : Y \to X$ be a continuous surjective map between topological spaces. Then the inverse multifunction $F : X \to \mathcal{P}(Y)$ given by $F(x) = f^{-1}(x)$ is closed.*

Next, we give a relationship between *u.s.c.* and closed multifunctions.

Theorem 1.35. *Let X be a topological space, Y a regular topological space and $F : X \to \mathcal{P}_{cl}(Y)$ an u.s.c. multifunction. Then F is closed.*

Proof. Let $y \in Y$, $y \notin F(x)$. Since Y is regular, there exist an open neighborhood $W(y)$ of the point y and an open neighborhood W_1 of the set $F(x)$ such that $W(y) \cap F(x) = \emptyset$. Let $V(x)$ be a neighborhood of x such that $F(V(x)) \subset W_1$. Then $F(v(x)) \cap W(y) = \emptyset$ and the statement follows from Theorem 1.33 part b). $\square$

Proposition 1.36. *A multifunction $F : X \to \mathcal{P}_{cp}(Y)$ is u.s.c. if and only if for every net $\{(x_\alpha, y_\alpha)\}_{\alpha \in J} \in \Gamma_F$ with $x_\alpha \to x$ in $X, \{y_\alpha\}_{\alpha \in J}$ has a accumulation point in $F(x)$.*

Proposition 1.37. *If $F : X \to \mathcal{P}_{cp}(Y)$ is u.s.c., then F is closed.*

Proof. Let $\{(x_\alpha, y_\alpha)\}_{\alpha \in J} \subset \Gamma_F$ such that $(x_\alpha, y_\alpha) \to (x, y)$ in $X \times Y$, we show that $y \in F(x)$. Suppose that $y \notin F(x)$, then $y \in Y \backslash F(x)$. Since $y_\alpha \to y$ there exists $\alpha_0 \in J$ such that

$$y_\alpha \in Y \backslash F(x), \text{ for all } \alpha \geq \alpha_0.$$

Then

$$F(x) \subset Y \backslash \{y_\alpha\}, \text{ for all } \alpha \geq \alpha_0.$$

Using the fact that F is *u.s.c.*, we can find $U_x^\alpha \in \mathcal{N}(x)$ such that

$$F(U_x^\alpha) \subset Y \backslash \{y_\alpha\}, \text{ for all } \alpha \geq \alpha_0.$$

Because $F(x)$ is compact, there exist finite sets $J_* \subset \{\alpha \in J : \alpha \geq \alpha_0\} := J_0$ such that

$$F(x) \subset \cup_{\alpha \in J_*} Y \backslash \{y_\alpha\}.$$

Let $V_x = \cap_{\alpha \in J_*} U_x^\alpha \in \mathcal{N}(x)$, then there exists $\beta_0 \in J_0$ such that

$$x_\beta \in V_x, \text{ for all } \beta \geq \beta_0 \Rightarrow F(x_\beta) \subset F(U_x^\beta).$$

Hence

$$y_\beta \in F(x_\beta) \subset Y \backslash \{y_\beta\}.$$

which is a contradiction.

$\square$

Proposition 1.38. *Let $F : X \to \mathcal{P}(Y)$ be an closed multivalued map and $K \in \mathcal{P}_{cp}(X)$, then $F(K)$ is closed in Y.*

Proof. Let $y \in \overline{F(K)}$. Then we can find a net $\{y_\alpha\}_{\alpha \in J} \subset F(K)$ such that $y_\alpha \to y$. For every $\alpha \in J$ there exists $x_\alpha \in K$ with $y_\alpha \in F(x_\alpha)$. Since K is compact, we can find a subnet $\{x_\beta\}_{\beta \in I}$ such that $x_\beta \to x \in K$. Then $\{(x_\beta, y_\beta)\}_{\beta \in I} \subset \Gamma_F$ and $(x_\beta, y_\beta) \to (x, y)$. Since Γ_F is closed in $X \times Y$, therefore

$$y \in F(x) \subset F(K) \Rightarrow y \in F(K).$$

$\square$

In the next result, we give sufficient conditions for a closed multifunction to be *u.s.c.* We need the following definition.

Definition 1.39. *A multifunction $F : X \to \mathcal{P}(Y)$ is said to be:*

a) *compact, if its range $F(X)$ is relatively compact in Y, i.e., $\overline{F(X)}$ is compact in Y;*

b) *locally compact, if every point $x \in X$ has a neighborhood $V(x)$ such that the restriction of F to $V(x)$ is compact;*

c) *quasicompact, if its restriction to every compact subset $A \subseteq X$ is compact.*

Remark 1.40. *It is clear that a) $\Longrightarrow$ b) $\Longrightarrow$ c).*

Theorem 1.41. *Let $F : X \to \mathcal{P}_{cl}(Y)$ be a closed multimap. If $A \subset X$ is a compact set then its image $F(A)$ is a closed subset of Y.*

Proof. For $F(A) = Y$ the result is trivial. Let $y \in Y \backslash F(A)$, then for any $x \in A$ there exist V_x and $W_x(y)$ be neighborhoods of x and y respectively such that

$$F(V_x) \cap W_x(y) = \emptyset.$$

Since A is compact, thus there exist collection of open $V_{x_1}, \ldots, V_{x_n}$ forming a finite covering of A and $W(y) = \cap_{i=1}^n W_{x_i}(y)$ such that

$$W(y) \cap F(A) = \emptyset \Rightarrow W(y) \subset Y \backslash F(A).$$

$\square$

Theorem 1.42. *Let $F : X \to \mathcal{P}_{cp}(Y)$ be an u.s.c. multimap. If $A \subset X$ is a compact set then its image $F(A)$ is a compact subset.*

Proof. Let $\{V_j\}_{j \in J}$ be an open covering of $F(A)$. Then for each $x \in A$ there exists a finite collection of open sets $\{V_{j_i}\}_{i=1}^{n(x)}$ covering $F(x)$. Set $U_x = \cup_{i=1}^{n(x)} V_{j_i}$. Since F is u.s.c. there exists $W_x \in \mathcal{N}(x)$ such that

$$F(W_x) \subset U_x.$$

Hence

$$A \subset \cup_{x \in A} W_x.$$

This implies that there exists $m \in \mathbb{N}$ such that

$$A \subset \cup_{i=1}^m W_x \Rightarrow F(A) \subset \cup_{i=1}^m U_{x_i}.$$

$\square$

Theorem 1.43. *Let $F : X \to \mathcal{P}(Y)$ be an u.s.c. multimap. If $A \subset X$ is a connected set then its image $F(A)$ is a connected subset.*

Proof. Let U_1, U_2 be two open sets such that

$$F(A) \subset U_1 \cup U_2 \text{ and } U_1 \cap U_2 = \emptyset.$$

Then

$$A \subset F_+^{-1}(U_1) \cup F_+^{-1}(U_2) \text{ and } F_+^{-1}(U_1) \cap F_+^{-1}(U_2) = \emptyset.$$

Since F is u.s.c. and A is connected we have

$$A \cap F_+^{-1}(U_1) = \text{ or } A \cap F_+^{-1}(U_2) = \emptyset \Rightarrow F(A) \subseteq U_1 \text{ or } F(A) \subseteq U_2.$$

$\square$

Theorem 1.44. *Let $F : X \to \mathcal{P}_{cp}(Y)$ be a closed locally compact multifunction. Then F is u.s.c.*

Proof. Let $x \in X$, let W be an open neighborhood of the set $F(x)$, and $V(x)$ be an open neighborhood of x such that the restriction of F to $V(x)$ is compact. Suppose that the set $Q = \overline{F(V(x))} \backslash W$ is nonempty. Since F is closed, for any $y \in Q$, there exist neighborhoods $\widetilde{W}(y)$ of y and $V_y(x)$ of x such that $F(V_y(x)) \cap \widetilde{W}(y) = \emptyset$. By virtue of the compactness of Q, we can find its finite covering

$$\widetilde{W}(y_1), \widetilde{W}(y_2), \ldots, \widetilde{W}(y_n).$$

But then if we consider the open neighborhood of x defined by

$$\widetilde{V}(x) = V(x) \cap (\cap_{i=1}^n V_{y_1}(x)),$$

we have

$$F(\widetilde{V}(x)) \subset W.$$

$\square$

Example 1.45. *The condition of local compactness is essential. The multifunction $F : [-1, 1] \to \mathcal{P}_{cp}(\mathbb{R})$ defined by*

$$F(x) = \begin{cases} \{\frac{1}{x}\}, & x \neq 0, \\ \{0\}, & x = 0, \end{cases}$$

is closed but loses its upper semicontinuity at $x = 0$.

Lemma 1.46. *If $G : X \to \mathcal{P}_{cp}(Y)$ is quasicompact and has a closed graph, then G is u.s.c.*

Proof. Assume that G is not *u.s.c.* at some point x. Then there exists an open neighborhood U of $G(x)$ in Y, a sequence $\{x_n\}$ which converges to x, and for every $l \in \mathbb{N}$ there exists $n_l \in \mathbb{N}$ such that $G(x_{n_l}) \not\subset U$. Then for each $l = 1, 2, \ldots$, there are y_{n^l} such that $y_{n_l} \in G(x_{n_l})$ and $y_{n_l} \notin U$; this implies that $y_{n_l} \in Y \backslash U$. Moreover $\{y_{n_l} : l \in \mathbb{N}\} \subset G(\overline{\{x_n : n \geq 1\}})$. Since G is compact, there exists a subsequence of $\{y_{n_l} : l \in \mathbb{N}\}$ which converges to y. G closed implies that $y \in G(x) \subset U$; but this is a contradiction to the assumption that $y_{n_l} \notin U$ for each n_l. $\qquad\square$

Proposition 1.47. *Let $F : X \to \mathcal{P}(Y)$ be an u.s.c. multivalued map. If $A \subseteq X$ is a connected set and $F(x)$ is connected for every $x \in A$ then the image $F(A)$ is a connected subset of Y.*

1.6 Upper and lower semicontinuous (u.s.c. and l.s.c.) functions and their relations

In this part we consider various upper semicontinuous properties of multivalued mappings which are the results of set-theoretic, topological or algebraic operations on multivalued mappings. Operations on multivalued mappings and their properties have been investigated by Berge [47, 48] and Kuratowski [193–195].

Theorem 1.48. *Let X, Y and Z be topological spaces. Then the following hold:*

a) *If $F, G : X \to \mathcal{P}(Y)$ are u.s.c., then so is $F \cup G$;*

b) *If $F, G : X \to \mathcal{P}(Y)$ are l.s.c., then so is $F \cup G$;*

c) *If $F, G : X \to \mathcal{P}_{cl}(Y)$ are u.s.c. and the space Y is normal, then so is $F \cap G$;*

d) *If $F : X \to \mathcal{P}_{cp}(Y)$ and $G : X \to \mathcal{P}_{cp}(Z)$ are u.s.c., then so is $F \times G$;*

e) *If $F, G : X \to \mathcal{P}(Y)$ and $G : X \to \mathcal{P}(Z)$ are l.s.c., then so is $F \times G$;*

f) *If $F : X \to \mathcal{P}(Y)$ and $G : Y \to \mathcal{P}(Z)$ are u.s.c., then so is $G \circ F$;*

g) *If $F : X \to \mathcal{P}(Y)$ and $G : Y \to \mathcal{P}(Z)$ are l.s.c., then so is $G \circ F$.*

Proof. Let $V \subset Y$ be an open. From Proposition 1.2, we have

$$(F \cup G)_+^{-1}(V) = F_+^{-1}(V) \cap G_+^{-1}(V) \quad \text{and} \quad (F \cup G)_-^{-1}(V) = F_-^{-1}(V) \cup G_-^{-1}(V),$$

so $a)$ and $b)$ hold.

Let $B \subset Y$ be closed set. From Proposition 1.2, we obtain

$$(F \cap G)_-^{-1}(B) \subseteq F_-^{-1}(B) \cap G_-^{-1}(B).$$

Now we show that $\overline{(F \cap G)_-^{-1}(B)} = (F \cap G)_-^{-1}(B)$. Indeed let $x \in \overline{(F \cap G)_-^{-1}(B)}$ then there exists a generalized sequence $\{x_\alpha\}_{\alpha \in J} \subset (F \cap G)_-^{-1}(B)$ such that x_α converges to x, hence $\{x_\alpha\}_{\alpha \in J} \subset F_-^{-1}(B)$ and $\{x_\alpha\}_{\alpha \in J} \subset G_-^{-1}(B)$. Since F and G are *u.s.c.*, then $x \in F_-^{-1}(B)$

and $x \in G_-^{-1}(B)$ are closed. If $F(x) \cap G(x) \cap B = \emptyset$, the normality of Y implies that there exist disjoint open sets U_1 and U_2 such that

$$F(x) \subset U_1, \quad G(x) \cap B \subset U_2.$$

Then, there exists $\alpha_0 \in J$ such that for each $\alpha \geq \alpha_0$, we have

$$x_\alpha \in U_1 \Rightarrow G(x_\alpha) \cap B = \emptyset \quad \text{for all } \alpha \geq \alpha_0,$$

which is a contradiction with $\{x_\alpha\}_{\alpha \in J} \subset G_-^{-1}(B)$, proving that the set $(F \cap G)_-^{-1}(B)$ is closed. So the statement c) holds. $\qquad\square$

Theorem 1.49. *Let X, Y and Z be topological spaces. Then the following hold:*

a) *If $F, G : X \to \mathcal{P}(Y)$ are closed, then so is $F \cup G$;*

b) *If $F, G : X \to \mathcal{P}(Y)$ are closed, then so is $F \cap G$;*

c) *If $F, G : X \to \mathcal{P}_{cl}(Y)$ are closed, then so is $F \times G$.*

Unfortunately, the composition of two closed multivalued mapping need not be closed.

Example 1.50. *Let $F, G : \mathbb{R} \to \mathcal{P}_{cp,cv}(\mathbb{R})$ defined by*

$$F(x) = \begin{cases} \{\frac{1}{x}\}, & x \neq 0, \\ \{0\}, & x = 0, \end{cases}$$

and

$$G(x) = \begin{cases} \{\frac{1}{x}\}, & x \neq 0, \\ \{1\}, & x = 0. \end{cases}$$

Then both F and G are closed, while the composition

$$(G \circ F)(x) = \begin{cases} \{x\}, & x \neq 0, \\ \{1\}, & x = 0, \end{cases}$$

is not.

Theorem 1.51. *Let $F : X \to \mathcal{P}_{cp}(Y)$ be an u.s.c. multimap and $G : Y \to \mathcal{P}_{cl}(Z)$ a closed multivalued map. Then the composition $G \circ F : X \to \mathcal{P}_{cl}(Z)$ is a closed multivalued map.*

Theorem 1.52. *Let X, Y and Z be topological spaces. Then the following holds:*

a) *If $F, G : X \to \mathcal{P}_{cl}(Y)$ are closed, then so is $F \cup G$;*

b) *If $F, G : X \to \mathcal{P}_{cl}(Y)$ are closed and the space Y is normal, then so is $F \cap G$;*

c) *If $F, G : X \to \mathcal{P}_{cl}(Y)$ are closed, then so is $F \times G$.*

The following are complementary theorems.

Theorem 1.53. *Let X, Y and Z be topological spaces. Then the following hold:*

a) *If $F : X \to \mathcal{P}_{cl}(Y)$ is closed and $G : X \to \mathcal{P}_{cp}(Y)$ is u.s.c. then, $F \cap G : X \to \mathcal{P}_{cp}(Y)$ is u.s.c.*

b) *If $F : X \to \mathcal{P}_{cp}(Y)$ is u.s.c and $G : Y \to \mathcal{P}_{cp}(Z)$ is closed, then, $F \cap G : X \to \mathcal{P}_{cl}(Y)$ is closed.*

Corollary 1.54. *Let X, Y and Z be topological spaces, $F : X \to \mathcal{P}_{cp}(Y)$ be an u.s.c. multifunction and $C \subset Y$ a closed subset such that*

$$F(x) \cap C \neq \emptyset, \text{ for all } x \in C.$$

Then the multimap $F_* : X \to \mathcal{P}_{cp}(Y)$, defined by $F_*(x) = F(x) \cap C$ is u.s.c.

1.7 Lower semicontinuity and open graphs

Definition 1.55. *A multivalued mapping $F : X \to \mathcal{P}(Y)$ is said to be quasiopen at the point $x \in X$ if $int\, F(x) \neq \emptyset$ and if for any $y \in int\, F(x)$ one can find a neighborhood $V(y) \subset Y$ and a neighborhood $U(x) \subset X$, such that*

$$V(y) \subseteq F(z) \text{ for all } z \in U(x).$$

If F is quasiopen at each point $x \in X$, then it will be called quasiopen.

Theorem 1.56. *Let $F : X \to \mathcal{P}(Y)$ be a multivalued map. The following statements are equivalent:*

(a) F is quasiopen;

(b) $int\, F(x) \neq \emptyset$ for all $x \in X$ and the graph of the multivalued mapping $int F : X \to \mathcal{P}(Y)$ defined by

$$(int\, F)(x) = int\, F(x)$$

is open in $X \times Y$.

We would like to stress that the intersection of two *l.s.c.* mappings does not have to be *l.s.c.*

Example 1.57. *Consider two multi-valued mappings $F, G : [0, \pi] \to \mathcal{P}(\mathbb{R}^2)$ defined by*

$$F(t) = \{(x, y) \in \mathbb{R}^2 | y \geq 0 \text{ and } x^2 + y^2 \leq 1\}, \text{ for every } t \in [0, \pi];$$

$$G(t) = \{(x, y) \in \mathbb{R}^2 | x = \lambda \cos t, \ y = \lambda \sin t, \ \lambda \in [-1, 1]\}.$$

Then F is a constant map hence continuous, G is a l.s.c. map but $F \cap G$ is no longer l.s.c. (to see this consider $t = 0$ or $t = \pi$).

Theorem 1.58. *Let X, Y be two topological spaces, $F_1 : X \to \mathcal{P}(Y)$ be lower semi continuous, $F_2 : X \to \mathcal{P}(Y)$ has an open graph, and for every $x \in X$ $(F_1 \cap F_2)(x) \neq \emptyset$, then $F_1 \cap F_2$ is l.s.c. at x.*

Proof. Let $V \subseteq Y$ be a nonempty open set, $x \in (F_1 \cap F_2)^{-1}(V)$ and let $y \in F_1(x) \cap F_2(x) \cap V$. Then $(x, y) \in \Gamma_{F_2} \cap X \times V$ is open in $X \times Y$, so we can find $U_1(x) \in \mathcal{N}(x)$ and $V_1(y) \in \mathcal{N}(y)$ such that $U_1(x) \times V_1(y) \subseteq \Gamma_{F_2} \cap X \times V$. It is clear that $F_2(x) \cap V_1(y) \neq \emptyset$. Since F_1 is lower semi continuous we can find $U_2(x)$ such that $F_1(z) \cap V_1(y) \neq \emptyset$ for every $z \in U_2(x)$. Set $U(x) = U_1(x) \cap U_2(x) \in \mathcal{N}(x)$. Thus for all $z \in U(x)$ we have $F_1(x) \cap V_1(y) \neq \emptyset$, while $U(x) \times V_1(y) \subseteq \Gamma_{F_2} \cap (X \times V)$. Thus for all $z \in U(x)$, $(F_1 \cap F_2)(x) \cap V \neq \emptyset$. This implies that $F_1 \cap F_2$ is l.s.c. $\qquad \square$

The following facts are essential, not surprising, yet tricky to prove if one does not go about it just right.

Lemma 1.59. *Let C be a convex subset of the normed space X. Then:*

(a) $\overline{C}$ is convex;

(b) $x \in \overline{C}, y \in int\, C \implies (x, y] \subset C$.

(c) $intC$ is convex,

(d) $intC \neq \emptyset \Longrightarrow \overline{C} = \overline{(intC)}$ and $(intC) = int(\overline{C})$.

Theorem 1.60. *Let X be a topological space and (Y,d) be a metric space. A multifunction $F : X \to \mathcal{P}_{cp}(Y)$ is l.s.c. at $x \in X$ if and only if for every $\epsilon > 0$ there exists a neighborhood $U(x)$ such that*

$$F(x) \subseteq V(F(x),\epsilon) = \{a \in Y : d(a, F(x')) < \epsilon\}, \quad \text{for all } x' \in U(x).$$

Proof. Let $\epsilon > 0$; then

$$F(x) \subseteq \bigcup_{y \in F(x)} B\left(y, \frac{\epsilon}{2}\right).$$

Thus there exist $y_1, \ldots, y_n \in F(x)$ such that

$$F(x) \subseteq \bigcup_{i=1}^{n} B\left(y_i, \frac{\epsilon}{2}\right).$$

Now, we show that

$$B\left(y_i, \frac{\epsilon}{2}\right) \subseteq V(F(x'),\epsilon), \quad i = 1, \ldots, n, \quad x' \in U_i(x).$$

Let $z \in B\left(y_i, \frac{\epsilon}{2}\right)$; then

$$d(z, F(x')) \leq d(z,a), \quad a \in F(x') \cap B\left(y_i, \frac{\epsilon}{2}\right).$$

Therefore,

$$d(z, F(x')) \leq d(z,a) \leq d(z, y_i) + d(y_i, a) < \epsilon \Longrightarrow z \in V(F(x'),\epsilon).$$

This implies that for each $x' \in V(x) = \bigcap_{i=1}^{n} U_i(x)$, we have

$$\bigcup_{i=1}^{n} B\left(y_i, \frac{\epsilon}{2}\right) \subset V(F(x'),\epsilon) \Longrightarrow F(x) \subset V(F(x'),\epsilon).$$

Conversely, let $W \subset Y$ be an open set such that $F(x) \cap W \neq \emptyset$. Let $y \in F(x) \cap W$; for any arbitrary $\epsilon > 0$ such that $B(y,\epsilon) \subset W$, there exists a neighborhood $U(x)$ of x such that, if $x' \in U(x)$, then

$$F(x) \subset V(F(x'),\epsilon).$$

Hence,

$$F(x') \cap B(y,\epsilon) \neq \emptyset \quad \text{for } x' \in U(x).$$

$\square$

Theorem 1.61. *Let X be a topological space, Y be a normed space and $F : X \to \mathcal{P}_{cl,cp}(Y)$ be a multivalued map. The following statements are equivalent:*

(a) F is quasiopen at x;

(b) $int\, F(x) \neq \emptyset$ and F is l.s.c. at x.

Proof. $a \Longrightarrow b$: Since F is quasiopen at x, then $int\, F(x) \neq \emptyset$. Let $W \subset Y$ be an open set such that $W \cap F(x) \neq \emptyset$. Now we show that $W \cap int\, F(x) \neq \emptyset$. Assume that $W \cap int\, F(x) = \emptyset$. For $y \in W \cap F(x)$ then $y \in F(x) \backslash int F(x)$. From Lemma 1.59 we have $y \in \overline{int F(x)}$. Thus there exists a sequence $(y_n)_{n \in \mathbb{N}} \subset int F(x)$ that converges to y. So there exists $n_0 \in \mathbb{N}$ such that

$$y_n \in W, \quad \text{for every } n \geq n_0.$$

This is in contradiction with $W \cap int\, F(x) = \emptyset$. Hence

$$W \cap int\, F(x) \neq \emptyset.$$

For an arbitrary $y \in W \cap int\, F(x)$, let $W(y) \subset Y$ and $V(x) \subset X$ be neighborhoods such that $W(y) \subset F(z)$ for all $z \in V(x)$. But $W \cap W(y) \neq \emptyset$ implies that $W \cap F(z) \neq \emptyset$ for all $x \in V(x)$, so F is l.s.c. at x.

$b \Longrightarrow a$: Let $y \in int F(x)$; then there exists $\delta > 0$ such that $B(y, \delta) \subseteq F(x)$. Choose $0 < \delta_1 < \delta$ so that $0 < \delta - \delta_1 = \eta$. Since F is *l.s.c.*, from Theorem 1.60, there exists a open neighborhood $U(x)$ such that

$$F(x) \subset V(F(x'), \eta), \quad \text{for all } x' \in U(x).$$

Then,

$$B(y, \eta) \subset B(y, \delta) \subset F(x) \subset V(F(x'), \eta), \quad \text{for every } x' \in U(x). \tag{1.5}$$

Now let now $y' \in B(y, \delta_1)$ with $y' \notin F(x')$ for some $x' \in U(x)$. Then there exists $a \in B(y', \eta)$ such that $d(a, F(x')) \geq \eta$. Assume that

$$d(a, F(x')) < \eta, \quad \text{for all } a \in B(y', \eta).$$

Since $F(x') \in \mathcal{P}_{cp}(Y)$ and $y' \notin F(x')$, there exists $r > 0$ with $d(y', F(x')) > r$ such that

$$B(y', r) \cap B(y', \eta) \cap F(x') = \emptyset. \tag{1.6}$$

Then,

$$d(y', F(x')) < \min(r, \eta).$$

Case 1: $\min(r, \eta) = \eta$. There exist $\epsilon > 0$ with $\eta - \epsilon > 0$ such that

$$d(y', F(x')) < \eta - \epsilon < \eta.$$

Then there exists $b_\epsilon \in F(x')$ such that

$$d(y', b_\epsilon) \leq d(a, F(x')) + \epsilon < \eta \Longrightarrow b_\epsilon \in B(y', r) \cap B(y', \eta),$$

which contradicts (1.6).

Case 2: $\min(r, \eta) = r$. Then there exists $\epsilon > 0$ such that

$$d(y', F(x')) < r - \epsilon < r < \eta.$$

Hence, there exists $b_r \in F(x')$ such that

$$d(y', b_r) < r < \eta \Longrightarrow b_r \in B(y', r) \cap B(y', \eta),$$

which again contradicts (1.6).

Thus,

$$d(y, a) \leq d(y, y') + d(y', a) < \delta \Longrightarrow a \in B(y, \delta) \subset F(x).$$

By (1.5), we have

$$\eta \le d(a, F(x')) < \eta,$$

which is a contradiction. Hence,

$$B(y, \delta_1) \subset F(x'), \quad \text{for all } x' \in U(x).$$

$\square$

Theorem 1.62. *Let X and Y be topological spaces. Let $F_0 : X \to \mathcal{P}(Y)$ be a multimap l.s.c. at $x \in X$ and let $F_1 : X \to \mathcal{P}(Y)$ be a multimap quasi-open at x_0, such that*

$$F_0(x) \cap F_1(x) \ne \emptyset \text{ for every } x \in X$$

and

$$F_0(x_0) \cap F_1(x_0) \subset F_0(x_0) \cap int \, F_1(x_0).$$

Then the intersection $F_0 \cap F_1$ is l.s.c. at x_0.

Proof. Let V be an open set such that $V \cap F_0(x_0) \cap F_1(x_0) \ne \emptyset$. From assumptions, there exists a point $y \in F_0(x_0) \cap F_1(x_0)$ which is an interior point of the set $F_1(x_0)$. Let $W(y)$ be a neighborhood of y such that

$$W(y) \subset F_1(x_0).$$

From the quasi-openness of the multimap F_1, there exists a neighborhood U_{x_0} of the point x_0 such that

$$W(y) \subset F_1(x) \text{ for all } x \in U_{x_0}.$$

Since $y \in F_0(x_0)$ and the multimap F_0 is l.s.c. there exists a neighborhood $U(x_0)$ of a point x_0 such that

$$F_0(x) \cap W(y), \quad \text{for all } x \in U(x_0).$$

Then for every $x \in U_{x_0} = U(x_0) \cap U_{x_0}$, we have

$$(F_0 \cap F_1)(x) \cap W(y) \ne \emptyset.$$

This implies that $F_0 \cap F_1$ is l.s.c.

$\square$

Proposition 1.63. *Let $\varphi \colon X \to \mathcal{P}(Y)$ be a l.s.c. map, and let $f \colon X \to Y$ and $\lambda \colon X \to (0, \infty)$ be continuous mappings. Assume further that, for every $x \in X$, we have:*

$$\varphi(x) \cap B(f(x), \lambda(x)) \ne \emptyset.$$

Then the map $\psi \colon X \to \mathcal{P}(Y)$ defined by

$$x \to \psi(x) = \overline{\varphi(x) \cap B(f(x), \lambda(x))}$$

is a l.s.c. map.

Proof. Let $x_0 \in X$ and V be an open set of Y such that $V \cap \psi(x_0) \ne \emptyset$. Let $y_0 \in V \cap (\varphi(x_0) \cap B(f(x_0), \lambda(x_0)))$ and let V_{y_0} be an open neighbourhood of y_0 in Y such that $V_{y_0} \subset V \cap B(f(x_0), \lambda(x_0))$. Now, continuity of f and λ implies that there is an open neighbourhood U_{x_0} of x_0 in X such that $V_{y_0} \subset B(f(x), \lambda(x))$ for every $x \in U_{x_0}$. Consequently, since φ is l.s.c. we choose an open neighbourhood W_{x_0} of x_0 in X such that $\varphi(x) \cap V_y \ne \emptyset$, for every $x \in W_{x_0}$. Let $U = U_{x_0} \cap W_{x_0}$. Then we get that $(\varphi(x) \cap B(f(x), \lambda(x))) \cap V_{y_0} \ne \emptyset$ which implies $\overline{(\varphi(x) \cap B(f(x), \lambda(x)))} \cap V \ne \emptyset$, for every $x \in U$ and the proof is completed. $\square$

1.8   Linear operations on multifunctions

Let X be a topological space, and let Y be a normed space.

Definition 1.64. *Let $F_1, F_2 : X \to \mathcal{P}(Y)$ be multimaps. The multimap $F_1 + F_2 : X \to \mathcal{P}(Y)$ defined by*

$$(F_1 + F_2)(x) = F_1(x) + F_2(x)$$

is said to be the sum of the multimaps F_1 and F_2.

Theorem 1.65. *If the multivalued maps $F_1, F_2 : X \to \mathcal{P}(Y)$ are l.s.c., then their sum $F_1 + F_2$ is l.s.c.*

Proof. From Theorem 1.48, the multivalued map $\widetilde{F} : X \to \mathcal{P}(Y \times Y)$ defined by

$$\widetilde{F}(x) = F_1(x) \times F_2(x) \quad \text{for all } x \in X,$$

is lower semi continuous and consider the continuous single map $f : Y \times Y \to Y$ defined by

$$f(x, y) = x + y \quad \text{for all } x, y \in Y.$$

Then

$$F_1 + F_2 = f \circ \widetilde{F}.$$

By Theorem 1.48 we conclude that $F_1 + F_2$ is lower semicontinuous. $\square$

Theorem 1.66. *If the multivaled maps $F_1, F_2 : X \to \mathcal{P}_{cp}(Y)$ are u.s.c., then their sum $F_1 + F_2$ is u.s.c.*

Example 1.67. *We consider the multimaps $F_1, F_2 : \mathbb{R}^2 \to \mathcal{P}(\mathbb{R}^2)$ defined in the following way:*

$$F_1(x_1, x_2) = \{(x_1, x_2)\}$$

and

$$F_2(x_1, x_2) = \{(y_1, y_2) : y_1.y_2 = 1, \ y_1 > 0\}.$$

They are continuous, but their sum

$$(F_1 + F_2)(x_1, x_2) = \{(z_1, z_2) : z_1 = x_1 + y_1, \ z_2 = x_2 + y_2; \ y_1.y_2 = 1 \ y_1 > 0\}$$

is not u.s.c., since

$$(F_1 + F_2)_+^{-1}(\mathbb{R}_+^2) = \overline{\mathbb{R}}_+^2,$$

where

$$\mathbb{R}_+^2 = \{(x_1, x_2) : \ x_1 > 0, \ x_2 > 0\}.$$

Theorem 1.68. *Let $f : X \to \mathbb{R}$ be a continuous map. Then*

(a) *If the multivalued map $F : X \to \mathcal{P}(Y)$ is lower semi continuous, then the the product $f \cdot F : X \to \mathcal{P}(Y)$, defined by*

$$(f \cdot F)(x) = f(x)F(x) \quad \text{for all } x \in X,$$

is lower semicontinuous.

(b) *If the multivalued map $F : X \to \mathcal{P}(Y)$ is upper semicontinuous, then then the product $f \cdot F : X \to \mathcal{P}_{cp}(Y)$, is upper semicontinuous.*

Proof. By Theorem 1.48 the multivalued map $f \times F : X \to \mathcal{P}(\mathbb{R} \times Y)$ defined by $(f \times F) = (f, F)$ is *l.s.c.* The single map $\phi : \mathbb{R} \times Y \to Y$ defined as $\phi(a, y) = ay$ is continuous. Also from Theorem 1.48 the multimap

$$f \cdot F = \phi \circ (f \times F),$$

is *l.s.c.* $\qquad\qquad\qquad\qquad\qquad\qquad\qquad\qquad\qquad\qquad\qquad\qquad\qquad\qquad\square$

Definition 1.69. *Let $F : X \to \mathcal{P}(Y)$ be a multimap. The multivalued map $coF : X \to \mathcal{P}_{cv}(Y)$, defined by $(coF)(x) = co(F(x))$, (and $(\overline{co}F)(x) = \overline{co}(F(x))$), is said to be the closed convex hull of $F(x)$, (is said to be the convex closure of the multimap F.)*

Proposition 1.70.   *(a) If $F : X \to \mathcal{P}(Y)$ is lower semi continuous, then so are $co(F)$ and $\overline{co}(F)$.*

   (b) If Y is a Banach space and $F : X \to \mathcal{P}_{cp}(Y)$ is u.s.c., then so is $\overline{co}(F)$.

1.9   Closed and proper multivalued maps

Definition 1.71. *A multivalued map $F : X \to \mathcal{P}(Y)$ is called closed if for every closed set $C \subset X$ the image $F(C)$ is closed in Y.*

Proposition 1.72. *The multivalued map $F : X \to \mathcal{P}(Y)$ is upper semicontinuous if and only if $F^{-1} : Y \to \mathcal{P}(X)$ is closed.*

Proof. For each $C \in \mathcal{P}_{cl}(X)$, we have

$$F^{-1}(C) = \{F^{-1}(y) : y \in C\} = F_-^{-1}(C).$$

Hence we conclude that F is u.s.c. if and only if F^{-1} is closed. $\qquad\qquad\qquad\square$

Definition 1.73. *A multivalued map $F : X \to \mathcal{P}(Y)$ is called proper if for every $K \in \mathcal{P}_{cp}(Y)$ we have $F^{-1}(K)$ is compact.*

Lemma 1.74. *Let $F : X \to \mathcal{P}(Y)$ be a multivalued map, then the following statements are equivalent:*

  (i) F is u.s.c., and $F(x) \in \mathcal{P}_{cp}(Y)$ for every $x \in X$.

  (ii) F^{-1} is closed, and $(F^{-1})^{-1}(x) \in \mathcal{P}_{cp}(Y)$ for every $x \in X$.

  (iii) F^{-1} is closed and proper.

Proof. • $i) \Longrightarrow ii)$
From Proposition 1.72, we deduce that F^{-1} is closed. Now we show that for $x \in X$ we have $(F^{-1})^{-1}(x) \in \mathcal{P}_{cp}(Y)$. Indeed let $(y_\alpha)_{\alpha \in J} \subset (F^{-1})^{-1}(x)$. Then

$$(y_\alpha)_{\alpha \in J} \subset F^{-1}(x) \Longrightarrow (y_\alpha)_\alpha \subset F(x).$$

Since $F(x) \in \mathcal{P}_{cp}(Y)$, there exists a subsequence $(y_\alpha)_{\alpha \in I}$, $I \subset J$, converging to $y \in F(x)$. This implies that $y \in (F^{-1}(x))^{-1}$.
    • $ii) \Longrightarrow iii)$

Let $K \in \mathcal{P}_{cp}(X)$. We show that $(F^{-1})^{-1}(K) \in \mathcal{P}_{cp}(Y)$. Let $(y_\alpha)_{\alpha \in J} \subset F(x)$, $x \in X$. Then $(y_\alpha)_{\alpha \in J} \subset F^{-1}(x)$. Since $F^{-1}(x) \in \mathcal{P}_{cp}(Y)$, thus there exists a subsequence $(y_\alpha)_{\alpha \in I}$, $I \subset J$, converging to $y \in F^{-1}(x)$. By Proposition 1.72, F is u.s.c., so $F(K) \in \mathcal{P}(X)$. Let $(y_\alpha)_{\alpha \in J} \subset (F^{-1})^{-1}(K)$, hence there exists $(x_\alpha)_{\alpha \in J} \subset K$ such that for every $\alpha \in J$ we have

$$y_\alpha \in F^{-1}(x_\alpha) \implies y_\alpha \in F(x_\alpha).$$

Therefore, $(y_\alpha)_{\alpha \in J} \subset F(K)$. Using the fact that K and $F(K)$ are compact subsets of X and Y, respectively, there are subsequences $(y_\alpha)_{\alpha \in I}$ and $(x_\alpha)_{\alpha \in I}$ converging to x and y, respectively. Since F has a closed graph we conclude that $y \in F(x)$, so F^{-1} is proper.

• *iii*) $\implies$ *i*)

By Proposition 1.72, F is u.s.c. Let $x \in X$ and $(y_\alpha)_{\alpha \in J} \subset F(x)$ be a sequence. Then $(y_\alpha)_{\alpha \in J} \subset F^{-1}(x)$. Since F^{-1} is proper then the compactness of $\{x\}$ implies that $F^{-1}(x)$ is compact. Hence there exists a subsequence $(y_\alpha)_{\alpha \in I}$ converging to $y \in F(x)$. □

Corollary 1.75. *A closed singular map $p : X \to Y$ is proper if and only if $p^{-1}(y) \in \mathcal{P}_{cp}(X)$ for every $y \in Y$.*

Definition 1.76. *A single-valued map of topological spaces is called perfect if it is continuous, proper, closed, and onto.*

Theorem 1.77. *Let $F : X \to \mathcal{P}_{cp}(Y)$ be u.s.c. Let the map $p : \Gamma_F \to X$ be defined by*

$$p(x, y) = x, \text{ for every } (x, y) \in \Gamma_F.$$

Then

$$p^{-1}(x) = \{x\} \times F(x) \text{ for every } x \in X,$$

and p is continuous, closed, and proper. The map p is perfect if and only if $dom(F) = X$.

Proof. Let $x \in X$, then

$$
\begin{aligned}
p^{-1}(\{x\}) &= \{(a, b) \in \Gamma_F : p(a, b) = x\} \\
&= \{(a, b) \in \Gamma_F : a = x\} \\
&= \{(x, y) \in X \times Y : y \in F(x)\} \\
&= \{x\} \times F(x).
\end{aligned}
$$

• p is continuous. Let $B \subset X$ be a closed set and $(x_\alpha, y_\alpha)_{\alpha \in J} \in p^{-1}(B)$ be a sequence converging to (x, y). Hence, $x \in B$ and

$$p(x_\alpha, y_\alpha) = x_\alpha, \text{for every } \alpha \in J \implies y_\alpha \in F(x_\alpha), \text{ or all } \alpha \in J.$$

It clear that $K = \{x_\alpha : \alpha \in J\} \cup \{x\}$ is compact in X, thus $F(K)$ is compact. Then there exists a subsequence $(y_\alpha)_{\alpha \in I}$, $I \subset J$, converging to y. Since F is u.s.c. and $F(\cdot) \in \mathcal{P}_{cp}(X)$, thus F has a closed graph. This implies that $y \in F(x)$ and $p(x, y) = x$. So $(x, y) \in p^{-1}(B)$.

• p is closed. Let $B \subset \Gamma_F$ be a closed subset and $(x_\alpha)_{\alpha \in J} \in p(B)$ be a sequence converging to x. Then for every $\alpha \in J$ there exists $y_\alpha \in F(x_\alpha)$ such that

$$p(x_\alpha, y_\alpha) = x_\alpha \text{ and } (y_\alpha)_{\alpha \in J} \in F(K).$$

Since F is u.s.c. and $F(\cdot) \in \mathcal{P}_{cp}(X)$, thus $F(K)$ is compact and F has a closed graph. So that there exists a subsequence $(y_\alpha)_{\alpha \in I}$, $I \subset J$, converging to $y \in F(x)$. Consequently,

$$(x_\alpha, y_\alpha) \to (x, y).$$

Because B is closed, (x, y) belongs to B, hence $p(x, y) = x \in p(B)$.

• p is proper. Let $K \in \mathcal{P}_{cp}(X)$, and $(x_\alpha, y_\alpha)_{\alpha \in J} \in p^{-1}(K)$ be a sequence. Then

$$p(x_\alpha, y_\alpha) = x_\alpha \in K \quad \text{for all } \alpha \in J.$$

Hence there exists a subsequence $(x_\alpha)_{\alpha \in I} \subset K$, $I \subset J$, converging to $x \in K$. Consequently, since F is u.s.c. and for every $x \in X$ we have $F(x) \in \mathcal{P}_{cp}(X)$, we deduce that there exists a subsequence $(y_\alpha)_{\alpha \in I} \subset F(\{x_\alpha : \alpha \in I\}$, $I \subset J$ converging to $y \in F(x)$. So there exists a subsequence $(x_\alpha, y_\alpha)_{\alpha \in I} \subset p^{-1}(K)$ converging to $(x, y) \in p^{-1}(K)$. $\qquad \square$

1.10   Open multivalued maps

Definition 1.78. *A multivalued map $F : X \to \mathcal{P}(Y)$ is called open if for every open set $C \subset X$ the image $F(C)$ is open in Y.*

Proposition 1.79. *The multivalued map $F : X \to \mathcal{P}(Y)$ is lower semicontinuous if and only if $F^{-1} : Y \to \mathcal{P}(X)$ is open.*

Proof. Since for each $C \in \mathcal{P}(X)$, we have

$$F^{-1}(C) = \{F^{-1}(y) : y \in C\} = F_-^{-1}(C),$$

we conclude that, F is l.s.c. if and only if F^{-1} is open. $\qquad \square$

Example 1.80. *Let $f : Y \to X$ be onto. Then f is open if and only if $f^{-1} : X \to \mathcal{P}(Y)$ is l.s.c.*

Lemma 1.81. *Let $F : X \to \mathcal{P}(Y)$ be a multivalued map. Then the following statements are equivalent:*

(i) F is l.s.c., and $F(x) \in \mathcal{P}_{cp}(Y)$ for every $x \in X$.

(ii) F^{-1} is open, and $(F^{-1})^{-1}(x) \in \mathcal{P}_{cp}(Y)$ for every $x \in X$.

Proof. • $i) \Longrightarrow ii)$
From Proposition 1.79, we deduce that F^{-1} is open. Now we show that for $x \in X$ we have $(F^{-1})^{-1}(x) \in \mathcal{P}_{cp}(Y)$. Indeed let $(y_\alpha)_{\alpha \in J} \subset (F^{-1})^{-1}(x)$, then

$$(y_\alpha)_{\alpha \in J} \subset F^{-1}(x) \Longrightarrow (y_\alpha)_\alpha \subset F(x).$$

Since $F(x) \in \mathcal{P}_{cp}(Y)$, there exists a subsequence $(y_\alpha)_{\alpha I}$, $I \subset J$, converging to $y \in F(x)$, this implies that $y \in (F^{-1}(x))^{-1}$.
• $ii) \Longrightarrow i)$
By Proposition 1.79, F is l.s.c. Let $x \in X$ and $(y_\alpha)_{\alpha \in J} \subset F(x)$ be a sequence, such that

$$(y_\alpha)_{\alpha \in J} \subset F^{-1}(x) \Longrightarrow (y_\alpha)_{\alpha \in J} \subset (F^{-1})^{-1}(x).$$

Since $(F^{-1})^{-1}(x) \in \mathcal{P}(Y)$, there exists a subsequence $(y_\alpha)_{\alpha \in I}$ converging to $y \in F(x)$. $\quad \square$

Theorem 1.82. *Let $F : X \to \mathcal{P}_{cp}(Y)$ be l.s.c. and have open graph. Let the map $p : \Gamma_F \to X$ be defined by*

$$p(x, y) = x, \text{ for every } (x, y) \in \Gamma_F.$$

Then

$$p^{-1}(x) = \{x\} \times F(x) \text{ for every } x \in X,$$

and p is continuous and open.

Proof. From Proposition 1.72, we have

$$p^{-1}(x) = \{x\} \times F(x) \text{ for every } x \in X.$$

• p is continuous. Let $U \subset X$ be an open set and $(x_0, y_0) \in p^{-1}(U)$. Then, $x_0 \in B$ and $p(x_0, y_0) = x_0$. Since Γ_F is open then there exists $V_{x_0} \times V_{y_0} \in \mathcal{N}(x_0) \times \mathcal{N}(y_0)$ and $W_{x_0} \in \mathcal{N}(x_0)$ such that $V_{x_0} \times V_{y_0} \subset \Gamma_F$ and $W_{x_0} \subset U$, hence $V_{x_0} \cap W_{x_0} \times V_{y_0} \subset \Gamma_F$ and $V_{x_0} \cap W_{x_0} \subset U$. This implies that $V_{x_0} \cap W_{x_0} \times V_{y_0} \subset p^{-1}(U)$. So $p^{-1}(U)$ is open. • p is open. Let $U \subset \Gamma_F$ be a open set. Then for every $(x, y) \in U$, there exists $V_x \times V_y \in \mathcal{N}(x) \times \mathcal{N}(y)$ such that

$$V_x \times V_y \subset U \subset \Gamma_F.$$

Therefore

$$p(a, b) = a, \text{ for every } (a, b) \in V_x \times V_y \Longrightarrow V_x \subset p(U).$$

Consequently, $p(U)$ is open. □

Theorem 1.83. *Let $F : X \to \mathcal{P}(Y)$ be a multivalued function.*

(a) *If F is a l.s.c,, then p_X is an open map, where $p_X : \Gamma_F \to X$ is defined by*

$$p_X(x, y) = x, \text{ for every } x \in X.$$

(b) *If F is open then p_Y is open, where $p_Y : \Gamma_F \to Y$ is defined by*

$$p_Y(x, y) = y, \text{ for every } y \in Y.$$

1.11 Weakly upper and lower semicontinuous functions

Definition 1.84. *Let X be a set and let $(Y_i, \tau_i)_{i \in I}$ be a family of topological spaces. For each $i \in I$ let $f_i : X \to Y_i$ be a mapping. We define the initial topology on X (associated with the mappings f_i) to be the coarsest topology on X for which all the mappings f_i are continuous. Its basic open sets are of the form*

$$\cap_{i \in J} f_i^{-1}(U_i)$$

with J a finite subset of I and $U_i \in \tau_i$ for each i. Also it is called the weak topology on X generated by the $(f_i)_{i \in I}$ and we denote it by $\sigma(X, (f_i)_{i \in I})$.

Proposition 1.85. *Let $(x_n)_{n \in \mathbb{N}}$ be a sequence in X. It converges in the topology $\sigma(X, (f_i)_{i \in I})$ to some $x \in X$ if and only if for each $i \in I$, $(f_i(x_n)_{n \in \mathbb{N}})$ converges to $f_i(x)$.*

Proof. Since for every $j \in I$, the function $f_j : (X, \sigma(X, (f_i)_{i \in I})) \to Y_j$ is continuous, we have

$$f_j(x_n) \to f_j(x) \text{ as } n \to \infty.$$

Conversely, suppose that there exists x in X such that

$$\forall i \in I, \ f_i(x_n) \to f_i(x) \text{ as } n \to \infty.$$

Let $U \in \sigma(X, (f_i)_{i \in I})$ be any open set containing x. By definition, there exists a finite subset J of I, and open sets $(U_j)_{j \in J}$ such that $U_j \subset Y_j$ for all $j \in J$, and such that

$$x \in \bigcap_{j \in J} f_j^{-1}(U_j),$$

which means that for each $j \in J$, we have

$$f_j(x) \in U_j.$$

Given $j \in J$, we know that the sequence $(f_j(x_n)_{n \in \mathbb{N}})$ converges to $f_j(x)$. Then, since U_j contains $f_j(x)$, there exists $n_j \in \mathbb{N}$ such that

$$\forall n \geq n_j, \quad f_j(x_n) \in U_j.$$

Letting $N = \max_{j \in J} n_j$, we have

$$\forall n \geq N, \ \forall j \in J, \ f_j(x_n) \in U_j.$$

In other words, for each $n \in \mathbb{N}$ such that $n \geq N$, we have

$$x_n \in \bigcap_{j \in J} f_j^-(U_j) \subset U.$$

So $(x_n)_{n \in \mathbb{N}}$ converges to x for the topology $\sigma(X, (f_i)_{i \in I})$. $\qquad\square$

Lemma 1.86. *For any topological space Z, a mapping $g : Z \to X$ is continuous when X is equipped with the initial topology if and only if each composition $f_i \circ g : Z \to Y_i$ is continuous*

Proof. Since for each $i \in I$, f_i is continuous, then $f_i \circ g$ is also continuous for every $i \in I$. Conversely, we show that $g^{-1}(U)$ is open in Z for every open set U in X. But we know that U has the form $U = \bigcup_{i \in I_*} \bigcap_{j \in J_i} f_{i_j}^{-1}(U_{i_j})$ where I_* is arbitrary and J_i are finite where U_{i_j} is open in Y_{i_j}. Therefore

$$g^{-1}(U) = g^{-1}\left(\bigcup_{i \in I_*} \bigcap_{j \in J_i} f_{i_j}^{-1}(U_{i_j}) \right) = \bigcup_{i \in I_*} \bigcap_{j \in J_i} (f_{i_j} \circ g)^{-1}(U_{i_j})$$

which is open in Z since every map $f_{i_j} \circ g$ is continuous. $\qquad\square$

An example of a weak topology is the topology of pointwise convergence. It is defined as follows: let A be any set and let X be the set of functions $A \to \mathbb{R}$. For every $a \in A$, define the function $\phi_a : X \to \mathbb{R}$ by

$$\phi_a(f) = f(a), \text{ for all } f \in X.$$

The topology of pointwise convergence is $\sigma(X, (f_a)_{a \in A})$.

1.12 The topology $\sigma(X, X^*)$

Let X be a Banach space (or locally convex space), X^* denote the dual space of X, and let $f \in X^*$. We denote by $\psi_f : X \to \mathbb{R}$ the linear functional $\psi_f(x) = \langle f, x \rangle = f(x)$. As f runs through X^* we obtain a collection $(\psi_f)_{f \in X^*}$ of maps from X into $\mathbb{R}$.

Definition 1.87. *The weak topology $\sigma(X, X^*)$ on X is the coarsest topology associated with the collection $(\psi_f)_{f \in X^*}$.*

The next result tells us the form of the neighborhoods of a point $x \in X$ for the weak topology.

Proposition 1.88. *The neighborhood basis at the point $x \in X$ is the collection of sets of the form*

$$V(\epsilon, f_1, f_2, \ldots, f_m, x) = \{y \in X : |f_i(y - x)| < \epsilon\},$$

where $n \in \mathbb{N}$, $\epsilon > 0$ and $f_1, \ldots, f_n \in X^*$.

Proof. Since $f_1, \ldots, f_m$ are continuous then

$$V(\epsilon, f_1, f_2, \ldots, f_m, x) = \bigcap_{i=1}^{m} f_i^{-1}((f_i(x) - \epsilon, f_i(x) + \epsilon))$$

is an open set. Furthermore it contains x since

$$|f_i(x - x)| = 0 < \epsilon, \quad i = 1, \ldots, m.$$

Now let O be any open set containing x. By definition of the topology $\sigma(X, X^*)$, O is a union of finite intersections of preimages of open sets U_i in $\mathbb{R}$ of bounded linear functionals on X, i.e.,

$$O = \bigcup_{n \in \mathbb{N}} \bigcap_{i=1}^{n} f_i^{-1}(O_i).$$

Since $x \in O$ there exist finitely many bounded linear functionals $f_1, \ldots, f_n \in X^*$ and open subsets $(O_i)_{i=1,\ldots,n}$ of $\mathbb{R}$ such that

$$x \in \bigcap_{i=1}^{n} f_i^{-1}(O_i) \implies f_i(x) \in O_i, \quad \text{for all } i = 1, \ldots, n.$$

Hence, for each $i \in \{1, \ldots, n\}$ there exists $\epsilon_i > 0$ such that

$$(f_i(x) - \epsilon_i, f_i(x) + \epsilon_i) \in O_i.$$

Let $\epsilon = \min_{1 \leq i \leq n} \epsilon_i$, then we have

$$(f_i(x) - \epsilon, f_i(x) + \epsilon) \in O_i, \quad i = 1, \ldots, n.$$

It follows that

$$x \in V(f_1, \ldots, f_n, \epsilon) \subset \bigcap_{i=1}^{n} f_i^{-1}(O_i) \subset O.$$

$\square$

Now we want to check is that the weak topology on X is Hausdorff, which will guarantee us the uniqueness of limits.

Proposition 1.89. *The topology $\sigma(X, X^*)$ is Hausdorff.*

From Proposition 1.88 we already know the weak open subsets of X and we can define the convergence of a net $(x_\alpha)_{\alpha \in \mathbb{N}}$ in the weak topology of X.

Definition 1.90. *Let X be a normed space. A set U is called weakly open in X if for each $x \in U$ there exists $\epsilon > 0$ and $f_1, \ldots, f_n \in X^*$ such that*

$$V(f_1, \ldots, f_n, \epsilon) \subset U.$$

Definition 1.91. *(neighborhood). We say that a set $U \subset X$ is a weak neighborhood of a point x in X if there is a weakly open set O such that $O \subset U$.*

Definition 1.92. *We say that the net $(x_\alpha)_\alpha \subset X$ converges weakly to $x \in X$ if for each $f \in X^*$ we have*

$$\lim f(x_\alpha) = f(x).$$

Remark 1.93. *If a net $(x_\alpha)_{\alpha \in J}$ in X converges to x in the weak topology $\sigma(X, X^*)$ we shall write*

$$x_\alpha \rightharpoonup x.$$

Now we prove a few easy facts comparing the weak topology and the norm (also called strong) topology on X.

Proposition 1.94. *Let X be a Banach space and $(x_n)_{n \in \mathbb{N}}$.*

(a) *A sequence $(x_n)_{n \in \mathbb{N}}$ converges weakly to x if and only if for every $f \in X^*$ we have*

$$f(x_n) \to f(x) \quad as \ \ n \to \infty.$$

(b) *If $x_n \to x$ as $n \to \infty$, then $x_n \rightharpoonup x$ as $n \to \infty$.*

(c) *If $(x_n)_{n \in \mathbb{N}}$ is a sequence in X converging weakly to x, then $(\|x_n\|)_{n \in \mathbb{N}}$ is bounded and*

$$\|x\| \le \liminf_{n \to \infty} \|x_n\|.$$

(d) *If If $(x_n)_{n \in \mathbb{N}}$ is a sequence in X converging weakly to x and $(f_n)_{n \in \mathbb{N}}$ is a sequence in X^* converging strongly to f, then*

$$f_n(x_n) \to f(x) \quad as \ \ n \to \infty.$$

Proposition 1.95. *If X and Y are normed spaces and $A : X \to Y$ is a linear operator, then A is continuous if and only if*

$$A : (X, \sigma(X, X^*)) \to (Y, \sigma(Y, Y^*))$$

is weakly continuous.

Proof. $\Rightarrow$ Suppose $x_n \rightharpoonup 0$ in X. Then for every $y* \in Y^*$, we have $y^* \circ A \in Y^*$ and $(y^* \circ A)(x_n) \to 0$, hence $\langle y^*, A(x_n) \rangle \to 0$, which means that

$$A(x_n) \rightharpoonup 0, \text{ as } n \to \infty.$$

This implies that A is weakly continuous.

$\Leftarrow$: Suppose A was unbounded. Thus we can find $(x_n)_{n \in N} \subset X$ with

$$\|x_n\| \le 1, \ \ \|A(x_n)\| \ge n^2, \text{ for each } n \in \mathbb{N}.$$

Then

$$\frac{x_n}{n} \to 0 \text{ as } n \to \infty \text{ in } X,$$

hence

$$\frac{x_n}{n} \rightharpoonup 0 \text{ as } n \to \infty,$$

and

$$A\left(\frac{x_n}{n}\right) \rightharpoonup 0 \text{ as } n \to \infty.$$

From Proposition 1.94(c) there exists $M > 0$ such that

$$\left\|A\left(\frac{x_n}{n}\right)\right\| \leq M, \quad \text{for every } n \in \mathbb{N}.$$

By the choice of the sequence $(x_n)_{n \in \mathbb{N}}$ we have

$$n \leq \left\|A\left(\frac{x_n}{n}\right)\right\| \leq M \quad \text{for every } n \in \mathbb{N},$$

which is a contradiction. This proves that A is continuous. $\qquad\square$

Another useful consequence of the Banach-Steinhaus theorem is the following. Here we let $L(X, Y)$ denote the set of all linear operators from X to Y.

Corollary 1.96. *Assume that X is a Banach space and Y is a normed space. If a sequence $(A_n) \in L(X, Y)$ satisfies $A_n(x) \rightharpoonup A(x)$ in Y for each $x \in X$, then A is a continuous operator.*

Proof. Clearly, the mapping $A : X \to Y$ defined by $A(x) = w - \lim_{n \to \infty} A_n(x)$ is a linear operator. Next, let $\mathcal{M} = \{A_n : n \in \mathbb{N}\}$. Since the sequence $(A_n(x))_{n \in \mathbb{N}}$ is weakly convergent for each x, we see that $(A_n(x))_{n \in \mathbb{N}}$ is a norm bounded sequence for each x. By the Banach-Steinhaus theorem, there exists some $M > 0$ such that

$$\|A_n(x)\| \leq M, \text{ for each } n \in \mathbb{N}.$$

Now note that if $\|x\| \leq 1$ and $y^* \in Y^*$, then for each $n \in \mathbb{N}$,

$$\begin{aligned} \langle A_n(x), y^* \rangle &\leq \|y^*\| \|A_n\| \|x\| \\ &\leq M \|y^*\| \|x\|. \end{aligned}$$

This implies

$$|\langle A(x), y^* \rangle| \leq M \|y^*\|, \text{ for each } \|x\| \leq 1 \text{ and all } y^* \in Y^*.$$

Therefore

$$\|A(x)\| = \sup_{\|y^*\| \leq 1} |\langle A(x), y^* \rangle| \leq M, \text{ for all } x \in X, \ \|x\| \leq 1.$$

Hence

$$\|A(x)\| \leq M \|x\|, \text{ for all } x \in X, \ \|x\| \leq 1.$$

This shows that $A \in L(X, Y)$. $\qquad\square$

In 1940 Šmulian proved that every weakly compact subset of a Banach space is weakly sequentially compact. In 1947 Eberlein showed the converse.

Theorem 1.97. *(Eberlein-Šmulian Theorem) A subset A of a normed space X is relatively weakly compact (respectively, weakly compact) if and only if every sequence in A has a weakly convergent subsequence in X (respectively, in A).*

Theorem 1.98. *(Krein-Šmulian Theorem) In a Banach space, both the convex circled hull and the convex hull of a relatively weakly compact set are relatively weakly compact sets.*

Assume now X and Y are normed spaces (or locally convex topological Hausdorff spaces) with their weak topologies $\sigma(X, X^*)$ and $\sigma(Y, Y^*)$, respectively.

Definition 1.99. *A multivalued map $F : X \to \mathcal{P}(Y)$ is weakly-weakly upper semicontinuous (w-w.u.s.c.) on X if for every weakly closed subset $C \subseteq Y$ the set*

$$F_-^{-1}(M) = \{x \in X : F(x) \cap C \neq \emptyset\}$$

is sequentially weakly closed in X.

Definition 1.100. *A multivalued map $F : X \to \mathcal{P}(Y)$ is weakly-strongly upper semicontinuous (w-s.u.s.c.) on X if for every weakly closed subset $C \subseteq Y$ the set*

$$F_-^{-1}(M) = \{x \in X : F(x) \cap C \neq \emptyset\}$$

is closed (in the norm topology) in X.

In similar way as in Section 1.1 we obtain the following results.

Theorem 1.101. *Let X, Y are two Banach spaces and let $F : X \to \mathcal{P}(Y)$. The following assertions are equivalent:*

i) F is w.-w.u.s.c. on X,

ii) for every weakly open set $U \subseteq Y$ the set

$$F_+^{-1}(U) = \{x \in X : F(x) \subset U\}$$

    is weakly open in X.

iii) for each sequence $(x_n)_{n \in \mathbb{N}}$ weakly converging to x and every weakly open $U \subset Y$ with $F(x) \subset U$, there exists $N \in \mathbb{N}$ such that

$$F(x_n) \subset U \text{ for all } n \geq N.$$

Theorem 1.102. *A multivalued map $F : X \to \mathcal{P}(Y)$ is w.-w.u.s.c. (w.-s.u.s.c.) on X if for every sequence $(x_n)_{n \in \mathbb{N}}$ of X weakly converging to x and every $(y_n)_{n \in \mathbb{N}} \subset Y$ with $y_n \in F(x_n)$, $n \in \mathbb{N}$ there exists a subsequence of $(x_n)_{n \in \mathbb{N}}$ weakly convergent to any $y \in F(x)$. Moreover, if there exists a weakly compact subset $C \subseteq Y$ such that*

$$F(X) \subseteq C,$$

then the above conditions are also necessary for F to be w.-w.u.s.c.

Theorem 1.103. *Let X and Y be Banach spaces and let C be a weakly compact subset of Y. Suppose that $F : X \to \mathcal{P}(Y)$ is w.-w.u.s.c. on X and such that $F(x) \subseteq C$. Then for every weakly compact set $B \subseteq X$ the set $F(B)$ is weakly compact in Y.*

Proof. Let $B \in \mathcal{P}_{wkcp}(X)$. Since $F(B) \subset C$ then $F(B)$ is weakly sequentially compact. Then by Smulian's theorem for each $y \in \overline{F(B)}^w$ there exist sequences $(y_n)_{n \in \mathbb{N}} \in F(B)$ and $(x_n)_{n \in \mathbb{N}} \subset B$ such that y_n converges weakly to y and $y_n \in F(x_n)$ for each $n \in \mathbb{N}$. Since B is weakly compact, there exist subsequences $(x_k)_{k \in \mathbb{N}}$ of $(x_n)_{n \in \mathbb{N}}$ and $(y_k)_{k \in \mathbb{N}}$ of $(y_n)_{n \in \mathbb{N}}$ with $y_k \in F(x_k)$, $k \in \mathbb{N}$, such that $(x_k)_{k \in \mathbb{N}}$ and $(y_k)_{k \in \mathbb{N}}$ converge weakly to x and y, respectively. By Theorem 1.101, we have $y \in F(x)$. Hence

$$\overline{F(B)}^w \subseteq F(B) \implies \overline{F(B)}^w = F(B).$$

$\square$

Proposition 1.104. *Let X be a topological space, E be a separable Banach space, and suppose that the multifunction $F : X \to 2^E$ is u.s.c. when E is endowed with its weak topology. Then $F_+^{-1}(V) \in B(X)$ whenever V is open in the strong topology of E.*

Chapter 2

Hausdorff-Pompeiu Metric Topology

The Hausdorff (more precisely, the Hausdorff-Pompeiu) metric topology is the oldest, and probably the most popular hyperspace topology. The Haudorff metric is defined on a metric space and is the main tool to quantify the distance between subsets of the given metric space. Let (X, d) be a metric space. In what follows, given $x \in X$ and $A \in \mathcal{P}(X)$, the distance of x from A, is defined by

$$d(x, A) = \inf\{d(x, a) : a \in A\}, \text{ where } A \in \mathcal{P}(X).$$

As usual, $d(x, \emptyset) = +\infty$.

Definition 2.1. *Let (X, d) be a metric space. The generalized diameter functional $\delta : \mathcal{P}(X) \cup \{\emptyset\} \to \mathbb{R}_+ \cup \{\infty\}$ is defined by:*

$$\delta(A) = \begin{cases} \sup\{d(a, b) : a, b \in A\}, & \text{if } A \neq \emptyset, \\ 0 & , \quad \text{if } A = \emptyset. \end{cases}$$

Definition 2.2. *The subset A of X is said to be bounded if and only if*

$$\delta(A) < \infty.$$

Lemma 2.3. *Let (X, d) be a metric space and $A, B \in \mathcal{P}_b(X)$. Then:*

   i) $\delta(A) = 0$ if and only if $A = \{a\}$.

   ii) If $A \subset B$ then $\delta(A) \leq \delta(B)$.

   iii) $\delta(A) = \delta(\overline{A})$.

   iv) If $A \cap B \neq \emptyset$, then $\delta(A \cup B) \leq \delta(A) + \delta(B)$.

   v) If X is a normed space then:
   a) $\delta(x + A) = \delta(A)$, for each $x \in X$.
   b) $\delta(\lambda A) = |\lambda|\delta(A)$, where $\lambda \in \mathbb{R}$.
   c) $\delta(A) = \delta(coA)$.
   d) $\delta(A) \leq \delta(A + B) \leq \delta(A) + \delta(B)$.

Proof. i) $\delta(A) = \sup\{d(x, y) : x, y \in A\} = 0$ if and only if $d(x, y) = 0$ for every $x, y \in A$, this equivalent to $A = \{a\}$.

   ii) $\delta(A) = \sup\{d(x, y) : x, y \in A\}$. Since $A \subseteq B$, then

$$\sup\{d(x, y) : x, y \in A\} \leq \sup\{d(x, y) : x, y \in B\}.$$

Therefore

$$\delta(A) \leq \delta(B).$$

iii) Since $A \subseteq \overline{A}$, then $\delta(A) \leq \delta(\overline{A})$. Let $x, y \in \overline{A}$, then there are sequences $(x_n)_{n \in \mathbb{N}} \subset A$ and $(y_n)_{n \in \mathbb{N}} \subset A$ such that

$$x_n \to x, \quad y_n \to y, \quad \text{as } n \to \infty.$$

Hence, for every $n \in \mathbb{N}$ we have

$$d(x_n, y_n) \leq \delta(A) \Longrightarrow d(x, y) \leq \delta(A).$$

So,

$$\delta(A) \leq \delta(\overline{A}) \Longrightarrow \delta(A) = \delta(\overline{A}).$$

iv) Let $x \in A, y \in B$ and $z \in A \cap B$. Then

$$d(x, y) \leq d(x, z) + d(z, y) \Longrightarrow d(x, y) \leq \delta(A) + \delta(B).$$

Hence

$$\delta(A \cup B) \leq \delta(A) + \delta(B).$$

v) a) Let $x \in X$, then

$$
\begin{aligned}
\delta(x + A) &= \sup\{\|x + a - x - b\| : a, b \in a\} \\
&= \sup\{\|a - b\| : a, b \in a\} \\
&= \delta(A).
\end{aligned}
$$

b) Let $\lambda \in \mathbb{R}$, thus

$$
\begin{aligned}
\delta(\lambda A) &= \sup\{\|\lambda x - \lambda y\| : x, y \in A\} \\
&= \sup\{|\lambda| \|x - y\| : x, y \in A\} \\
&= |\lambda| \sup\{\|x - y\| : x, y \in A\} = |\lambda| \delta(A).
\end{aligned}
$$

c) $A \subseteq co(A)$, then

$$\delta(A) \leq \delta(co(A)).$$

For every $x, y \in co(A)$, there exist $\lambda_i, \bar{\lambda}_i, \in [0, 1]$ and $x_i, y_i \in A$, $i = 1, \ldots n$, with $\displaystyle\sum_{i=1}^{n} \lambda_i = 1$ and $\displaystyle\sum_{i=1}^{n} \bar{\lambda}_i = 1$ such that

$$x = \sum_{i=1}^{n} \lambda_i x_i \text{ and } y = \sum_{i=1}^{n} \bar{\lambda}_i y_i.$$

Then

$$
\begin{aligned}
\|x - y\| &= \left\| \sum_{i=1}^{n} \lambda_i x_i - \sum_{i=1}^{n} \bar{\lambda}_i y_i \right\| \\
&= \left\| \sum_{i=1}^{n} \bar{\lambda}_i \sum_{i=1}^{n} \lambda_i x_i - \sum_{i=1}^{n} \lambda_i \sum_{i=1}^{n} \bar{\lambda}_i y_i \right\| \\
&= \left\| \sum_{i,j=1}^{n} \bar{\lambda}_j \lambda_i x_i - \sum_{j,i=1}^{n} \lambda_i \bar{\lambda}_j y_j \right\|
\end{aligned}
$$

$$= \left\| \sum_{i,j=1}^{n} \bar{\lambda}_j \lambda_i (x_i - y_j) \right\|$$

$$\leq \sum_{i,j=1}^{n} \bar{\lambda}_j \lambda_i \|x_i - y_j\|$$

$$\leq \sum_{i,j=1}^{n} \bar{\lambda}_j \lambda_i \delta(A)$$

$$\leq \delta(A).$$

Consequently,

$$\delta(A) = \delta(co(A)).$$

d) Let $x \in B$, thus

$$\begin{aligned}
\delta(A) &= \delta(x + A) \\
&\leq \delta(A + B) \\
&= \sup\{\|x + y - \bar{x} - \bar{y}\| : x, \bar{x} \in A, y, \bar{y} \in B\} \\
&\leq \sup\{\|x - \bar{x}\| : x, \bar{x} \in A\} + \sup\{\|y - \bar{y}\| : y, \bar{y} \in B\} \\
&\leq \delta(A) + \delta(B).
\end{aligned}$$

$\square$

Definition 2.4. *Let $A, B \in \mathcal{P}(X)$, we define*

- $H_d^*(A, B) = \sup\{d(a, B) : a \in A\}$,

- $H_d^*(B, A) = \sup\{d(A, b) : b \in B\}$,

- $H_d(A, B) = \max(H_d^*(A, B), H_d^*(B, A))$ *(the Hausdorff (or Hausdorff-Pompeiu) distance between A and B).*

Remark 2.5. *Given $\epsilon > 0$, let*

$$A_\epsilon = \{x \in X : d(x, A) < \epsilon\}.$$

Then from the above definitions we have

$$H_d^*(A, B) = \inf\{\epsilon > 0 : A \subset B_\epsilon\}, \quad H_d^*(B, A) = \inf\{\epsilon > 0 : A \subset B_\epsilon\}.$$

From the definition we can easily prove the following properties:

- $H_d(A, A) = 0$, for all $A \in \mathcal{P}(X)$,

- $H_d(A, B) = H_d(B, A)$, for all $A, B \in \mathcal{P}(X)$,

- $H(A, B) \leq H_d(A, C) + H_d(C, B)$, for all $A, B, C \in \mathcal{P}(X)$.

Hence $H_d(\cdot, \cdot)$ is an extended pseudo-metric on $\mathcal{P}(X)$ (i.e., is a pseudo-metric which can also take the value $+\infty$). Moreover, we can prove that

$$H_d(A, B) = 0, \text{ if and only if } \overline{A} = \overline{B}.$$

So $\mathcal{P}_{cl}(X)$ furnished with the Hausdorff distance (H_d-distance), $H_d(\cdot, \cdot)$, becomes a metric space.

Lemma 2.6. $d(a, A) = 0$ *if and only if* $a \in \overline{A}$.

Proof. Let $x \in \overline{A}$, then for each $\epsilon > 0$ and every $B(x, \epsilon) \subset X$, we have

$$A \cap B(x, \epsilon) \neq \emptyset.$$

Then for each $\epsilon > 0$ there exists $x_\epsilon \in A$ such that

$$d(x, x_\epsilon) < \epsilon \Rightarrow d(x, A) < \epsilon.$$

Hence

$$d(x, A) = 0 \Rightarrow x \in \{a \in X : d(a, A) = 0\}.$$

Let $b \in X$ such that $d(a, A) = 0$, then for every $\epsilon > 0$ there exists $x_\epsilon \in A$ such that

$$d(a, x_\epsilon) < \epsilon \Rightarrow A \cap B(a, \epsilon) \neq \emptyset.$$

It follows that

$$\overline{A} = \{a \in X : d(a, A) = 0\}.$$

$\square$

Theorem 2.7. *Let* (X, d) *be a metric space. Then the pair* $(\mathcal{P}_{b,cl}(X), H_d)$ *is a metric space.*

Lemma 2.8. *Let* (X, d) *a metric space and* $A \in \mathcal{P}(X)$. *Then we have:*

    i) $d(\cdot, A) : (X, d) \to \mathbb{R}_+$, $x \to d(x, A)$ *is nonexpansive.*

    ii) $d(x, \cdot) : (\mathcal{P}_{cl}(X), H_d) \to \mathbb{R}_+$, $A \to d(x, A)$ *is nonexpansive.*

Proof. i) Let $x, y \in X$, then for each $a \in A$ we have

$$d(x, a) \leq d(x, y) + d(y, a) \Rightarrow d(x, A) \leq d(x, y) + d(y, A).$$

and

$$d(y, A) \leq d(x, y) + d(x, A).$$

    Then

$$d(x, A) - d(y, A) \leq d(x, y).$$

and

$$-d(x, y) \leq d(x, A) - d(y, A).$$

Hence

$$|d(x, A) - d(y, A)| \leq d(x, y).$$

(*ii*) Let $A, B \in \mathcal{P}(X)$, then for every $a \in A$ and $b \in B$ we have

$$d(x, A) \leq d(x, b) + d(b, A) \Rightarrow d(x, A) - d(x, B) \leq H_d(A, B)$$

and

$$d(x, B) \leq d(x, a) + d(a, B) \Rightarrow d(x, B) - d(x, A) \leq H_d(A, B).$$

Therefore

$$|d(x, A) - d(x, B)| \leq H_d(A, B).$$

$\square$

Proposition 2.9. *For every* $A, B \in \mathcal{P}_{cl}(X)$ *we have*

$$H_d(A, B) = \inf\{\epsilon > 0 : B \subset A_\epsilon, \ A \subset B_\epsilon\}.$$

Proof. Let $A, B \in \mathcal{P}_{cl}(X)$, and $\epsilon > 0$ be such that

$$A \subset B_\epsilon \text{ and } B \subset A_\epsilon.$$

Then, for every $a \in A$ and $b \in B$, we have

$$d(a, B) < \epsilon \Longrightarrow H_d^*(A, B) \leq \epsilon$$

and

$$d(b, A) < \epsilon \Longrightarrow H_d^*(B, A) \leq \epsilon.$$

Hence

$$H_d(A, B) \leq \epsilon \Longrightarrow H_d(A, B) \leq \inf\{\epsilon > 0 : A \subset B_\epsilon, \ B \subset A_\epsilon\}.$$

Next, assume that $H_d(A, B) < \inf\{\epsilon > 0 : A \subset B_\epsilon, \ B \subset A_\epsilon\}$, then there exists $\alpha > 0$ such that

$$H_d(A, B) < \alpha < \inf\{\epsilon > 0 : A \subset B_\epsilon, \ B \subset A_\epsilon\}.$$

Thus we have

$$A \subset B_\alpha \text{ and } B \subset A_\alpha \Longrightarrow \inf\{\epsilon > 0 : A \subset B_\epsilon, \ B \subset A_\epsilon\} \leq \alpha.$$

Hence

$$\alpha < \inf\{\epsilon > 0 : A \subset B_\epsilon, \ B \subset A_\epsilon\} \leq \alpha.$$

This is a contradiction and so

$$H_d(A, B) = \inf\{\epsilon > 0 : A \subset B_\epsilon, \ B \subset A_\epsilon\}.$$

$\square$

Lemma 2.10. *Let X be a Banach space. Then:*

i) $H_d(A_1 + \cdots + A_n, B_1 + \cdots + B_n) \leq H_d(A_1, B_1) + \cdots + H_d(A_n, B_n)$ *xfor all A_i, $B_i \in \mathcal{P}(X)$, $i = 1, 2, \ldots, n$,*

ii) $H_d(A + B, A + C) \leq H_d(B, C)$ *for all $A, B, C \in \mathcal{P}(X)$,*

iii) $H_d(A + B, A + C) = H_d(B, C)$ *for all $A \in \mathcal{P}_b(X)$ and for all $B, C \in \mathcal{P}_{b,cl,cv}(X)$,*

iv) $H_d(coY, coZ) \leq H_d(Y, Z)$ *for all $Y, Z \in \mathcal{P}_b(X)$,*

v) $H_d(\overline{co}Y, \overline{co}Z) \leq H_d(Y, Z)$ *for all $Y, Z \in \mathcal{P}_{b,cl}(X)$.*

Lemma 2.11. *Let (X, d) be a metric space. Then the generalized functional $\delta : (\mathcal{P}_{cl}(X), H_d) \to \mathbb{R}_+ \cup \{\infty\}$ is continuous.*

Lemma 2.12. *If $\{A_n, A\} \in \mathcal{P}_{cl}(X)$ and $A_n \to A$, then*

$$A = \bigcap_{n \geq 1} \overline{\bigcup_{m \geq n} A_m} = \bigcap_{\epsilon \geq 0} \bigcup_{n \geq 1} \bigcap_{m \geq n} (A_m)_\epsilon.$$

Proof. Let $\epsilon > 0$ be given. Since by hypothesis $A_n \to A$, we can find $n_0(\epsilon) \geq 0$ such that for $m \geq n_0(\epsilon)$, we have $A \subset (A_m)_\epsilon$ and $A_m \subset A_\epsilon$. From this inclusion, we have that

$$A \subset \bigcap_{\epsilon > 0} \bigcup_{n \geq 1} \bigcap_{m \geq n} (A_m)_\epsilon$$

and

$$\bigcap_{n \geq 1} \overline{\bigcup_{m \geq n} A_m} \subset A.$$

Hence,

$$\bigcap_{n \geq 1} \overline{\bigcup_{m \geq n} A_m} \subset A \subset \bigcap_{\epsilon > 0} \bigcap_{n \geq 1} \bigcup_{m \geq n} (A_m)_\epsilon.$$

Finally, let $x \in \bigcap_{\epsilon \geq 0} \bigcup_{n \geq 1} \bigcap_{m \geq n} (A_m)_\epsilon$. Then for all $\epsilon \geq 0$, there is $n_0(\epsilon) \geq 1$ such that, for $m \geq n_0(\epsilon)$, we have $x \in (A_m)_\epsilon$. Let $n \geq 1$ be given. Then there is $m \geq \max(n, n_0(\epsilon))$ such that $x \in (A_m)_\epsilon \subset (\bigcup_{m \geq n} A_m)_\epsilon$. Since $\epsilon > 0$ was arbitrary, we can deduce that $x \in \overline{\bigcup_{m \geq n} A_m}$, and so $x \in \bigcap_{n \geq 1} \overline{\bigcup_{m \geq n} A_m}$. Thus,

$$A = \bigcap_{n \geq 1} \overline{\bigcup_{m \geq n} A_m} = \bigcap_{\epsilon \geq 0} \bigcap_{n \geq 1} \bigcup_{m \geq n} (A_m)_\epsilon.$$

$\square$

Now we will check the completeness of the metric space $(\mathcal{P}_{cl}(X), H_d)$.

Theorem 2.13. *If (X, d) is a complete metric space, then so is $(\mathcal{P}_{cl}(X), H_d)$.*

Proof. Let $\{A_n\}_{n \in \mathbb{N}}$ be a Cauchy sequence in $(\mathcal{P}_{cl}(X), H_d)$. The previous Lemma 2.12 identifies the only possible candidate for a limit of $\{A_n\}_{n \in \mathbb{N}}$. Namely, let $A = \bigcap_{n \geq 1} \overline{\bigcup_{m \geq n} A_m}$. We will now show that

$$A \in \mathcal{P}_{cl}(X) \text{ and } A_n \to A \text{ as } n \to +\infty.$$

First, it is clear that A being the intersection of closed sets, is closed, yet possibly empty. Let $\epsilon > 0$. Then for every $k \geq 0$, we can find $N_k \geq 1$ such that $H(A_n, A_m) \leq \frac{\epsilon}{2^{k+1}}$ for all $n, m \geq N_k$. Pick $n_0 \geq N_0$ and $x_0 \in A_{n_0}$. Then choose $n_1 > \max(n_1, N_1)$ and $x_1 \in A_{n_1}$ with $d(x_0, x_1) < \frac{\epsilon}{2}$ (this is possible, since $d(x_0, A_{n_1}) \leq H(A_{n_0}, A_{n_1}) < \frac{\epsilon}{2}$). Then, if $\{n_k\}_{k \geq 0}$ is a strictly increasing sequence with $n_k \geq N_k$, inductively, we can generate a sequence $\{x_k\}_{k \geq 0} \subset X$ such that $x_k \in A_k$ and $d(x_k, x_{k+1}) < \frac{\epsilon}{2^{k+1}}$. So $\{x_k\}_{k \geq 0}$ is a Cauchy sequence in X and since X is complete, we have that $x_k \to x \in X$. Because $\{n_k\}_{k \geq 0}$ is strictly increasing, given $n \geq 1$, we can find $k_n \geq 1$ such that $n_{k_n} \geq n$. Hence $x_k \in \bigcup_{m \geq n} A_m$ for all $k \geq k_n$ and so $x \in \overline{\bigcup_{m \geq n} A_m}$ for all $n \geq 1$. Thus $x \in A$, which shows that $A \in \mathcal{P}_{cl}(X)$. In addition, we have

$$d(x, x_0) = \lim_{n \to +\infty} d(x_n, x_0) \leq \lim_{n \to +\infty} \sum_{k=1}^{n} d(x_k, x_{k-1}) < \epsilon.$$

So for all $n_0 \geq N_0$ and all $x_0 \in A_{n_0}$, we have obtained an $x \in A$ such that $d(x, x_0) < \epsilon$. Therefore $A_{n_0} \subset A_\epsilon$. We need to show that $A \subset (A_n)_\epsilon$ for all $n \geq N_0$. So let $x \in A$. Then $x \in \overline{\bigcup_{m \geq n_0} A_m}$, and we can find $m \geq N_0$ and $y \in A_m$ such that $d(x, y) < \frac{\epsilon}{2}$. Also, if $n \geq N_0$, we have $d(x, A_n) \leq d(x, A_m) + d(A_m, A_n) < \frac{\epsilon}{2} + \frac{\epsilon}{2} = \epsilon$. So $H_d^*(A, A_n) < \epsilon$ and this implies that $A \subset (A_n)_\epsilon$ for $n \geq N_0$. Therefore, we conclude that $A_n \to A$. $\square$

Lemma 2.14. *If (X, d) is a complete metric space, then $\mathcal{P}_{cp}(X)$ is a closed subset of $(\mathcal{P}(X), H_d)$; hence $(\mathcal{P}(X), H_d)$ is a complete metric space.*

Proof. Let $\{A_n\}_{n \geq 1} \subset \mathcal{P}_{cp}(X)$ and assume that $A_n \to A$. Then given $\epsilon > 0$, we can find $n_0(\epsilon) \geq 1$ such that for all $n \geq n_0(\epsilon)$, $H(A_n, A) < \epsilon$ and so $A \subset (A_n)_\epsilon$. But by hypothesis, A_n is compact, and so is totally bounded. Thus, we can find a finite set $F \subset X$ such that $A_n \subset F_\epsilon$; hence, $(A_n)_\epsilon \subset F_{2\epsilon}$. Therefore, $A \subset F_{2\epsilon}$ which shows that A is totally bounded and closed, and so $A \in \mathcal{P}_{cp}(X)$. From Lemma 2.12, we can easily show that $\mathcal{P}_{cp}(X)$ is a complete metric space. $\qquad\square$

In general, there is no simple relationship between the Hausdorff pseudometric (respectively, metric) topology $\hat{\tau}_{H_d}$ and the Vietoris topology $\hat{\tau}_V$, defined on $\mathcal{P}(X)$ (respectively, on $\mathcal{P}_{cl}(X)$). However, if we restrict ourselves to $\mathcal{P}_{cp}(X)$, then we have the following result.

Lemma 2.15. *([173]) If (X,d) is a metric space, then on $\mathcal{P}_{cp}(X)$, the Haudorff metric topology $\hat{\tau}_H$ and the Vietoris topology $\hat{\tau}_V$ coincide.*

The next lemma is obvious.

Lemma 2.16. $\mathcal{P}_{cl,b}(X)$ *is a closed subset of $(\mathcal{P}_{cl}(X), H_d)$. Therefore if (X,d) is complete metric space, then so is $\mathcal{P}_{cl,b}(X) = \mathcal{P}_{cl}(X) \cap \mathcal{P}_b(X)$.*

Now we assume that the underlying metric space is a normed space.

Lemma 2.17. *If X is a normed space, then $\mathcal{P}_{cl,cv}(X) = \mathcal{P}_{cl}(X) \cap \mathcal{P}_{cv}(X)$ is a closed subset of $(\mathcal{P}_{cl}(X), H_d)$.*

Proof. Let $\{A_n, A\}_{n \geq 1} \subset \mathcal{P}_{cl}(X)$, where A_n is convex for every $n \geq 1$ and assume that $A_n \overset{H}{\to} A$. Then from Lemma 2.12, we know that

$$A = \bigcap_{\epsilon > 0} \bigcup_{n \geq 1} \bigcap_{m \geq n} (A_m)_\epsilon.$$

Observe that for every $m \geq 1$, $(A_m)_\epsilon \subset \mathcal{P}_{cl,cv}$ is convex, hence $C_n^\epsilon = \bigcap_{m \geq n}(A_m)_\epsilon$ is convex. The sequence $\{C_n^\epsilon\}_{n \geq 1} \subset \mathcal{P}_{cl,cv}(X)$ is increasing for $n \geq 1$ for every $\epsilon > 0$. Therefore, $\bigcup_{n \geq 1} C_n^\epsilon = C^\epsilon$ is convex. So finally, $A = \bigcap_{\epsilon > 0} C^\epsilon$ is convex; i.e., $A \in \mathcal{P}_{cl,cv}(X)$. $\qquad\square$

Combining the previous three Lemmas, we can summarize the situation in a normed space.

Proposition 2.18. *If X is a normed space, then $\mathcal{P}_{cp,cv}(X) \subset \mathcal{P}_{cl,b,cv}(X) \subset \mathcal{P}_{cl,cv}(X)$ and $\mathcal{P}_{cp}(X) \subset \mathcal{P}_{cl,b}(X)$ are closed subspaces of $(\mathcal{P}_{cl}(X), H_d)$.*

Remark 2.19. *If X is a Banach space, all the above subsets are complete subspaces of the metric space $(\mathcal{P}_{cl}(X), H_d)$.*

Next, we derive two formulas for the Hausdorff distance. The first formula, also known as "Härmondar's formula," concerns the sets in $\mathcal{P}_{cl,b,cv}(X)$ and involves the supremum of the support functions of these sets.

Definition 2.20. *Let $(X, \|\cdot\|)$ be a Banach space, X^* its topological dual, and $A \in \mathcal{P}(X)$. The support function $\sigma(\cdot, A)$ of A is a function from X^* into $\overline{\mathbb{R}} = \mathbb{R} \cup \{+\infty\}$ defined by*

$$\sigma(x^*, A) = \sup\{\langle x^*, a \rangle : a \in A\},$$

where $\langle \cdot, \cdot \rangle$ denotes the duality bracket for the the pair (X^*, X).

Lemma 2.21. *([173]) If X is normed space and $A, B \in \mathcal{P}_{cl,b,cv}(X)$, then*

$$H_d(A, B) = \sup\{|\sigma(x^*, A) - \sigma(x^*, B)| : \|x^*\| \le 1\}.$$

Proof. Let $a \in A$ and $x^* \in X^*$, $\|x^*\| \le 1$, then for every $b \in B$, we get

$$\langle x^*, a \rangle = \langle x^*, b \rangle + \langle x^*, a - b \rangle \implies \langle x^*, a \rangle \le \langle x^*, b \rangle + \|a - b\|.$$

Thus

$$\langle x^*, a \rangle \le \sigma(x^*, B) + \|a - b\|.$$

For each $\epsilon > 0$ there exists $b_\epsilon \in B$ such that

$$\|a - b_\epsilon\| \le d(a, B) + \epsilon \implies \|a - b_\epsilon\| \le H_d^*(A, B) + \epsilon.$$

It follows that

$$\langle x^*, a \rangle \le \sigma(x^*, B) + H_d^*(A, B) + \epsilon \implies \sigma(x^*, A) - \sigma(x^*, B) \le H_d(A, B) + \epsilon.$$

By a similar argument as above, we get

$$\sigma(x^*, B) - \sigma(x^*, A) \le H_d(A, B) + \epsilon.$$

Since ϵ is arbitrary positive number, we have

$$|\sigma(x^*, A) - \sigma(x^*, A)| \le H_d(A, B).$$

So

$$\sup\{|\sigma(x^*, A) - \sigma(x^*, B)| : \|x^*\| \le 1\} \le H_d(A, B).$$

On the other hand, if $\epsilon = \sup\{|\sigma(x^*, A) - \sigma(x^*, B)| : \|x^*\| \le 1\} > 0$, then we have

$$A \subseteq B + \overline{B(0, \epsilon)} \text{ and } B \subseteq A + \overline{B(0, \epsilon)}.$$

So $H_d(A, B) \le \epsilon$. It follows that

$$H_d(A, B) = \sup\{|\sigma(x^*, A) - \sigma(x^*, B)| : \|x^*\| \le 1\} \le H_d(A, B).$$

$\square$

The second formula for the Hausdorff distance concerns nonempty subsets of an arbitrary metric space and involves the distance functions from the sets.

Lemma 2.22. *If (X, d) is a metric space and $A, B \in \mathcal{P}(X)$, then*

$$H_d(A, B) = \sup\{|d(x, A) - d(x, B)| : x \in X\}.$$

Proof. For every $x \in X$, we have

$$d(x, A) \le d(x, B) + H_d(A, b) \text{ and } d(x, B) \le d(x, A) + H_d(A, b).$$

Then

$$|d(x, A) - d(x, B)| \le H_d(A, b),$$

hence

$$\sup\{|d(x, A) - d(x, B)| : x \in X\} \le H_d(A, B).$$

Suppose that there exists $\alpha > 0$ such that

$$\sup\{|d(x, A) - d(x, B)| : x \in X\} < \alpha < H_d(A, B).$$

Thus, for every $a \in A$ and $b \in B$, we get

$$d(a, B) < \alpha, \quad d(b, A) < \alpha \Longrightarrow A \subseteq B_\alpha \text{ and } B \subseteq A_\alpha,$$

and so, we obtain

$$H_d(A, B) \leq \alpha \Longrightarrow \alpha < H_d(A, B) \leq \alpha.$$

This is a contradiction. Hence

$$H_d(A, B) = \sup\{|d(x, A) - d(x, B)| : x \in X\}.$$

$\square$

2.1 Hausdorff continuity

When Y is a metric space, by using the Hausdorff pseudometric, we can define three new continuity concepts that, in general, are distinct from ones considered in the previous section. Throughout this section, X is a Hausdorff topological space and Y is a metric space.

Definition 2.23. *A multifunction $F : X \to \mathcal{P}(Y)$ is said to be:*

a) *H_d-upper semicontinuous at $x_0 \in X$, if $H_d^*(F(x), F(x_0))$ is continuous at x_0; i.e.,*

$$\forall \epsilon > 0, \exists U_\epsilon \in \mathcal{N}(x_0) : \forall x \in U_\epsilon \Longrightarrow H_d^*(F(x), F(x_0)) < \epsilon,$$

where $\mathcal{N}(x)$ is a neighborhood filter of of x.

b) *H_d-lower semicontinuous at $x_0 \in X$, if $H_d^*(F(x_0), F(x))$ is continuous at x_0; i.e.,*

$$\forall \epsilon > 0, \exists U_\epsilon \in \mathcal{N}(x_0) : \forall x \in U_\epsilon \Longrightarrow H_d^*(F(x_0), F(x)) < \epsilon.$$

c) *H_d-continuous at x_0, if it is both H_d-upper semicontinuous and H_d-lower semicontinuous at x_0.*

We start by comparing these continuity concepts with the Vietoris ones studied earlier.

Proposition 2.24. *If $F : X \to \mathcal{P}(Y)$ is u.s.c., then $F(\cdot)$ is H_d-u.s.c.*

Proof. Since $F(\cdot)$ is upper semicontinuous, given $\epsilon > 0$ and $x \in X$, we have that $F^+((F(x))_\epsilon) = U \in \mathcal{N}(x)$. So for every $x' \in U$, we have $F(x') \subseteq (F(x))_\epsilon$. Hence, $H_d^*(F(x'), F(x)) < \epsilon$ for all $x' \in U$, and thus we conclude that $F(\cdot)$ is H_d-upper semicontinuous. $\square$

Example 2.25. *A single valued mapping $f : \mathbb{R} \to \mathbb{R}$ is H_d-u.s.c. (H_d-l.s.c.) if the set valued mapping F defined by $F(t) = [0, f(t)]$ is upper (lower) semicontinuous.*

Example 2.26. *The converse of Proposition 2.24 is not in general true. We consider the counterexample* $F : [0,1] \to \mathcal{P}(\mathbb{R})$ *defined by*

$$F(x) = \begin{cases} [0,1], & x \in [0,1), \\ [0,1), & x = 1. \end{cases}$$

It easy to check that $F(\cdot)$ *is* H_d*–upper semicontinuous but not upper semicontinuous at* $x = 1$*. Indeed note that* $F_+^{-1}((-1,1)) = \{1\}$ *is not an open set.*

The second example involves a closed-valued multifunction.

Example 2.27. *In the following counterexample, let* $F : \mathbb{R} \to \mathcal{P}_{cl}(\mathbb{R}^2)$ *be defined by*

$$F(x) = \begin{cases} \{[0,z] : z \geq 0\}, & x = 0, \\ \{[x,z] : 0 \leq z \leq \frac{1}{z}\}, & x \neq 0. \end{cases}$$

Then $F(\cdot)$ *is* H_d*-upper semicontinuous but not upper semicontinuous, since for* $C = \{[\frac{1}{n}, n] : n \geq 1\} \subset \mathbb{R}^2$ *is closed, but* $F_-^{-1}(C)$ *is not closed in* $\mathbb{R}$*.*

Proposition 2.28. *If* $F : X \to \mathcal{P}_{cl}(Y)$ *is* H_d*-u.s.c., then* $F(\cdot)$ *is closed.*

Proof. Let $(x,y) \in X \times Y$ and $(x_n, y_n)_{n \in \mathbb{N}} \subset \Gamma_F$ such that

$$(x_n, y_n) \to (x,y) \quad \text{as} \quad n \to \infty.$$

Since F is H_d–u.s.c., then

$$H_d^*(F(x_n), F(x)) \to 0 \quad \text{as} \quad n \to \infty.$$

On the other hand,

$$d(y, F(x)) \leq d(y, y_n) + d(y_n, F(x)) \Rightarrow d(y, F(x)) \leq d(y, y_n) + H_d^*(F(x_n), F(x)).$$

Hence

$$d(y, F(x)) \leq d(y, y_n) + H_d^*(F(x_n), F(x)) \to 0 \quad \text{as} \quad n \to \infty.$$

It follows that

$$d(y, F(x)) = 0 \Rightarrow y \in \overline{F(x)} = F(x).$$

$\square$

2.2 H_d-u.s.c, l.s.c., and single-valued u.s.c. and l.s.c. functions

This section deals with relations between H_d-u.s.c, l.s.c. and multi-valued and single-valued l.s.c. and u.s.c. functions.

Proposition 2.29. *Let* X *be a metric space and* $f : X \to \mathbb{R}$*. Then the following hold true:*

(i) if f *is a continuous function, then* $-f$ *is also a continuous function;*

(ii) if f *is a continuous function and* $\alpha \in \mathbb{R}$*, then the set*

$$\{x \in X : f(x) < \alpha\}$$

is an open set.

Remark 2.30. *From above Proposition 2.29 we observe that*

$$\{x \in X : \ f(x) > \alpha\},$$

is also an open set.

Corollary 2.31. *Let X be a metric space and $f : X \to \mathbb{R}$ be a continuous function. Then*

$$\{x \in X : \ f(x) \leq \alpha\}, \ \{x \in X : \ f(x) \geq \alpha\}$$

and

$$\{x \in X : \ f(x) = \alpha\}$$

are closed sets.

Definition 2.32. *(upper and lower semi-continuous functions). Let $f : X \to \mathbb{R}$ and $x_0 \in X$. Then*

(i) *f is said to be upper semi-continuous (u.s.c.) at x_0 if for every $\epsilon > 0$ there exists $\delta(\epsilon, x_0)$ such that*

$$f(x) - f(x_0) < \epsilon, \ \text{for all } x \in B(x_0, \delta(\epsilon, x_0)).$$

Moreover, f is said to be an upper semi-continuous function on X if f is upper semi-continuous at every $x \in X$;

(ii) *f is said to be lower semi-continuous (l.s.c.) if $-f$ is upper semi-continuous.*

Proposition 2.33. *Let X be a metric space and $f : X \to \mathbb{R}$. Then the following hold true:*

(i) *if f is a u.s.c. and $\alpha \in \mathbb{R}$,, then*

$$\{x \in X : \ f(x) > \alpha\}$$

is an open set.

(ii) *if f is a l.s.c. and $\alpha \in \mathbb{R}$, then the set*

$$\{x \in X : \ f(x) < \alpha\}$$

is an open set.

Corollary 2.34. *A real valued continuous function is both lower and upper semi-continuous.*

Now, we state some interesting results.

Proposition 2.35. *If $F : X \to \mathcal{P}_{cl}(Y)$ is H_d-u.s.c., then for every $v \in Y$, $x \to \phi_v(x) = d(v, F(x))$ is lower semicontinuous.*

Proposition 2.36. *If $F : X \to \mathcal{P}_{cl}(Y)$ is H_d-u.s.c., then $F(\cdot)$ is lower semicontinuous.*

Theorem 2.37. *([173]) Let $F : X \to \mathcal{P}_{cp}(Y)$. The following conditions are equivalent:*

a) *F u.s.c. (resp. F l.s.c.).*

b) *H_d-upper semicontinuous (resp. H_d-lower semicontinuous).*

Definition 2.38. *A multivalued operator* $N : X \to \mathcal{P}(X)$ *is called*

    a) L-Lipschitz if and only if there exists $L > 0$ *such that*

$$H_d(N(x), N(y)) \leq Ld(x, y), \quad \text{for each } x, \ y \in X,$$

    b) a contraction if and only if it is L-Lipschitz with $0 \leq \gamma < 1$.

Remark 2.39. *It clear that, if* N *is L-Lipschitz, then* N *is* H_d*−continuous.*

Lemma 2.40. *Let* $F : X \to \mathcal{P}(Y)$ *be a multivalued map with Lipschitz constant* L. *If* $A, B \in \mathcal{P}(X)$, *then*

$$H_d(F(A), F(B)) \leq L H_d(A, B).$$

Theorem 2.41. *Let* $F : X \to \mathcal{P}(Y)$ *be a multivalued map with Lipschitz constant* L *and let* $G : Y \to \mathcal{P}(Z)$ *be a multivalued map with Lipschitz constant* L'. *Then* GoF *is a multivalued map with Lipschitz constant* LL'.

Theorem 2.42. *Let* $F : X \to \mathcal{P}(Y)$ *be a multivalued map with Lipschitz constant* L *and let* $F_* : \mathcal{P}(X) \to \mathcal{P}(Y)Y$ *be given by* $F_*(A) = F(A)$. *Then* F_* *is a Lipschitz mapping with Lipschitz constant* L.

Theorem 2.43. *Let* $F : X \to \mathcal{P}_{cl,b}(Y)$ *be a multivalued Lipschitz map with Lipschitz constant* L_1 *and let* $G : X \to \mathcal{P}_{cl,b}(Y)$ *be a m.v.l.m. with Lipschitz constant* L_1. *Then* $F \cup G : X \to \mathcal{P}_{cl,b}(Y)$ *is a m.v.l.m. with Lipschitz constant* $L = \max(L_1, L_2)$.

Example 2.44. *Let* $F : [0, 1] \times [0, 1] \to \mathcal{P}_{cp}([0, 1] \times [0, 1])$ *be a multivalued map defined by:* $F(x, y)$ *is the line segment in* $[0, 1] \times [0, 1]$ *from the point* $(\frac{x}{2}, 0)$ *to the point* $(\frac{x}{2}, 1)$ *for each* $(x, y) \in [0, 1] \times [0, 1]$. *For each* $(x, y) \in [0, 1] \times [0, 1]$ *let* $G : [0, 1] \times [0, 1] \to \mathcal{P}_{cp}([0, 1] \times [0, 1])$ *be defined by* $G(x, y)$ *being the line segment in* $[0, 1] \times [0, 1]$ *from the point* $(\frac{x}{2}, 0)$ *to the point* $(\frac{x}{3}, 1)$. *It is clear that* F *and* G *are both multivalued contraction mappings, but* $F \cap G$, *which is given by*

$$(F \cap G)(x, y) := \begin{cases} \{(\frac{x}{2}, 1)\}, & x \neq 0, \\ \{(x, y) \in [0, 1] \times [0, 1] : x = 0\}, & x = 0, \end{cases}$$

for all $(x, y) \in [0, 1] \times [0, 1]$ *is not continuous.*

2.3 Fixed point theorems for multi-valued contractive mappings

Definition 2.45. *Recall that a point* $x \in X$ *is called a fixed point of a multi-valued operator* $F : X \to \mathcal{P}(X)$ *if*

$$x \in F(x).$$

Theorem 2.46. *Let* (X, d) *be a complete metric space, and* $F : X \to \mathcal{P}_{cl}(X)$ *be a contraction multi-valued mapping. Then* F *has a fixed point in* X.

Proof. We employ the standard iterative procedure for contracting mappings. Let $L \in (0, 1)$ such that

$$H_d(F(x), F(y)) \leq Ld(x, y) \quad \text{for all } x, y \in X.$$

Let $x_0 \in X$, be fixed and choose $x_1 \in F(x_0)$ such that

$$d(x_1, x_0) \leq d(x_0, F(x_0)) + L.$$

By the definition of the Hausdorff distance, we find $x_2 \in F(x_1)$ such that

$$d(x_1, x_2) \leq d(x_1, F(x_1)) + L \Rightarrow d(x_1, x_2) \leq H_d(F(x_0), F(x_1)) + L.$$

Similarly, we find $x_3 \in F(x_2)$ such that

$$d(x_3, x_2) \leq H_d(F(x_2), F(x_1)) + L^2.$$

Continuing this process, we find a sequence $(x_n)_{n \in \mathbb{N}}$ in X such that $x_{n+1} \in (x_n)$ and

$$d(x_{n+1}, x_n) \leq H_d(F(x_n), F(x_{n-1})) + L^n.$$

For fixed $k \in \mathbb{N}$ we get

$$
\begin{aligned}
d(x_k, x_{k+1}) &\leq H_d(F(x_k), F(x_{k-1})) + L^k \\
&\leq L d(x_k, x_{k-1}) + L^k \\
&\leq L H_d(F(x_{k-1}), F(x_{k-2})) + 2L^k \\
&\leq L^2 d(x_{k-1}, x_{k-2}) + 2L^k \\
&\leq L^2 (H_d(F(x_{k-2}), F(x_{k-3})) + L^{k-2}) + 2L^k \\
&\leq L^3 d(x_{k-2}, x_{k-3}) + 3L^k \\
&\vdots \\
&\leq L^k d(x_1, x_0) + k L^k.
\end{aligned}
$$

Consequently,

$$
\begin{aligned}
d(x_n, x_{n+p}) &\leq \sum_{k=n}^{n+p-1} d(x_k, x_{k+1}) \\
&\leq \sum_{k=n}^{n+p-1} L^k d(x_1, x_0) + \sum_{k=n}^{n+p-1} k L^k.
\end{aligned}
$$

This shows that $(x_n)_{n \in \mathbb{N}}$ is a Cauchy sequence, and hence $x_n \to x$ for some $x \in X$. Since F satisfies a Lipschitz condition, then F is H_d-u.s.c.

$$H_d(F(x_n), F(x)) \to 0 \text{ as } n \to \infty.$$

From Proposition 2.28, F has closed graph. So $x \in F(x)$. $\qquad\square$

In what follows we present the local version of the above result.

Theorem 2.47. *Let $B(x_0, r)$ be a closed ball of radius r centered at a point x_0 in a complete metric space X, and suppose $F : B(x_0, r) \to \mathcal{P}_{cl}(X)$ is a contraction multivalued map with Lipschitz constant $0 \leq L < 1$. Suppose that*

$$H_d(x_0, F(x_0)) < (1 - L)r.$$

Then F has at least one fixed point in $B(x_0, r)$.

Proof. Let $0 < r_1 < r$ such that

$$H_d(x_0, F(x_0)) \leq (1 - L)r_1 < (1 - L)r.$$

Set

$$K(x_0, r_1) = \overline{B(x_0, r_1)}.$$

Then $K(x_0, r_1)$ is complete metric space. In view of Theorem 2.46, for the proof it is sufficient to show that

$$F(K(x_0, r_1)) \subseteq K(x_0, r_1), \quad \text{for every } x \in K(x_0, r_1).$$

Let $x \in K(x_0, r_1)$, then for each $y \in F(x)$ and $a \in F(x_0)$, we have

$$d(x_0, y) \leq d(x_0, a) + d(a, y).$$

For every $\epsilon > 0$ there exists $a_\epsilon(y) \in F(x_0)$ such that

$$d(a_\epsilon(y), y) \leq d(y, F(x_0)) + \epsilon.$$

Then

$$d(x_0, y) \leq d(x_0, a_\epsilon(y)) + d(y, F(x_0)) + \epsilon,$$

hence

$$d(x_0, y) \leq H_d(x_0, F(x_0)) + H_d(F(x), F(x_0)) + \epsilon.$$

Letting $\epsilon \to 0$ we get

$$d(x_0, y) \leq H_d(x_0, F(x_0)) + H_d(F(x), F(x_0)) \leq (1 - L)r_1 + Lr_1 = r_1.$$

This implies that $y \in K(x_0, r_1)$, and it follows that

$$F(K(x_0, r_1)) \subseteq K(x_0, r_1).$$

$\square$

Now, let U be an open subset of Banach space E and let $F : U \to \mathcal{P}_{cl,b}(E)$ be an $L-$contraction. We set

$$\phi : U \to \mathcal{P}_{cl,b}(E)$$

where

$$\phi(x) = x - F(x) = \{x - y : y \in F(x)\}.$$

Theorem 2.48. *(Invariance of a domain for contraction mappings) Under the above assumptions, the multivalued map $\phi : U \to \mathcal{P}_{cl,b}(E)$ is an open map.*

Proof. Let $x_0 \in U$ then there exists $r > 0$ such that $B(x_0, r) \subseteq U$. Now, we show that $\phi(B(x_0, r))$ is an open set. Set

$$O_{(1-L)r}(\phi(x_0)) = \{x \in U : H_d(x, \phi(x_0)) < (1 - L)r\}.$$

Let $y_0 \in O_{(1-L)r}(\phi(x_0))$. We define $G : B(x_0, r) \to \mathcal{P}_{cl}(X)$ by putting

$$G(x) = y_0 + F(x), \quad x \in B(x_0, r).$$

Let $x, y \in B(x_0, r)$. Then

$$\begin{aligned}
d(G(x), G(y)) &\leq H_d(y_0 + F(x_0), y_0 + F(y)) \\
&= H_d(F(x), F(y)) \\
&\leq L d(x, y)
\end{aligned}$$

and

$$\begin{aligned}
H_d(G(x_0), x_0) &= H_d(y_0 + F(x_0), x_0) \\
&= H_d(y_0 + F(x_0), \phi(x_0) + F(x_0)) \\
&\leq H_d(y_0, \phi(x_0)) \\
&\leq (1 - r)L.
\end{aligned}$$

Then all the conditions of Theorem 2.47 hold. So there exists $z \in B(x_0, r)$ such that

$$z \in G(z) = y_0 + F(z) \implies y_0 \in \phi(B(x_0, r)).$$

Consequently

$$O_{(1-L)r}(\phi(x_0)) \subseteq \phi(B(x_0, r)),$$

and the proof is complete. $\square$

From the above Theorem 2.48, we have a couple of corollaries.

Corollary 2.49. *If U is a domain (i.e., U is open and connected) and $F : U \to \mathcal{P}_{b,cl}(X)$ is a contraction with connected values, then $\phi(U)$ is domain as well, where*

$$\phi(x) = x - F(x) \quad x \in U.$$

Corollary 2.50. *Let E be a Banach space. If $F : E \to \mathcal{P}_{b,cl}(E)$ is a contraction, then $\phi(E) = E$, where*

$$\phi(x) = x - F(x) \quad x \in E.$$

Proposition 2.51. *Let X be a complete metric space and $F, G : X \to \mathcal{P}(X)$ be two contraction mappings each having Lipschitz constant $0 < L < 1$, i.e.*

$$H_d(F(x), F(y)) \leq Ld(x,y) \text{ and } H_d(G(x), G(y)) \leq Ld(x,y) \quad \text{for all } x, y \in X.$$

Then

$$H_d(Fix(F), Fix(G)) \leq \frac{1}{1-L} \sup_{x \in X} H_d(G(x), F(x)).$$

Proof. Let $\epsilon > 0$ and $x_0 \in Fix(G)$. Then there exists $x_1 \in F(x_0)$ such that

$$d(x_1, x_0) \leq d(x_0, F(x_0)) + \epsilon \Rightarrow d(x_1, x_0) \leq H_d(G(x_0), F(x_0)) + \epsilon.$$

Also there exists $x_2 \in F(x_1)$ such that

$$\begin{aligned} d(x_1, x_2) &\leq d(x_1, F(x_1)) + \frac{c\epsilon L}{1-L} \\ &\leq d(F(x_0), F(x_1)) + \frac{c\epsilon L}{1-L}, \end{aligned}$$

where $c > 0$ such that

$$c \sum_{k=0}^{\infty} kL^k < 1.$$

Since $H_d(F(x_0), F(x_1)) \leq Ld(x_1, x_0)$, then

$$d(x_1, x_2) \leq Ld(x_0, x_1) + \frac{c\epsilon L}{1-L}.$$

For $\frac{c\epsilon L^2}{1-L} > 0$ there exists $x_3 \in F(x_2)$ such that

$$d(x_2, x_3) \leq Ld(x_2, x_1) + \frac{c\epsilon L^2}{1-L}.$$

Define $(x_n)_{n \in \mathbb{N}}$ inductively by

$$\begin{aligned} d(x_k, x_{k+1}) &\leq Ld(x_k, x_{k-1}) + \frac{c\epsilon L^k}{1-L} \\ &\leq L^2 d(x_{k-1}, x_{k-2}) + \frac{2c\epsilon L^k}{1-L} \\ &\leq L^3 d(x_{k-2}, x_{k-3}) + \frac{3c\epsilon L^k}{1-L} \\ &\vdots \\ &\leq L^k d(x_1, x_0) + \frac{ck\epsilon L^k}{1-L}. \end{aligned}$$

Consequently,

$$\begin{aligned} d(x_n, x_{n+p}) &\leq \sum_{k=n}^{n+p-1} d(x_k, x_{k+1}) \\ &\leq \sum_{k=n}^{n+p-1} L^k d(x_1, x_0) + \sum_{k=n}^{n+p-1} \frac{k\epsilon L^k}{1-L}. \end{aligned}$$

This shows that $(x_n)_{n \in \mathbb{N}}$ is a Cauchy sequence, and hence $x_n \to x$ for some $x \in X$. Since F satisfies a Lipschitz condition, then F is H_d-u.s.c., and so

$$H_d(F(x_n), F(x)) \to 0 \text{ as } n \to \infty.$$

From Proposition 2.28, F has a closed graph. Hence $x \in F(x)$, so $x \in Fix(F)$. Observe that

$$
\begin{aligned}
d(x_0, x) &\leq \sum_{k=0}^{\infty} d(x_k, x_{k+1}) \\
&\leq \sum_{k=0}^{\infty} L^k d(x_1, x_0) + \sum_{k=0}^{\infty} \frac{ck\epsilon L^k}{1-L} \\
&\leq (1-L)^{-1} d(x_1, x_0) + \sum_{k=0}^{\infty} \frac{ck\epsilon L^k}{1-L} \\
&\leq (1-L)^{-1} H_d(F(x_0), G(x_0)) + \tfrac{\epsilon}{1-L} + \sum_{k=0}^{\infty} \frac{ck\epsilon L^k}{1-L}.
\end{aligned}
$$

Therefore

$$\sup_{a \in Fix(G)} d(a, Fix(F)) \leq (1-L)^{-1} H_d(F(x_0), G(x_0)) + \frac{\epsilon}{1-L} + \sum_{k=0}^{\infty} \frac{ck\epsilon L^k}{1-L}.$$

Letting $\epsilon \to 0$, we get

$$H_d^*(Fix(G), Fix(F)) \leq (1-L)^{-1} H_d(F(x_0), G(x_0)). \tag{2.1}$$

Interchanging the roles of F and G, we conclude, for each $y_0 \in Fix(F)$, there exist $y_1 \in F(y_0)$ and $u \in Fix(G)$ such that

$$d(y_0, u) \leq (1-L)^{-1} H_d(F(y_0), G(y_0)) + \frac{\epsilon}{1-L} + \sum_{k=0}^{\infty} \frac{ck\epsilon L^k}{1-L}.$$

Because ϵ is arbitrary, we conclude that

$$H_d^*(Fix(F), Fix(G)) \leq (1-L)^{-1} H_d(F(y_0), G(y_0)). \tag{2.2}$$

From (2.1) and (2.2), we obtain

$$H_d(Fix(F), Fix(G)) \leq \frac{1}{1-L} \sup_{x \in X} H_d(G(x), F(x)).$$

$\square$

Theorem 2.52. *Let X be a complete metric space and $F_n : X \to P_{cl,b}(X)$, $n \in \mathbb{N}$, be a sequence of multivalued mappings. Suppose that there exists $0 < L < 1$ such that*

$$H_d(F_n(x), F_n(y)) \leq Ld(x, y), \text{ for all } x, y \in X \text{ and } n \in \mathbb{N}.$$

If $\lim_{n \to \infty} H_d(F_n(x), F_0(x)) = 0$ uniformly for $x \in X$, then

$$\lim_{n \to \infty} H_d(Fix(F_n), Fix(F_0)) = 0.$$

Proof. From Proposition 2.51, we have

$$H_d(Fix(F_n), Fix(F_0)) \leq \frac{1}{1-L} \sup_{x \in X} H_d(F_n(x), F(x)) \to 0 \quad \text{as} \quad n \to \infty.$$

$\square$

Theorem 2.53. *Let E be a Banach space, $Y \subseteq E$ be a nonempty convex compact subset of E and $F : X \to \mathcal{P}_{cl}(Y)$ be a multivalued map such that*

$$H_d(F(x), F(y)) \leq d(x, y), \quad \text{for each } x, y \in X.$$

Then there exists $x \in X$, such that $x \in F(x)$.

Proof. Let $x_0 \in Y$. For every $m \in \mathbb{N}$ we consider the mapping $F_m : Y \to \mathcal{P}_{cl}(Y)$ defined by

$$F_m(x) = \left(1 - \frac{1}{2^m}\right) F(x) + \frac{1}{2^m} x_0 \in Y \quad \text{for all } x \in Y.$$

Then

$$H_d(F_m(x), F_m(y)) \leq \frac{1}{2^m} d(x, y) \quad \text{for all } x, y \in Y.$$

From Theorem 2.46 there exists $x_m \in Y$ such that

$$x_m \in F_m(x_m), \quad m \in \mathbb{N}.$$

Since Y is compact, then there exists a subsequence of $(x_m)_{m \in \mathbb{N}}$ converging to $x \in Y$. Now we show that $x \in F(x)$.

$$
\begin{aligned}
d(x, F(x)) &\leq d(x, x_m) + d(x_m, F(x_m)) + H_d(F(x_m), F(x)) \\
&\leq 2d(x, x_m) + d(x_m, F(x_m))
\end{aligned}
$$

and

$$
\begin{aligned}
d(x_m, F(x_m)) &\leq d(x_m, F(x)) + H_d(F(x), F(x_m)) \\
&\leq d(x_m, z_m) + d(x, x_m) \\
&\leq d(x_m, x) + \frac{1}{2^m} d(z_m, x_0)
\end{aligned}
$$

where

$$z_m \in F(x) \text{ and } x_m = (1 - \frac{1}{2^m}) z_m + \frac{1}{2^m} x_0.$$

Since $x_m \to x$ as $m \to \infty$, then

$$z_m \to x \text{ as } m \to \infty.$$

Hence

$$d(x, F(x)) \leq 3d(x, x_m) + \frac{1}{2^m} d(z_m, x_0) \to 0 \text{ as } m \to \infty.$$

$\square$

Definition 2.54. *The map F from a metric space (X, d) into the subsets of a metric space Y is pseudo-Lipschitz around $(x_0, y_0) \in \Gamma_F$ with constant L if there exist positive constants ϵ and δ such that $H_d^*(F(z) \cap B(x_0, \epsilon), F(y)) \leq Ld(x, y)$ for all $x, y \in B(x_0, \delta)$.*

Remark 2.55. *If $\epsilon = \infty$ then the map F is Lipschitz in $B(x_0, \delta)$ (with respect to the Hausdorff metric). When F is single-valued, this corresponds to the usual concept of Lipschitz continuity.*

Theorem 2.56. *Let (X, d) be a complete metric space; $F : X \to \mathcal{P}_{cl}(X)$ be a multivalued map, $x_0 \in X$, and let r and $0 < L < 1$ be such that,*

(a) $d(x_0, F(x_0)) < r(1 - L)$,

(b) $H_d^*(F(x) \cap B(x_0, r), F(y)) \leq L d(x, y)$ for all $x, y \in B(x_0, r)$.

Then F has a fixed point in $B(x_0, r)$.

Proof. From assumption (a) there exists $x_1 \in F(x_0)$ such that

$$d(x_1, x_0) < r(1 - L).$$

By the assumption (b) we have $H_d^*(F(x_1) \cap B(x_0, \epsilon), F(x_0)) \leq L d(x_0, x_1)$. Then, there exists $x_2 \in F(x_1) \cap B(x_0, r)$ such that

$$d(x_2, x_1) \leq r(1 - L)L.$$

Proceeding by induction, suppose that there exists $x_{k+1} \in F(x_{k+1}) \cap B(x_0, r)$, $k = 1, 2, \ldots, n - 1$ such that

$$d(x_{k+1}, x_k) \leq r(1 - L)L^k.$$

By assumption (b),

$$
\begin{aligned}
d(x_n, F(x_n)) &\leq H_d^*(F(x_{n-1}) \cap B(x_0, r), F(x_n)) \\
&\leq L d(x_n, x_{n-1}) < r(1 - L)L^n.
\end{aligned}
$$

This implies that there exists $x_{n+1} \in F(x_n)$ such that

$$d(x_{n+1}, x_n) < r(1 - L)L^n.$$

By the triangle inequality, we get

$$
\begin{aligned}
d(x_{n+1}, x_0) &\leq \sum_{k=0}^{n} d(x_{k+1}, x_k) \\
&< r(1 - L) \sum_{k=0}^{n} L^k < r.
\end{aligned}
$$

Hence,

$$x_{n+1} \in F(x_n) \cap B(x_0, r).$$

For $n > m$, we have

$$
\begin{aligned}
d(x_n, x_m) &\leq \sum_{k=m}^{n-1} d(x_{k+1}, x_k) \\
&< r(1 - L) \sum_{k=m}^{n-1} L^k < rL^m.
\end{aligned}
$$

Thus $(x_n)_{n \in \mathbb{N}}$ is a Cauchy sequence converging to some $x \in B(x_0, r)$. From (b), we get

$$d(x_n, F(x)) \leq H_d^*(F(x_n) \cap B(x_0, r), F(x)) \leq Ld(x_{n-1}, x).$$

So,

$$\begin{aligned}
d(x, F(x)) &\leq d(x, x_n) + d(x_n, F(x)) \\
&\leq d(x, x_n) + Ld(x_{n-1}, x) \to 0 \quad \text{as } n \to \infty.
\end{aligned}$$

Since $F(x)$ is closed we conclude that $x \in F(x)$.

$\square$

Chapter 3

Measurable Multifunctions

Apart from semi-continuous multivalued mappings, multivalued measurable mappings will be of great importance in what follows. Throughout this section, assume that X is a separable metric space and $(\Omega, \mathcal{U}, \mu)$ is a complete σ–finite measurable space, i.e. a set Ω equipped with σ-algebra $\mathcal{U}$ of subsets and a countably additive measure μ on $\mathcal{U}$. A typical example is when Ω is a bounded domain in the Euclidean space $\mathbb{R}^k$ equipped with the Lebesgue measure.

Definition 3.1. *A multivalued map $F : \Omega \to \mathcal{P}(X)$ is said to be:*

a) *measurable if for every closed subset $C \subseteq X$, we have*

$$F_-^{-1}(C) = \{\omega \in \Omega : F(\omega) \cap C \neq \emptyset\} \in \mathcal{U},$$

b) *weakly measurable (or Effros measurable) if for every open subset $U \subseteq X$, we have*

$$F_-^{-1}(U) = \{\omega \in \Omega : F(\omega) \cap U \neq \emptyset\} \in \mathcal{U},$$

c) *K-measurable if for every compact subset $K \subseteq X$, we have*

$$F_-^{-1}(K) = \{\omega \in \Omega : F(\omega) \cap K \neq \emptyset\} \in \mathcal{U},$$

d) *graph measurable if*

$$\Gamma(F) = \{(\omega, x) \in \Omega \times X : x \in F(\omega)\} \in \mathcal{U} \otimes \mathcal{B}(X),$$

where $\mathcal{B}(X)$ is the σ-algebra generated by the family of open all sets from X.

e) *Borel measurable, if $F_-^{-1}(B) \in \mathcal{U}$ for each Borel subset $B \in \mathcal{B}(X)$.*

Another way of defining measurability is by requiring the measurability of the graph Γ_φ of φ in the product $\Omega \times Y$, equipped with the minimal σ-algebra $\mathcal{U} \otimes \mathcal{B}(Y)$ generated by the sets $A \times B$ with $A \in \mathcal{U}$ and $B \in \mathcal{B}(Y)$, where $\mathcal{B}(Y)$ denotes the family of all Borel subsets of Y.

We collect some relationships between these definitions in the following proposition.

Proposition 3.2. *Assume that $\varphi, \psi \colon \Omega \to \mathcal{P}(Y)$ are two multivalued mappings. Then the following hold true:*

(3.2.1) *φ is measurable if and only if $\varphi_-^{-1}(A) \in \mathcal{U}$ for each closed $A \subset Y$,*

(3.2.2) *φ is weakly measurable (Effros measurable or simply measurable) if and only if $\varphi_-^{-1}(V) \in \mathcal{U}$ for each open $V \subset Y$,*

(3.2.3) *if φ is measurable then φ is also weakly measurable,*

(3.2.4) *φ is weakly measurable if and only if the distance function $f_y : \Omega \to \mathbb{R}$, $f_y(x) = d(y, \varphi(x))$ is measurable for all $y \in Y$,*

(3.2.5) *if φ is weakly measurable then the graph Γ_φ of φ is product measurable,*

(3.2.6) *if φ and ψ are measurable then so is $\varphi \cup \psi$,*

(3.2.7) *if φ and ψ are measurable then so is $\varphi \cap \psi$,*

(3.2.8) *if φ and ψ are measurable then so is $\varphi \times \psi$.*

The proof of Proposition 3.2 is straightforward and therefore is left to the reader.

Of course, the composition of two measurable multivalued mappings need not be measurable.

Example 3.3. *Let $\Omega = [0,1]$ be equipped with the Lebesgue measure and let $f: \Omega \to \mathbb{R}$ be a strictly increasing Cantor function which of course is measurable. It is well known that one may find a measurable set $\mathcal{D} \subset \mathbb{R}$ such that $f^{-1}(\mathcal{D})$ is not measurable. If we define $\varphi: \Omega \to \mathcal{P}(\mathbb{R})$ and $\psi: \mathbb{R} \to \mathcal{P}(\mathbb{R})$ by*

$$\varphi(t) = \{f(t)\} \quad for\ t \in \Omega, \qquad \psi(u) = \left\{ \begin{array}{ll} \{1\} & if\ u \in \mathcal{D}, \\ \{0\} & if\ u \notin \mathcal{D}, \end{array} \right.$$

then both φ and ψ are measurable, but $\psi \circ \varphi$ is not.

For further reference, we collect the results and counterexamples given so far on the conservation of semi-continuity or measurability properties in the following table where φ and ψ are assumed to have compact values.

φ, ψ	$u.s.c.$	$l.s.c.$	measurable
$\varphi \cup \psi$	yes	yes	yes
$\varphi \cap \psi$	yes	no	yes
$\varphi \times \psi$	no	no	yes
$\varphi \circ \psi$	yes	yes	no

Next, we present some important properties of measurability of multifunctions.

Proposition 3.4. *Let $F : \Omega \to \mathcal{P}(X)$ be a measurable multi-valued map. Then F is weakly measurable.*

Proof. Let U be a open set in X. Then

$$U = \cup_{n \in \mathbb{N}} V_n, \quad V_n = \left\{ x \in: d(x, X \backslash U) \geq \frac{1}{n} \right\}, \quad n \in \mathbb{N}.$$

Thus,

$$F_-^{-1}(U) = \cup_{n \in \mathbb{N}} V_n \in \mathbb{U},$$

and so F is measurable. □

Proposition 3.5. *The mapping $F : \Omega \to \mathcal{P}(X)$ is weakly measurable if and only if for every $x \in X$, $\omega \to d(x, F(\omega)) = \inf\{d(x, x') : x' \in F(\omega)\}$ is a measurable $\overline{\mathbb{R}}_+ = \mathbb{R}_+ \cup \{\infty\}-valued function.*

Proof. Let $x \in X$; define $h_x : \Omega \to \overline{\mathbb{R}}_+$ by

$$h_x(\omega) = d(x, F(\omega)), \ \omega \in \Omega.$$

For $\gamma > 0$, we have

$$
\begin{aligned}
h_x^{-1}([0, \gamma)) &= \{\omega \in \Omega : d(x, F(\omega)) < \gamma\} \\
&= F_-^{-1}(B(x, \gamma)).
\end{aligned}
$$

If F is measurable, then $F_-^{-1}(B(x, \gamma)) \in \mathcal{U}$, and hence h_x is measurable. Thus, necessity holds.

Now let U be an open set in X. Since X is a separable space,

$$
U = \bigcup_{n \in \mathbb{N}} B(x_n, \gamma_n), \ \gamma_n > 0, x_n \in X, \ n \in \mathbb{N}.
$$

Thus,

$$
\begin{aligned}
F_-^{-1}(U) &= \bigcup_{n \in \mathbb{N}} F_-^{-1}(B(x_n, \gamma_n)) \\
&= \bigcup_{n \in \mathbb{N}} h_{x_n}^{-1}([0, \gamma_n)).
\end{aligned}
$$

For every $n \in \mathbb{N}$, h_{x_n} is measurable, so we have $\bigcup_{n \in \mathbb{N}} h_{x_n}^{-1}([0, \gamma_n)) \in \mathcal{U}$. Therefore F is measurable. $\qquad\square$

Definition 3.6. *A function $f : \Omega \times X \to X$ is a Carathéodory function if*

(a) *the function $\omega \mapsto f(\omega, x)$ is measurable for each $x \in X$;*

(b) *for every (or a.e.) $\omega \in \Omega$, the map $x \mapsto f(\omega, x)$ is continuous.*

Theorem 3.7. *(Carathéodory) Let X be a separable metric space, Y be a metric space, and $f : \Omega \times X \to X$ be a Carathédory function. Then the map $(\omega, x) \to f(\omega, x)$ is jointly measurable.*

Proof. Since X is a separable space, there exists a countable dense subset D of X. Let $C \subseteq X$ be a closed set. Then $f(\omega, x) \in C$ if and only if for every $n \in \mathbb{N}$ there exists $v \in D$ such that $d(x, v) < \frac{1}{n}$ and $f(\omega, v) \in C_n$, where

$$
C_n = \left\{ y \in Y : d_Y(y, C) < \frac{1}{n} \right\}.
$$

Hence,

$$
f^{-1}(C) = \bigcap_{n \in \mathbb{N}} \bigcup_{v \in D} [\{\omega \in \Omega : f(\omega, v) \in C_n\}
$$

$$
\times \left\{ y \in Y : d_Y(y, C_n) < \frac{1}{n} \right\}] \in \mathcal{U} \otimes \mathcal{B}(X).
$$

This implies that f is jointly measurable. $\qquad\square$

Lemma 3.8. *Let X and Y be complete separable metric spaces and $\phi : \Omega \times X \to Y$ be a Carathéodory map. Then for every measurable $f : \Omega \to X$, the map $\omega \to \phi(\omega, f(\omega))$ is measurable.*

Proof. Since f is a measurable function, there exists a measurable sequence $(f_n)_{n\in\mathbb{N}} : \Omega \to X$ such that

$$\lim_{n\to\infty} f_n(\omega) = f(\omega), \quad \omega \in \Omega.$$

Then,

$$\lim_{n\to\infty} \phi(\omega, f_n(\omega)) = \phi(\omega, f(\omega)), \quad \omega \in \Omega.$$

By Theorem 3.7, $\phi(\cdot, \cdot)$ is measurable, so $\omega \to \phi(\omega, f(\omega))$ is measurable. $\qquad\square$

Proposition 3.9. *If $F : \Omega \to \mathcal{P}_{cl}(X) \cup \{\emptyset\}$ is weakly measurable, then F is graph measurable.*

Proof. Note that since F is closed valued, then

$$
\begin{aligned}
\Gamma(F) &= \{(\omega, x) \in \Omega \times X : d(x, F(\omega)) = 0\} \\
&= f^{-1}(0),
\end{aligned}
$$

where $f : \Omega \times X \to \mathbb{R}$ is the function defined by

$$f(\omega, x) = d(x, F(\omega)), \quad (\omega, x) \in \Omega \times X.$$

From Lemma 2.8, for every $\omega \in \Omega$, the function $x \to f(\omega, x)$ is continuous. Since F is measurable, by Proposition 3.5, for each $x \in X$, the function $\omega \to f(\omega, x)$ is measurable. Thus, f is a Carathédory function. By Theorem 3.7, f is jointly measurable. This implies that

$$\Gamma(F) = f^{-1}(\{0\}) \in \mathcal{U} \times \mathcal{B}(X).$$

$\qquad\square$

Proposition 3.10. *Let $F : \Omega \to \mathcal{P}(X)$ be a multifunction such that for every $x \in X$ and every $r > 0$, $F_-^{-1}\left(\overline{B(x,r)}\right) \in \mathcal{U}$. Then F is weakly measurable.*

Proof. Since X is a separable space, there exists $\{x_n : n \in \mathbb{N}\}$ and $\lambda_n > 0$, $n \in \mathbb{N}$, such that

$$U = \cup \overline{B(x_n, \lambda_n)}$$

Then

$$F_-^{-1} B(x_n, \lambda_n)$$

by

$$F_-^{-1}(\overline{B(x_n, \lambda_n)}).$$

$\qquad\square$

Lemma 3.11. *For every open set $U \subset X$ and set $A \subset X$, then $A \cap U \neq \emptyset$ if and only if $\overline{A} \cap U \neq \emptyset$.*

Proposition 3.12. *The mapping $F : \Omega \to \mathcal{P}(X) \cup \{\emptyset\}$ is weakly measurable if and only if $\overline{F} : \Omega \to \mathcal{P}_{cl}(X) \cup \{\emptyset\}$ is weakly measurable.*

Proof. Let $V \subset X$ be a open set. Then from Lemma 3.11 we have

$$
\begin{aligned}
\overline{F}_-^{-1}(V) &= \{\omega \in \Omega : \overline{F(\omega)} \cap V \neq \emptyset\} \\
&= \{\omega \in \Omega : F(\omega) \cap V \neq \emptyset\} \\
&= F_-^{-1}(V).
\end{aligned}
$$

$\qquad\square$

Proposition 3.13. *Let $F : \Omega \to \mathcal{P}_{cp}(X) \cup \{\emptyset\}$ is weakly measurable, if and only if F is measurable.*

Proof. First note that measurable implies weakly measurable. Now assume that F is weakly measurable. Let $C \subseteq X$ be a closed set. Since X is a metrizable space,

$$U = X \backslash C = \cup_{n \in \mathbb{N}} C_n,$$

where

$$C_n = \left\{ x \in X : d(x, C) \geq \frac{1}{n} \right\}, \quad n \in \mathbb{N}.$$

It is clear that

$$F_-^{-1}(C) = \Omega \backslash F_+^{-1}(U) = \Omega \backslash F_+^{-1}(\cup_{n \in \mathbb{N}} C_n).$$

From Proposition 1.2 we have

$$\cup_{n \in \mathbb{N}} F_+^{-1}(C_n) \subset F_+^{-1}(\cup_{n \in \mathbb{N}} C_n).$$

We will show that

$$F_+^{-1}(\cup_{n \in \mathbb{N}} C_n) \subseteq \cup_{n \in \mathbb{N}} F_+^{-1}(C_n).$$

Let $\omega \in F_+^{-1}(\cup_{n \in \mathbb{N}} C_n)$ such that

$$F(\omega) \cap X \backslash C_n \neq \emptyset, \quad n \in \mathbb{N},$$

and let $x_n \in F(\omega) \cap X \backslash C_n \neq \emptyset, \quad n \in \mathbb{N}$. Since $F(\omega)$ is compact, there exists a subsequence of $(x_n)_{n \in \mathbb{N}}$ such that $x_n \to x \in F(\omega) \subset U$. On the other hand, $x_n \in X \backslash C_n, n \in \mathbb{N}$, so

$$d(x_n, C) < \frac{1}{n}, \quad n \in \mathbb{N}.$$

Hence,

$$d(x_n, C) < \frac{1}{n} \to 0 \text{ as } n \to \infty.$$

So $d(x, C) = 0$, and this implies that $x \in C$, which is a contradiction. Therefore,

$$\cup_{n \in \mathbb{N}} F_+^{-1}(C_n) = F_+^{-1}(\cup_{n \in \mathbb{N}} C_n).$$

Finally, we have

$$F_-^{-1}(C) = \Omega \backslash \left(\Omega \backslash \cup_{n \in \mathbb{N}} F_-^{-1}(X \backslash C_n) \right) \in \mathcal{U}.$$

$\square$

Definition 3.14. *The Hilbert cube $\mathcal{H}$, is the product of countably many copies of the closed unit interval with the product topology, i.e., $\mathcal{H} = [0, 1]^{\mathbb{N}}$ with the product topology.*

Theorem 3.15. *Every separable metrizable space X is homeomorphic to a subset of the Hilbert cube $\mathcal{H}$.*

Proposition 3.16. *Let $F : \Omega \to \mathcal{P}_{cl}(X) \cup \{\emptyset\}$ be weakly measurable. Then F is $K-$measurable.*

Proof. In view of Theorem 3.15, we can view X as a dense subspace of a compact metric space Z. Let $G : \Omega \to \mathcal{P}_{cp}(X)$ be defined by

$$G(\omega) = \overline{F(\omega)}, \quad \text{for every } \omega \in \Omega.$$

From Propositions 3.12 and 3.13, G is measurable. Let $K \in \mathcal{P}_{cp}(X)$, then

$$
\begin{aligned}
\overline{F}_-^{-1}(K) &= \{\omega \in \Omega : F(\omega) \cap K \neq \emptyset\} \\
&= \{\omega \in \Omega : G(\omega) \cap K \neq \emptyset\} \\
&= G_-^{-1}(K).
\end{aligned}
$$

Since for each $\omega \in \Omega$ we have $G(\omega) \in \mathcal{P}_{cp}(X)$, by Proposition 3.13, it follows that

$$
F_-^{-1}(K) = G_-^{-1}(K) \in \mathcal{U}.
$$

$\square$

Next we prove the following proposition when the metric space is σ-compact.

Proposition 3.17. *If X is a $\sigma-$compact metric space and $F : \Omega \to \mathcal{P}_{cl}(X) \cup \{\emptyset\}$ is a multivalued map, then the following statements are equivalent:*

  a) *F is measurable,*

  b) *F is weakly measurable,*

  c) *F is $K-$measurable.*

Theorem 3.18. *(Weak measurability versus measurable graph) Consider a nonempty-valued mulivalued map $F : \Omega \to \mathcal{P}(X)$, where X is a separable metrizable space. If F is weakly measurable, then its closure $\overline{F}$ has a measurable graph, that is, $\Gamma(\overline{F}) \in \mathcal{U} \otimes \mathcal{B}(X)$.*

Lemma 3.19. *Let $F_n : \Omega \to \mathcal{P}(X)$ be a sequence of multivalued maps. We have the following.*

  1) *The union multivalued map $F : \Omega \to \mathcal{P}(X)$ defined by*

$$
F(x) = \bigcup_{n=1}^{\infty} F_n(\omega), \quad \omega \in \Omega
$$

  is
  a) *weakly measurable, if every F_n is weakly measurable,*
  b) *measurable, if every F_n is measurable, and*
  c) *Borel measurable, if every F_n is Borel measurable.*

  2) *if $\{X_n\}_{n \in \mathbb{N}}$ are separable metric spaces and $F_n : \Omega \to \mathcal{P}_{cl}(X_n)$ are weakly measurable (graph measurable), then $F(\omega) := \prod_{n=1}^{\infty} F_n(\omega)$, $\omega \in \Omega$, is weakly measurable (graph measurable).*

  3) *If X is a separable metrizable space and each $F_n : \Omega \to \mathcal{P}_{cl}(X)$ is weakly measurable, and for each $\omega \in \Omega$ there is some k such that $F_k(\omega)$ is compact, then the intersection multifunction $F : \Omega \to \mathcal{P}(X)$, defined by*

$$
F(\omega) = \bigcap_{n=1}^{\infty} F_n(\omega),
$$

  is measurable (and hence weakly measurable).

Proposition 3.20. *If $(\Omega, \mathcal{U})$ is a complete measurable space, X is a Polish space and $F : \Omega \to \mathcal{P}(X)$ is graph measurable, then*

$$
\omega \to \partial F(\omega), \quad \omega \in \Omega,
$$

is measurable.

Theorem 3.21. ([185]) *Let X be a separable metric space, Y be a metric space, $f : \Omega \times X \to X$ be a Carathéodory function, and U be an open subset of Y. Then the multivalued map $F : \Omega \to \mathcal{P}(X)$ defined by*

$$F(\omega) = \{\omega \in \Omega : f(t, x) \in U\}$$

is measurable. In particular, if f is real valued, then

$$F_*(\omega) = \{\omega \in \Omega : f(\omega, x) > \lambda\} \quad and \quad \widetilde{F}(\omega) = \{\omega \in \Omega : f(\omega, x) < \lambda\}$$

are measurable.

3.1 Measurable selection

In the study of multi-valued mappings, the notion of a selection function is an important concept.

Definition 3.22. *Let $F : X \to \mathcal{P}(Y)$ be a multi-valued map. A single-valued map $f : X \to Y$ is said to be a selection of f, and we write $f \subset F$, if $f(x) \in F(x)$ for every $x \in X$.*

In what follows, we shall use the following Kuratowski–Ryll–Nardzewski selection theorem (see [33], [34], [185].)

Theorem 3.23. *Let Y be a separable complete space. Then every measurable $\varphi \colon \Omega \to \mathcal{P}(Y)$ has a selection.*

Proof. Without loss of generality we can change the metric of Y into an equivalent metric, preserving completeness and separability, so that Y becomes a bounded (say, with diameter M) complete metric space. Now, let us divide the proof into two steps.

Step 1. Let C be a countable dense subset of Y. Set $\varepsilon_0 = M$, $\varepsilon_i = M/2^i$. We claim that we can define a sequence of mappings $s_m \colon \Omega \to C$ such that:

(3.23.1) s_m is measurable,

(3.23.2) $s_m(x) \in O_{\varepsilon_m}(\varphi(x))$,

(3.23.3) $s_m(x) \in B(s_{m-1}(x), \varepsilon_{m-1})$, $m > 0$.

In fact, arrange the points of C into a sequence $\{c_j\}_{j=0,1,\ldots}$ and define s_0 by putting:

$$s_0(x) = c_0, \quad \text{for every } x \in \Omega.$$

Then (3.23.1) and (3.23.2) are clearly satisfied.

Assume we have defined functions s_m satisfying (3.23.1) and (3.23.2) up to $m = p - 1$, and define s_p satisfying (3.23.1)–(3.23.3) as follows. Set

$$A_j = \varphi_+^{-1}(B(c_j, \varepsilon_p)) \cap s_{p-1}^{-1}(B(c_j, \varepsilon_{p-1})),$$
$$E_0 = A_0, \quad E_j = A_j \setminus (E_0 \cup \ldots \cup E_{j-1}).$$

We claim that

$$\Omega = \bigcup_{j=0}^{\infty} E_j.$$

Of course E_j, $j = 0, 1, \ldots$, is measurable (see Proposition 3.2). In fact, let $x \in \Omega$ and consider $s_{p-1}(x)$ and $\varphi(x)$. By (3.23.2) $s_{p-1}(x) \in O_{\varepsilon_{p-1}}(\varphi(x))$; by the density of C there is a c_j such that at once $s_{p-1}(x) \in B(c_j, \varepsilon_{p-1})$ and $\varphi(x) \cap B(c_j, \varepsilon_{p-1}) \neq \emptyset$, i.e. $x \in A_j$. Finally, either $x \in E$, or it is in some E_i, $i < j$. In either case $x \in \bigcup_{j=0}^{\infty} E_j$. We define $s_p : \Omega \to C$ by putting:

$$s_p(x) = c_j \quad \text{whenever } x \in E_j.$$

Then s_p satisfies (3.23.1)–(3.23.3). Condition (3.23.3) implies that $\{s_m(x)\}$ is a Cauchy sequence for every $x \in \Omega$.

We let $s : \Omega \to Y$ be defined as follows:

$$s(x) = \lim_{m \to \infty} s_m(x), \quad x \in \Omega.$$

Since φ has closed values by (ii) we deduce that $s(x) \in \varphi(x)$ for every $x \in \Omega$.

Step 2. It remains to show that s is measurable. This is equivalent to proving that counter images of closed sets are measurable. Let K be a closed subset of Y. Then each set $s_m^{-1}(O_{\varepsilon_m}(K))$ is measurable. We shall complete the proof showing that $s^{-1}(K) = \bigcap s_m^{-1}(O_{\varepsilon_m}(K))$. In fact on the one hand, when $x \in s^{-1}(K)$, $s(x) \in K$ and since $d(s_m(x), s(x)) < \varepsilon_m$, $s_m \in O_{\varepsilon_m}(K)$, for every m. On the other hand, when $x \in s_m^{-1}(O_{\varepsilon_m}(K))$ for all m, $s_m(x) \in O_{\varepsilon_m}(K)$ and since $\{s_m(x)\}$ converges to $s(x)$ and K is closed we get $s(x) \in K$. The proof of Theorem 3.23 is completed. $\qquad\square$

Theorem 3.24. *Let $(\Omega, \sum)$, Y be a separable metric space and $F : \Omega \to \mathcal{P}_{cl}(Y)$ be a weakly measurable multivalued map. Then F has a measurable selection.*

The Kuratowski-Ryll-Nardzewski selection theorem was first published in 1965. In 1966, Castaing observed that it is possible to represent measurable multi-valued maps by the union of single-valued measurable maps. Recall the definition.

Definition 3.25. *A multi-valued map $G : \Omega \to \mathcal{P}(X)$ has a Castaing representation if there exists a family measurable single-valued maps $g_n : \Omega \to X$ such that*

$$G(\omega) = \overline{\{g_n(\omega) : n \in \mathbb{N}\}}.$$

The following result is due to Castaing (see [79]).

Theorem 3.26. *Let X be a separable metric space. Then the multivalued map $G : \Omega \to \mathcal{P}_{cl}(X)$ is measurable if and only if G has a Castaing representation.*

Proof. Let $D = \{x_n : n \in \mathbb{N}\}$ be such that $\overline{D} = X$. Assume that G is measurable. For each $n, k \in \mathbb{N}$, define the following family of multi-valued maps $G_{n,k} : \Omega \to \mathcal{P}(X)$ by

$$G_{n,k}(\omega) = \begin{cases} G(\omega) \cap B\left(x_n, \dfrac{1}{2^k}\right), & \omega \in G_-^{-1}(B(x_n, \tfrac{1}{2^k})), \\[2mm] G(\omega), & \omega \notin G_-^{-1}\left(B(x_n, \dfrac{1}{2^k})\right). \end{cases}$$

Since G is weakly measurable, $G_-^{-1}\left(B\left(x_n, \dfrac{1}{2^k}\right)\right) \in \mathcal{U}$. Let $V \subset X$ be an open set. Then

$$(G_{n,k}^{-1})_-(V) = G_-^{-1}\left(B(x_n, 2^{-k}) \cap V\right) \cup G_-^{-1}(V) \cap \Omega \backslash G_-^{-1}\left(B(x_n, 2^{-k})\right).$$

Hence,

$$\omega \to G_{n,k}(\omega)$$

is a measurable multifunction and from Proposition 3.2, the multi-function

$$\omega \to \overline{G_{n,k}(\omega)}$$

is measurable. Then from Theorem 3.24, 3.4 there exists a family of single-valued measurable maps $g_{n,k}$ defined Ω to X such that

$$g_{n,k}(\omega) \in \overline{G_{n,k}(\omega)}, \quad \text{for all} \quad \omega \in \Omega.$$

Now, we shall prove that

$$G(\omega) = \overline{\{g_{n,k}(\omega) : \ n, k \geq 1\}}.$$

Let $x \in G(\omega)$. For every $\varepsilon > 0$, there exist $n, k \geq 1$ such that $\dfrac{1}{2^{k-1}} < \varepsilon$ and $x \in B(x_n, 2^{-k})$. Hence

$$\omega \in G_-^{-1}(B(x_n, 2^{-k})) \text{ and } g_{n,k}(\omega) \in \overline{B(x_n, 2^{-k})}.$$

Then

$$d(g_{n,k}(\omega), x) \leq d(g_{n,k}(\omega), x_n) + d(x_n, x) \leq \frac{1}{2^{k-1}} \leq \varepsilon.$$

This implies that for every $\varepsilon > 0$, we have

$$\{g_{n,k}(\omega) : n, k \geq 1\} \cap B(x, \varepsilon) \neq \emptyset.$$

Hence

$$G(\omega) = \overline{\{g_{n,k}(\omega) : n, k \geq 1\}}.$$

Conversely, let $V \subset X$ be an open set. Then

$$
\begin{aligned}
G_-^{-1}(V) &= \{\omega \in \Omega : G(\omega) \cap V \neq \emptyset\} \\
&= \bigcup_{n \geq 1} \{\omega \in \Omega : g_n(\omega) \in V\} \\
&= \bigcup_{n \geq 1} g_n^{-1}(V) \in \mathcal{U}.
\end{aligned}
$$

Thus G is weakly measurable. $\qquad \square$

Next, we have the Castaing representation for measurable multifunctions.

Theorem 3.27. *Let X be a topological space and $F : \Omega \to \mathcal{P}_{cl}(X)$ be a multifuction.*

1) *If there exists a sequence $\{f_n\}_{n \in \mathbb{N}}$ of measurable selectors G satisfying*

$$G(\omega) = \overline{\{f_1(\omega), f_2(\omega), \dots, \}}$$

for each $\omega \in \Omega$, then G is weakly measurable.

2) *If there exists a sequence $\{f_n\}_{n \in \mathbb{N}}$ of measurable selectors G converging to a measurable function f and satisfying*

$$G(\omega) = \{f_1(\omega), f_2(\omega), \dots, \}$$

for each $\omega \in \Omega$, then G is measurable.

Proof. For 1) we refer to the proof of Theorem 3.26.

2) Now, we show that G is measurable. Since $\{f_n\}_{n\in\mathbb{N}}$ converge to f, then for every $\omega \in \Omega$, we have

$$
\begin{aligned}
G(\omega) &= \overline{\{f_1(\omega), f_2(\omega), \dots, \}} \\
&= \{f_1(\omega), f_2(\omega), \dots, \} \cup \{f(\omega)\}.
\end{aligned}
$$

Conversely, let $V \subseteq X$ be a closed set. Then then for every $\omega \in \Omega$, we have

$$
\begin{aligned}
G_-^{-1}(V) &= \{\omega \in \Omega : G(\omega) \cap V \neq \emptyset\} \\
&= \cup_{n\in\mathbb{N}}\{\omega \in \Omega : f_n(\omega) \in V\} \cup \{\omega \in \Omega : f(\omega) \in V\} \\
&= \cup_{n\in\mathbb{N}} f_n^{-1}(V) \cup f^{-1}(V) \in \mathcal{U}.
\end{aligned}
$$

□

Using Theorem 3.26, we can prove the following superposition result for multifunctions.

Proposition 3.28. *Let X be a Polish space, Y be a metric space, $U : \Omega \to \mathcal{P}_{cl}(X) \cup \{\emptyset\}$ be weakly measurable, $f : \Omega \times X \to Y$ be a Carathédory function and we define $F : \Omega \to \mathcal{P}(Y)$ by*

$$
F(\omega) = g(\omega, U(\omega)), \quad \omega \in \Omega.
$$

Then F is weakly measurable.

Proposition 3.29. *Let X be a separable Banach space and $F : \Omega \to \mathcal{P}_{cl}(X)$ be weakly measurable. Then so are the multifunctions*

$$
F_1(\omega) = co(F(\omega)), \quad \omega \in \Omega
$$

and

$$
F_2(\omega) = \overline{co}(F(\omega)), \quad \omega \in \Omega.
$$

Proposition 3.30. *If $(\Omega, \mathcal{U})$ is a complete measurable space, X is a separable Banach space and $F : \Omega \to \mathcal{P}_{cl}(X)$ is graph measurable, then,*

$$
F_*(\omega) = \overline{co}(F(\omega)), \quad \omega \in \Omega,
$$

is measurable.

The following important result is due to J.R. Aummann (see [86, 170, 185]).

Theorem 3.31. *Let $(\Omega, \mathcal{U})$ be a complete measurable space, X be a Polish space. If a multifunction $F : \Omega \to \mathcal{P}(X)$ has measurable graph, then it has a measurable selection. Moreover, there exists a countable family of measurable selections $f_n : \Omega \to X$ for F, such that*

$$
F(\omega) \subseteq \overline{\{f_n(\omega) : \omega \in \Omega\}}, \quad \text{for each } \omega \in \Omega.
$$

For the proof of Aummann's Theorem, we need a very useful projection property enjoyed by complete measurable spaces.

Theorem 3.32. *Let X be a complete separable metric space and $G \in \mathcal{U} \otimes \mathcal{B}(X)$. Then its projection is measurable:*

$$
\Pi_\Omega(G) := \{\omega \in \Omega | \exists x \in X, \ (\omega, x) \in G\} \in \mathcal{U}.
$$

Proof of Theorem 3.31. Let C be a closed set in X, then

$$
G_+^{-1}(C) = \Pi_\Omega(\Gamma(G) \cap \Omega \times C).
$$

Hence G is a measurable multifunction. From Theorem 3.23, G has a measurable selection.

□

3.2 Scalar measurable

In this section we the study the measurability of Banach space valued multifunctions.

Definition 3.33. *A multifunction $F : \Omega \to \mathcal{P}(X)$ is said to be scalarly measurable if for every $x^* \in X^*$ the function $\omega \to \sigma(x^*, F(\omega))$ is measurable.*

Definition 3.34. *A single valued function $f : \Omega \to X$ is scalarly measurable if the composition $x^* \circ f$ is measurable for every $x^* \in X^*$.*

Proposition 3.35. *Let $F : \Omega \to 2^X$ be a multifunction such that for each V nonempty weakly open subset of X we have $F_-^{-1}(V) \in \mathcal{U}$. Then F is scalarly measurable.*

Proof. Let $(x^*, \lambda) \in X^* \times \mathbb{R}$ and define

$$U(x^*, \lambda) = \{x \in X : \langle x^*, x \rangle > \lambda\}.$$

It is clear that $U(x^*, \lambda)$ is weakly open in X, so $F_-^{-1}(U(x^*, \lambda)) \in \mathcal{U}$. Observe that

$$
\begin{aligned}
F_-^{-1}(U(x^*, \lambda)) &= \{\omega \in \Omega : F(\omega) \cap U(x^*, \lambda) \neq \emptyset\} \\
&= \{\omega \in \Omega : \lambda < \sigma(x^*, F(\omega))\} \in \mathcal{U}.
\end{aligned}
$$

$\square$

Proposition 3.36. *Let $(\Omega, \mathcal{U})$ be a complete measurable space, X be a separable Banach space and $F : \Omega \to \mathcal{P}(X)$ be graph measurable. Then F is scalarly measurable.*

Proof. From Theorem 3.31 there exists a sequence of measurable selections f_n such that

$$F(\omega) \subseteq \overline{\{f_n(\omega) : n \in \mathbb{N}\}}, \quad \text{for every} \omega \in \Omega.$$

Then for every $x^* \in X^*$ we have

$$\sigma(x^*, F(\omega)) = \sup\{\langle x^*, f_n(\omega) \rangle : n \in \mathbb{N}\}.$$

Hence the function $\omega \to \sigma(x^*, F(\omega))$ is measurable. $\square$

Theorem 3.37. *Let $(\Omega, \mathcal{U})$ be a measurable space, X be a separable Banach space and $F : \Omega \to \mathcal{P}_{wcp,cv}(X)$ be weakly measurable. Then F is measurable if and only if F is scalarly measurable.*

Proof. If F is measurable, then from Theorem 3.18, it has a measurable graph. So from Proposition 3.36 F is a scalarly measurable. For the converse, we show that F is measurable. Let $(x_n^*)_{n \in \mathbb{N}}$ be a dense set of unit balls of X^*. By our assumption, for each $n \in \mathbb{N}$, $\sigma(x_n^*, F(\cdot))$ is measurable. Since $F(\omega)$ is weakly compact and convex, then for fixed $y \in X$, we get

$$
\begin{aligned}
d(y, F(\omega)) &= d(0, F(\omega) - y) \\
&= -\inf_{\|x^*\| \leq 1} \sigma(x^*, F(\omega) - y) \\
&= \sup_{\|x^*\| \leq 1} (\langle x^*, y \rangle - \sigma(x^*, y)) \\
&= \sup_{n \in \mathbb{N}} (\langle x_n^*, y \rangle - \sigma(x_n^*, y)).
\end{aligned}
$$

Hence $d(x, F(\cdot))$ is measurable. Consequently F is an Effros measurable multifunction. $\square$

3.2.1   Scalarly measurable selection

Throughout this subsection $(\Omega, \sum, \mu)$ is a complete finite measure space (or $\sigma-$finite measure space) and X is a real Banach space. We denote

$$\sum{}^{+} = \left\{ A \in \sum : \mu(A) > 0 \right\}, \quad \sum{}^{+}_{A} = \left\{ B \in \sum{}^{+} : B \subseteq A \right\}.$$

Definition 3.38. *We say that a multi-function* $F : \Omega \to 2^{X}$ *satisfies property* (P) *if for each* $\epsilon > 0$ *and each* $A \in \sum^{+}$ *there exist* $B \in \sum^{+}_{A}$ *and* $D \subset X$ *with* $diam(D) \leq \epsilon$ *such that* $B \subseteq F_{-}^{-1}(D)$.

Now, we present the first example of a function with property (P) in the single valued case.

Theorem 3.39. *Let* $f : \Omega \to X$ *be a single valued function. Then the following statements are equivalent:*

(i) *f satisfies property (P).*

(ii) *For each $A \in \sum^{+}$ there exists $B \in \sum^{+}_{A}$ with $diam(f(B)) \leq \epsilon$.*

(iii) *f is strongly measurable.*

For the next result we prove the relation between Effros measurable multifunctions and functions satisfying property (P); see Cascals *et al* [77].

Theorem 3.40. *Suppose X is a separable Banach space. Let $F : \Omega \to \mathcal{P}$ be an Effros measurable multifunction. Then F satisfies the property (P).*

Proof. Fix $\epsilon > 0$ and $A \in \sum^{+}$. Using the fact that X is a separable Banach space, we can write $X = \bigcup_{n \in \mathbb{N}} B_{n}$, where each B_{n} is an open ball with $diam(B_{n}) \leq \epsilon$. Since F is measurable then $C_{n} = F_{-}^{-1}(B_{n}) \in \sum$ and $\Omega = \bigcup_{n \in \mathbb{N}} C_{n}$. Since $\mu(A) > 0$ then there exists $n_{0} \in \mathbb{N}$ such that $C = A \cap B_{n_{0}} \in \sum^{+}_{A}$. Then for all $\omega \in B_{n_{0}} := D$ we have $F(\omega) \cap D \neq \emptyset$. $\square$

Theorem 3.41. *Every scalarly measurable multi-function $F : \Omega \to \mathcal{P}_{w,cp}(X)$ admits a scalary measurable selection.*

3.3   Lusin's theorem type

A famous relation between measurability and continuity of single-valued functions is established by Lusin's theorem, which states, roughly speaking, that $f : \Omega \to Y$ is measurable if and only if f is continuous on subsets of Ω of arbitrarily small measure. It is not surprising that this result has an analogue for multivalued mappings (for details see [28], [118]) which we shall sketch below.

In this section we assume throughout that Ω is also a metric space.

Definition 3.42. *We will say that a multivalued map $\varphi : \Omega \to \mathcal{P}_{cl}(Y)$ has the Lusin property if, given $\delta > 0$, one may find a closed subset $\Omega_{\delta} \subset \Omega$ such that $\mu(\Omega \setminus \Omega_{\delta}) \leq \delta$ and the restriction $\varphi|_{\Omega_{\delta}}$ of φ to Ω_{δ} is continuous.*

Lemma 3.43. *[176] Let $F : \Omega \to \mathcal{P}_{cl}(X)$ be a measurable multifunction. Then for every $\varepsilon > 0$, there exists a compact set $\Omega_\varepsilon \subset \Omega$, with $\mu(\Omega \backslash \Omega_\varepsilon) < \varepsilon$ such that F restricted to Ω_ε has a closed graph.*

Proof. Fix $\varepsilon > 0$ and let $D = \{x_n\}_{n \in \mathbb{N}}$ be a dense subset in X. Then each mapping $d_n(\omega) = d(x_n, F(\omega))$ is measurable and by Lusin's Theorem, there exists a compact set $\Omega_n \subset \Omega$, with $\mu(\Omega \backslash \Omega_n) \leq \dfrac{\varepsilon}{2^n}$ such that d_n restricted to Ω_n is continuous. Take $\Omega_\varepsilon = \cap_{n=1}^{\infty} \Omega_n$. It is clear that Ω_ε is compact, $\mu(\Omega \backslash \Omega_\varepsilon) \leq \varepsilon$ and for each d_n restricted to Ω_ε is continuous. We claim that F restricted to Ω_ε has a closed graph. Let $(\omega_m, y_m) \in \Gamma(F) \cap \Omega_\varepsilon \times X$ be a sequence which converges to (ω, y). For every fixed $\delta > 0$ there exists $x_n \in X$ such that $d(x_n, y) < \delta$. Then for sufficiently large $m \geq m(n)$, one has $d(x_n, F(\omega_m)) \leq d(x_n, y_m) < \delta$. Thus, by the continuity of d_n, we have $d_n(z_n, F(\omega)) \leq \delta$. Therefore $d(y, F(\omega)) \leq 2\delta$. Since δ is arbitrary, we have $d(y, F(\omega)) = 0$ which implies that $(\omega, y) \in \Gamma(F)$. $\square$

A similar fact holds for the lower semi-continuity.

Lemma 3.44. *Let $F : \Omega \to \mathcal{P}_{cl}(X)$ be a measurable multi-valued map. Then for every $\varepsilon > 0$, there exists a compact map $\Omega_\varepsilon \subset \Omega$ with $\mu(\Omega \backslash \Omega_\varepsilon) \leq \varepsilon$ such that F restricted to Ω_ε is l.s.c.*

Proof. Since F is measurable and X is separable space, there exists a sequence $f_n : \Omega \to X$ of measurable single-valued maps such that

$$F(\omega) = \overline{\{f_n(\omega) : n \in \mathbb{N}\}}.$$

By Lusin's Theorem, for every $\varepsilon > 0$, there exists a compact set $\Omega_\varepsilon \subset \Omega$ with $\mu(\Omega \backslash \Omega_\varepsilon) \leq \varepsilon$ such that each f_n restricted to Ω_ε is continuous. Now we prove that F is l.s.c. on Ω_ε. Let $C \subset X$ ba a closed subset; then

$$F_+^{-1}(C) = \{\omega \in \Omega_\varepsilon : F(\omega) \cap C \neq \emptyset\} = \bigcap_{n=1}^{\infty} f_n^{-1}(C \cap \Omega_\varepsilon).$$

Using the fact that the f_n are continuous functions, we deduce that $F_+^{-1}(C)$ is a closed set. $\square$

Theorem 3.45. *[176, 180] A multifunction $F : \Omega \to \mathcal{P}_{cl}(X)$ is measurable if and only if F has the Lusin property.*

Proof. From Lemmas 3.43 and 3.44, we have that if F is measurable, then F has the Lusin property. Conversely, let $C \subset X$ be a closed set in X. For arbitrary $\varepsilon > 0$, we have a closed set $\Omega_\varepsilon \subset \Omega$ such that $\mu(\Omega \backslash \Omega_\varepsilon) \leq \varepsilon$ and the restriction of F on Ω_ε is continuous. Then $F_+^{-1}(C)$ consists of a closed set $F_+^{-1}(C) \cap \Omega_\varepsilon$ and a set $F_+^{-1}(C) \cap \Omega \backslash \Omega_\varepsilon$ whose outer measure is less than or equals ε and therefore $F_+^{-1}(C)$ is measurable. $\square$

Remark 3.46. *The notion of a measurable multivalued map used in this book is called a strong measurable multifunction (see for example [98, 180]).*

Lemma 3.47. *[180] Let E be a Banach space, $J \subset \mathbb{R}$ be an interval and $F : J \to \mathcal{P}_{cp}(E)$ be a measurable multivalued map. Then F is almost separable (i.e. there is a subset $I \subset J$ with $\mu(I) = 0$ such that $\bigcup \{F(t) : t \in J \backslash I\}$ is separable).*

Lemma 3.48. *(see [296], Lemma 3.2) Let E be a separable Banach space, $G : [a, b] \to \mathcal{P}_{cl}(E)$ be a measurable multifunction, and $u : [a, b] \to E$ a measurable function. Then for any measurable $v : [a, b] \to \mathbb{R}^+$, there exists a measurable selection g of G such that for a.e. $t \in [a, b]$,*

$$\|u(t) - g(t)\| \leq d(u(t), G(t)) + v(t).$$

Proof. By Theorem 3.26, there is a sequence of measurable selections $\{g_n : n \in \mathbb{N}\}$ of G such that

$$G(t) = \overline{\{g_n(t) : n \in \mathbb{N}\}}, \text{ for all } t \in [a, b].$$

Set

$$T_n = \{t \in [a, b] | \; \|g_n(t) - u(t)\| \le d(u(t), G(t)) + r(t)\}.$$

Consider a single-valued map $\Psi_n : [a, b] \to \mathbb{R}^+$ defined by

$$\Psi_n(t) = \|g_n(t) - u(t)\| - d(u(t), G(t)) + r(t), \; t \in [a, b].$$

It clear that Ψ_n is a measurable map; then

$$\Psi_n^{-1}((-\infty, 0]) = \{t \in [a, b] : \|g_n(t) - u(t)\| \le d(u(t), G(t)) + r(t)\} = T_n.$$

Then the T_n, $n \in \mathbb{N}$, are measurable and we can easily show that $[a, b] = \cup_{n=1}^{\infty} T_n$ up to a negligible set. Let $E_1 = T_1$, $E_2 = T_2 \backslash E_1, \ldots, E_n = T_n \backslash \cup_{i=1}^{n-1} E_i, \ldots$. Then $[a, b] = \cup_{i=1}^{\infty} E_i$ up to a negligible set and $\{E_i\}_{i=1}^{\infty}$ is a disjoint sequence of measurable sets. Set

$$g(t) = \sum_{n=1}^{\infty} \chi_{E_n}(t) g_n(t),$$

where χ_{E_n} represents the characteristic function of the set E_n. Then g is a measurable selection of G satisfying the requirement of the lemma. $\qquad\square$

Corollary 3.49. *Let $G : [0, b] \to \mathcal{P}_{cp}(E)$ be a measurable multifunction and $g : [0, b] \to E$ be a measurable function. Then there exists a measurable selection u of G such that*

$$\|u(t) - g(t)\| \le d(g(t), G(t)).$$

Proof. Let $v_\varepsilon : [0, b] \to \mathbb{R}^+$ be defined by $v_\varepsilon(t) = \varepsilon > 0$. From Lemma 3.48, there exists a measurable selection u_ε of G such that

$$\|u_\varepsilon(t) - g(t)\| \le d(g(t), G(t)) + \varepsilon.$$

Take $\varepsilon = 1/n$, $n \in \mathbb{N}$; hence for every $n \in \mathbb{N}$, we have

$$\|u_n(t) - g(t)\| \le d(g(t), G(t)) + 1/n.$$

Using the fact that G has compact values, we may pass to a subsequence if necessary to get that $u_n(\cdot)$ converges to measurable function u in E. Then

$$\|u(t) - g(t)\| \le d(g(t), G(t)).$$

$\qquad\square$

Corollary 3.50. *Let E be a reflexive Banach space, $G : [0, b] \to \mathcal{P}_{cl,cv}(E)$ be a measurable multifunction, and suppose there exists $k \in L^1([0, b], E)$ such that*

$$G(t) \subseteq k(t)B(0, 1), \; t \in [0, b],$$

where $B(0, 1)$ denotes the closed ball in E and $g : [0, b] \to E$ is a measurable function. Then there exists a measurable selection u of G such that

$$\|u(t) - g(t)\| \le d(g(t), G(t)).$$

Lemma 3.51. *Let X be a separable metric space with $\{x_k\}$ a countable dense subset of X and Y be a Banach space. Let $F : X \to \mathcal{P}_{cp,cv}(Y)$ be a u.s.c. mapping; then the mapping $G : X \to \mathcal{P}(Y)$ defined by*

$$G(x) = \cap_{n=1}^{\infty} \overline{co}\left(\cup\{F(x_k) : d(x_k) < 1/n\}\right), \quad x \in X$$

satisfies the conditions:

(i) *For any $x \in X$, we have $\emptyset \neq G(x) \subset F(x)$.*

(ii) *G is u.s.c.*

Proof. Consider the family of multi-valued maps $G_n : X \to \mathcal{P}_{cl,cv}(Y)$, $n \in \mathbb{N}$ defined by

$$G_n(x) = \overline{co}\left\{\cup F(x_k) : d(x_k, x) < 1/n\right\}, \ x \in X,$$

and let

$$G(x) = \cap_{n=1}^{\infty} G_n(x).$$

First, we show that $G(x)$ is nonempty for every $x \in X$. For any n, we can take $k_n \in \mathbb{N}$ such that $d(x_{k_n}, x) < 1/n$. Then

$$\overline{co}\left(\bigcup_{i \geq n} F(x_{k_i})\right) \subset G_n(x).$$

Since $\{x_{k_n} : n \in \mathbb{N}\} \cup \{x\}$ is compact and F is u.s.c., then $\bigcup_{n=1}^{\infty} F(x_{k_n}) \cup F(x)$ is compact; hence $\overline{co}\left(\cup_{i \geq n} F(x_{k_i})\right)$ is compact. Thus

$$G(x) = \cap_{i=1}^{\infty} G_n(x) \supset \cap_{n=1}^{\infty} \overline{co}\left(\bigcup_{i \geq n} F(x_{k_i})\right) \neq \emptyset.$$

Using the fact F is u.s.c., we can easily show that $G(x) \subset F(x)$. Indeed for every $\varepsilon > 0$, and for n sufficiently large, we have that $d(x_k, x) < 1/n$ implies $F(x_k) \subset (F(x))_\varepsilon$. Since $(F(x))_\varepsilon$ is convex, then

$$G(x) \subset G_n(x) \subset \overline{(F(x))_\varepsilon},$$

which yields $G(x) \subset F(x)$. Now we prove that G is u.s.c. Let C be any closed subset of Y and $\{z_q\}$ be a sequence of $G_+^{-1}(C)$ converging to some limit $z \in X$. For each n, choose z_q such that $d(z_q, z) < 1/2n$. If $d(x_k, z_q) < 1/2n$, then $d(x_k, z) < 1/n$; hence $G_{2n}(z_q) \subset G_n(z)$ and $\emptyset \neq G_{2n}(z_q) \cap C \subset G_n(z) \cap C$. Since F is u.s.c. there exists $j_n > n$ for which $d(x_k, z) < \dfrac{1}{j_n}$ implies that $F(x_k) \subset (F(z))_{1/n}$. Thus $\emptyset \neq G_{j_n}(z) \cap C \subset \overline{(F(z))_{1/n}}$. Then there exists $y_n \in G_{j_n}(z) \cap C$ such that $d(y_n, F(z)) \leq 1/n$. Since $F(z)$ is compact, then some subsequence (y_m) of (y_n) converges to an element y of C. If $j_m > n$, then

$$G_{j_m}(z) \cap C \subset G_n(z) \cap C.$$

Hence

$$y \in G_n(z) \cap C, \ \text{ for all } \ n.$$

This implies that

$$y \in \cap_{n=1}^{\infty} G_n(z) \cap C = G(z) \cap C$$

and $z \in G_+^{-1}(C)$. Hence $G_+^{-1}(C)$ is closed and G is u.s.c. $\qquad \square$

Now, we shall be concerned with multivalued mappings which are defined on the topological product of some measurable set with the Euclidean space $\mathbb{R}^n$. We are particularly interested in Carathéodory multivalued mappings and Scorza–Dragoni multivalued mappings. Apart from their fundamental importance in all fields of multivalued analysis, such multivalued mappings are useful in differential inclusions.

Let $\Omega = [0, a]$ be equipped with the Lebesgue measure and $Y = \mathbb{R}^n$.

Definition 3.52. *A map* $\varphi \colon [0, a] \times \mathbb{R}^n \to \mathcal{P}_{cp}(\mathbb{R}^n)$ *is called u-Carathéodory (resp. l-Carathéodory; resp. Carathéodory) if it satisfies:*

(3.52.1) $t \to \varphi(t, x)$ *is measurable for every* $x \in \mathbb{R}^n$,

(3.52.2) $x \to \varphi(t, x)$ *is u.s.c. (resp. l.s.c.; resp. continuous) for almost all* $t \in [0, a]$,

(3.52.3) $\|y\| \leq \mu(t)(1 + \|x\|)$, *for every* $(t, x) \in [0, a] \times \mathbb{R}^n$, $y \in \varphi(t, x)$, *where* $\mu \colon [0, a] \to [0, +\infty)$ *is an integrable function.*

As before, by $\mathcal{U} \otimes \mathcal{B}(\mathbb{R}^n)$, we denote the minimal σ-algebra generated by the Lebesgue measurable sets $A \in \mathcal{U}$ and the Borel subsets of $\mathbb{R}^n$, and then the term "product-measurable" means measurability with respect to $\mathcal{U} \otimes \mathcal{B}(\mathbb{R}^n)$.

Proposition 3.53. *Let* $\varphi \colon [0, a] \times \mathbb{R}^n \to \mathcal{P}(\mathbb{R}^m)$ *be a Carathéodory multivalued map. Then* φ *is product-measurable.*

Proof. Consider the countable dense subset $\mathbb{Q}^n \subset \mathbb{R}^n$ of rationals. For closed $A \subset \mathbb{R}^n$, $a \in \mathbb{Q}^n$ and k, the set

$$G_k(A, a) = \{t \in [0, a] \mid \varphi(t, a) \cap O_{1/k}(A) \neq \emptyset\} \times B(a, 1/k)$$

belongs to $\mathcal{U} \otimes \mathcal{B}(\mathbb{R}^n)$. Since φ is *l.s.c.* in the second variable, we have:

$$\varphi_+^{-1}(A) \subseteq \bigcap_{k=1}^{\infty} \bigcup_{a \in Q^n} G_k(A, a),$$

while the *u.s.c.* of φ implies the reverse inclusion. The proof is completed. $\qquad\square$

The following example shows that an *l*-Carathéodory multivalued map needs not to be product measurable.

Example 3.54. *Let* $\varphi \colon [0, 1] \times \mathbb{R} \to \mathcal{P}(\mathbb{R})$ *be defined as follows:*

$$\varphi(t, u) = \begin{cases} \{0\}, & \text{if } u = 0, \\ [0, 1], & \text{otherwise.} \end{cases}$$

Then φ *is l-Carathéodory but not u-Carathéodory.*

An analogous example can be constructed for *u*-Carathéodory mappings.

Let $\varphi \colon [0, a] \times \mathbb{R}^n \to \mathcal{P}(\mathbb{R}^n)$ be a fixed multivalued map. We are interested in the existence of Carathéodory selections, i.e. Carathéodory functions $f \colon [0, a] \times \mathbb{R}^n \to \mathbb{R}^n$ such that $f(t, u) \in \varphi(t, u)$ for almost all $t \in [0, a]$ and all $u \in \mathbb{R}^n$. It is evident that, in the case when φ is *u*-Carathéodory, this selection problem does not have a selection in general (the reason is exactly the same as in Michael's selection theorem). For *l*-Carathéodory multivalued maps φ, however, this is an interesting problem.

In order to study this problem, we shall use the following notation:

$$C(\mathbb{R}^n, \mathbb{R}^n) = \{f \colon \mathbb{R}^n \to \mathbb{R}^n \mid f \text{ is continuous}\}.$$

We shall understand that $C(\mathbb{R}^n, \mathbb{R}^n)$ is equipped with the topology on uniform convergence on compact subsets of $\mathbb{R}^n$. In fact this topology is metrizable. Moreover, as usual by $L_1([0, a], \mathbb{R}^n)$, we shall denote the Banach space of Lebesgue integrable functions.

There are two ways, essentially, to deal with the above selection problem. Let $\varphi \colon [0, a] \times \mathbb{R}^n \to \mathcal{P}(\mathbb{R}^n)$ be an l-Carathéodory mapping. On the one hand, we may show that the multivalued map

$$\Phi \colon [0, a] \to \mathcal{P}(C(\mathbb{R}^n, \mathbb{R}^n)),$$
$$\Phi(t) = \{u \colon \mathbb{R}^n \to \mathbb{R}^n : u(x) \in \varphi(t, u(x)) \text{ and } u \text{ is continuous}\}$$

is measurable.

Then, if we assume that φ has convex values, in view of the Michael selection theorem, we obtain that $\Phi(t) \neq \emptyset$ for every t. Moreover, let us observe that every measurable selection of Φ will give rise to a Carathéodory selection of φ. On the other hand, we may show that the multivalued map:

$$\Psi \colon \mathbb{R}^n \to \mathcal{P}(L^1([0, a], \mathbb{R}^n)),$$
$$\Psi(x) = \{u \colon [0, a] \to \mathbb{R}^n \mid u(t) \in \varphi(t, u(t)), \text{ for almost all } t \in [0, a]\}$$

is a *l.s.c.* mapping. Consequently, continuous selections of Ψ will give rise to Carathéodory selections of φ. Hence, our problem can be solved by using Michael and Kuratowski–Ryll–Nardzewski selection theorems. Let us formulate, only for informative purposes, the following result due to A. Cellina.

Theorem 3.55. *Let $\varphi \colon [0, a] \times \mathbb{R}^n \to \mathcal{P}_{cp,cv}(\mathbb{R}^n)$ be a multivalued map. If $\varphi(\,\cdot\,, x)$ is u.s.c. for all $x \in \mathbb{R}^n$ and $\varphi(t, \,\cdot\,)$ is l.s.c. for all $t \in [0, a]$, then φ has a Carathéodory selection.*

Proposition 3.56. *Let X be a separable metric space with $\{x_k\}$ a countable dense subset of X and Y a separable Banach space. Let $F \colon \Omega \times X \to \mathcal{P}_{cp,cv}(Y)$ be an upper Carathéodory multifunction. Then the mapping $G \colon \Omega \times X \to \mathcal{P}_{cl,cv}(Y)$ defined by*

$$G(\omega, x) = \cap_{n=1}^{\infty} \overline{co} \{F(\omega, x_k) : d(x_k, x) < 1/n\}$$

satisfies the following conditions:

(3.56.1) For each $\omega \in \Omega$ and $x \in X$, $\emptyset \neq G(\omega, x) \subset F(\omega, x)$.

(3.56.2) For each $\omega \in \Omega$, $G(\omega, .)$ is u.s.c.

(3.56.3) G is $\mathcal{L} \otimes \mathcal{B}(X)$- measurable.

Proof. Let $G_n \colon \Omega \times X \to \mathcal{P}(Y)$ be the multivalued map defined by

$$G_n(\omega, x) = \cup\{F(\omega, x_k) : d(x_k, x) < 1/n\}, \quad (\omega, x) \in \Omega \times X.$$

Together with (3.56.1) and (3.56.2), we apply Lemma 3.51; then we have

$$F(t, x) \supset H_n(\omega, x) \neq \emptyset, \text{ for every } (\omega, x) \in \Omega \times X,$$

and $H_n(\omega, \cdot)$ is *u.s.c.* Now, we prove that $H_n(\cdot, \cdot)$ is $\mathcal{L} \otimes \mathcal{B}(X)$- measurable. Indeed, for any open subset V of Y, we have

$$\begin{aligned}
H_n^{-1}{}_{+}(V) &= \{(\omega, x) \in \Omega \times X : H_n(\omega, x) \cap V \neq \emptyset\} \\
&= \cup_{n=1}^{\infty} \{\omega \in \Omega : F(\omega, x) \cap V \neq \emptyset\} \times \\
&\quad \{x \in X : d(x, x_k) < 1/n\} \in \mathcal{L} \otimes \mathcal{B}(X).
\end{aligned}$$

Then the multi-map $G_n : \Omega \times X \to \mathcal{P}(Y)$ defined by

$$G_n(\omega, x) = \overline{co\,(H_n(\omega, x))}$$

is measurable (see [34]). If we show that

$$G_+^{-1}(C) = \cap_{n=1}^{\infty} (G_n)_+^{-1}(C_{n-1}),$$

where

$$C = \cap_{n=1}^{\infty} C_{n-1}$$

and

$$C_{n-1} = \{y \in Y \,|\, d(y, C) < 1/n\},$$

then we can conclude that G is $\mathcal{L} \otimes \mathcal{B}(X)$- measurable. It is obvious that

$$G_+^{-1}(C) \subset \cap_{n=1}^{\infty} (G_n)_+^{-1}(C_{n-1}).$$

Conversely, let

$$(\omega, x) \in \cap_{n=1}^{\infty} (G_{n-1})_+^{-1}(C_{n-1})$$

then $G_n(\omega, x) \cap C_{n-1} \neq \emptyset$ for all $n \in \mathbb{N}$. Since $F(\omega, \cdot)$ is *u.s.c.*, by the same way as in the proof of Lemma 3.51, we have

$$\emptyset \neq \cap_{n=1}^{\infty} G_n(\omega, x) \cap \overline{C_{n-1}} = G(\omega, x) \cap C.$$

Hence

$$\cap_{n=1}^{\infty} (G_n)_+^{-1}(C_{n-1}) \subset G_+^{-1}(C).$$

$$\square$$

Proposition 3.57. *[79, 296] Let X be a complete separable Banach space, $G : [t_0, b] \to \mathcal{P}_{cl}(X)$ be a Lebesgue measurable multivalued map (i.e. for every open subset $V \subset X$, the set $G_+^{-1}(V)$ is Lebesgue measurable) and $f : [t_0, b] \to X$, and $g : [t_0, b] \to \mathbb{R}^+$ be measurable single-valued maps. Then the maps*

$$t \to \overline{co}\,G(t), \quad t \to \overline{B}(f(t), g(t)), \quad t \to d(f(t), G(t))$$

and

$$\widetilde{\Pi}_{G(t)}(g(t) = \{x \in G(t)\,|\, d_X(x, f(t)) = d_Y(f(t), G(t))\}$$

are measurable. Consequently, if

$$\{v \in G(t)\,|\, \|v - g(t)\| \leq k(t)\} \neq \emptyset, \quad a.e. \ in \ [t_0, b],$$

then there exists a measurable selection $u(t) \in G(t)$ such that for almost every $t \in [t_0, b]$, we have

$$\|u(t) - f(t)\| \leq k(t).$$

Proof. Since G is a measurable multifunction and X is a separable Banach space, from Theorem 3.32, there is a sequence of measurable selections $\{g_n(\cdot)\,|\, n \geq 1\}$ such that

$$G(t) = \overline{\{g_n(t)\,|\, n \in \mathbb{N}\}}.$$

Let $\{\lambda_n\}$ be a sequence of nonnegative rational numbers such that there are only finitely many $\lambda_n \neq 0$ and $\sum_{n=1}^{\infty} \lambda_n = 1$. The set

$$\left\{ \sum_{n=1}^{\infty} g_n(\cdot)\,|\, (\lambda_n)_{n \geq 1} \in \mathbb{Q}_+ \right\}$$

is a countable family of measurable functions. Using the fact that

$$\{g_n(t)|\ n \in \mathbb{N}\} \subset G(t),$$

then

$$\left\{ \sum_{n=1}^{\infty} g_n(\cdot) : (\lambda_n)_{n \geq 1} \in \mathbb{Q}_+ \right\} \subset \overline{co}\, G(t)$$

and

$$\overline{co}\, G(t) \subset \overline{\left\{ \sum_{n=1}^{\infty} g_n(\cdot)|\, (\lambda_n)_{n \geq 1} \in \mathbb{Q}_+ \right\}}.$$

Hence

$$\overline{co}\, G(t) = \overline{\left\{ \sum_{n=1}^{\infty} g_n(\cdot)|\, (\lambda_n)_{n \geq 1} \in \mathbb{Q}_+ \right\}};$$

we conclude that $\overline{co}\, G$ is measurable.

Now, we show that $t \to \overline{B}(f(t), g(t))$ is a measurable multifunction. We can easily verify that

$$\overline{B}(f(t), g(t)) = f(t) + g(t)\overline{B}(0,1), \quad t \in [t_0, b].$$

Since X is a separable space, then there exists $\{x_n : n \in \mathbb{N}\}$, a countable subset in $\overline{B}(0,1)$, such that

$$\overline{\{x_n|\ n \geq 1\}} = \overline{B}(0,1).$$

Set

$$\{f(t) + g(t)x_n|, \quad n \geq 1\} \subset \overline{B}(f(t), g(t)).$$

Hence

$$\overline{B}(f(t), g(t)) = \overline{\{f(t) + g(t)x_n|, \quad n \geq 1\}}.$$

This implies that $\overline{B}(f(\cdot), g(\cdot))$ is measurable.

The multi-map $t \to d(f(t), G(t))$ is measurable. Let $r > 0$ and

$$\{t \in [t_0, b]|\, d(f(t), G(t)) < r\} \quad = \quad \bigcup_{n=1}^{\infty} \{t \in [t_0, b]|\ \|f(t) - f_n(t)\| < r\}$$

$$= \quad \bigcup_{n=1}^{\infty} \psi_n(r, \infty),$$

where $\psi_n(t) = \|f_n(t) - f(t)\|$. We can easily prove that the ψ_n are measurable functions, then $\bigcup_{n=1}^{\infty} \psi_n((r, \infty))$ is a measurable set; we conclude that $d(f(\cdot), G(\cdot))$ is a measurable single-valued function

$$\widetilde{\Pi}_{G(t)}(g(t)) = G(t) \cap \{v \in Y|\, d_X(v, f(t)) = d_Y(f(t), G(t))\}.$$

□

Theorem 3.58. *[34] Let $(\Omega, \mathcal{A}, \mu)$ be a complete $\sigma-$finite measurable space, X be a complete separable metric space, and $F : \Omega \to \mathcal{P}(X)$ be a measurable multivalued map with closed images. Consider a Carathéodory multivalued map G from $\Omega \times X$ to a complete separable metric space Y. Then the map*

$$\Omega \ni \omega \to \overline{G(\omega, F(\omega))} \in \mathcal{P}(Y)$$

is measurable.

While this result characterizes the measurability, the following lemma is a measurable selection result (Filippov's Theorem). It is crucial in the proof that the control system coincide with the differential inclusion problem.

Lemma 3.59. *(see [34], Thm. 8.2.10) Consider a complete $\sigma-$finite measurable space (Ω, A, μ) (A is a $\sigma-$algebra and μ is a positive measure). Let X, Y be two complete separable metric spaces. Let $F : X \to \mathcal{P}(Y)$ be a measurable multivalued map with closed nonempty values and $g : \Omega \times X \to Y$ a Carathéodory map. Then for every measurable map $h : \Omega \to Y$ satisfying*

$$h(\omega) \in g(\omega, F(\omega)) \quad \text{for almost all } \omega \in \Omega,$$

there exists a measurable selection $f(\omega) \in F(\omega)$ such that

$$h(\omega) = g(\omega, f(\omega)) \quad \text{for almost all } \omega \in \Omega.$$

Proof. Define the multi-valued map $H : \Omega \to \mathcal{P}(X)$ by letting

$$H(\omega) = F(\omega) \cap \bigcap_{n=1}^{\infty} \overline{\{x \in X| \ d_y(g(\omega, x), h(\omega)) < 1/n\}}.$$

Let $\phi : \Omega \times X \to \Omega \times Y$ be a measurable function defined by

$$\phi(\omega, x) = (\omega, g(\omega, x)) \quad (\omega, x) \in \Omega \times X$$

and the multifunction $G : \Omega \to \mathcal{P}(X)$ defined by

$$G(\omega) = \overline{\{x \in X| \ d_y(g(\omega, x), h(\omega)) < 1/n\}}.$$

Observe that

$$\mathcal{G}r(G) = \phi^{-1}(\mathcal{G}r(\widetilde{G})),$$

where $\widetilde{G}$ is a multi-valued map defined by

$$\widetilde{G}(\omega) = \overline{B}\left(h(\omega), 1/n\right).$$

Now, we show that $g(\cdot, \cdot)$ is a measurable function. Since X is a separable Banach space, then there exists a set $D = \{x_n : n \in \mathbb{N}\} \subset X$ such that $\overline{D} = X$, and let C be a closed subset in Y. Then

$$g^{-1}(C) = \bigcap_{n=1}^{\infty} \bigcup_{v \in D} \{\omega \in \Omega| \ g(\omega, v) \in C_n\} \times$$
$$\{x \in X| \ d_X(x, v) < 1/n\} \in A \otimes \mathcal{B}(X),$$

where

$$C_n = \{y \in Y| \ d_Y(y, C) < 1/n\}.$$

Then g is a measurable single-valued map which implies that ϕ is measurable. Then $\mathcal{G}r(G) \in A \otimes \mathcal{B}(X)$. From Proposition 3.2, G is measurable; also H is a measurable multifunction. Hence by Theorem 3.23, H has a measurable selection f. Then for every $n \in \mathbb{N}$, we have

$$d_Y(g(\omega, f(\omega)), h(\omega)) \leq 1/n.$$

Hence

$$h(\omega) = g(\omega, h(\omega)), \quad \text{for almost every } \omega \in \Omega.$$

$\square$

Lemma 3.60. *[275] Let X, Y be complete separable metric spaces and $F : [0, b] \times X \to \mathcal{P}_{cl}(Y)$ be a $\mathcal{L} \otimes \mathcal{B}(X)$-measurable multifunction. Then for any continuous function $x : [0, b] \to Y$ the multifunction $t \to F(t, x(t))$ is measurable and has a strongly measurable selection.*

Proof. Let $B \subset Y$ be a closed set and put

$$\mathcal{C} = \{t \in [0, b] : F(t, x(t)) \cap B \neq \emptyset\}.$$

Let

$$V = \{(t, u) \in [0, b] \times X : F(t, u) \cap B \neq \emptyset\}.$$

Then the set V is $\mathcal{L} \otimes \mathcal{B}(X)$-measurable, hence

$$W = \{(t, u) \in V : u = x(t)\}$$

is measurable. From Theorem 3.5 and Proposition 2.2 in [170], it follows that the set

$$\mathcal{C} = \{t \in [0, b] : (t, x(t)) \in W\}$$

is measurable. This implies that $t \to F(t, x(t))$ is measurable. By Theorem 5.6 in [170], we obtain that there exists a strong measurable selection. $\square$

Lemma 3.61. *[173] Let (Ω, Σ) is a measurable space, X, Y are separable metric spaces and $F : \Omega \times X \to \mathcal{P}_{cl}(Y)$ be a multifunction. Assume that for every $x \in X$*

$$t \to F(t, x) \text{ is measurable}$$

and for every $t \in \Omega$, we have

$$x \to F(t, x) \text{ is continuous or } H_d - \text{continous.}$$

Then

$$(t, x) \to F(t, x) \text{ is } \Omega \otimes B(X) \text{ measurable.}$$

3.4 Hausdorff-measurable multivalued maps

In this section, we study of some Hausdorrff measurability properties of multivalued maps. Let (Ω, Σ) be a measurable space and let μ be a non-negative measure on Ω. Moreover we say that the measurable space (Ω, Σ) is complete if the σ-algebra Σ coincides with $\Omega*$ the Lebesgue completion of Ω with respect to μ and X a metric space.

Definition 3.62. *A multifunction $F : \Omega \to \mathcal{P}(X)$ is said to be:*

(3.62.1) d-measurable if for every $x \in X$, the function $\omega \to d(x, F(\omega))$ is measurable on Ω;

(3.62.2) H_d-measurable if for every $C \in \mathcal{P}(X)$, the functions $\omega \to H_u(F(\omega), C)$ and $\omega \to H_l(C, F(\omega))$ are measurable on Ω;

(3.62.3) $\widetilde{H}_{\max}$-measurable if for every $C \in \mathcal{P}(X)$ the function $\omega \to H_d(F(\omega), C)$ is measurable on Ω.

Proposition 3.63. *Let $F : \Omega \to \mathcal{P}(X)$ be a multifunction. We have:*

(3.63.1) F is H_d-measurable if and only if $\overline{F}$ is h-measurable.

(3.63.2) F is H_d-measurable (resp. $\widetilde{H}_{\max}$-measurable) if and only if for every $C \in \mathcal{P}(X)$, $\omega \to H_u(F(\omega), C)$ and $\omega \to H_l(C, F(\omega))$ are measurable on Ω(resp. $\omega \to H_d(F(\omega), C)$ is measurable on Ω).

(3.63.3) F is H_d-measurable, which implies that F is $\widetilde{H}_{\max}$-measurable.

(3.63.4) F is H_d-measurable, which implies that F is d-measurable.

Definition 3.64. *A multivalued map $F : \Omega \to \mathcal{P}(X)$ is called simple if there is an admissible partition $\{\Omega_k\}$ (i.e., there is a countable family $\{\Omega_k\}$ of nonempty measurable pairwise disjoint subsets Ω_k of Ω, whose union is Ω) and such that F restricted to each Ω_k is constant. An analogous notion applies to single-valued maps.*

Remark 3.65. *Each simple multivalued map $F : \Omega \to \mathcal{P}(X)$ is weakly measurable and H_d-measurable.*

The following proposition is a variant of the theorem of Kuratowski and Ryll-Nardzewski, Theorem 3.23.

Proposition 3.66. *Let $F : \Omega \to \mathcal{P}_{cl,b}(X)$ be an $\widetilde{H}_{\max}$-measurable multifunction, whose range $F(\Omega)$ is a separable subset of $\mathcal{P}_{cl,b}(X)$. Then we have:*

(3.66.1) there is a sequence $\{F_n : n \in \mathbb{N}\}$ of simple multivalued maps $F_n : \Omega \to \mathcal{P}_{cl,b}(X)$, converging to F uniformly on Ω;

(3.66.2) if X is complete, then F has a measurable selection.

Proposition 3.67. *Let $F : \Omega \to \mathcal{P}_{cl,b}(X)$ be an $\widetilde{H}_{\max}$-measurable multifunction, whose range $F(\Omega)$ is a separable subset of $\mathcal{P}_{cl,b}(X)$. Then we have:*

(3.67.1) If X is separable then

$$F \text{ is } H_d - \text{measurable} \iff F \text{ is weakly measurable}$$
$$\iff F \text{ is } d - \text{measurable};$$

(3.67.2) If $F(\Omega)$ is a separable subset of $\mathcal{P}_{cl,b}(X)$, then

$$F \text{ is } \widetilde{H}_{\max} - \text{measurable} \Leftrightarrow F \text{ is weakly measurable}.$$

Corollary 3.68. *Let $(\Omega, \mathcal{L})$ be a Borel space, where Ω is a metric space. Then each H_d-u.s.c. or H_d-l.s.c. multivalued map $F : \Omega \to \mathcal{P}_{cl,b}(X)$ is H_d-measurable.*

Proposition 3.69. *Let Ω be a complete separable metric space. Let μ be a non-negative finite measure defined on the completion $\mathcal{L}$ of the Borel Ω σ-algebra $\sum$. Let X be a metric space. For a multivalued $F : \Omega \to \mathcal{P}_{cl,b}(X)$, the following statements are equivalent:*

(3.69.1) F is Lusin measurable.

(3.69.2) F is H_d-measurable, and there exists a set $\Omega_0 \in \mathcal{L}$ with $\mu(\Omega_0) = 0$ such that $F(\Omega \backslash \Omega_0)$ is a separable subset of $\mathcal{P}_{cl,b}(X)$.

More details on the above results may be found in [96].

3.5 The Scorza-Dragoni property

We now introduce mappings having the Scorza-Dragoni property. First, we recall some definitions. Recall that a map $\varphi \colon [a, b] \times \mathbb{R}^n \to \mathcal{P}(\mathbb{R}^n)$ is said to be *integrably bounded* if there exists an integrable function $\mu \in L^1([a, b])$ such that $\|y\| \leq \mu(t)$ for every $x \in \mathbb{R}^n$, $t \in [a, b]$ and $y \in \varphi(t, x)$.

We say that φ has *linear growth* if there exists an integrable function $\mu \in L^1([a, b])$ such that

$$\|y\| \leq \mu(t)(1 + \|x\|)$$

for every $x \in \mathbb{R}^n$, $t \in [a, b]$ and $y \in \varphi(t, x)$.

Now we extend Definition 3.52.

Definition 3.70. *Let X and Y be two complete separable Banach spaces. $G : J \times X \to \mathcal{P}(Y)$ is called a multivalued upper Carathéodory (u-Carathéodory) or lower Carathéodory (l-Carathéodory) function if*

(3.70.1) The function $t \mapsto G(t, z)$ is measurable for each $z \in X$.

(3.70.2) For a.e. $t \in J$, the map $z \mapsto G(t, z)$ is upper semi-continuous (u.s.c.) or lower semi-continuous (l.s.c.), respectively.

It is further an $L^1 - Carathéodory$ if it is locally integrable bounded, i.e. for each positive real number r, there exists some $h_r \in L^1(J, \mathbb{R}^+)$ such that

$$\|G(t, z)\|_{\mathcal{P}} \leq h_r(t) \ \text{ for a.e. } \ t \in J \ \text{ and all } \ \|z\| \leq r,$$

where $J = [a, b]$ is a compact interval in $\mathbb{R}$ or $J = \mathbb{R}$.

Definition 3.71. *We say that a multivalued map $\varphi \colon [0, a] \times \mathbb{R}^n \to \mathcal{P}_{cl}(\mathbb{R}^n)$ has the u-Scorza–Dragoni property (resp. l-Scorza–Dragoni property; resp. Scorza–Dragoni property) if, given $\delta > 0$, one may find a closed subset $A_\delta \subset [0, a]$ such that the measure $\mu([0, a] \setminus A_\delta) \leq \delta$ and the restriction $\widetilde{\varphi}$ of φ to $A_\delta \times \mathbb{R}^n$ is u.s.c. (resp. l.s.c.; resp. continuous).*

Let us observe that the Scorza–Dragoni property plays the same role for multivalued mappings of two variables as the Lusin property for multivalued mappings of one variable. In addition, there is a close connection between Carathéodory multivalued mappings and multivalued mapping having the Scorza–Dragoni property.

Proposition 3.72. *Let $\varphi \colon [0, a] \times \mathbb{R}^m \to \mathcal{P}_{cp}(\mathbb{R}^n)$ be a multivalued map. Then we have:*

(3.72.1) φ is Carathéodory if and only if φ has the Scorza–Dragoni property,

(3.72.2) if φ has the u-Scorza–Dragoni property then φ is u-Carathéodory,

(3.72.3) if φ has the l-Scorza–Dragoni property then φ is l-Carathéodory,

(3.72.4) if φ is product-measurable l-Carathéodory then φ has the l-Scorza–Dragoni property.

Assume further that φ satisfies the Filippov condition, i.e. for every open U, $V \subset \mathbb{R}^n$ the set $\{t \in [0, a] : \varphi(t, U) \subset V\}$ is Lebesgue measurable; then:

(3.72.5) φ is u-Carathéodory multivalued map if and only if φ has the u-Scorza–Dragoni property.

Proposition 3.72 is taken from [28]. All proofs are rather technical and sometimes need long calculations. Therefore we shall present below only two examples showing that l-Carathéodory (u-Carathéodory) maps need not have the l-Scorza–Dragoni (u-Scorza–Dragoni) property.

Example 3.73. *Let $\varphi \colon [0,1] \times \mathbb{R} \to \mathcal{P}(\mathbb{R})$ be the map defined by*

$$\varphi(t,u) = \begin{cases} \{0\}, & \text{if } u = t \text{ and } t \in [0,1] \setminus A, \\ \{1\}, & \text{if } u = t \text{ and } t \in A, \\ [0,1], & \text{otherwise}, \end{cases}$$

where A is a nonmeasurable subset of $[0,1]$. Then obviously φ is l-Carathéodory but does not have l-Scorza–Dragoni property. Moreover, φ is not product measurable.

Example 3.74. *Let $\varphi \colon [0,1] \times \mathbb{R} \to \mathcal{P}(\mathbb{R})$ be defined by*

$$\varphi(t,u) = \begin{cases} [0,1], & \text{if } t = u \text{ and } t \in A, \\ \{0\}, & \text{otherwise}, \end{cases}$$

where A is a nonmeasurable subset of $[0,1]$. It is not hard to see that φ is u-Carathéodory but does not have the u-Scorza–Dragoni property.

For the remainder of this section, X is a metric separable space and Ω is a complete measure space. We also assume that $\varphi \colon \Omega \times X \to \mathcal{P}_{cp}(X)$ is a product-measurable multivalued mapping.

First we shall prove:

Proposition 3.75. *If $\varphi \colon \Omega \times X \to \mathcal{P}(X)$ is product-measurable, then the function $f \colon \Omega \times X \to [0, +\infty)$ defined by the formula,*

$$f(\omega, x) = d(x, \varphi(\omega, x)),$$

is also product measurable.

Proof. We have:

$$\{(\omega, x) \in \Omega \times X \mid f(\omega, x) < r\} = \{(\omega, x) \in \Omega \times X \mid \varphi(\omega, x) \cap O_r(\{x\}) \neq \emptyset\}.$$

Therefore, our assertion follows from the assumption that φ is measurable. $\qquad\square$

Theorem 3.76. *(Aumann) If $\varphi \colon \Omega \to \mathcal{P}_{cp}(X)$ is a multivalued map such that the graph Γ_φ of φ is measurable, then φ possesses a measurable selection.*

The following Scorza-Dragoni type result describes possible regularization of Carathéodory maps. For the proof, we refer to [178, 255].

Theorem 3.77. *Let X be a compact subset of $\mathbb{R}^n$ and $\varphi \colon [0,a] \times X \to \mathcal{P}_{cp,cv}(\mathbb{R}^n)$ be an upper-Carathéodory map. Then there exists a u-Scorza–Dragoni map $\psi \colon [0,a] \times X \to \mathcal{P}_{cp,cv}(\mathbb{R}^n)$ such that:*

(3.77.1) $\psi(t,x) \subset \varphi(t,x)$ for every $(t,x) \in [0,a] \times X$,

(3.77.2) if $\Delta \subset [0,a]$ is measurable, $u \colon \Delta \to \mathbb{R}^n$ and $v \colon \Delta \to X$ are measurable maps and $u(t) \in \varphi(t,v(t))$ for almost all $t \in \Delta$, then $u(t) \in \psi(t,v(t))$ for almost all $t \in \Delta$.

The following definitions and the next theorem can be found, for example, in [24, 33, 105, 137, 138, 202]. Let (X,d) and (Y,d') be two metric spaces.

Definition 3.78. *We say that a map $F : X \to \mathcal{P}(Y)$ is $\sigma - Ca$-selectionable if there exists a decreasing sequence of compact valued u.s.c. maps $F_n : X \to \mathcal{P}(Y)$ satisfying:*

(a) F_n has a Carathédory selection, for all $n \geq 0$ (F_n are called Ca-selectionable),

(b) $F(x) = \bigcap\limits_{n \geq 0} F_n(x)$, for all $x \in X$.

Definition 3.79. *A single-valued map $f : [0, a] \times X \to Y$ is said to be measurable-locally-Lipschitz (mLL) if $f(\cdot, x)$ is measurable for every $x \in X$ and for every $x \in X$, there exists a neighborhood V_x of $x \in X$ and an integrable function $L_x : [0, a] \to [0, \infty)$ such that*

$$d'(f(t, x_1), f(t, x_2)) \leq L_x(t)d(x_1, x_2) \text{ for every } t \in [0, a] \text{ and } x_1, x_2 \in V_x.$$

Definition 3.80. *A multi-valued mapping $F : [0, a] \times X \to \mathcal{P}(Y)$ is mLL-selectionable if it has an mLL-selection.*

Definition 3.81. *We say that a multi-valued map $\phi : [0, a] \times E \to \mathcal{P}(E)$ with closed values is upper-Scorza-Dragoni (u-Scorza-Dragoni) if, given $\delta > 0$, there exists a closed subset $A_\delta \subset [0, 1]$ such that the measure $\mu([0, a] \setminus A_\delta) \leq \delta$ and the restriction ϕ_δ of ϕ to $A_\delta \times E$ is u.s.c.*

Now, we prove:

Theorem 3.82. *Let E, E_1 be two separable Banach spaces and $\varphi : [a, b] \times E \to \mathcal{P}_{cp,cv}(E_1)$ be a u-Scorza–Dragoni map. Then φ is σ-Ca-selectionable. The maps $\varphi_k : [a, b] \times E \to \mathcal{P}(E_1)$ are u-Scorza–Dragoni and we have*

$$\varphi_k(t, x) \subset \left(\bigcup_{x \in E} \varphi(t, x) \right).$$

Moreover, if φ is integrably bounded, then φ is σ-mLL-selectionable.

Proof. Consider the family $\{B(y, r_k)\}_{y \in E}$, where $r_k = (1/3)^k$, $k = 1, 2, \ldots$. Using Stone's theorem for every $k = 1, 2, \ldots$, we get a locally finite subcovering $\{U_i^k\}_{i \in I_k}$ of $\{B(y, r_k)\}_{y \in E}$. For every $i \in I_k$, $k = 1, 2, \ldots$, we fix the center $y_i^k \in E$ such that $U_i^k \subset B(y_i^k, r_k)$. Now, let $\eta_i^k : E \to [0, 1]$ be a locally Lipschitz partition of unity subordinated to $\{U_i^k\}_{i \in I^k}$.

Define $\psi_i^k : [0, a] \to \mathcal{P}(E)$ and $f_i^k : [0, a] \to E$ as follows:

$$\psi_i^k(t) = \overline{co} \left(\bigcup_{y \in B(y_i^k, 2r_k)} \varphi(t, y) \right),$$

and let f_i^k be a measurable selection of ψ_i^k which exists in view of the Kuratowski–Ryll–Nardzewski theorem.

Finally, define $\varphi_k : [a, b] \times E \to \mathcal{P}(E_1)$ and $f_k : [a, b] \times E \to E_1$ as follows:

$$\varphi_k(t, z) = \sum_{i \in I_k} \eta_i^k(z) \cdot \psi_i^k(t), \qquad f_k(t, z) = \sum_{i \in I_k} \eta_i^k(z) \cdot f_i^k(t).$$

Then $f_k \subset \varphi_k$. Fix $t \in [a, b]$. If $\varphi(t, \cdot)$ is u.s.c. then $\varphi(t, z) = \bigcap_{k=1}^{\infty} \varphi_k(t, z)$ and $\varphi_{k+1}(t, z) \subset \varphi_k(t, z)$, for every $z \in E$. By the assumptions on φ, the map $\varphi(t, \cdot)$ is u.s.c. for almost all $t \in [0, a]$, and the first part of Theorem 3.82 is proved. The second claim is an immediate consequence of the first one. $\square$

The following approximation lemma can be proved as in De Belasi [95, Proposition 4.1]] or Papageorgiou [234, Lemma]].

Lemma 3.83. *Assume that X is a Polish space, Y is a separable Banach space, and $F : [0, b] \times X \to \mathcal{P}_{cl,cv}(Y)$ satisfies*

(3.83.1) $t \to F(t, x)$ is measurable;

(3.83.2) $x \to F(t, x)$ is H_d-u.s.c.;

(3.83.3) $\|F(t, x)\|_{\mathcal{P}} = \sup\{|v| : v \in F(t, x)\} \le p(t)$ a.e. on $[0, b]$ with $p \in L^p([0, b], \mathbb{R}^+)$, $1 \le p < \infty$.

Then, there exists a sequence of $F_n : [0, b] \times X \to \mathcal{P}_{cl,cv}(Y)$ such that for every $n \in \mathbb{N}$ and $x \in X$ there exists $l_n(x) > 0$ and $\epsilon_n > 0$ such that if $x_1, x_2 \in B(x, \epsilon_n)$, then

$$H_d(F_n(t, x_1), F_n(t, x_2)) \le l_n(x)p(t)|x_1 - x_2| \quad \text{a.e. on } [0, b],$$

$$F(t, x) \subset \ldots F_n(t, x) \subset F_{n-1}(t, x) \ldots, \quad \|F_n(t, x)\|_{\mathcal{P}} \le p(t) \quad \text{a.e. on } [0, b],$$

$$H_d(F_n(t, x), F(t, x)) \to 0 \quad \text{as } n \to \infty \text{ for all } (t, x) \in [0, b] \times X,$$

and there exists $u_n : [0, b] \times X \to Y$, measurable in $t \in [0, b]$, and locally Lipschitz in $x \in X$, and $u_n(t, x) \in F_n(t, x)$ for $(t, x) \in [0, b] \times X$ and $n \in \mathbb{N}$.

Moreover, if $x \to F(\cdot, x)$ is H_d-continuous, then for every $n \in \mathbb{N}$, $t \to F_n(t, x)$ is measurable.

The following result is in part the multivalued version of a well-known theorem of Scorza-Dragoni, and of Lusin's theorem, which states that every measurable function is almost continuous.

Theorem 3.84. *[255] Let X and Y be separable Banach spaces and $F : J \times X \to \mathcal{P}_{cl}(Y)$ with J a measurable subset of $\mathbb{R}$. Assume that $Graph(F(t, \cdot))$ is closed in $X \times Y$ for almost $t \in J$. Then there exists $\widehat{F} : J \times X \to \mathcal{P}(Y)$ such that*

(3.84.1) for almost all $t \in J$, $\widehat{F}(t, x) \subset F(t, x)$ for all $x \in X$;

(3.84.2) if $\Delta \subset J$ is measurable and $u : \Delta \to X$ and $v : \Delta \to Y$ are measurable functions with $v(t) \in F(t, u(t))$ a.e in Δ, then $v(t) \in \widehat{F}(t, u(t))$ a.e. in Δ;

(3.84.3) for every $\varepsilon > 0$, there exists a closed $J_\varepsilon \subset J$ with $\mu(J \backslash J_\varepsilon) < \varepsilon$ and the graph of $\widehat{F}|_{J_\varepsilon \times X}$ is closed in $J \times X \times Y$.

If E is a separable Banach space we present the following result of Scorza-Dragoni type (essentially due to Rzézuchowski [255]).

Theorem 3.85. *[255] Let $F : [0, b] \times D \to \mathcal{P}_{cl}(E)$ with D a closed convex subset of E. Assume that F satisfies the following conditions:*

(3.84.1) F is an upper-Carathéodory map.

(3.84.2) F maps compact subsets of $J \times D$ into compact ones.

Then there is a map $F_0 : [0, b] \times D \to \mathcal{P}(E) \cup \{\emptyset\}$ such that

(a) for all $t \in [0, b]$ and $x \in D$, $F_0(t, x) \subset F(t, x)$;

(b) If $\Delta \subset [0, b]$ is measurable, $u, v : \Delta \to D$, then $v(t) \in F_0(t, u(t))$ a.e. in Δ;

(c) for any $\varepsilon > 0$, there is a closed $\Delta_\varepsilon \subset [0,b]$ such that F_0 restricted to $\Delta_\varepsilon \times D$ has nonempty values and is (jointly) upper semi-continuous.

For *u.s.c.* multivalued maps, there exists a useful approximation result similar to the Lasota-Yorke Lemma [201].

Lemma 3.86. *[[97] Lemma 2.2] Let Ω be a subset of a Banach space, E be a Banach space and $F : \Omega \to \mathcal{P}_{cl,cv}(E)$ be a multivalued map. Let $r_n = 3^{-n}$, $\{U_\lambda\}_{\lambda \in \Lambda}$ be a locally finite refinement of $\Omega = \bigcup_{\omega \in \Omega} B(\omega, r_n)$, $\{\phi_\lambda\}_{\lambda \in \Lambda}$ be a locally Lipschitz partition of unity subordinate to $\{U_\lambda\}_{\lambda \in \Lambda}$, pick $\omega_\lambda \in U_\lambda \subset B(\omega_\lambda, r_n) \cap \Omega$, and let*

$$F(\omega) = \sum_{\lambda \in \Lambda} \phi_\lambda(\omega) C_\lambda, \ \ C_\lambda = \overline{co} \, F(B(\omega_\lambda, 2r_n) \cap \Omega).$$

Then

(3.86.1) $F(\omega) \subset F_{n+1}(\omega) \subset F_n(\omega) \subset \overline{co} \, F(B(\omega, 3r_n) \cap \Omega)$ on Ω.

(3.86.2) If F is uniformly locally bounded, then F_n is locally Lipschitz for large n.

(3.86.3) If F is $\varepsilon - \delta - u.s.c.$, then $H_d(F_n(\omega), F(\omega)) \to 0$ on Ω as $n \to \infty$.

Lemma 3.87. *Let E be a Banach space and $F : [0,b] \times E \to \mathcal{P}_{cl,cv}(E)$ be a multivalued map. Let $r_n = 3^{-n}$, $\{U_\lambda\}_{\lambda \in \Lambda}$ be a locally finite refinement of $E = \bigcup_{x \in E} B(x, r_n)$, $\{\phi_\lambda\}_{\lambda \in \Lambda}$ be a locally Lipschitz partition of unity subordinated to $\{U_\lambda\}_{\lambda \in \Lambda}$, pick $\omega_\lambda \in U_\lambda \subset B(x_\lambda, r_n)$ and let*

$$F(t,x) = \sum_{\lambda \in \Lambda} \phi_\lambda(x) C_\lambda(t), \ \ C_\lambda(t) = \overline{co} \, F(t, B(x_\lambda, 2r_n)).$$

Then

(3.86.1) $F(t,x) \subset F_{n+1}(t,x) \subset F_n(t,x) \subset \overline{co} \, F(t, B(x, 3r_n))$ on E.

(3.86.2) If $F(t, \cdot)$ is locally bounded, then $F_n(t, \cdot)$ is locally Lipschitz for large n.

(3.86.3) If $F(t, \cdot)$ is $\varepsilon - \delta - u.s.c.$, then $H_d(F_n(t,x), F(t,x)) \to 0$ on Ω as $n \to \infty$.

Finally, we present some interesting approximations of upper semicontinuous multivalued maps.

Theorem 3.88. *[96] Let Ω be a complete separable metric space. Let μ be a non-negative finite measure on the completion $\mathcal{L}$ of the Borel σ-algebra Σ. Let X be a complete separable metric space with Borel σ-algebra $\mathcal{B}(X)$, and let E be a separable Banach space. Suppose that $F : \Omega \times X \to \mathcal{P}_{cl,cv,b}(E)$ is a bounded multivalued map such that*

(3.88.1) F is $\mathcal{L} \otimes \mathcal{B}(X)$ weakly measurable;

(3.88.2) for each $\omega \in \Omega$, $x \to F(\omega, x)$ is H_d-u.s.c..

Then there exists a sequence $\{F_n : n \in \mathbb{N}\}$ of multivalued maps $F_n : \Omega \times X \to \mathcal{P}_{cv,cl,b}(E)$ and a sequence $\{f_n : n \in \mathbb{N}\}$ of single functions $f_n : \Omega \times X \to E$ satisfying, for every $n \in \mathbb{N}$, the following conditions:

(a$_1$) for each $x \in X$, $\omega \to F(\omega, x)$ is $\mathcal{L}$ weakly measurable;

(a$_2$) for each $\omega \in \Omega$, $x \to F(\omega, x)$ is locally Lipschitz;

(a_3) *for each* $(\omega, x) \in \Omega \times X$,

$$F(t, x) \subset F_{n+1}(t, x) \subset F_n(t, x) \subset \overline{co}\left(\bigcup_{x \in X} (F(t, x))\right), \text{ for all } n \in \mathbb{N};$$

(a_4) *for each* $(t, x) \in \Omega \times X$, $\lim\limits_{n \to \infty} H_d(F_n(t, x), F(t, x)) = 0$;

(a_5) f_n *is a Carathéodory-Lipschitz selector of* F_n.

Theorem 3.89. *[96] Let Ω be a complete separable metric space. Let μ be a nonnegative finite measure on the completion $\mathcal{L}$ of the Borel σ-algebra Σ. Let X be a complete separable metric space with Borel σ-algebra $\mathcal{B}(X)$, and let E be a separable Banach space. Suppose that $F : \Omega \times X \to \mathcal{P}_{cl,cv,b}(E)$ is a bounded multivalued map such that*

(3.89.1) F is $\mathcal{L} \otimes \mathcal{B}(X)$ H_d- measurable;

(3.89.2) for each $\omega \in \Omega$, $x \to F(\omega, x)$ is H_d-u.s.c..(resp. H_d-l.s.c.);

(3.89.3) $F(\Omega \times X)$ is a separable subset of $\mathcal{P}_{cl,b}(X)$.

Then for every $\varepsilon > 0$, there exists a compact set $K_\varepsilon \subset \Omega$, with $\mu(\Omega \backslash K_\varepsilon) < \varepsilon$ such that F restricted to $K_\varepsilon \times X$ is H_d-u.s.c..(resp. H_d-l.s.c.).

Corollary 3.90. *[96] Let Ω, X, $\mathcal{L}$, $\sum$, and μ be as in Theorem 3.89 and let Y be a separable metric space. Let $F : \Omega \times X \to \mathcal{P}_{cl,b,cp}(Y)$ be a multivalued map such that the conditions (3.89.1), (3.89.2) of Theorem 3.89 hold. Then for every $\varepsilon > 0$, there exists a compact set $K_\varepsilon \subset \Omega$, with $\mu(\Omega \backslash K_\varepsilon) < \varepsilon$ such that F restricted to $K_\varepsilon \times X$ is H_d-u.s.c..(resp. H_d-l.s.c.).*

Theorem 3.91. *[96] Let $\Omega, X, \mathcal{L}, \sum$, and μ be as in Theorem 3.89, let Y be a metric space, and let $F : \Omega \times X \to \mathcal{P}_{cl,b,cp}(Y)$ be a multivalued map such that*

(3.91.1) for each $x \in X$, $t \to F(\omega, x)$ is $\mathcal{L}$ H_d-measurable and has $F(\Omega \times \{x\})$ separable $\mathcal{P}_{cl,b}(Y)$;

(3.91.2) for each $t \in \Omega$, $x \to F(\omega, x)$ is H_d-continuous.

Then for every $\varepsilon > 0$, there exists a compact set $K_\varepsilon \subset \Omega$ with $\mu(\Omega \backslash K_\varepsilon) < \varepsilon$ such that F restricted to $K_\varepsilon \times X$ is H_d-continuous.

3.6 L^p selection

In this section we assume that $(\Omega, \mathcal{U}, \mu)$ is $\sigma-$finite measurable space and X is separable Banach space.

Definition 3.92. *A subset $K \subset L^0(\Omega, X)$ is called decomposable if for each $(f_1, f_2, A) \in K \times K \times \mathcal{U}$, we have*

$$f_1 \chi_A + f_2 \chi_{\Omega \backslash A} \in K.$$

Definition 3.93. *Let $F : \Omega \to \mathcal{P}(X)$ be a multivalued map.*

- *For $p = 0$, we define the set*

$$S(F) = \{f \in L^0(\Omega, X) : f(\omega) \in F(\omega), \ \mu - a.e.\},$$

i.e. $S(F)$ is the set of measurable selections.

- *For $1 \leq p \leq \infty$, we define*

$$S_F^p = \{f \in L^p(\Omega, X) : f(\omega) \in F(\omega), \ \mu - a.e.\}$$

i.e. S_F^p is the set of all L^p-selections.

Now, we give a lemma that provides a necessary and sufficient condition for the nonemptiness of S_F^p.

Lemma 3.94. *Let $F : \Omega \to \mathcal{P}(X)$ be a multivalued map, which has a measurable graph, and let $1 \leq p \leq \infty$. Then $S_F^p \neq \emptyset$ if and only if for some $h \in L^p(\Omega)$ we have*

$$\inf\{\|x\| : x \in F(\omega)\} \leq h(\omega), \ \mu - a.e. \ \omega \in \Omega.$$

Proof. The first implication is easy.

$\Leftarrow$ From Theorem 3.31, we know that there exists a sequence of measurable selections $f_n : \Omega \to X, \ n \in \mathbb{N}$ such that

$$F(\omega) \subseteq \overline{\{f_n(\omega), \ \mu.a.e., \ n \in \mathbb{N}\}}.$$

Hence,

$$\omega \to m(\omega) = \inf_{n \in \mathbb{N}} \|f_n(\omega)\|$$

is a measurable function and $m \in L^p(\Omega)$. Let $\epsilon \in L^p(\Omega, \mathbb{R}\backslash\{0\})$. Define $L_\epsilon : \Omega \to \mathcal{P}(X)$ by

$$L_\epsilon(\omega) = \{x \in F(\omega) : \|x\| \leq m(\omega) + \epsilon(\omega), \ \mu - a.e.\}.$$

Since F has a measurable graph, the multivalued map $F_*(\cdot) = \overline{B(0, m(\cdot) + \epsilon(\cdot))}$ has a measurable graph and $\Gamma(L_\epsilon) = \Gamma(F) \cap \Gamma(F_*) \in \mathcal{U} \bigotimes \mathcal{B}(X)$. By Theorem 3.31 there exists a measurable selection $f : \Omega \to X$ such that

$$f(\omega) \in L_\epsilon(\omega), \quad \omega \in \Omega \Rightarrow f(\omega) \in F(\omega), \quad \omega \in \Omega.$$

Hence $S_F^p \neq \emptyset$. $\qquad\square$

Proposition 3.95. *Let $F : \Omega \to \mathcal{P}(X)$ be a multivalued map with measurable graph and $S_F^p \neq \emptyset$,. Then there exists a sequence $(f_n)_{n \in \mathbb{N}} \subset S_F^p$ such that*

$$F(\omega) \subseteq \overline{\{f_n(\omega), \ \mu.a.e., \ n \in \mathbb{N}\}}.$$

Corollary 3.96. *Let $F, G : \Omega \to \mathcal{P}_{cl}(x)$ be measurable multivalued maps and $S_F^p = S_G^p \neq \emptyset$, then $F(\omega) = G(\omega) \ \mu \ a.e.$*

Chapter 4

Continuous Selection Theorems

4.1 Partitions of unity

Partitions of unity play an important role in the existence of continuous selections of some lower semicontinuous classes and in approximation of upper multivalued maps.

Definition 4.1. *Let $\{U_\lambda : \lambda \in \Lambda\}$ and $\{V_\beta : \beta \in \Lambda'\}$ be two coverings of a space. $\{U_\lambda : \lambda \in \Lambda\}$ is said to refine (or be a refinement of) $\{V_\beta : \beta \in \Lambda'\}$ if for each U_λ, there is some V_β with $U_\lambda \subset V_\beta$.*

Definition 4.2. *Let $\{U_\lambda : \lambda \in \Lambda\}$ be a covering of X. If Λ' is contained in Λ and $\{U_\lambda : \lambda \in \Lambda'\}$ is again a covering, it is called a subcovering.*

Definition 4.3. *A covering $\{U_\lambda : \lambda \in \Lambda\}$ of a topological space X is called locally finite if for every $x \in X$, there exists a neighborhood V of x such that $U_\lambda \cap V \neq \emptyset$ only for a finite number of indexes.*

Definition 4.4. *A Hausdorff (T_2 separated space) space is called paracompact if each open covering has a locally finite open refinement.*

Lemma 4.5. *A closed subset of a paracompact space is paracompact.*

Theorem 4.6. *(A. H. Stone) [107] (see also [60]) Every metric space is paracompact.*

Definition 4.7. *Let X be a Hausdorff space. A family $\{\phi_\lambda : \lambda \in \Lambda\}$ of continuous maps $\phi_\lambda : X \to [0,1]$ is called a partition of unity on X if:*

(4.7.1) the support (i.e. $supp(\phi_\lambda) = \overline{\{x \in X : \phi_\lambda(x) \neq 0\}}$) of ϕ_λ is closed and locally finite,

(4.7.2) for each $x \in X$, $\displaystyle\sum_{\lambda \in \Lambda} \phi_\lambda(x) = 1$.

If $\{U_\lambda : \lambda \in \Lambda\}$ is a given open covering of X, we say that a partition $\{\phi_\lambda : \lambda \in \Lambda\}$ of unity is subordinated to $\{U_\lambda : \lambda \in \Lambda\}$ if for every $\lambda \in \Lambda$, $supp(\phi_\lambda) \subset U_\lambda$.

Theorem 4.8. *[33] Let X be a metric space. To any locally finite open covering $\{U_\lambda : \lambda \in \Lambda\}$ of X, we can associate a locally Lipschitz partition of unity subordinated to it.*

4.2 Michael's selection theorem

The most famous continuous selection theorem is the following result proved by Michael in 1956 (see [138]).

Theorem 4.9. *Let X be a metric space, E a Banach space and $\varphi\colon X \to \mathcal{P}_{cl,cv}(E)$ a l.s.c. map. Then there exists $f\colon X \to E$, a continuous selection of φ ($f \subset \varphi$), i.e. $f(x) \in \varphi(x)$ for every $x \in X$.*

Proof. Step 1. Let us begin by proving the following claim: given any convex (not necessarily closed) valued *l.s.c.* map $\Phi\colon X \to \mathcal{P}(E)$ and every $\varepsilon > 0$, there exists a continuous $g\colon X \to E$ such that $d(g(x), \Phi(x)) \leq \varepsilon$, i.e. $g(x) \in O_\varepsilon(\Phi(x))$, for every $x \in X$.

In fact, for every $x \in X$, let $y_x \in \Phi(x)$ and let $\delta_x > 0$ be such that $B(y_x, \varepsilon) \cap \Phi(x') \neq \emptyset$, for x' in $B(x, \delta_x)$. Since X is metric, it is paracompact by Stone's Theorem. Hence there exists a locally finite refinement $\{U_x\}_{x \in X}$ of $\{B(x, \delta_x)\}_{x \in X}$. Let $\{L_x\}_{x \in X}$ be a partition of unity subordinate to it. The mapping $g\colon X \to E$ defined by,

$$g(u) = \sum_{x \in X} L_x(u) \cdot y_x,$$

is continuous since it is locally a finite sum of continuous functions. Fix $n \in X$. Whenever $L_x(u) > 0$, $n \in B(x, \delta_x)$, hence $y_x \in O_\varepsilon(\Phi(u))$. Since this latter set is convex, any convex combination of such y's belongs to it.

Step 2. Next we claim that we can define a sequence $\{f_n\}$ of continuous mappings from X to E with the following properties

$$d(f_n(u), \varphi(u)) \leq \frac{1}{2^n}, \quad n = 1, 2, \ldots, \quad u \in X. \text{(4.1.1)}$$

$$\|f_n(u) - f_{n-1}(u)\| \leq \frac{1}{2^{n-2}}, \quad n = 2, 3, \ldots, \quad u \in X. \text{(4.1.2)}$$

$$(4.1)$$

For $n = 1$ it is enough to take in the Step 1, $\Phi = \varphi$ and $\varepsilon = 1/2$.

Assume we have defined mappings f_n satisfying (4.1.1) up to $n = k$. We shall define f_{k+1} satisfying (4.1.1) and (4.1.2) as follows.

Consider the set $\Phi(u) = B(f_k(u), 1/2^k) \cap \varphi(u)$. By (4.1.1), it is not empty, and it is a convex set. By Proposition 1.63, the map Φ is *l.s.c.*; so by the claim in Step 1 there exists a continuous g such that

$$d(g(x), \Phi(x)) < \frac{1}{2^{n+1}}.$$

Set $f_{k+1}(u) = g(u)$. Then $d(f_{k+1}(u), \varphi(u)) < 1/2^{k+1}$, proving (a). Also

$$f_{k+1}(u) \in O_{1/2^{k+1}}(\Phi(u)) \subset B\left(f_k(u), \frac{1}{2^k} + \frac{1}{2^{k+1}}\right),$$

i.e.

$$\|f_{k+1}(u) - f_n(u)\| \leq \frac{1}{2^{k-1}}$$

proving (4.1.2).

Step 3. Since the series $\sum(1/2^n)$ converges, $\{f_n\}$ is a Cauchy sequence, uniformly converging to a continuous f. Since the values of φ are closed, by (4.1.1), f is a selection of F. The proof is completed. $\qquad\square$

The next result is also quite interesting.

Theorem 4.10. *Let X be a paracompact space, Y be a topological vector space and let $F\colon X \to \mathcal{P}_{cv}(Y)$. If for each $y \in Y$, $F_-^{-1}(y)$ is open, then F has a continuous selection.*

Proof. The family $\{F^{-1}(y) : y \in \bigcup_{x \in X} F(x)\}$ is an open cover of X. From the paracompactness of X, it follows that there exists a locally finite continuous partition of unity $\{f_y\}_{y \in F(X)}$ such that $f_y(x) = 0$ for each $x \notin F^{-1}(y)$. In particular, notice that $f_y(x) > 0$ implies $y \in F(x)$. Now for each $x \in X$ let $g(x) = \sum_{x \in F(X)} f_y(x) y$ and note that the local finiteness of $\{f_y\}_{y \in F(X)}$ in conjunction with the convexity of $F(x)$ guarantees that $g(x) \in F(x)$ for all $x \in X$. Now it remains to observe that the formula $g(x)$ defines a function $g : X \to Y$ that is continuous and hence it is a continuous selection from F. $\square$

Chapter 5

Linear Multivalued Operators

Definition 5.1. *Let X and Y be vector spaces over the field $\mathbb{K} = \mathbb{R}$ or $\mathbb{C}$. A multivalued map (multimap) $A : X \to \mathcal{P}(Y)$ is is said to be a multivalued linear operator (MLO)(or linear relation) if:*

i) $A(x) + A(y) = A(x + y), \quad \forall x, y \in D(A),$

ii) $A(\lambda x) = \lambda A(x), \quad \forall \lambda \in \mathbb{K} \backslash \{0\}, \ \forall x \in D(A).$

The class of multivalued linear operators will be denoted by $ML(X, Y)$. We write $MLO(X, X) := MLO(X)$.

Corollary 5.2. *Let $A \in ML(X, Y)$, then $D(A)$ is a linear subspace of X.*

Corollary 5.3. *Let $A \in MLO(X, Y)$ such that $D(A^{-1}) = R(A)$, then $A^{-1} \in MLO(Y, X)$.*

Proof. Let $y, z \in Y$, then

$$A^{-1}(y + z) = \{x \in X : y + z \in A(x)\}.$$

Since $D(A^{-1}) = R(A)$, thus there exists $x_1 \in X$ such that

$$z \in A(x_1) \Rightarrow y \in A(x - x_1) \text{ and } x_2 \in A^{-1}(z).$$

Hence

$$x - x_1 \in A^{-1}(y) \text{ and } x_1 \in A^{-1}(z) \implies x \in A^{-1}(y) + A^{-1}(z).$$

Therefore

$$A^{-1}(y + z) \subseteq A^{-1}(y) + A^{-1}(z). \tag{5.1}$$

Let $a \in A^{-1}(y)$ and $b \in A^{-1}(z)$, then

$$y \in A(a), \ z \in A(b) \implies y + z \in A(a + b) = A(a) + A(b),$$

this implies that

$$a + b \in A^{-1}(y + z).$$

By (5.1), we deduce that

$$A^{-1}(y + z) = A^{-1}(y) + A^{-1}(z), \quad \text{for all } y, z \in Y.$$

Now we show that

$$A^{-1}(\lambda y) = \lambda A^{-1}(y) \text{ for every } \lambda \in \mathbb{K}, \ y \in Y.$$

Indeed, let $y \in Y$, $\lambda \in \mathbb{K}$ and $x \in \lambda A^{-1}(y)$, then there exists $a \in A^{-1}(y)$ such that $x = \lambda a$, thus for each $\lambda \in \mathbb{K} \backslash \{0\}$ we get

$$y \in A(\lambda^{-1} x) \implies y \in \lambda^{-1} A(x).$$

So
$$\lambda y \in A(x) \Longrightarrow x \in A^{-1}(\lambda y).$$
$$\lambda A^{-1}(y) \subseteq A^{-1}(\lambda y). \qquad (5.2)$$

Let $x \in A^{-1}(\lambda y)$, thus
$$\lambda y \in A(x) \Longrightarrow y \in A(\lambda^{-1}x).$$

Therefore, we also have $\lambda^{-1}x \in A^{-1}(y)$, and hence $x \in \lambda A^{-1}(y)$. Consequently, we get
$$A^{-1}(\lambda y) \subseteq \lambda A^{-1}(y).$$

From (5.2), we deduce that
$$A^{-1}(\lambda y) = \lambda A^{-1}(y).$$

$\square$

Corollary 5.4. *Let* $A \in MLO(X,Y)$. *Then* $A(0)$ *and* $A^{-1}(0)$ *are linear subspaces.*

Proposition 5.5. *Let* $A : X \to \mathcal{P}(Y)$ *be a multivalued map.* $A \in MLO(X,Y)$ *if and only if the following properties are satisfied:*

(i) $A(\lambda x) = \lambda A(x)$ *for every scalar* $\lambda \in \mathbb{K}$ *and every* $x \in X$.

(ii) $A(x) + A(y) \subseteq A(x+y)$, *for every* $x, y \in X$.

Proposition 5.6. *Let* $A : X \to \mathcal{P}(Y)$ *be a multivalued map.* $A \in MLO(X,Y)$ *if and only if*
$$A(\alpha x) + A(\beta y) = A(\alpha x + \beta y), \text{for every } x, y \in X \text{ and } \alpha, \beta \in \mathbb{K}\backslash\{0\}.$$

Proposition 5.7. *Let* $A : X \to \mathcal{P}(Y)$ *be a multivalued map.* $A \in MLO(X,Y)$ *if and only if*
$$A(\alpha x) + A(\beta y) \subseteq A(\alpha x + \beta y), \text{for every } x, y \in X \text{ and } \alpha, \beta \in \mathbb{K}\backslash\{0\}.$$

Proposition 5.8. *Let* $A \in MLO(X,Y)$. *The following properties are equivalent:*

(i) A *is a multivalued linear operator.*

(ii) $Gr(A)$ *is a linear subspace of* $X \times Y$.

(iii) A^{-1} *is a multivalued linear operator.*

(iv) $Gr(A^{-1})$ *is a linear subspace of* $Y \times X$.

Proof.   • $(i) \Longrightarrow (ii)$: Let $(x_1, y_1), (x_2, y_2) \in Gr(A)$ and $\alpha, \beta \in \mathbb{K}$. Then
$$\alpha y_1 + \beta y_2 \in \alpha A(x_1) + \beta A(x_2) = A(\alpha x_1 + \beta x_2).$$

So
$$\alpha(x_1, y_1) + \beta(x_2, y_2) \in Gr(A).$$

Similarly, we can prove that $(ii) \Longrightarrow (iii)$.

• $(iii) \Longrightarrow (i)$: Let $x_1, x_2 \in X, \alpha, \beta \in \mathbb{K}, y_1 \in A(x_1)$ and $y_2 \in A(x_2)$, then
$$\alpha(y_1, x_1) + \beta(y_2, x_2) \in Gr(A^{-1}) \Longrightarrow \alpha y_1 + \beta y_2 \in A(\alpha x_1 + \beta x_2).$$

This implies that
$$\alpha A(x_1) + \beta A(x_2) \subseteq A(\alpha x_1 + \beta x_2),$$
and by Proposition 5.7, $A \in MLO(X,Y)$.

□

Proposition 5.9. *Let* $A : X \to \mathcal{P}(Y)$ *be a multivalued linear operator. Then for every* $y \in A(x)$ *we have*

$$A(x) = y + A(0).$$

Proof. Let $y, z \in A(x)$, then

$$y + A(0) \subseteq A(x) + A(0) \Longrightarrow y + A(0) \subseteq A(x)$$

and

$$z - y \in A(x) - A(x) \Longrightarrow z \in y + A(0).$$

Consequently, we have

$$A(x) = y + A(0).$$

□

Proposition 5.10. *Let* $A \in MLO(X, Y)$, *then:*

(a) $A(\alpha M) = \alpha A(M)$ $(M \subset X, \alpha \in \mathbb{K} \backslash \{0\})$,

(b) $A(M) + A(N) \subset A(M + N)$ $(M, N \subset X)$,

(c) $A(M + N) = A(M) + A(N)$ $(M \subset X, \ N \subset D(A))$,

(d) $AA^{-1}(M) = M \cap R(A) + A(0)$ ($M \subset Y$),

(e) $A^{-1}A(M) = M \cap D(A) + A^{-1}(0)$ $(M \subset X)$,

(f) $A^{-1}(0) \times \{0\} = Gr(A) \cap (X \times \{0\}) = Gr(A) \cap Gr(0)$,

(g) $\{0\} \times A(0) = Gr(A) \cap (\{0\} \times Y)$,

(h) $X \times R(A) = Gr(A) + (X \times \{0\}) = Gr(A) + Gr(0)$,

(i) $D(A) \times Y = Gr(A) + (\{0\} \times Y)$.

Proof. (a) Let $y \in A(\alpha x)$ where $x \in M$ then $y \in \alpha A(x)$, so $A(\alpha M) \subseteq \alpha A(M)$. Conversely, let $y \in \alpha A(x)$ where $x \in M$, thus $y \in A(\alpha M)$, therefore $\alpha A(M) \subseteq A(\alpha M)$. This implies that

$$A(\alpha M) = \alpha A(M).$$

(b) Let $(x, y) \in M \times N$, then

$$A(x) + A(y) = A(x + y) \Longrightarrow A(x) + A(y) \in A(M + N).$$

Hence

$$A(M) + A(N) \subseteq A(M + N).$$

(c) Let $(x, y) \in M \times N$. If $x + y \notin D(A)$, then

$$A(x + y) = \emptyset \Longrightarrow A(x + y) \subset A(M) + A(N).$$

If $x + y \in D(A)$, thus $y \in D(x)$, therefore

$$A(x + y) = A(x) + A(y) \Longrightarrow A(x + y) \subset A(M) + A(N).$$

We deduce from (b) that

$$A(M) + A(N) = A(M + N).$$

(d) Let $y \in A(A^{-1}(M))$, then there exist $a \in M$ and $b \in A^{-1}(a)$ such that $y \in A(b)$. It follows that $a \in A(b)$ and by proposition 5.9 we have

$$A(b) = a + A(0) \Rightarrow y \in a + A(0) \subseteq M \cap R(A) + A(0).$$

Hence

$$A(A^{-1}(M)) \subseteq M \cap R(A) + A(0).$$

Let $y \in M \cap R(A) + A(0)$, then there exist $a \in M \cap R(A)$, $b \in X$ and $c \in A(0)$ with $a \in A(b)$ such that

$$y = a + c \Rightarrow y \in A(A^{-1}(M)) + A(A^{-1}(0)),$$

by proposition 5.9 we get

$$y \in A(A^{-1}(M)) \Rightarrow M \cap R(A) + A(0) \subseteq A(A^{-1}(M)).$$

Therefore

$$A(A^{-1}(M)) = M \cap R(A) + A(0).$$

(e) Let $y \in A^{-1}(A(M))$, then there exist $a \in M$ and $b \in A(a)$ such that $y \in A^{-1}(b)$. Then $a \in M \cap D(A)$ and $a \in A^{-1}(b)$ by proposition 5.9 we have

$$A^{-1}(b) = a + A^{-1}(0) \Rightarrow y \in a + A^{-1}(0) \subseteq M \cap D(A) + A^{-1}(0).$$

This implies that

$$A^{-1}(A(M)) \subseteq M \cap D(A) + A^{-1}(0).$$

Let $y \in M \cap D(A) + A^{-1}(0)$, then there exist $a \in M \cap D(A)$, $b \in Y$ and $c \in A^{-1}(0)$ with $b \in A(a)$ such that

$$y = a + c \Rightarrow y \in A^{-1}(A(M)) + A^{-1}(0),$$

by Proposition 5.9, we obtain

$$y \in A^{-1}(A(M)) \Rightarrow M \cap D(A) + A^{-1}(0) \subseteq A^{-1}(A(M)).$$

Hence

$$A^{-1}(A(M)) = M \cap D(A) + A^{-1}(0).$$

(f) Let $(x, 0) \in A^{-1}(0) \times \{0\}$, then $0 \in A(x)$, so $(x, 0) \in Gr(A)$. Hence

$$A^{-1}(0) \times \{0\} \subseteq Gr(A) \cap X \times \{0\}.$$

For every $(x, y) \in Gr(A) \cap (X \times \{0\})$ we have

$$y = 0 \text{ and } 0 \in A(x),$$

hence

$$x \in A^{-1}(0) \Rightarrow (x, 0) \in A^{-1}(0) \times \{0\}.$$

So

$$Gr(A) \cap (X \times \{0\}) \subseteq A^{-1}(0) \times \{0\}.$$

Thus

$$Gr(A) \cap (X \times \{0\}) = A^{-1}(0) \times \{0\}.$$

Moreover

$$Gr(0) = \{(x, y) \in X \times Y : y = 0\} = X \times \{0\}.$$

Hence
$$Gr(A) \cap (X \times \{0\}) = A^{-1}(0) \times \{0\} = Gr(A) \cap Gr(0).$$

(g) Let $(0, y) \in \{0\} \times A(0)$ thus $(0, y) \in Gr(A)$. Therefore
$$(0, y) \in Gr(A) \cap \{0\} \times Y.$$

Then we have
$$\{0\} \times A(0) \subseteq Gr(A) \cap \{0\} \times Y.$$

Conversely, let $(x, y) \in Gr(A) \cap \{0\} \times Y$, then $x = 0$ and $y \in A(0)$, thus we obtain that
$$(0, y) \in \{0\} \times A(0) \Rightarrow Gr(A) \cap \{0\} \times Y \subseteq \{0\} \times A(0).$$

Hence we conclude that
$$Gr(A) \cap \{0\} \times Y = \{0\} \times A(0).$$

(h) Let $(x, y) \in X \times R(A)$, thus there exists $a \in X$ such that
$$y \in A(a) \Rightarrow (a, y) \in Gr(A).$$

It follows that
$$(x, y) = (a, y) + (x - a, 0) \in Gr(A) + X \times \{0\}.$$

From above we have
$$X \times R(A) \subseteq Gr(A) + X \times \{0\}.$$

Conversely, let $(x, y) \in Gr(A) + X \times \{0\}$, then there exist $a_1, a_2 \in X$, $b \in Y$ with $b \in A(a_1)$ such that
$$(x, y) = (a_1, b) + (a_2, 0) \Rightarrow (x, y) = (a_1 + a_2, b) \in X \times R(A).$$

It remains to show that
$$Gr(A) + X \times \{0\} \subseteq X \times R(A).$$

Therefore
$$Gr(A) + X \times \{0\} = X \times R(A).$$

Moreover, since $Gr(0) = X \times \{0\}$ we get
$$Gr(A) + X \times \{0\} = X \times R(A) = Gr(A) + Gr(0).$$

(i) Let $(x, y) \in D(A) \times Y$, then there exists $z \in Y$ such that
$$z \in A(x) \Rightarrow (x, z) \in Gr(A),$$

and hence
$$(x, y) = (x, z) + (0, y - z) \in Gr(A) + \{0\} \times Y.$$

Thus
$$D(A) \times Y \subseteq Gr(A) + \{0\} \times Y.$$

On the other hand for each $(x, y) \in Gr(A) + \{0\} \times Y$ there exist $a_1 \in X$, $b_1, b_2 \in Y$ with $b_1 \in A(a_1)$ such that
$$(x, y) = (a_1, b_1) + (0, b_2) \Rightarrow (x, y) = (a_1, b_1 + b_2) \in D(A) \times Y.$$

It follows that
$$Gr(A) + \{0\} \times Y \subseteq D(A) \times Y.$$

Consequently
$$Gr(A) + \{0\} \times Y = D(A) \times Y.$$

$\square$

Corollary 5.11. *Let $A \in MLO(X, Y)$ and $B \in MLO(Y, Z)$. Then $B \circ A \in MLO(X, Z)$.*

Proof. Let $x, y \in X$ and α, β $\mathbb{K}$, then by proposition 5.10 we have

$$
\begin{aligned}
(B \circ A)(\alpha x + \beta y) &= B(A(\alpha x + \beta y)) \\
&= B(\alpha A(x) + \beta A(y)) \\
&= \alpha B(A(x)) + \beta B(A(y)) \\
&= \alpha(B \circ A)(x) + \beta(B \circ A)(y).
\end{aligned}
$$

This implies that $B \circ A \in MLO(X, Z)$. $\qquad\square$

5.1 Uniform boundedness principle

Firstly, we turn our attention to linear operators between normed spaces.

Proposition 5.12. *If X and Y are two normed spaces and $A : X \to \mathcal{P}_b(Y)$ is a linear operator, then the following statements are equivalent:*

(a) A is u.s.c. at the origin;

(b) A is linear bounded(i.e. $\|A\|_* = \sup\{\|y\| : y \in A(x) \text{ and } \|x\| \leq 1\} < \infty$);

(c) There exists $M > 0$ such that

$$\|A(x)\|_{\mathcal{P}} = \sup\{\|y\| : y \in A(x)\} \leq M\|x\| \quad \text{for all } x \in X\backslash\{0\};$$

(d) The set $A(B(0, 1))$ is bounded in Y, where

$$B(0, 1) = \{x \in X : \|x\| < 1\};$$

(f) A is u.s.c.

Proof. • $(a) \Rightarrow (b)$ Let $r > 0$ such that $A(0) \subset B(0, r)$. Since A is u.s.c. at 0, then there exists $r_1 > 0$ such that

$$A(B(0, r_1)) \subset B(0, r).$$

For every $x \in X\backslash\{0\}$ we have $z = \frac{r_1 x}{2\|x\|} \in B(0, r_1)$, thus

$$\frac{r_1}{2\|x\|} A(x) \subset B(0, r) \Longrightarrow \|y\| \leq \frac{2r\|x\|}{r_1}, \ \forall \, y \in A(x).$$

It follows that

$$\|A\|_* \leq \max\left(\frac{2r}{r_1}, r\right) < \infty.$$

• $(b) \Rightarrow (c)$ For each $x \in X\backslash\{0\}$ and $y \in A(x)$, we have

$$\frac{y}{\|x\|} \in A\left(\frac{y}{\|x\|}\right) \Longrightarrow \|y\| \leq \|A\|_*\|x\|.$$

Hence

$$\|A(x)\|_{\mathcal{P}} \leq \|A\|_*\|x\|, \quad \text{for every } x \in X\backslash\{0\}.$$

- $(c) \Rightarrow (d)$ Note for $x \in X \backslash \{0\}$ and $\|x\| \leq 1$ we have

$$\|A(x)\|_{\mathcal{P}} \leq M \Longrightarrow A(x) \subset B(0, M+1). \tag{5.3}$$

On the other hand there exists $r > 0$ such that

$$A(0) \subset B(0, r). \tag{5.4}$$

Combining (5.3) and (5.4), one obtains

$$A(B(0,1)) \subset B(0, \max\{M+1, r\}).$$

- $(d) \Rightarrow (f)$ Let $x_0 \in X$, $A(x_0) \subset V$ where V is open in Y. From proposition 5.9, we get

$$A(x_0) = y + A(0) \quad \text{for every } y \in A(x_0).$$

Then there exist $r, r_* > 0$ such that

$$y + B(0, r) \subset V \text{ and } A(B(0, 1)) \subset B(0, r_*). \tag{5.5}$$

Set $U_{x_0} = x_0 + \frac{r}{r_*} B(0, 1)$. Since $A \in MLO(X, Y)$ and $D(A) = X$, it follows from Proposition 5.10 that

$$A(U_{x_0}) = A(x_0 + \frac{r}{r_*} B(0, 1)) \tag{5.6}$$

$$= A(x_0) + \frac{r}{r_*} A(B(0, 1)) \tag{5.7}$$

$$= y + A(0) + \frac{r}{r_*} A(B(0, 1)) \tag{5.8}$$

$$= y + \frac{r}{r_*} A(B(0, 1)). \tag{5.9}$$

Combining (5.5) and (5.6) we obtain

$$A(U_{x_0}) \subset y + B(0, r) \subset V.$$

- $(f) \Rightarrow (a)$. This is obvious. □

Theorem 5.13. *(Uniform boundedness principle) If X is a Banach space, Y is a normed space, I is an arbitrary index set and $\{A_i\}_{i \in I} \subset BMLO(X, Y)$. Suppose that for each $x \in X$ we have*

$$\sup_{i \in I} \|A_i(x)\|_{\mathcal{P}} < \infty. \tag{5.10}$$

Then there exists $M > 0$ such that

$$\sup_{i \in I} \|A_i\|_* \leq M,$$

where

$$\|A_i\|_* = \sup\{\|A_i(x)\|_{\mathcal{P}} : \|x\| \leq 1\}.$$

Proof. For every $n \in \mathbb{N}$, let

$$X_n = \{x \in X : \|A_i(x)\|_{\mathcal{P}} \leq n, \ \forall \, i \in I\}.$$

Let $(x_p)_{p \in \mathbb{N}} \subset X_n$ be a sequence converging to x in X. Then

$$A_i(x_p) \subset \overline{B(0, n)}, \quad \text{for all } p \in \mathbb{N}.$$

For every $\epsilon > 0$ we have

$$A_i(x_p) \subset B(0, n + \epsilon), \quad \text{for all } p \in \mathbb{N}.$$

Since A_i is u.s.c., then for every $y \in A_i(x)$ there exists a sequence $(y_p)_{p \in \mathbb{N}}$ such that $y_p \in A_i(x_p)$ and $(y_p)_{p \in \mathbb{N}}$ converges to y. Thus

$$y \in \overline{B(0, n + \epsilon)} \Longrightarrow \|y\| \le n + \epsilon.$$

It follows that

$$\|y\| \le n + \epsilon, \text{ as } \epsilon \to 0 \Longrightarrow \|y\| \le n.$$

Thus

$$\|A_i(x)\|_{\mathcal{P}} \le n \Longrightarrow x \in X_n.$$

We conclude that X_n is closed and by (5.10) we have

$$X = \bigcup_{n \in \mathbb{N}} X_n.$$

Since X is Baire space then there exists $n_0 \in \mathbb{N}$ such that $int(X_{n_0}) \ne \emptyset$. Pick x_0 and $r > 0$ such that

$$B(x_0, r) \subset X_{n_0} \Longrightarrow \|A_i(x_0 + rz)\|_{\mathcal{P}} \le n_0, \ \forall \ i \in I, \ \forall \ z \in \overline{B(0,1)}. \tag{5.11}$$

On the other hand, since $A \in MLO(X, Y)$, we have

$$A_i(x_0 + rz) = A_i(x_0) + rA_i(z), \ \forall \ i \in I, \ \forall \ z \in \overline{B(0,1)}.$$

We deduce from (5.11) that

$$r\|A_i(z)\|_{\mathcal{P}} \le \|A_i(x_0)\|_{\mathcal{P}} + n_0, \ \forall \ i \in I, \ \forall \ z \in \overline{B(0,1)},$$

so

$$\|A_i\|_* \le \frac{2n_0}{r}, \ \forall \ i \in I.$$

We conclude that

$$\sup_{i \in I} \|A_i\|_* \le \frac{2n_0}{r} := M.$$

$\square$

Here is an important consequence of the uniform bounded principle.

Corollary 5.14. *Let X and Y be two Banach spaces. Let $(A_n)_{n \in \mathbb{N}}$ be a sequence of u.s.c. multivalued linear operators from X into $\mathcal{P}_b(Y)$ such that $D(A_n) = X$, for every $n \in \mathbb{N}$ and for each $x \in X$,*

$$H_d(A_n(x), A(x)) \to 0 \text{ as } n \to \infty.$$

Then we have:

i) $\sup_{n \in \mathbb{N}} \|A_n\|_* < \infty$.

ii) If for each $n \in \mathbb{N}$, we have $A_n : X \to \mathcal{P}_{cp}(Y)$, then $A \in MLO(X, Y)$.

5.2   Norm of linear multivalued operators

Let X and Y be normed spaces and M be a closed subspace of X spaces.

Definition 5.15. *(Quotient Space). If M is a subspace of a vector space X, then the quotient space X/M is*

$$X/M = \{x + M : x \in X\}.$$

Definition 5.16. *If M is a subspace of a vector space X, then the canonical projection or the canonical mapping of X onto X/M is $\pi : X \to X/M$ defined by*

$$\pi(x) := [x] = x + M, \quad x \in X.$$

Corollary 5.17. *Let M be a subspace of X. If $E \subset X$ the inverse image of $\pi(E)$ is*

$$\pi^{-1}(\pi(E)) = E + M = \{x + m : x \in E, \ m \in M\}.$$

Proposition 5.18. *Let $A \in MLO(X,Y)$ and $\pi : X \to X/\overline{A(0)}$ be a canonical mapping, then $\pi \circ A$ is single map.*

Proof. Let $y_1, y_2 \in \pi(A(x))$, then there exist $a_1, a_2 \in A(x)$ such that $y_1 = \pi(a_1)$ and $\pi(a_2)$. From Proposition 5.9

$$A(x) = a_1 + A(0) \text{ and } A(x) = a_2 + A(0).$$

Hence

$$\pi(A(x) = \pi(a_1) + \pi(A(0)) \text{ and } \pi(A(x) = \pi(a_2) + \pi(A(0)).$$

It is clear that $\pi(A(0)) \subseteq \pi(\overline{A(0)}) = 0$. So $y_1 = y_2$.   □

Definition 5.19. *Let X and Y be two normed spaces and $A \in MLO(X,Y)$. We define*

$$\|A(x)\| = \|\pi(A(x))\| = \inf\{\|a\| : a \in \pi(A(x))\}$$

and

$$\|A\| = \|\pi \circ A\|,$$

called the norm of Ax and A, respectively.

Chapter 6

Fixed Point Theorems

In this chapter we discuss a variety of fixed point theorems that are useful in the study of differential inclusions. This includes the Schauder-Tychonoff theorem, Fan's theorem, and various forms of Krasnosel'skii's theorem. We also discuss fixed point theorems in topological vector spaces.

6.1 Approximation methods and fixed point theorems

For a fairly general class of multifunctions with compact and nonconvex values, approximate continuous selections have been constructed by Cellina [80], Górniewicz, Granas and Kryszewski [141], and Górniewicz and Lassonde [142]. Apparently, the idea of constructing such approximations seems to be an old one and goes back to works of J. von Neumann [280]; later it was studied by many authors ([25, 26, 31, 44, 58, 140, 141, 202]). Below we shall present most general approximation results proved recently in [140] and [141] (see also [31, 59] and [192]).

If (X, d_X) and (Y, d_Y) are two metric spaces, then in the Cartesian product $X \times Y$ we consider the max-metric $d_{X \times Y}$, i.e.,

$$d_{X \times Y}((x, y), (u, v)) = \max\{d_X(x, u), d_Y(y, v)\}, \text{ for } x, u \in X \text{ and } y, v \in Y.$$

Secondly, we shall use the following approximation result.

Proposition 6.1. *Let K be a compact subset of X and let $f : X \to Y$ be a continuous map. Then for each $\varepsilon > 0$, there exists $\eta > 0$ such that $d_2(f(x), f(u)) < \varepsilon$, provided $d_1(u, x) < \eta$ and $x, u \in O_\eta(K)$.*

Proof. Assume on the contrary that there exists $\varepsilon > 0$ such that for every $n = 1, 2, \ldots$, there are $x_n, u_n \in O_{1/n}(K)$ such that for every $n \in \mathbb{N}$

$$d(x_n, u_n) < 1/n \quad \text{and} \quad d(f(x_n), f(u_n)) \geq \varepsilon.$$

Since $x_n, u_n \in O_{1/n}(K)$, we can find $\widetilde{x}_n, \ \widetilde{u}_n \in K$ such that

$$d_1(x_n, \widetilde{x}_n) < 1/n \text{ and } d_1(u_n, \widetilde{u}_n) < 1/n. \tag{6.1}$$

Then we deduce that

$$d_1(\widetilde{u}_n, \widetilde{x}_n) < \frac{3}{n} \text{ and } d_1(u_n, \widetilde{u}_n) < 1/n. \tag{6.2}$$

Now, since K is compact, we can assume that the sequences $\widetilde{x}_n$ and $\widetilde{u}_n$ are convergent. So, in view of (6.2), we have

$$\lim_{n \to \infty} \widetilde{x}_n = \lim_{n \to \infty} \widetilde{u}_n = x. \tag{6.3}$$

Consequently, from (6.1) we get

$$\lim_{n\to\infty} x_n = \lim_{n\to\infty} u_n = x.$$

Then

$$\lim_{n\to\infty} f(u_n) = \lim_{n\to\infty} f(x_n) = f(x),$$

and this contradicts the fact that

$$d_2(f(u_n), f(x_n)) \geq \varepsilon, \quad \text{for every } n.$$

$\square$

Definition 6.2. *Let $F : X \to \mathcal{P}(Y)$ be a multivalued mapping, $Z \subset X$, and $\varepsilon > 0$. A mapping $f : Z \to Y$ is called an $\varepsilon-$approximation (on the graph) of F if*

$$\Gamma(f) \subset O_\varepsilon(\Gamma(F)).$$

If $Z = X$ and f is an $\varepsilon-$approximation (on the graph of G), then we write $f \in a(G, \varepsilon)$. Some important properties about approximation of multi-valued maps are included in the next theorem.

Theorem 6.3. *(Approximation selection)* ([185]) *Let X be a normed space, Y be a Banach space, and $F : X \to \mathcal{P}_{cv}(Y)$ be an u.s.c. multivalued map. Then, for every $\epsilon > 0$, there exists a locally Lipschitz function $f_\epsilon : X \to Y$ such that*

$$f_\epsilon(X) \subseteq co\, F(X)$$

and

$$\Gamma(f_\epsilon) \subseteq \Gamma(F) + \epsilon B(0, 1),$$

where $B(0,1)$ is the open unit ball in $X \times Y$.

Proof. Fix $\varepsilon > 0$ and $x \in X$; let $\delta = \delta(x)$, $\delta < \varepsilon$, be such that

$$F(B(x, \delta(x))) \subset O_\varepsilon(F(x)).$$

The family of balls $\{B(x, \eta(x))\}_{x \in X}$, where $\eta(x) = \dfrac{1}{4}\delta(x)$ covers the paracompact space X. Using Stone's Theorem, this cover has a locally finite refinement $\{V_i\}_{i \in I}$ and $\{\phi\}_{i \in I}$ a locally Lipschitzean corresponding partition of unity. Now, choosing for each $i \in I$ an arbitrary point $y_i \in F(V_i)$, define the map $f_\varepsilon : X \to Y$ by

$$f_\varepsilon(x) = \sum_{i \in I} \phi_i(x) y_i.$$

The map f_ε is the desired one. Let $x \in X$ belong to all members of the family $\{V_i\}_{i=1}^n$ from the covering $\{v_i\}_{i \in I}$. Every V_i, $i = 1, \ldots, n$ is contained in some ball $B(x_i, \eta(x_i))$, therefore $x \in \cap_{i=1}^n B(x_i, \eta(x_i))$. Let $k \in \{1, \ldots, n\}$ be such that $\eta_k = \max_{1 \leq i \leq n} \eta(x_i)$. Take $x' = x_k$, then we have $x_i \in B(x, \eta_k)$, hence $x_i \in B(x', 2\eta_k)$ for all $i = 1, \ldots, n$. Thus

$$B(x_i, \eta(x_i)) \subset B(x', 4\eta_k), \quad i = 1, \ldots, n.$$

Then

$$y_i \in F(x_i) \subset F(B(x_i, \eta(x_i)) \subset F(B(x', 4\eta_k) \subset O_\varepsilon(F(x')), \text{ for all } i = 1, \ldots, n.$$

Using the fact that $O_\varepsilon(F(x'))$ is convex, then $f_\varepsilon(x) \in O_\varepsilon(F(x'))$. Since $x \in V_i$, $i = 1, \ldots, n$, we have also

$$F(x) \subset O_\varepsilon(F(x')).$$

Since $\{\phi_i\}_{i \in I}$ are locally Lipschitzean, then we can easily prove the continuity of f_ε. $\square$

By the Schauder fixed point theorem and Theorem 6.3 we obtain the following Kakutani's fixed point theorem.

Theorem 6.4. *(1941) Let X be a Banach space, $C \in \mathcal{P}_{cp,cv}(X)$ and $F : C \to \mathcal{P}_{cp,cv}(C)$ be u.s.c. on C. Then there is an $x \in C$ such that $x \in F(x)$.*

Proof. Let $B(0,1)$ be an open ball of $X \times X$ and $(f_n)_{n \in \mathbb{N}}$ be a sequence of continuous mappings from C to C such that

$$\Gamma(f) \subseteq \Gamma(F) + \epsilon_n B(0,1),$$

where $(\epsilon_n)_{n \in \mathbb{N}}$ is a real sequence converging to zero. By Schauder's fixed point theorem there exists $x_n \in C$ such that $x_n = f(x_n)$, for each $n \in \mathbb{N}$. Since C is compact there is a subsequence, say $(x_k)_{k \in \mathbb{N}}$ of $(x_n)_{n \in \mathbb{N}}$ converging to any $x \in C$. Thus

$$d((x_k, f_k(x_k)), \Gamma(F)) \le \epsilon_k \to 0, \text{ as } k \to \infty,$$

then $(x,x) \in \overline{\Gamma(F)}$. By Proposition 1.37 it follows that $(x,x) \in \Gamma(F)$ i.e. $x \in F(x)$. $\square$

Kakutani proved this theorem for $X = \mathbb{R}^n$. The generalization is due to Ky Fan (1952) and Glicksberg (1952).

Theorem 6.5. *(Bohnenlust and Karlin (1950)) Let X be a normed space, $K \in \mathcal{P}_{cl,b}(X)$ and $F : K \to \mathcal{P}_{cl,cv}(K)$ be a compact, u.s.c. multi-map. Then F has at least one fixed point.*

Proof. Set $C = \overline{co}(F(K))$, so $C \subseteq K$. It is clear that $F(C) \subseteq C$. Then by Theorem 6.4, the operator F has at least one fixed point. $\square$

Now, we present the classical multivalued version of the nonlinear alternative of Leray and Schauder type.

Theorem 6.6. *Let E be a normed linear space, and $F : E \to \mathcal{P}_{cv,cp}(E)$ be an upper semi-continuous multi-valued map. Suppose there exists $r > 0$ such that*

$$x \in \lambda F(x) \ (0 \le \lambda \le 1) \Rightarrow \|x\| \le r$$

holds and F is completely continuous. Then F has a fixed point in $B(0,r)$.

Proof. Let $C = \{x \in E : x \in \lambda F(x), \ \lambda \in (0,1)\}$; then C is a bounded set in E. Hence there exists $r_* > 0$ such that

$$F(C) \subseteq B(0, r_*) = \{x \in E : \|x\| \le r_*\}.$$

Set

$$K = \sup\{\|y\| : y \in F(\overline{B}(0, 2r_*))\}, \ k = \max(K, 2r_* + 1).$$

Finally, consider the multivalued operator $G : E \to \mathcal{P}(E)$ defined by

$$G(x) = \begin{cases} F(x) \cap \overline{B}(0, 2r_*), & \text{if } F(x) \cap \overline{B}(0, 2r_*) \ne \emptyset, \\ \dfrac{2r_*}{k} F(x), & \text{if } F(x) \cap \overline{B}(0, 2r_*) = \emptyset. \end{cases}$$

Now we can easily prove that $G(\overline{B}(0, 2r_*)) \subseteq \overline{B}(0, 2r_*)$, $G(\cdot) \in \mathcal{P}_{cp,cv}(\overline{B}(0, 2r_*))$ and G is u.s.c. Then by Theorem 6.5, there exists $x_* \in \overline{B}(0, 2r_*)$ such that $x_* \in G(x_*)$. Assume that $x_* \in \dfrac{2r_*}{k} F(x_*)$ with $F(x_*) \cap \overline{B}(0, 2r_*) = \emptyset$; then there exists $y_* \in F(x_*)$ such that

$$x_* = \frac{2r_*}{k} y_* \Rightarrow y_* = \frac{k}{2r_*} x_* \Rightarrow 2r_* < \|y_*\| \le k \Rightarrow \frac{2r_*}{k} < 1.$$

Then
$$x_* \in C \Rightarrow y_* \in F(C) \Rightarrow \|y_*\| \leq r_* \Rightarrow 2r_* < r_*.$$
This is a contradiction with $F(x_*) \cap \overline{B}(0, 2r_*)$; hence $x_* \in F(x_*)$.   □

We end this section with a nonlinear alternative.

Lemma 6.7. *Let X be a normed space and $F\colon X \to \mathcal{P}_{cl,cv}(X)$ be a compact, u.s.c. multi-map. Then either one of the following conditions holds:*
(a) F has at least one fixed point,
(b) the set $C := \{x \in X,\ x \in \lambda F(x),\ \lambda \in (0,1)\}$ is unbounded.

6.2   Schauder-Tychonoff fixed point theorem

In 1934, Tychonoff proved the version of Schauder's fixed point theorem in a locally convex space. This version is known as the Schauder-Tychonoff fixed point theorem. Firstly, we present the generalization of Schauder approximation theorem for locally convex space.

Theorem 6.8. *Let X be a locally convex space, K a compact subset of X and C a convex subset of X with $K \subseteq C$. Then given an open neighborhood U of the zero element of X, there exists a continuous mapping $f_u : K \to C$, with*

$$f_u(K) \subseteq L$$

and
$$(f_u(x) - x) \in U, \quad \text{for } x \in K,$$
where L is a finite dimensional subspace of X.

Proof. Without any loss of generality, assume that U is convex and balanced. Let

$$p_u(x) = \inf\{\alpha > 0 : x \in \alpha U\}$$

be the Minkowski functional associated with U. Obviously, p_u is a continuous seminorm on X and
$$U = \{x \in X : p_u(x) < 1\}.$$
Since K is compact, there exists a finite set $\{x_1, \ldots, x_n\} \subseteq K$ such that

$$K \subseteq \bigcup_{i=1}^{n} U(x_i),$$

where $U(x) = U + x$, $x \in X$. Let

$$f_u(x) = \frac{\displaystyle\sum_{i=1}^{n} \phi_i(x) x_i}{\displaystyle\sum_{i=1}^{n} \phi_i(x)}, \quad x \in K,$$

where
$$\phi_i(x) = \max(0, 1 - p_u(x - x_i)), \ i = 1, \ldots, n.$$

Since p_u is a continuous function on X, we have that ϕ_i, $i = 1, \ldots, n$, are also continuous functions on X. In addition, for $i = 1, \ldots, n$

$$0 \leq \phi_i(x) \leq 1, \text{ for all } x \in X,$$

and

$$\phi_i(x) = \begin{cases} 0, & \text{if } x \notin U(x_i), \\ > 0, & \text{if } x \in U(x_i). \end{cases}$$

Hence f_u is well defined and is continuous on K. Now we show that $f_u(K) \subseteq L$ where L, is a linear subspace generated by $\{x_1, \ldots, x_n\}$. Since $K \subseteq C$ and C is convex, we have

$$f_u(x) \in C, \quad \text{for each } x \in K.$$

Therefore

$$f_u(x) \in C \cap L, \quad \text{for each } x \in K.$$

Furthermore, we have

$$f_u(x) - x = \frac{\displaystyle\sum_{i=1}^{n} \phi_i(x)(x_i - x)}{\displaystyle\sum_{i=1}^{n} \phi_i(x)}, \quad x \in K,$$

and so

$$p_u(f_u(x) - x) = \frac{\displaystyle\sum_{i=1}^{n} \phi_i(x) p_u(x_i - x)}{\displaystyle\sum_{i=1}^{n} \phi_i(x)} < 1, \quad x \in K.$$

This immediately yields

$$f_u(x) - x \in U, \quad \text{for each } x \in K.$$

$\square$

Now, from Theorem 6.8 and Brouwer's fixed point theorem, we get the following.

Theorem 6.9. *Let X be a locally convex space, K a compact subset of X and C a convex subset of X with $K \subseteq C$ and $f : C \to X$ a continuous mapping such that*

$$f(C) \subseteq K \subseteq C.$$

Then f has at least one fixed point.

Proof. Let U be an open, convex, balanced neighbourhood of $0 \in X$, then from Theorem 6.8 there exists $f_u : K \to C$ continuous function and L a finite dimensional subspace such that

$$f_u(K) \subseteq L$$

and

$$(f_u(x) - x) \in U, \quad \text{for } x \in K.$$

Define a function $\widetilde{f}_u : C \to C$ by

$$\widetilde{f}_u(x) = f_u(f(x)), \quad \text{for all } x \in C.$$

Since f_u takes values in the space L (defined in Theorem 6.8), we can show that

$$\widetilde{f_u}(L \cap C) \subseteq f_u(K) \subseteq L \cap C.$$

Indeed, let $x \in L \cap C$, then $f(x) \in K$, and then

$$\widetilde{f_u}(x) = f_u(f(x)) \in L \cap C, \quad \text{for all } x \in L \cap C.$$

Let $K_* = co(f_u(K))$, the convex hull of the compact set $f_u(K)$ in L. Notice K_* is compact in L and

$$K_* \subseteq C \Rightarrow f(K_*) \subseteq f(C) \subseteq K \subseteq C.$$

Then

$$\widetilde{f_u}(K_*) = f_u(f(K_*)) \subseteq f_u(K) \subseteq K_*.$$

By Brouwer's fixed point theorem, we can deduce that there exists $x \in K_*$ such that $x = \widetilde{f_u}(x)$. This implies that $x = f_u(f(x))$. From Theorem 6.8, we get

$$f_u(f(x)) - f(x) \in U.$$

Hence

$$x - f(x) \in U. \tag{6.4}$$

We conclude that for each $U \in \mathcal{N}(0)$, there corresponds at least one $x \in K_* \subseteq C$ such that (6.4) is satisfied.

Suppose now, $x \neq f(x)$ for all $x \in C$. By the continuity of f, and since X is a Hausdorff space, there exist $V_x, W_x \in \mathcal{N}(0)$ such that

$$f(C \cap V_x(x)) \subset W_x(f(x)), \tag{6.5}$$

and

$$V_x(x) \cap W_x(f(x)) = \emptyset. \tag{6.6}$$

Choose $U_x \in \mathcal{N}(0)$ such that

$$2U_x \subset V_x \cap W_x.$$

Since K is compact, there exists a finite set $\{a_i : i = 1, \ldots, n\} \subseteq K$ such that

$$K \subset \bigcup_{i=1}^{n} U_{a_i}(a_i),$$

where $U_{a_i}(a_i) = U_{a_i} + a_i, i = 1, \ldots, n$. We claim that for every $x \in C$ there exists $j \in \{1, \ldots, n\}$ such that

$$x - f(x) \in U_{a_j} \tag{6.7}$$

cannot hold. Fix $x \in C$. Since $y = f(x) \in K$ then there exists $j \in \{1, \ldots, n\}$ such that $y \in U_{a_j}$. It then follows that

$$U_{a_j}(y) \subset V_{a_i}(a_i). \tag{6.8}$$

To see this notice that

$$y = u + a_j \text{ for some } u \in U_{a_j}.$$

Therefore for $z \in U_{a_j}(y)$, then there exists $w \in U_{a_j}$ such that

$$z = w + y = w + u + a_j \Rightarrow z \in 2U_{a_j} + a_j \subset V_{a_j}(a_j).$$

Assume that (6.7) is not true. Then for any $x \in C$ we have that $x \in U_{a_j}(y)$ with $y = f(x)$, and therefore from (6.8), we see that $x \in V_{a_j}(a_j)$. Now (6.4) guarantees that

$$y = f(x) \in W_{a_j}(f(a_j)).$$

However, $y \in W_{a_j}(f(a_j))$ and (6.6) lead to $y \notin V_{a_j}(a_j)$, which contradicts (6.8). Therefore, (6.7) cannot be true. Choosing U such that

$$U \subseteq \bigcap_{i=1}^{n} U_{a_i},$$

it follows that

$$x - f(x) \notin U, \quad \text{for all } x \in C.$$

This however contradicts (6.4). Consequently there exists $x \in C$ with $x = f(x)$. $\square$

Theorem 6.10. *Let C be a convex subset of a Hausdorff locally convex linear topological space X. Suppose that $f : C \to C$ is a continuous, compact map. Then F has at least one fixed point in C.*

Proof. Since $f(C)$ is compact, then $K_* = co(f(C))$ is also a compact and convex subset in X. It is clear that

$$f(C) \subset K_* \subset C.$$

Hence, by Theorem 6.9, the function f has at least one fixed point. $\square$

Immediately we have the following version of Schauder's fixed point theorem.

Theorem 6.11. *Let C be a convex weakly compact subset of a Banach space X. Then every sequentially weakly continuous operator $A : C \to C$ has a fixed point.*

Proof. Let X have the weak topology. Then X becomes a locally convex space. Now for each closed subset C_* of X, $A^{-1}(C_* \cap C)$ is sequentially closed in C, hence weakly compact by the Ebertein-Smulian theorem, so $A^{-1}(C_* \cap C)$ is weakly closed. Hence A is weakly continuous. From Theorem 6.10, the operator A has at least one fixed point. $\square$

Corollary 6.12. *Let X be a reflexive, separable Banach space, C be a closed bounded convex subset in X and $A : C \to C$ be a weakly sequentially continuous operator. Then A has a fixed point.*

6.3 Fan's fixed point theorem

In this section we present the multivalued version of the Schauder fixed point theorem in locally convex spaces.

Theorem 6.13. *[273] ([185]) Let X be a Hausdorff locally convex space and M be a compact convex subset of X. If $F : M \to \mathcal{P}_{cp,cv}(M)$ is a u.s.c. multivalued operator, then there exists $x \in M$ such that $x \in F(x)$.*

Proof. Let $\{U_i : i \in I\}$ denote a neighborhood base at 0 in X consisting of open convex circle sets. For each $i \in I$ there exists a finite set

$$\{x_{ij:\ j\in J(i)}\} \subseteq M,$$

such that

$$M \subseteq \bigcup U_{j\in J(i)}(x_{ij} + U_i).$$

There exists a continuous partition of unity subordinate to this covering, i.e. for $j \in J(i)$ there is a continuous function $\phi_{ij} : M \to \mathbb{R}$ such that

$$\phi_{ij}(x) \geq 0 \text{ for } x \in M, \ \phi_{ij}(x) = 0 \text{ for } x \notin x_{ij} + U_i$$

and

$$\sum_{j\in J(i)} \phi_{ij}(x) = 1, \qquad x \in M.$$

Choose $y_{ij} \in F(x_{ij})$ and define the continuous function $f_i : M \to X$ by

$$f_i(x) = \sum \phi_{ij}(x)y_{ij}, \quad x \in M.$$

Let $C_i = co\{y_{ij} : j \in J(i)\} \subseteq M$. Clearly $f_i(C_i) \subseteq C_i$. By Brouwer's fixed point theorem we may choose $x_i \in C_i$ such that $x_i = f_i(x_i)$. The neighborhood base $\{U_i : i \in I\}$ is directed by $\subset$. Let $x \in M$ be a cluster point of the corresponding net $\{x_i : i \in I\}$ in M, and suppose $x \notin F(x)$. By separation there is a closed convex neighborhood W of $F(x)$ with $x \notin W$. Since $F(x)$ is upper semicontinuous there exists a neighborhood V of x with $V \cap W = \emptyset$ such that $x \in V \cap M$ implies $F(x) \subset W$. Choose $m \in I$ with $U_m + U_m \subset V - x$. There exists an $i \in I$ with $U_i \subset U_m$ such that $x_i \in x + U_i$, and then $x_i + U_i \subset V$ holds. For any $j \in J(i)$ with $\phi_{ij}(x_i) \neq 0$ we have $x_i \in x_{ij} + U_i$, hence $x_{ij} \in V$ which implies $y_{ij} \in W$. Then

$$x_i = f_i(x_i) = \sum_{j\in J(i)} \phi_{ij}(x_i)y_{ij} \in W,$$

a contradiction that $x_i \in V$. Therefore $x \in F(x)$.

$\square$

Theorem 6.14. ([185]) *Let X be a Banach space and M be a weakly compact convex subset of X. If $F : M \to \mathcal{P}_{wcl,cv}(M)$ is a $w. - w.u.s.c.$ multivalued operator, then there exists $x \in M$ such that $x \in F(x)$.*

Proof. Since X is a Banach space, then $(X, \sigma(X, X^*))$ is a locally convex topological Hausdorff space. Now we show that F is u.s.c. with the weak topology $\sigma(X, X^*)$. Let B be a weakly closed subset of M. Since F is $w.w.u.s.c.$, then $F_-^{-1}(B)$ is sequentially weakly closed. But $F_-^{-1}(B) \subset M$ and M is weakly compact. Then $F_-^{-1}(B)$ is sequentially compact and by Eberlein-Smulian's theorem, $\overline{F_-^{-1}(B)}^w$ is weakly compact. Let $x \in \overline{F_-^{-1}(B)}^w$. By Smulian's theorem there exists a subsequence, say $(x_n)_{n\in\mathbb{N}}$ of $F_-^{-1}(B)$ weakly converging to x. Since $F_-^{-1}(B)$ is weakly closed, then $x \in F^-(B)$, i.e. $\overline{F_-^{-1}(B)}^w \subseteq F_-^{-1}(B)$. Thus for every weakly closed set $B \subseteq M$, we have $F_-^{-1}(B)$ is weakly closed. Then F is u.s.c. and for each $x \in M$, $F(x) \in \mathcal{P}_{wcp,cv}(M)$. Thus by Fan's fixed point theorem there exists at least one fixed point of F.

$\square$

6.4 Krasnosel'skii-type fixed point theorems

Many problems arising from diverse areas of natural science, when modeled from a mathematical point of view, involve the study of solutions of nonlinear differential equations or inclusions of the respective forms,

$$B(u) + A(u) = u, \quad u \in M, \tag{6.9}$$

or

$$u \in B(u) + G(u), \quad u \in M, \tag{6.10}$$

where G is a multivalued map and M is a closed convex subset of a Banach space X. In particular, many integral equations and inclusions can be formulated in terms of (6.9) or (6.10); see, for example, [228]. In 1958, Krasnosel'skii [188] established that equation (6.9) has a solution in M if A and B satisfy:

(i) $A(x) + B(y) \in M$ for all $x, y \in M$;

(ii) A is continuous on M and $\overline{A(M)}$ is a compact set in X;

(iii) B is a $k-$contraction on X.

That result combined the Banach contraction principle and Schauder's fixed point theorem. The existence of fixed points for the sum of two operators has attracted tremendous interest, and their applications occur frequently in nonlinear analysis. Over time, many improvements of Krasnosel'skii's theorem have appeared in the literature by modifying the above assumptions; see, for example, [36, 37, 41, 42, 67, 68, 106, 126–128].

Fixed point theory for multivalued mappings is an important topic in set-valued analysis. Several well-known fixed point theorems for single-valued mappings such as those of Banach and Schauder have been extended to multivalued mappings in Banach spaces; see, for example, the monographs of Górniewicz et al. [24, 138].

Recently, multivalued analogues of Krasnosel'skii's fixed point theorem were obtained by Boriceanu [57] and Petruşel [241]. Several authors have extended the classical Krasnosel'skii fixed point theorem [36, 41, 106, 127, 128]. All the results of this section can be found in Graef *et al.* [143].

6.4.1 Krasnosel'skii-type fixed point theorem for weakly-weakly u.s.c.

In this subsection, we use Theorem 6.14 to obtain a multivalued version of the Krasnosel'skii theorem presented by Barroso [41]. We let $\mathcal{L}(X)$ denote the space of continuous, linear operators on X.

Theorem 6.15. *Let X be a Banach space and M be a nonempty weakly compact convex subset of X. Assume that $G : M \to \mathcal{P}_{wcl,cv}(X)$ is $w. - w.$u.s.c. and $B \in \mathcal{L}(X)$ satisfy*

$(\mathcal{H}_1)$ *$\|B^p\| < 1$ for some $p \in \mathbb{N}$,*

$(\mathcal{H}_2)$ *$x \in B(x) + G(y)$ and $y \in M$ implies $x \in M$.*

Then there exists $y \in M$ such that $y \in B(y) + G(y)$.

Proof. From $(\mathcal{H}_1)$, we can prove that $I-B$ is invertible and $(I-B)^{-1} = (I-B^p)^{-1} \displaystyle\sum_{k=0}^{k=p-1} B^k$.
This operator is well defined and $(I-B)^{-1} \in \mathcal{L}(X)$. Moreover, by Proposition 1.95, $(I-B)^{-1}$ is weakly continuous. Let us define $N : M \to \mathcal{P}_{wcl,cv}(M)$ by

$$y \to N(y) = (I - B)^{-1}G(y).$$

Since $G(\cdot) \in \mathcal{P}_{wcl,cv}(X)$ and $(I - B)^{-1} \in \mathcal{L}(X)$, we see that $N(\cdot) \in \mathcal{P}_{wcl,cv}(X)$. Now we show that $N(\cdot)$ is $w. - w.u.s.c.$ Let $x \in M$ and $h \in (I - B)^{-1}(G(x))$. Then there exists $y \in G(x)$ such that

$$h = (I - B)^{-1}(y) \quad \text{implies} \quad h = B(h) + y \subseteq B(h) + G(x),$$

and thus $(\mathcal{H}_2)$ implies that $h \in M$. Hence, $N(M) \subset M$.

Next, we show that for every $y \in M$, we have $N(y) \in \mathcal{P}_{wcl}(M)$. Let $\{y_n : n \in \mathbb{N}\} \subset M$ be a sequence converging weakly to some z in M and $y_n \in N(y)$, $n \in \mathbb{N}$. Then there exists $x_n \in G(y)$ such that

$$(I - B)(y_n) = x_n, \ n \in \mathbb{N}.$$

Since G is $w. - w.u.s.c.$, there exists a subsequence of $\{x_n\}$ converging weakly to some $x \in G(y)$. Then $(I - B)^{-1}(x_n)$ converges weakly to $(I - B)^{-1}(x) \in N(y)$. Hence, there exists a subsequence of $\{y_n\}$ converging weakly to $z = (I - B)^{-1}(x) \in N(y)$.

Finally, we show that N has a weakly closed graph. Let $h_n \in N(y_n)$ be such that $\{h_n\}_{n \in \mathbb{N}}$ and $\{y_n\}_{n \in \mathbb{N}}$ converge weakly to h and y respectively. Then, $(I - B)(h_n) \in G(y_n)$, $n \in \mathbb{N}$. It is clear that $(I - B)(h_n)$ converges weakly to $(I - B)(h)$. Using the fact that $G(\cdot)$ is $w. - w.u.s.c.$, we conclude that $(I - B)(h) \in G(y)$. Hence, $h \in N(y)$. Now M is weakly compact, so by Theorem 1.102, N is $w. - w.u.s.c.$ Thus, from Theorem 6.14, there exists $y \in M$ such that $y \in (I - B)^{-1}G(y)$, and therefore $B + G$ has a fixed point in M. □

Now, we can easily prove the following result.

Theorem 6.16. *Let X be a Banach space and M be a nonempty weakly closed bounded convex subset of X. Assume that $B \in \mathcal{L}(X)$, $G : M \to \mathcal{P}_{wcl,cv}(X)$ is $w.-w.u.s.c.$, conditions $(\mathcal{H}_1)$-$(\mathcal{H}_2)$ are satisfied, and*

$(\mathcal{H}_3)$ *$G(M)$ is weakly relatively compact and for each $y \in co\, G(M)$,*

$$x \in B(x) + G(y) \quad \text{implies} \quad x \in co\, G(M).$$

Then the operator $B + G$ has at least one fixed point.

Proof. Let $\widetilde{M} = \overline{co}G(M)$ be weakly compact convex. Now we prove only that $N(\widetilde{M}) \subseteq \widetilde{M}$, where N is defined in the proof of Theorem 6.15. Indeed, let $x \in N(\widetilde{M})$. Then there exists $y \in \widetilde{M}$ such that

$$x \in N(y).$$

Hence,

$$x = (I - B)^{-1}z, \ z \in G(y) \quad \text{implies} \quad x \in \overline{co}\, G(M).$$

Thus,

$$N(\widetilde{M}) \subseteq \widetilde{M}.$$

So, by Theorem 6.14, there exists $x \in X$ that is a fixed point of N. □

Next we present the case where B^p is a non-expansive operator.

Theorem 6.17. *Let X be a Banach space and M be a weakly compact convex, subset of X. Assume that*

$(\mathcal{H}_\alpha)$ $\alpha \in (0,1)$, $x \in \alpha B(x) + G(y)$, *and* $y \in M$ *imply* $x \in M$.

Let $B \in \mathcal{L}(X)$ with $\|B^p\| \le 1$, and let $G : M \to \mathcal{P}_{wcl,cv}(X)$ be a $w. - w.u.s.c.$ multivalued operator. Then $B + G$ has at least one fixed point.

Proof. Let $\alpha_n \in (0,1)$, $n \in \mathbb{N}$ with $\alpha_n \to 1$ as $n \to \infty$, such that

$$\|B^p_{\alpha_n}\| := \|\alpha_n B^p\| < 1, \ n \in \mathbb{N}.$$

From Theorem 6.15 there exists $x_n \in M$ such that

$$x_n \in \alpha_n B(x_n) + G(x_n), \ n \in \mathbb{N} \Rightarrow x_n \in B(\alpha_n x_n) + G(x_n), \ n \in \mathbb{N}.$$

Since M is weakly compact, there exists a subsequence of $\{x_n\}$ converging weakly to some $x \in M$. Let $L : [0,1] \times M \to \mathcal{P}_{wcl}(X)$ be the multivalued operator defined by

$$(\alpha, x) \to L(\alpha, x) = B(\alpha x) + G(x).$$

Using the fact that $B \in \mathcal{L}(X)$, G is $w. - w.u.s.c$, and M is weakly compact, we have that L has a weakly sequentially closed graph (see Theorem 1.102). Then

$$x_n \in B(\alpha x_n) + G(x_n) \to x \in B(x) + G(x), \text{ as } n \to \infty.$$

Hence, $B + G$ has at least one fixed point. $\qquad\square$

Now, we present a Leary-Shauder alternative type of the Krasnosel'skii fixed point theorem.

Theorem 6.18. *Let X be a Banach space, $\Omega \subset X$ be a closed convex set, U be a weakly open subset of Ω with $0 \in U$, $B \in \mathcal{L}(X)$, and let $G : U \to \mathcal{P}_{wcl,cv}(X)$ be a $w. - w.u.s.c.$ multivalued operator such that $G(\overline{U})$ is relatively weakly compact. Assume that $(\mathcal{H}_1)$ holds and*

$(\overline{\mathcal{H}}_1)$ $x \in B(x) + G(y)$ *and* $y \in \overline{U}$ *implies* $x \in \Omega$.

Then, either

(a) $x \in B(x) + \lambda G(x)$ has a solution for $\lambda = 1$,

or

(b) there exists $x \in \partial_\Omega U$ (the weakly boundary of U in Ω) and some $\lambda \in (0,1)$ such that $x \in B(x) + \lambda G(x)$.

Proof. Assume that (b) does not hold and $B + G$ has no fixed point in $\partial_\Omega U$. Let

$$D = \{x \in \overline{U} : x \in \lambda(I - B)^{-1}G(x), \ \lambda \in [0,1]\}.$$

For $\lambda = 0$, we have $0 \in D$. We first we show that D is weakly compact. Let $(x_n)_{n \in \mathbb{N}}$ be a sequence in D; then $x_n \in \lambda_n(I - B)^{-1}G(x_n)$ for some $\lambda_n \in [0,1]$. Thus, for each $n \in \mathbb{N}$,

$$x_n = \lambda_n(I - B)^{-1}(z_n)$$

for some $z_n \in G(x_n)$. There exists a subsequence of (λ_n), again denoted by (λ_n), such that $\lambda_n \to \lambda$.

Now the set

$$\{G(x_n) : \ n \in \mathbb{N}\} = G(\{x_n : \ n \in \mathbb{N}\})$$

is weakly relatively compact, so there exists a subsequence of $(z_n)_{n \in \mathbb{N}}$, also denoted as $(z_n)_{n \in \mathbb{N}}$, such that

$$z_n \rightharpoonup z, \quad \text{as } n \to \infty.$$

Since $(I - B)^{-1}$ is continuous,

$$(I - B)^{-1}(z_n) \rightharpoonup (I - B)^{-1}(z), \quad \text{as } n \to \infty.$$

Thus, there exists a subsequence of $x_n = \lambda_n (I - B)^{-1}(z_n)$, again denoted by $(x_n)_{n \in \mathbb{N}}$, that weakly converges to $x = \lambda(I - B)^{-1}(z)$. Since G has a weakly closed graph, we have $z \in G(x)$. This implies that $x \in \lambda(I - B)^{-1}G(z)$. Consequently, D is weakly compact.

Notice that $\partial U \cap D = \emptyset$. Since $(X, \sigma(X, X^*))$ is a Tychonoff space, there exists a continuous mapping $\mu : \overline{U} \to [0, 1]$ with $\mu(D) = 1$ and $\mu(\partial U) = 0$. Let

$$N(x) := \begin{cases} \mu(x)(I - B)^{-1}G(x), & \text{if } x \in \overline{U}, \\ 0, & \text{if } x \in \Omega \backslash \overline{U}. \end{cases}$$

By the weak compactness of G and the continuity of $(I - B)^{-1}$, we see that the set $C = \overline{co}\,((I-B)^{-1}(G(U)) \cup \{0\})$ is weakly compact. Also, $N : C \to \mathcal{P}_{wcl,cv}(C)$ is $w. - w.u.s.c$ (see Theorem 1.102). Then, from Theorem 6.14, there exists $x \in X$ such that $x \in N(x)$.   □

The following corollary is a direct consequence of the above theorem.

Corollary 6.19. *Let X be a reflexive Banach space, $B \in \mathcal{L}(X)$, $G : X \to \mathcal{P}_{wcl,cv}(X)$ be a $w. - w.u.s.c.$ multivalued operator. Assume that $(\mathcal{H}_1)$ holds and*

$(\overline{\mathcal{H}}_2)$ *For each $D \in \mathcal{P}_b(X)$, $G(B)$ is weakly relatively compact.*

Then, either

(a) *$x \in B(x) + \lambda G(x)$ has a solution for $\lambda = 1$,*

or

(b) *the set $\mathcal{M} = \{x \in X : x \in B(x) + \lambda G(x) \quad \text{for some } \lambda \in [0, 1]\}$ is unbounded.*

6.4.2   Krasnosel'skii-type fixed point theorem for u.s.c.

In this section we use a technique of approximation proved recently for a version of a Krasnosel'skii-type fixed point result.

Theorem 6.20. *Let X be a Banach space, M be a compact convex subset of X, $G : M \to \mathcal{P}_{cp,cv}(X)$ be an u.s.c. multivalued map, and $B : M \to M$ be a single-valued map. Assume that G and B satisfy the following conditions:*

$(\mathcal{H}_4)$ *B is a contraction mapping;*

$(\mathcal{H}_5)$ *$B(M) + G(M) \subset M$.*

Then the inclusion $x \in B(x) + G(x)$ has a solution.

Proof. From Theorem 6.3, given $\epsilon > 0$, there exists a continuous map $f_\epsilon : M \to X$ such that

$$\Gamma(f_\epsilon) \subseteq \Gamma(G) + \epsilon B(0, 1) \tag{6.11}$$

and

$$f_\epsilon(M) \subseteq co\, G(M).$$

From $(\mathcal{H}_5)$ and the convexity of M, we obtain

$$B(M) + f_\epsilon(M) \subseteq M.$$

For fixed $y \in M$, we consider $F_\epsilon^y : M \to M$ defined by

$$F_\epsilon^y(x) = B(x) + f_\epsilon(y), \quad x \in M.$$

By the Banach fixed point theorem, there exists a unique $x_\epsilon(y) \in M$ such that

$$x_\epsilon(y) = B(x_\epsilon(y)) + f_\epsilon(y).$$

From $(\mathcal{H}_4)$, the mapping $I - B : M \to (I - B)(M)$ is a homeomorphism. We define the operator $N_\epsilon : M \to M$ by

$$N_\epsilon(x) = (I - B)^{-1} f_\epsilon(x).$$

It easy to see that N_ϵ satisfies the conditions of Schauder's fixed point theorem, and so there exists $x_\epsilon \in M$ such that

$$x_\epsilon = B(x_\epsilon) + f_\epsilon(x_\epsilon).$$

Let $\{\epsilon_n : n \in \mathbb{N}\}$ be such that

$$\epsilon_n \to 0 \text{ as } n \to \infty,$$

and for each $n \in \mathbb{N}$, choose x_{ϵ_n} so that $x_{\epsilon_n} = B(x_{\epsilon_n}) + f_\epsilon(x_{\epsilon_n})$. Since M is compact, there exists a subsequence of x_{ϵ_n} converging to some $x \in M$, and so

$$f_{\epsilon_n}(x_{\epsilon_n}) = (I - B)(x_{\epsilon_n}) \to (I - B)(x), \text{ as } n \to \infty. \tag{6.12}$$

Hence, from (6.11), we have

$$d((x_{\epsilon_n}, f_{\epsilon_n}(x_{\epsilon_n})), \Gamma(G)) \le \epsilon_n, \text{ for every } n \in \mathbb{N}.$$

It is clear that G has a closed graph in $X \times X$ and so $(I - B)(x) \in G(x)$. This implies

$$x \in B(x) + G(x)$$

and completes the proof of the theorem. $\qquad \square$

Using Theorem 6.20 we can easily prove our next result.

Theorem 6.21. *Let X be a Banach space and M be a closed bounded convex subset of X. Assume that $G : M \to \mathcal{P}_{cp,cv}(X)$ is u.s.c., B satisfies $(\mathcal{H}_4)$, and*

$(\mathcal{H}_6)$ *$G(M)$ is compact and $B(M) + G(M) \subseteq co\, G(M)$.*

Then the operator $B + G$ has at least one fixed point.

Proof. Let $\widetilde{M} = \overline{co}\, G(M)$. Then, $\widetilde{M}$ is compact convex. It suffices to show that $N_\epsilon(\widetilde{M}) \subseteq \widetilde{M}$, where N_ϵ is defined in the proof of Theorem 6.20.

To this end, let $x \in \widetilde{M}$ be such that $y = N_\epsilon(x)$; then

$$y = (I - B)^{-1} f_\epsilon(x) \quad \text{implies} \quad x \in \overline{co}\, G(M).$$

Thus,

$$N_\epsilon(\widetilde{M}) \subseteq \widetilde{M}.$$

By Theorem 6.20, there exists $x \in X$ that is a fixed point of $B + G$. $\qquad\square$

Next, we examine the case where B is a non-expansive operator.

Theorem 6.22. *Let X be a Banach space and M be a compact convex subset of X. Assume that the condition $(\mathcal{H}_\alpha)$ of Theorem 6.17 holds, B is a non-expansive operator, and $G : M \to \mathcal{P}_{cp,cv}(X)$ is an u.s.c. multivalued operator. Then $B + G$ has at least one fixed point.*

Proof. Let $\alpha_n \in (0,1)$, $n \in \mathbb{N}$, with $\alpha_n \to 1$ as $n \to \infty$ such that $\alpha_n B^p$, $n \in \mathbb{N}$, is a contraction. From Theorem 6.15, there exists $x_n \in M$ such that

$$x_n \in \alpha_n B(x_n) + G(x_n), \ n \in \mathbb{N}, \quad \text{implies} \quad x_n \in B(\alpha_n x_n) + G(x_n), \ n \in \mathbb{N}.$$

Since M is compact, there exists a subsequence of $\{x_n\}$ converging to some $x \in M$. Let $L_* : [0,1] \times M \to \mathcal{P}_{cp}(X)$ be a multivalued operator defined by

$$(\alpha, x) \to L_*(\alpha, x) = \alpha B(x) + G(x).$$

Using the facts that B is a continuous operator, G is u.s.c., and M is compact, we see that L_* has closed graph, and so

$$x_n \in \alpha_n B(x_n) + G(x_n) \to x \in B(x) + G(x), \ \text{as } n \to \infty.$$

Hence, $B + G$ has at least one fixed point. $\qquad\square$

We next present our global multivalued version of the Krasonsel'skii fixed point theorem.

Theorem 6.23. *Let X be a Banach space and $G : X \to \mathcal{P}_{cp,cv}(X)$ be an u.s.c. multivalued operator. Assume condition $(\mathcal{H}_4)$ holds and G is compact. Then, either*

(a) $x \in \lambda B(\frac{x}{\lambda}) + \lambda G(x)$ has a solution for $\lambda = 1$,

or

(b) the set $\{x \in X : x \in \lambda B(\frac{x}{\lambda}) + \lambda G(x), \ \lambda \in (0,1)\}$ is unbounded.

Proof. Assume that $\mathcal{M} = \{x \in X : x \in \lambda B(\frac{x}{\lambda}) + \lambda G(x), \ \lambda \in (0,1)\}$ is bounded. Then there exists $K > 0$ such that

$$|x| \leq K \quad \text{for every } x \in \mathcal{M} \text{ and } \lambda \in [0,1].$$

Set

$$U = \{x \in X : |x| < K + 1\}.$$

It clear that U is an open bounded convex subset of X. Moreover, $\overline{G(U)}$ is compact and so $\overline{co}\, G(U)$ is compact.

Let $f_\epsilon : X \to X$ be an ϵ-approximation selection of G such that

$$f_\epsilon(X) \subseteq co\, G(X).$$

Set

$$D_\epsilon = \{x \in \overline{U} : x = \lambda(I - B)^{-1}f_\epsilon(x), \ \lambda \in [0, 1]\}.$$

It clear that $\partial U \cap D_\epsilon = \emptyset$. Therefore, by Urysohn's lemma, there exists a continuous function $\mu : \overline{U} \to [0, 1]$ with $\mu(D_\epsilon) = 1$ and $\mu(\partial U) = 0$. Given $\epsilon > 0$, we define

$$\overline{N}_\epsilon(x) := \begin{cases} \mu(x)(I - B)^{-1}f_\epsilon(x), & \text{if } x \in \overline{U}, \\ 0, & \text{if } x \in X \backslash \overline{U}. \end{cases}$$

By the compactness of G and the continuity of $(I - B)^{-1}$, we see that the set $C = \overline{co}\,((I - B)^{-1}(co\,G(U)) \cup \{0\})$ is compact. Also, $\overline{N}_\epsilon : C \to C$ is continuous and compact. Then, by Schauder's fixed point theorem, there exists $x_\epsilon \in X$ such that $x_\epsilon = \overline{N}_\epsilon(x_\epsilon)$. By the same technique used in the proof of Theorem 6.20, we conclude that $B + G$ has at least one fixed point. $\qquad\square$

6.4.3   Expansive Krasnosel'skii type fixed point theorem

In this section, we present a multivalued version of an expansive Krasnosel'skii-type fixed point theorem. Our result extends some of those in [290] for which the mapping was a single-valued map.

Definition 6.24. *Let* (X, d) *be a metric space and* M *be a subset of* X. *The mapping* $B : M \to X$ *is said to be expansive, if there exists a constant* $k > 1$ *such that*

$$d(B(x), B(y)) \geq kd(x, y) \quad \text{for all } x, y \in M.$$

We will need the following auxiliary results.

Theorem 6.25. ([290]) *Let* X *be a complete metric space and* M *be a closed subset of* X. *Assume that* $B : M \to X$ *is expansive and* $M \subseteq B(M)$. *Then there exists a unique point* $x \in M$ *such that* $x = B(x)$.

Proof. Since B is expansive, there exists $h > 1$ such that

$$d(B(x), B(y)) \geq hd(x, y) \quad \text{for every } x, y \in M.$$

Then B is injective. By $M \subseteq B(M)$ we get B is bijective. Now, we show that $B^{-1} : M \to M$ is a contraction operator. Let $y_1, y_2 \in M$, then there exist $x_1, x_2 \in M$ such that

$$y_1 = B(x_1), \quad y_2 = B(x_2) \Rightarrow x_1 = B^{-1}(y_1) \quad \text{and} \quad x_2 = B^{-1}(y_2).$$

Hence

$$d(y_1, y_2) = d(B(x_1), B(x_2)) \geq hd(x_1, x_2).$$

Therefore

$$d(B^{-1}(y_1), B^{-1}(y_2)) \leq \frac{1}{h}d(y_1, y_2), \quad \text{for each } y_1, y_2 \in M.$$

So, by the Banach fixed point theorem, the equation $B^{-1}(x) = x$ has unique solution on M which is unique fixed point of B. $\qquad\square$

Lemma 6.26. *Assume there exists* $n \in \mathbb{N}$ *such that* $B^n : X \to X$ *is expansive and* M *is a closed subset of* X *with* $M \subset B(M)$. *Then there exists a unique fixed point of* B.

Proof. Since B^n is an expansive map and $M \subset B^n(M)$, by Theorem 6.25, there exists a unique fixed point of B^n, call it $x \in M$. Using the fact that B^n is an expansive map, there exists $k > 1$ such that

$$d(B^n(x), B^n(y)) \geq kd(x, y) \text{ for all } x, y \in M.$$

Hence,

$$d(x, B(x)) = d(B^n(x), B^{n+1}(x)) \geq kd(x, B(x)) \Rightarrow d(x, B(x)) = 0.$$

Therefore, B has a unique fixed point in M. □

Lemma 6.27. ([290]) *Let X be a normed linear space, $M \subseteq X$, and assume that the mapping $B : M \to X$ is expansive with constant $h > 1$. Then the inverse of $I - B : M \to (I - B)(M)$ exists and*

$$|(I - B)^{-1}(x) - (I - B)^{-1}(y)| \leq \frac{1}{h - 1}|x - y| \text{ for } x, y \in (I - B)(M).$$

Now we are ready to give our main results in this section.

Theorem 6.28. *Let X be a Banach space, M be a compact convex subset of X, $G : M \to \mathcal{P}_{cp,cv}(X)$ be an u.s.c. multivalued map, and $B : M \to X$ be a single-valued mapping. Assume that G and B satisfy the following conditions:*

($\mathcal{C}_1$) B is a continuous expansive mapping;

($\mathcal{C}_2$) For every $z \in co\, G(M)$, we have

$$M \subseteq z + B(M).$$

Then the inclusion $x \in B(x) + G(x)$ has a solution.

Proof. Given $\epsilon > 0$, there exists a continuous map $f_\epsilon : M \to X$ such that

$$\Gamma(f_\epsilon) \subset \Gamma(G) + \epsilon B_*. \tag{6.13}$$

For fixed $y \in M$, consider $F_\epsilon^y : M \to X$ defined by

$$F_\epsilon^y(x) = B(x) + f_\epsilon(y), \ x \in M.$$

By Theorem 6.25, there exists a unique fixed point $x_\epsilon(y) \in M$ such that

$$x_\epsilon(y) = B(x_\epsilon(y)) + f_\epsilon(y).$$

Since B is expansive, there exists $h > 1$ such that for $y_1, y_2 \in M$,

$$|B(x_\epsilon(y_1)) - B(x_\epsilon(y_2))| \geq h|x_\epsilon(y_1) - x_\epsilon(y_2)|.$$

Then,

$$|x_\epsilon(y_1) - x_\epsilon(y_2)| \leq \frac{1}{h - 1}|f_\epsilon(y_1) - f_\epsilon(y_2)|.$$

Since $f_\epsilon(\cdot)$ is continuous, $x_\epsilon(\cdot) : G(M) \to M$ is continuous as well. Let $L_\epsilon : M \to M$ be defined by $L_\epsilon(z) = x_\epsilon(f_\epsilon(z))$. It is clear that L_ϵ is continuous, and by Schauder's fixed point theorem, there exists $z_\epsilon \in M$ such that $x_\epsilon(f_\epsilon(z_\epsilon)) = z_\epsilon$. Hence,

$$B(z_\epsilon) + f_\epsilon(z_\epsilon) = z_\epsilon.$$

By the same method used in the proof of Theorem 6.20, we can show that there exists $x \in M$ such that $x \in B(x) + G(x)$, and this completes the proof of the theorem. □

Using an analogous argument, we will prove the following result.

Theorem 6.29. *Let X be a Banach space, M be a compact convex subset of X, $G : M \to \mathcal{P}_{cp,cv}(X)$ be an u.s.c. multivalued map, and $B : M \to X$ be a single-valued mapping. Assume that G and B satisfy $(\mathcal{C}_1)$ and $(\mathcal{C}_2)$. Then the inclusion $x \in B(I - G)(x)$ has a solution.*

Proof. Since B is expansive, we see that $B^{-1} : B(M) \to M$ is contractive, and since B is continuous, it is easy to see that $B(M)$ is closed. Let $\epsilon > 0$ be given. Then there exists an ϵ-approximate selection f_ϵ of F. For fixed $y \in M$, by the Banach contraction principle, the equation

$$B^{-1}(x) + f_\epsilon(y) = x$$

has a unique solution $x_\epsilon(y) \in B(M)$. Let y_1, $y_2 \in M$; then

$$|x_\epsilon(y_1) - x_\epsilon(y_2)| \le \frac{1}{h}|x_\epsilon(y_1) - x_\epsilon(y_2)| + |f_\epsilon(y_1) - f_\epsilon(y_2)|.$$

Therefore,

$$|x_\epsilon(y_1) - x_\epsilon(y_2)| \le \frac{h}{h-1}|f_\epsilon(y_1) - f_\epsilon(y_2)|.$$

Hence, $L_\epsilon(y) = x_\epsilon(f_\epsilon(\cdot)) : M \to M$ is continuous. By Schauder's fixed point theorem, there exists $z_\epsilon \in M$ such that $x_\epsilon(f_\epsilon(z_\epsilon)) = z_\epsilon$, and hence

$$B^{-1}(z_\epsilon) + f_\epsilon(z_\epsilon) = z_\epsilon.$$

By the same method used in the proof of Theorem 6.20, we can show that there exists $x \in M$ such that

$$x \in B^{-1}(x) + G(x) \text{ implies } x \in B^{-1}(I - G)(x),$$

and this proves the theorem. $\qquad\square$

6.4.4 Expansive Krasnosel'skii-type fixed point theorem for weakly continuous maps

In this section, we will provide some expansive Krasnosel'skii-type fixed point theorem. Our result extends some results presented in [290].

Theorem 6.30. *Let X be a Banach space and M be a nonempty weakly compact convex subset of X. Assume that $A : M \to X$ is weakly continuous and $B \in \mathcal{L}(X)$ satisfy*

$(\widetilde{\mathcal{H}}_1)$ *B^p is expansive and $\|B^p\| > 1$ for some $p \in \mathbb{N}$,*

$(\widetilde{\mathcal{H}}_2)$ *for each $y \in M$ such that*

$$x = B(x) + A(y) \Rightarrow x \in M.$$

$(\widetilde{H}_3)$ *For every $y \in M$ we have*

$$M \subseteq B^p(M) + A(y)$$

Then there exists $y \in M$ such that $y = B(y) + A(y)$.

Proof. Let $y \in M$. Let $F_y : M \to X$ be a operator defined by

$$F_y(x) = B(x) + A(y), \quad x \in M.$$

From Theorem 6.25 there exist unique $x(y) \in M$ such that

$$x(y) = B(x(y)) + A(y).$$

By $(\widetilde{\mathcal{H}}_1)$, we can prove that $I - B$ is invertible and $(I - B)^{-1} = (I - B^p)^{-1} \sum_{k=0}^{k=p-1} B^k$. This operator is well defined and $(I - B)^{-1} \in \mathcal{L}(X)$. Moreover, by Proposition 1.95, $(I - B)^{-1}$ is weakly continuous. Let us define $N : M \to M$ by

$$y \to N(y) = (I - B)^{-1}A(y).$$

Let $x \in M$ and $h = (I - B)^{-1}(A(x))$. Then

$$h = (I - B)^{-1}(A(x)) \Rightarrow h = B(h) + A(x),$$

and thus $(\widetilde{\mathcal{H}}_2)$ implies that $h \in M$. Let $\{y_n : n \in \mathbb{N}\} \subset M$ be a sequence converging weakly to y in M. We show that $N(y_n)$ converges weakly to $N(y)$. Set $x_n = (I - B)^{-1}A(y_n)$, then $(I - B)(x_n) = A(y_n)$, $n \in \mathbb{N}$.

Since M is weakly compact, there exists a subsequence of $\{x_n\}$ converging weakly for some $x \in M$. Then $(I - B)(x_n)$ converges weakly to $(I - B)(x)$. Hence there exists a subsequence of $\{y_n\}$ converging weakly to $(I - B)(x)$. Then $N(y_n)$ converge weakly to $N(y)$. Hence from Theorem 6.11, there exists $y \in M$ such that $y = (I - B)^{-1}A(y)$, and we deduce that $B + G$ has a fixed point in M. □

Now we are ready to state our results in this part.

Theorem 6.31. *Let X be a Banach space, M be a weakly compact convex subset of X, $A : M \to X$ be an weakly continuous map and $B \in \mathcal{L}(X)$ be a linear continuous operator single-valued mapping. Assume that G and B satisfy the following hypotheses:*

$(\mathcal{C}_1)$ *there exists $p \in \mathbb{N}$ such that $\|B^p\| > 1$ is an expansive mapping.*

$(\mathcal{C}_2)$ *for each $y \in coA(M)$ such that*

$$x = B(x) + A(y) \Rightarrow x \in coA(M).$$

$(\mathcal{C}_\ni)$ *For every $y \in M$ we have*

$$M \subseteq B^p(M) + A(y).$$

Then the abstract equation $x = B(x) + A(x)$ has a solution.

Proof. Let $\widetilde{M} = \overline{co}A(M)$ be weakly compact convex. Now we prove only that $N(\widetilde{M}) \subseteq \widetilde{M}$, where N is defined in the proof of Theorem 6.30. Indeed, let $x \in N(\widetilde{M})$. Then there exists $y \in \widetilde{M}$ such that

$$x = N(y).$$

Hence

$$x = (I - B)^{-1}A(y) \Rightarrow x \in \overline{co}A(M).$$

Then

$$N(\widetilde{M}) \subseteq \widetilde{M}.$$

So, by Theorem 6.30, there exists $x \in X$ which is fixed point of N. □

6.4.5 Expansive Krasnosel'skii-type fixed point theorem for weakly-weakly u.s.c.

In this section, we use Theorem 6.14 to obtain a multivalued version of the Krasnosel'skii theorem presented by Xiang and Yuan [290].

Theorem 6.32. *Let X be a Banach space and M be a nonempty weakly compact convex subset of X. Assume that $G : M \to \mathcal{P}_{wcl,cv}(X)$ is $w. - w.u.s.c.$ and $B \in \mathcal{L}(X)$ satisfy*

$(\overline{\mathcal{C}}_1)$ B *is expansive and $\|B\| > 1$*

$(\overline{\mathcal{C}}_2)$ *for each $y \in M$ such that*

$$x \in B(x) + G(y) \Rightarrow x \in M.$$

(C_*) *For every $y \in M$ and $z \in G(y)$ we have*

$$M \subseteq B(M) + z.$$

Then there exists $y \in M$ such that $y \in By + G(y)$.

Proof. Let $y \in M$. Let $F_y : M \to \mathcal{P}(X)$ be the multivalued operator defined by

$$F_y(x) = B(x) + G(y), \quad x \in M.$$

Since $G(y) \in \mathcal{P}_{wk,cp,cv}(X)$ and G is $w. - w.u.s.c.$, then $F_y(x) \in \mathcal{P}_{wcp,cv}(X)$. By $(\overline{\mathcal{C}}_2)$ we have $F_y(M) \subseteq M$. Now we show that F_y is $w. - w.u.s.c.$ Let $(x_n, y_n) \in \Gamma(F_y) = \{(x, z) \in M \times M : z \in F_y(x)\}$ be a sequence such that

$$y_n \in F_y(x_n), \ x_n \rightharpoonup x \text{ and } y_n \rightharpoonup y_*.$$

Thus there exists $z_n \in G(y)$ such that

$$y_n = B(x_n) + z_n, \quad n \in \mathbb{N}.$$

Since $B \in \mathcal{L}(X)$, then, by Proposition 1.95, $B(x_n)$ converges weakly to $B(x)$. This implies that $z_n \rightharpoonup y_* - B(x) \in G(y)$. Hence F_y has a weakly closed graph. By $(\overline{\mathcal{C}}_2)$ and Theorem 1.102, we deduced that F_y is $w. - w.u.s.c.$ From Theorem 6.14, there exists $x(y) \in M$ such that

$$x(y) \in B(x(y)) + G(y).$$

By $(\overline{\mathcal{C}}_1)$, we can prove that $I - B$ is invertible and $(I - B)^{-1} \in L(X)$. Moreover, from Proposition 1.95, $(I - B)^{-1}$ is weakly continuous. Let us define $N : M \to \mathcal{P}_{wcl,cv}(M)$ by

$$y \to N(y) = (I - B)^{-1} G(y).$$

Since $G(\cdot) \in \mathcal{P}_{wcl,cv}(X)$ and $(I - B)^{-1} \in L(X)$, then $N(\cdot) \in \mathcal{P}_{wcl,cv}(X)$. Now we show that $N(\cdot)$ is $w. - w.u.s.c.$ Let $x \in M$ and $h \in (I - B)^{-1}(G(x))$. Then there exists $y \in G(x)$ such that

$$x = (I - B)^{-1}(y) \Rightarrow x = Bx + y \subseteq B(x) + G(y),$$

and thus $(\overline{\mathcal{C}}_2)$ implies that $x \in M$. Let $\{y_n : n \in \mathbb{N}\} \subset M$ be a sequence converging weakly to x in M and $y_n \in N(y)$, $n \in \mathbb{N}$. Then there exists $x_n \in G(y)$ such that

$$(I - B)(y_n) = x_n, \ n \in \mathbb{N}.$$

Since G is $w. - w.u.s.c.$, there exists a subsequence of $\{x_n\}$ converging weakly for some $x \in G(y)$. Then $(I - B)^{-1}(x_n)$ converges weakly to $(I - B)^{-1}(x) \in N(y)$. Hence there exists a subsequence of y_n converging weakly to $(I - B)^{-1}(x)$. By Theorem 1.102, N is $w. - w.u.s.c.$ Hence from Theorem 6.14, there exists $y \in M$ such that $y \in (I - B)^{-1}G(y)$, and we deduce that $B + G$ has a fixed point in M. $\qquad\square$

Now, we can easily prove the next result.

Theorem 6.33. *Let X be a Banach space and M be a nonempty weakly closed bounded convex subset of X. Assume that $G : M \to \mathcal{P}_{wcl,cv}(X)$ is $w.-w.u.s.c.$, that B satisfies $(\overline{\mathcal{C}}_1)$ and $(\overline{\mathcal{C}}_*)$, and the condition*

$(\overline{\mathcal{C}}_3)$ *$G(M)$ is weakly relatively compact and for each $y \in coG(M)$ such that*

$$x \in B(x) + G(y) \Rightarrow x \in coG(M).$$

Then the operator $B + G$ has at least one fixed point.

Proof. Let $\widetilde{M} = \overline{co}G(M)$ be weakly compact convex. Now we prove only that $N(\widetilde{M}) \subseteq \widetilde{M}$, where N is defined in the proof of Theorem 6.32. Indeed, let $x \in N(\widetilde{M})$. Then there exists $y \in \widetilde{M}$ such that

$$x \in N(y).$$

Hence

$$x = (I - B)^{-1}z, \ z \in G(y) \Rightarrow x \in \overline{co}G(M).$$

Then

$$N(\widetilde{M}) \subseteq \widetilde{M}.$$

So, by Theorem 6.32, there exists $x \in X$ which is fixed point of N. $\qquad\square$

6.4.6 Krasnosel'skii type in a Fréchet space

First, we present a multivalued version of the Covitz and Nadler fixed point theorem in Fréchet spaces. Let X be a Fréchet space with the topology generated by a countable family of semi-norms $\{\|\cdot\|_n : n \in \mathbb{N}\}$ with the corresponding distances $d_n(x, y) = \|x - y\|_n$, $n \in \mathbb{N}$. Suppose that $\{\|\cdot\|_n : n \in \mathbb{N}\}$ is sufficient, *i.e.*, for every $x \in X$, $x \neq 0$, there exists $n \in \mathbb{N}$ such that $\|x\|_n \neq 0$. Every space $(X, \|\cdot\|_n)$ endowed with a countable and sufficient family of semi-norms can be considered as a metric space, and so we set

$$d(x, y) = \sum_{n=1}^{\infty} \frac{1}{2^n} \frac{d_n(x, y)}{1 + d_n(x, y)}. \tag{6.14}$$

The convergence determined by the metric (6.14) can be characterized by the semi-norms, i.e.,

$x_n \to x$ as $n \to \infty$ if and only if for each $n \in \mathbb{N}$, $\|x_m - x\|_n \to 0$ as $m \to \infty$.

Two families of semi-norms, $|\cdot|_n$ and $\|\cdot\|_n$, are *equivalent* if and only if they define the same metric topology. Clearly, if $(X, |\cdot|_n)$ is complete, then each equivalent family of semi-norms is also complete.

We remark that for every family of semi-norms $|\cdot|_n$, there is an equivalent family of semi-norms that are ordered in the sense that

$$|x|_n \le |x|_{n+1} \text{ for every } n \in \mathbb{N} \text{ and } x \in X.$$

It is easy to characterize continuity of a mapping and the compactness of a set through the notion of semi-norms.

Definition 6.34. *Let $F : X \to \mathcal{P}(X)$ be a multivalued map. We call F an admissible contraction on X if:*

(*i*) *There exists* $\alpha_n \in [0,1)$, $n \in \mathbb{N}$, *such that*

$$H_{d_n}(F(x), F(y)) \leq \alpha_n d_n(x, y) \ \text{ for each } x, y \in X,$$

where H_{d_n} *is the Hausdorff-Pompeiu distance on* $(X, |\cdot|_n)$;

(*ii*) *For every* $x \in X$ *and every* $\epsilon_n \in [0, \infty)$, *there exists* $y \in F(x)$ *such that*

$$d_n(x, y) \leq d_n(x, F(y)) + \epsilon_n \ \text{ for every } n \in \mathbb{N}.$$

We will make use of the following two theorems in extending the results in Section 6.4.1 to the case where X is a Fréchet space.

Theorem 6.35. ([124]) *Let* $(X, |\cdot|_n)$ *be a Fréchet space and let* $M \subseteq X$ *be a closed subset of* X. *Let* $F : M \to \mathcal{P}_{cl}(M)$ *be an admissible multivalued map. Then there exists* $x \in M$ *such that* $x \in F(x)$.

Theorem 6.36. ([228, Theorem 2.2]) *Let* X *be a metrizable locally convex linear space and* M *be a weakly compact, convex subset of* X. *Suppose* $F : M \to \mathcal{P}_{cl}(M)$ *has weakly sequentially closed graph. Then* F *has a fixed point.*

Our main result in this direction is contained in the following theorem.

Theorem 6.37. *Let* $(X, |\cdot|_n)$ *be a Fréchet space and let* M *be a weakly compact, convex subset of* X, *let* $G : M \to \mathcal{P}_{cl,cv}(X)$ *have a weakly sequentially closed graph, and let* $B \in \mathcal{L}(X)$ *be a contraction operator with respect to a family of semi-norms* $\|\cdot\|_n$ *that are equivalent to the family* $|\cdot|_n$. *Assume that the following conditions are satisfied:*

($\mathcal{S}_1$) $x \in B(x) + G(y)$ *and* $y \in M$ *implies* $x \in M$.

($\mathcal{S}_2$) *For each fixed* $y \in M$, *the multifunction* $x \to B(x) + G(y)$ *is an admissible function.*

Then there exists $x \in M$ *such that*

$$x \in B(x) + G(x).$$

Proof. Let $y \in M$ and consider the multivalued operator $F_y : M \to \mathcal{P}_{cl,cv}(X)$ defined by

$$F_y(x) = B(x) + G(y), \quad x \in M.$$

Let $\{x_m\} \subseteq B(x) + G(y)$ be a sequence converging weakly to x_*; then there exists $y_m \in G(y)$ such that $x_m = B(x) + y_m$ for each $m \in \mathbb{N}$ and so $x_m - B(x) = y_m$ for $m \in \mathbb{N}$. Hence, y_m converges weakly to $y_* = B(x) - x_*$, and since $G(\cdot)$ has closed values, we conclude that

$$x_* \in B(x) + G(y).$$

By ($\mathcal{S}_2$), the multifunction $F_y(\cdot)$ is admissible. Hence, from Theorem 6.35, there exists $x(y) \in M$ such that

$$x(y) \in B(x(y)) + G(y).$$

Since B is a contraction, $(I - B)^{-1}$ exists and is continuous. So define the mapping $N : M \to \mathcal{P}_{wcl,cv}(X)$ by

$$y \to N(y) = (I - B)^{-1} G(y).$$

Since $(I - B)^{-1} \in \mathcal{L}(X)$, $N(\cdot)$ has a weakly sequentially closed graph. From ($\mathcal{S}_1$) it is easy to see that $N(M) \subset M$ and $N : M \to \mathcal{P}_{wcl,cv}(M)$ has a weakly sequentially closed graph. Then, by Theorem 6.36, N has at least one fixed point. $\qquad\square$

6.4.7 Measure of noncompactness and Krasnosel'skii's theorem

In this section, we present a Kranosel'skii type fixed point theorem by using the measure of noncompactness combined with an approximation method. First, we give definitions and main properties of a measure of noncompactness. For more details, we refer the reader to [16, 180, 274] and the references therein.

Definition 6.38. *Let E be a Banach space and $(\mathcal{A}, \leq)$ be a partially ordered set. A map $\beta \colon \mathcal{P}(E) \to \mathcal{A}$ is called a measure of noncompactness (MNC) on E if*

$$\beta(\overline{co}\,\Omega) = \beta(\Omega)$$

for every bounded $\Omega \in \mathcal{P}(E)$.

Notice that if D is dense in Ω, then $\overline{co}\,\Omega = \overline{co}\,D$ and hence $\beta(\Omega) = \beta(D)$.

Definition 6.39. *A measure of noncompactness β is called:*

(a) *Monotone if $\Omega_0, \Omega_1 \in \mathcal{P}(E)$ with $\Omega_0 \subset \Omega_1$ implies $\beta(\Omega_0) \leq \beta(\Omega_1)$;*

(b) *Nonsingular if $\beta(\{a\} \cup \Omega) = \beta(\Omega)$ for every $a \in E$ and $\Omega \in \mathcal{P}(E)$;*

(c) *Invariant with respect to the union of compact sets if $\beta(K \cup \Omega) = \beta(\Omega)$ for every relatively compact set $K \subset E$ and $\Omega \in \mathcal{P}(E)$;*

(d) *Real if $\mathcal{A} = \overline{\mathbb{R}}_+ = [0, \infty]$ and $\beta(\Omega) < \infty$ for every bounded Ω;*

(e) *Semi-additive if $\beta(\Omega_0 \cup \Omega_1) = \max(\beta(\Omega_0), \beta(\Omega_1))$ for every $\Omega_0, \Omega_1 \in \mathcal{P}(E)$;*

(f) *Lower-additive if β is real and $\beta(\Omega_0 + \Omega_1) \leq \beta(\Omega_0) + \beta(\Omega_1)$ for every $\Omega_0, \Omega_1 \in \mathcal{P}(E)$;*

(g) *Regular if the condition $\beta(\Omega) = 0$ is equivalent to the relative compactness of Ω.*

As an example of a MNC, consider the Kuratowski measure defined by $\alpha(A) = \inf D$ where

$$D = \{\varepsilon > 0 : A = \bigcup_{i=1}^{i=n} \Omega_i, \ \mathrm{diam}\,(\Omega_i) \leq \varepsilon \ \text{for all } i = 1, \ldots, n\}.$$

The Hausdorff MNC is defined by

$$\chi(\Omega) = \inf\{\varepsilon > 0 \colon \Omega \ \text{has a finite} \ \varepsilon - net\}.$$

(Recall that a bounded set $A \subset E$ has a finite $\varepsilon-$net if there exist ε and a finite subset $S \subset E$ such that $A \subset S + \varepsilon\overline{B}$, where $\overline{B}$ is a closed unit ball in E.) Other examples of MNCs are given by the following measures of noncompactness defined on the space of continuous functions $C([0, b], E)$ with values in a Banach space E:

(i) the modulus of fiber noncompactness

$$\varphi(\Omega) = \sup_{t \in [0,b]} \chi_E(\Omega(t)),$$

where χ_E is the Hausdorff MNC in E and $\Omega(t) = \{y(t) : \ y \in \Omega\}$;

(ii) the modulus of equicontinuity

$$\mathrm{mod}_C(\Omega) = \lim_{\delta \to 0} \sup_{y \in \Omega} \max_{|\tau_1 - \tau_2| \leq \delta} |y(\tau_1) - y(\tau_2)|.$$

(It should be mentioned that these examples of $MNCs$ satisfy all above-mentioned properties in Definition 6.39 except regularity (g).)

Definition 6.40. *Let X, Y be two normed spaces. The multivalued map $F : X \to \mathcal{P}(Y)$ is called a $k-$set contraction with respect to β if there exists $k \in [0, 1)$ such that for every $D \in \mathcal{P}_b(X)$, we have*

$$\beta(F(D)) \leq k\beta(D).$$

The next result applies to both $\beta-$condensing (see [145, Definition 2.93]) and $k-$contractive *u.s.c.* multivalued maps.

Theorem 6.41. ([16]) *Let V be a bounded closed convex subset of E and let $N : \overline{V} \to V$ be a β-condensing (or $k-$contractive) continuous mapping, where β is a nonsingular measure of noncompactness defined on subsets of E. Then the set of fixed points of N,*

$$Fix(N) = \{x \in V : x \in N(x)\},$$

is nonempty.

Theorem 6.42. ([138, 180]) *Let $V \subset E$ be a bounded open neighborhood of zero and $N : \overline{V} \to \mathcal{P}_{cp,cv}(V)$ be a $k-$contractive u.s.c. multivalued map, where β is a nonsingular measure of noncompactness defined on subsets of E. Then the set*

$$Fix(N) = \{x \in V : x \in N(x)\}$$

is nonempty.

Next, we recall the definition of an expansive mapping.

Definition 6.43. *A mapping $A : D(A) \subseteq X \to X$ is said to be $\phi-$expansive if there exists a function $\phi : [0, \infty) \to [0, \infty)$ such that*

$$|A(x) - A(y)| \geq \phi(|x - y|) \quad \text{for all } x, y \in D(A),$$

and ϕ satisfies:

(i) $\phi(0) = 0$;

(ii) $\phi(r) > 0$ for $r > 0$;

(iii) Either ϕ is continuous or ϕ is nondecreasing.

In what follows, for any mapping Q, we let $\mathcal{R}(Q)$ denote the image of Q. We also need the following lemma.

Lemma 6.44. ([128, Lemma 3.1]) *Let M be a nonempty bounded closed subset of a Banach space X and let $B : M \to X$ be a $\phi-$expansive mapping. Then B is injective and the mapping $B^{-1} : \mathcal{R}(M) \to M$ is uniformly continuous.*

Our first result in this direction is the following.

Theorem 6.45. *Let X be a Banach space, M be a closed bounded convex subset of X, $G : M \to \mathcal{P}_{cp,cv}(X)$ be an u.s.c. multivalued mapping, and $B : M \to X$ be a continuous mapping such that:*

($\mathcal{S}_3$) G is compact;

($\mathcal{S}_4$) B is $k-$contractive;

(S_5) $(I - B)^{-1} : \mathcal{R}(I - B) \to M$ *exists and is uniformly continuous;*

(S_6) $B(M) + G(M) \subseteq M.$

Then there exists $x \in M$ such that

$$x \in B(x) + G(x).$$

Proof. Let $\epsilon > 0$ be given. Then, by Theorem 6.3 on approximate selections, there exists $f_\epsilon : M \to X$ such that

$$\Gamma(f_\epsilon) \subseteq \Gamma(F) + \epsilon B(0, 1),$$

and

$$f_\epsilon(M) \subset co\, G(M).$$

Hence, by (S_6) and the convexity of M, we obtain

$$B(M) + f_\epsilon(M) \subseteq B(M) + co\, G(M) \subseteq co\, (B(M) + G(M)) \subset co\, (M) = M.$$

For fixed $y \in M$, consider the operator $F_\epsilon^y : M \to M$ defined by

$$F_\epsilon^y(x) = B(x) + f_\epsilon(y).$$

Let K be a bounded subset of M; then

$$\beta(F_\epsilon^y(K)) = \beta(B(K)) \le k\beta(K).$$

Since B is continuous and $k-$contractive, Theorem 6.41 ensures the existence of an $x_\epsilon(y) \in M$ such that

$$x_\epsilon(y) = B(x_\epsilon(y)) + f_\epsilon(y)$$

so

$$f_\epsilon(y) = (I - B)(x_\epsilon(y)) \in \mathcal{R}(I - B). \tag{6.15}$$

Hence, from (6.15) and (S_5), we can define the operators $\widetilde{L}_\epsilon : M \to M$ by

$$\widetilde{L}_\epsilon(y) = (I - B)^{-1} f_\epsilon(y) = x_\epsilon(y).$$

Again by (S_5) and the continuity of f_ϵ, we see that L_ϵ is continuous.

Next, we want to show that L_ϵ is compact, so let $K \subset M$ be a bounded set. Then

$$f_\epsilon(K) \subset co\, G(M) \quad \text{implies} \quad \overline{f_\epsilon(K)} \subset \overline{co\, G(M)}.$$

Since G is compact, so is $\overline{f_\epsilon(K)}$. Moreover, by the continuity of $(I - B)^{-1}$, we obtain that $\widetilde{L}_\epsilon$ is compact. Hence, from the Schauder fixed point theorem, there exists $y_\epsilon \in M$ such that

$$y_\epsilon = B(y_\epsilon) + f_\epsilon(y_\epsilon).$$

By the same methods used in the proof of Theorem 6.20, we can prove that $B + G$ has at least one fixed point. $\qquad\square$

Analogous to Theorem 6.23, it is straightforward to prove the following global multivalued version of Krasnosel'skii's fixed point theorem.

Theorem 6.46. *Let X be a Banach space, M be a closed and convex subset of X with $0 \in M$, and $G : M \to \mathcal{P}_{cp,cv}(X)$ be an u.s.c. multivalued operator. In addition to conditions (S_3)–(S_4), assume that*

$(\mathcal{S}_7)$ $(I - B)^{-1} : \mathcal{R}(I - B) \to X$ *exists, is uniformly continuous, and* $B(X)$ *is bounded.*

Then, either

(a) $x \in B(x) + \lambda G(x)$ *has a solution for* $\lambda = 1$,

or

(b) the set $\{x \in X : x \in \lambda B(\frac{x}{\lambda}) + \lambda G(x), \ \lambda \in (0, 1)\}$ *is unbounded.*

6.5 Fixed point theorems for sums of two multivalued operators

We begin this section by establishing some results about fixed points and continuous selections of multivalued maps.

Definition 6.47. *Let* X *and* Y *be two normed spaces. The multivalued map* $F : X \to \mathcal{P}(Y)$ *is weakly lower semicontinuous at* $x \in X$, *(w-l.s.c.), if for every* $\epsilon > 0$ *and every neighborhood* V *of* x *there is a point* $x' \in V$ *such that, for every* $z \in F(x')$, *there is a neighborhood* U_z *of* x *such that*

$$z \in \bigcap \{F(a + \epsilon B(0, 1)) : a \in U_z\},$$

where B_* *is the closed unit ball in* X.

Definition 6.48. *Let* X *be a topological space and* Y *be a normed space. A multifunction* $F : X \to \mathcal{P}(Y)$ *is weakly Hausdorff lower semicontinuous (w-H_d-l.s.c.) at* $x_0 \in X$ *if for every* $\epsilon > 0$ *and every neighborhood* V *of* x_0 *there are a neighborhood* U *of* x_0 *($U \subset V$) and a point* $x' \in U$ *such that*

$$F(x') \subset F(x) + \epsilon S(0, 1) \quad \text{for every } x \in U,$$

where

$$S(0, 1) = \{x \in Y : \|x\| < 1\}.$$

If F *is w-H_d-l.s.c. at each* $x_0 \in X$, *then* F *is called w-H_d-l.s.c.*

Remark 6.49. *If* $x' = x_0$, *we obtain the definition of a Hausdorff lower semi-continuous multifunction* $(H_d - l.s.c.)$. *If* F *is l.s.c., then* F *is* $w - l.s.c.$, *and if* F *is* $w - H_d - l.s.c.$ *then* F *is* $w - l.s.c.$

Definition 6.50. *Let* X *and* Y *be two normed spaces. A multivalued map* $F : X \to \mathcal{P}(Y)$ *is midconvex (resp., convex) if*

$$\frac{1}{2}(F(x) + F(y)) \subset F\left(\frac{x+y}{2}\right) \quad (resp., \lambda F(x) + (1 - \lambda)F(y) \subset F(\lambda x + (1 - \lambda)y)$$

for all $x, y \in X$ *(and* $\lambda \in [0, 1]$).

Lemma 6.51. *([226, pp. 29-30, Lemma 3.1, Remark 3.1]) Let* X *be a real vector space,* $D \subset X$ *be closed and convex, and let* Y *be a topological vector space. If a multifunction* $F : D \to \mathcal{P}_{cl,b}(Y)$ *is midconvex, then*

$$\lambda F(x) + (1 - \lambda)F(y) \subset F(\lambda x + (l - \lambda)y)$$

for all $x, y \in D$ *and all* $\lambda \in [0, 1]$.

We can easily prove the following corollary.

Corollary 6.52. *Let X be a real topological vector space, $D \subset X$ be convex, and let Y be a topological vector space. If a multifunction $F : D \to \mathcal{P}(Y)$ has a closed graph and is midconvex, then*

$$\lambda F(x) + (1 - \lambda)F(y) \subset F(\lambda x + (l - \lambda)y)$$

for all x, $y \in D$ and all rational numbers $\lambda \in [0, 1]$.

We will need the following theorem in the proof of our main result in this section.

Theorem 6.53. *([254]) Let X be a paracompact and perfectly normal topological space and Y be a closed subset of a Banach space Y. Assume that $F : X \times Y \to \mathcal{P}_{cl,cv}(Y)$ is a multivalued map such that*

$$H_d(F(x, y_1), F(x, y_2)) \leq L\|y_1 - y_2\| \quad \text{for each } x \in X \quad \text{and} \quad y_1, y_2 \in Y,$$

and for every $y \in Y$, the multifunction $F(\cdot, y)$ is $w - l.s.c.$ Then there exists a continuous mapping $f : X \times Y \to Y$ such that $f(x, y) \in P_F(x)$ for every $(x, y) \in X \times Y$, where

$$P_F(x) = \{y \in Y : y \in F(x, y)\}.$$

Our first fixed point result in this section is in keeping with the spirit of the preceding results and is related to Theorem 3.2 in [57].

Theorem 6.54. *Let X be a Banach space, $M \subseteq X$ be a closed convex compact subset of X, and $G : M \to \mathcal{P}_{wcl,cv}(X)$ and $B : M \to \mathcal{P}_{cl,cv,b}(X)$ be two multivalued maps satisfying:*

$(\mathcal{K}_1)$ *G is w-l.s.c.;*

$(\mathcal{K}_2)$ *$B(M) + G(M) \subseteq M$;*

$(\mathcal{K}_3)$ *B is a λ-contraction.*

Then $B + G$ has at least one fixed point.

Proof. Let $y \in M$. Then from $(\mathcal{K}_3)$, the multivalued function $F_y : M \to \mathcal{P}(M)$ defined by

$$F_y(x) = B(x) + G(y), \quad x \in M,$$

is a λ-contraction. From the theorem of Covitz and Nadler [86], there exists $x(y) \in M$ such that $x(y) \in F_y(x(y))$. Define the multivalued function

$$\bar{L} : M \times M \to \mathcal{P}_{cl,cv}(M)$$

by

$$\bar{L}(x, y) = G(x) + B(y).$$

By Theorem 6.53, there exists a continuous selection $f : M \times M \to M$ for L satisfying

$$f(x, y) \in B(f(x, y)) + G(x) \quad \text{for } (x, y) \in M \times M.$$

Let $h(y) = Fix(F_y)$ and consider $h_* : M \to M$ defined by $h_*(x) = f(x, x)$. It is clear that h_* is a continuous map and $h_*(x) \in B(f(x, x)) + G(x) \in M$, so $h_*(M) \subset M$. By Schauder's fixed point theorem, there exists $x \in M$ such that $x = h_*(x)$, which is a fixed point of $B + G$. $\qquad\square$

Our second fixed point theorem concerns midconvex maps.

Theorem 6.55. *Let X be a Banach space and let $M \subseteq X$ be a convex weakly compact subset of X. Let $G : M \to \mathcal{P}_{cl,cv}(X)$ have a weakly sequentially closed graph, and $B : M \to \mathcal{P}_{wcp,cv}(X)$ be a multivalued operator. If*

($\mathcal{R}_1$) B is $k-$contractive and midconvex

and

($\mathcal{R}_2$) $B(M) + G(M) \subset M$,

then there exists $x \in M$ such that $x \in B(x) + G(x)$.

Proof. Let $y \in M$ and define the multivalued operator $F_y : M \to \mathcal{P}_{cl}(X)$ by

$$F_y(x) = B(x) + G(y), \quad x \in M.$$

Since $G(y)$ is convex,

$$H_d(F_y(x), F_y(x')) = H_d(B(x), B(x')) \le kd(x, x') \quad \text{for all} \quad x, x' \in M.$$

By the theorem of Covitz and Nadler [86], there exists $x(y) \in M$ such that

$$x(y) \in B(x(y)) + G(y).$$

Consider $L_* : M \to \mathcal{P}(M)$ defined by

$$L_*(y) = Fix(F_y) = \{x \in M : x \in B(x) + G(y)\}.$$

We now show that for every $y \in M$, we have $L_*(y) \in \mathcal{P}_{cl}(M)$.

Let $\{x_n\} \subset L_*(y)$ be a sequence converging weakly to some x. For each $n \in \mathbb{N}$, there exists $h_n \in B(x)$ and $z_n \in G(y)$ such that

$$x_n = h_n + z_n.$$

Since $z_n \in G(y) \subset \overline{G(M)}$, there exists a subsequence of z_n converging weakly to some z. Since G has a closed graph, we obtain that $z \in G(y)$. Also, $x_n - z_n = h_n \in B(x_n)$, $n \in \mathbb{N}$, and it is clear that $x_n - z_n$ converges weakly to $x - z$. Using the fact that $B(\cdot) \in \mathcal{P}_{cv}(X)$ and is H_d-continuous, combined with Mazur's Lemma, we have that $x - z \in B(x)$ implies $x \in B(x) + G(y)$, which in turn implies $x \in L_*(y)$. Hence,

$$L_*(y) \in \mathcal{P}_{cl}(M).$$

Now L_* has a weakly sequentially closed graph, so let $(y_n, x_n) \in \Gamma(L_*)$ be such that y_n, x_n converges weakly to y, x, respectively. Thus, there exists $h_n \in B(x_n)$ and $r_n \in G(y_n)$ such that

$$x_n = h_n + r_n \quad \text{for each } n \in \mathbb{N}.$$

Since $r_n \in G(y_n) \subset \overline{G(M)}$, there exists a subsequence of r_n converging weakly to some r, and since G has a closed graph, we obtain that $r \in G(y)$. From Mazur's Lemma, there exists a subsequence $\widetilde{x}_n \in co\{x_n : n \in \mathbb{N}\}$ converging strongly to x, and $\widetilde{y}_n \in co\{y_n : n \in \mathbb{N}\}$ converging strongly to y. Since B is midconvex, $\widetilde{y}_n \in B(\widetilde{x}_n)$. Using the fact that $B(\cdot) \in \mathcal{P}_{cl,cv}(X)$ and is H_d-continuous, we see that B has closed graph. Then, by Mazur's Lemma, we have $x \in B(x) + G(y)$. Therefore, the operator $L_* : M \to \mathcal{P}_{cl,cv}(M)$ has a sequentially closed graph. Thus, L_* satisfies the hypotheses of Theorem 6.36, and so there exists $x \in M$ such that $x \in L_*(x)$. This implies that $x \in B(x) + G(x)$ and completes the proof of the theorem. $\square$

6.6   Kakutani fixed point theorem type in topological vector spaces

Definition 6.56. *Let X and Y be two topological vector spaces. A multifunction $F : X \to \mathcal{P}(Y)$ is said to be convex u.s.c. at $x \in X$ if for any open neighborhood V such that $F(x) \subset V$ there exists an open neighborhood W_x of x such that $co(F(W_x)) \subset V$, where $co(F(W_x))$ represents the convex hull of $F(W_x)$.*

Proposition 6.57. *Let X and Y be two Hausdorff topological vector spaces. If a multifunction $F : X \to \mathcal{P}_{cp}(Y)$ is convex u.s.c., then F is closed.*

Proof. Let $\{x_\alpha\}_{\alpha \in J}$ and $\{y_\alpha\}_{\alpha \in J}$ be two generalized sequences, for which there are $x \in X$ and $y \in Y$, with $x_\alpha \to x$, $y_\alpha \to y$, and $y_\alpha \in F(x_\alpha)$ where J is a directed set. We show that $y \in F(x)$. Assume $y \notin F(x)$, then for each $z \in F(x)$ there exist neighborhoods $W_z(y)$ of y and $V_y(z)$ of z such that $W(y) \cap V_y(z) = \emptyset$. By virtue of the compactness of $F(x)$, we can find its finite covering $V_y(z_1), V_y(z_2), \ldots, V_y(z_n)$. We consider the open neighborhood of y defined by

$$W(y) = \cap_{i=1}^n W_{z_i}(y) \Longrightarrow W(y) \bigcap \bigcup_{i=1}^n V_y(z_i) = \emptyset.$$

Since F is convex u.s.c., then there exists $V(x)$ an open neighborhood of x such that

$$coF(V_x) \subseteq \bigcup_{i=1}^n V_y(z_i) \Longrightarrow F(V_x) \subseteq \bigcup_{i=1}^n V_y(z_i).$$

From $x_\alpha \to x$, $y \in Y$, $y_\alpha \to y$, there exists $\alpha_0 \in J$ such that

$$x_\alpha \in V_x \text{ and } y_\alpha \in F(V_x) \subseteq \bigcup_{i=1}^n V_y(z_i), \text{ for every } \alpha \in J, \ \alpha \geq \alpha_0,$$

which is a contradiction with $W(y) \cap \bigcup_{i=1}^n V_y(z_i) = \emptyset$. $\qquad\square$

Lemma 6.58. *Let X be a Hausdorff topological vector space, $G : X \to \mathcal{P}(X)$ be a convex u.s.c. multivalued map and $B : X \to X$ be a linear continuous mapping. Then $B \circ G$ is convex u.s.c.*

Proof. Let $x_0 \in E$ and let V be an open neighborhood of 0 such that

$$(B \circ G)(x_0) \in (B \circ G)(x_0) + V.$$

Since B is continuous, then $B^{-1}(V)$ is neighborhood of 0, we have

$$G(x_0) \in G(x_0) + B^{-1}(V).$$

Then there exists W an open neighborhood of 0 such that

$$co(G(x_0 + W)) \subset G(x_0) + B^{-1}(V) \Rightarrow co(B(G(x_0 + W))) \subset B(G(x_0)) + V.$$

This implies that $B \circ G$ is convex u.s.c. $\qquad\square$

When X is a locally convex space, it turns out that these two concepts are equivalent.

Proposition 6.59. *Let X be a locally convex space. Then $F : X \to \mathcal{P}_{cv}(X)$ is convex u.s.c. if and only if F is u.s.c.*

Definition 6.60. *Let X be a topological vector space. A mapping $A : X \to X$ is said to be convex continuous at $x_0 \in X$ if for any open neighborhood $V(A(x_0))$ of $A(x_0)$, there exists an open neighborhood $V(x_0)$ of x_0 such that $co(A(V(x_0))) \subset V(A(x_0))$.*

Lemma 6.61. *Let X be a Hausdorff topological vector space, $A : X \to X$ be a convex continuous operator and $B : X \to X$ be a linear continuous mapping. Then $B \circ A$ is convex continuous.*

Proof. Let $x_0 \in E$ and let V be an open neighborhood of 0 such that

$$(B \circ A)(x_0) = B(A(x_0)) \in B(A(x_0)) + V, \quad A(x_0) \in A(x_0) + B^{-1}(V).$$

B is continuous, then $B^{-1}(V)$ is neighborhood of 0, we have

$$A(x_0) \in A(x_0) + B^{-1}(V).$$

Then there exists W an open neighborhood of 0 such that

$$co(A(x_0 + W) \subset A(x_0) + B^{-1}(V) \Rightarrow co(B(A(x_0 + W))) \subset B(A(x_0)) + V.$$

This implies that $B \circ A$ is convex continuous. $\qquad\qquad\square$

In the following, we assume that X is a Hausdorff topological vector space with property (W); that is,

X **has property** (W) iff X has a local base $\{W_i\}_{i \in I}$ of 0, where I is an index set with a partial order '$\subset$', such that $W_i \subset W_j$ if $i < j$ and for any $i_1, i_2, \ldots, i_k$, there is an $l \in \{1, 2, \ldots, k\}$, such that $i_l = \max\{i_1, i_2, \ldots, i_k\}$ and $W_{i_j} \subset W_{i_l}$ for $j = 1, 2; \ldots, k, j \neq l$. For example, if X is first countable (equivalently, metrizable), then such a base exists. The following result is due to Chen [82].

Theorem 6.62. *Let X be a Hausdorff topological vector space with property (W), and $C \subset X$ a convex compact subset. Suppose $A : C \to C$ is a convex continuous mapping. Then A has a fixed point in C.*

Here is our first main result in this section.

Theorem 6.63. *Let X be a Hausdorff topological vector space with property (W), and let $M \subset X$ be a convex compact subset. Suppose $F : M \to \mathcal{P}_{cv,cp}(M)$ is convex u.s.c. Then F has a fixed point in M.*

Proof. Let $\{W_i : i \in I\}$ be the local base of 0 such that property (W) holds. We assume that $\{W_i : i \in I\}$ is a symmetric base. (Otherwise, we put $\{W_i' = W_i \cap (-W_i) : i \in I\}$). For any open V of 0 and $x \in M$, by the convex u.s.c. of F, there exists $W_{i_x} \in \{W_i : i \in I\}$ such that

$$W_{i_x} \subset V \quad \text{and} \quad co(F(x + W_{i_x})) \subset F(x) + V.$$

Since X is a topological vector space, then there exists $W_{i_x}' \in \{W_i : i \in I\}$ such that $W_{i_x}' + W_{i_x}' \subset W_{i_x}$. It is clear that $M \subset \bigcup_{x \in M}(x + W_{i_x}')$, so there exists $n \in \mathbb{N}$ such that

$$M \subset \bigcup_{j=1}^{n}(x_j + W_{i_{x_j}}').$$

Let $\{\phi_i\}_{i=1}^{n}$ be a continuous partition of unity subordinate to the covering of

$$\{x_j + W_{i_{x_j}}' : j = 1, \ldots, n\}, \quad \sum_{j=1}^{n} \phi_j(x) = 1 \quad \text{for all } x \in M.$$

For each $j \in \{1, \ldots, k\}$, we choose $z_j \in F(x_j)$ and define $F_V : M \to M$ by

$$F_V(x) = \sum_{j=1}^{n} \phi_j(x)z_j, \ x \in M.$$

Let $X_* = span\{z_1, \ldots, z_n\}$ and $\widetilde{M} = \{\sum_{i=1}^{n} \lambda_j z_j : 0 \leq \lambda_j \leq 1\}$. It is clear that X_* is a vector space with finite dimension, and $\widetilde{M}$ is a convex closed subset of X_*. Also $F_V : \widetilde{M} \to \widetilde{M}$ is a continuous function. By the Brouwer fixed point theorem there exists $x_v \in M$ such that $x_v = F_V(x_v)$. Now we show that there exists $z_v \in M$ such that $x_v - z_v \in V$ and $x_v - F(z_v) \subseteq V$. Since $x_v \in M$ then there exists $j \in \{1, \ldots, n\}$ such that $x_v \in x_j + W'_{i_x}$. We assume that $\phi_i(x_v) \neq 0$, $i = 1, \ldots, n$; then

$$
\begin{aligned}
x_j &\in & x_v + W'_{i_{x_j}} &\subseteq & x_{j_0} + W'_{i_{x_{j_0}}} + W'_{i_{x_j}} \\
&\in & x_{j_0} + W'_{i_{x_j}} &\subseteq & x_{j_0} + W_{i_{x_{j_0}}},
\end{aligned}
$$

where $i_{x_{j_0}} = \max\{i_{x_1}, \ldots, i_{x_n}\}$ and $x_v \in x_{i_{x_i}} + W'_{i_x}$, $i = 1, \ldots, n$. Hence

$$F_V(x) = \sum_{j=1}^{n} \phi_j(x)z_j \in coF(x_{j_0} + W_{i_{x_{j_0}}}) \subseteq F(x_{j_0}) + V.$$

Let $z_v = x_{j_0}$. Thus $x_v - z_v \in V$ and $x_v - F(z_v) \subseteq V$. Let $\mathcal{V}$ be a neighborhood base of 0. Using the fact that M is compact, $\{x_v, V \in \mathcal{V}\}$ has subnet $\{x'_v, V \in \mathcal{V}\}$ converging to some $x \in M$. Then $\{z'_v : V \in \mathcal{V}\}$ converges to x. By Proposition 6.57 we have $x \in F(x)$. $\qquad \square$

Definition 6.64. *Let X and Y be two Hausdorff topological vector spaces. A multivalued mapping $F : X \to \mathcal{P}(Y)$ has a convex continuous approximation mapping if for every $V \in \mathcal{V}$ there exists a convex continuous map $f_V : X \to Y$ such that*

$$F(x) - f_V(x) \in V \ for \ every \ \ x \in X.$$

Theorem 6.65. *Let X be a Hausdorff topological space satisfying property (W), $M \subseteq X$ be a compact convex subset and $F : M \to \mathcal{P}_{cp,cv}(M)$ be a u.s.c. multivalued operator which has a continuous approximation mapping. Then there exists $x \in M$ such that $x \in F(x)$.*

Proof. Let $V \in \mathcal{V}$. Then there exists $f_V : M \to M$, a convex continuous map, such that

$$F(x) - f_V(x) \in V \ for \ every \ \ x \in M.$$

From Theorem 6.62 there exists $x_v \in C$ such that $x_v = f_V(x_v)$, and thus $F(x_v) - x_v \in V$. Thus there exists $y_v \in F(x_v)$ such that

$$y_v - x_v \in V. \tag{6.16}$$

Since M is compact, there exist subnets $\{x_{V'} : V \in \mathcal{V}\}$ of $\{x_V : V' \in \mathcal{V}\}$ and $\{y_{V'} : V' \in \mathcal{V}\}$ of $\{y_V : V \in \mathcal{V}\}$ such that $x_{V'}, y_{V'}$ converges to $x, y \in M$, respectively. It is clear that F is closed, and then $y \in F(x)$. From (6.16) we have $y = x$. Hence $x \in F(x)$. $\qquad \square$

6.7 Krasnosel'skii-type fixed point theorem in topological vector spaces

Theorem 6.66. *Let X be a Hausdorff topological vector space with property (W), and $M \subset X$ be a symmetric convex compact subset. Assume that $A : M \to X$ is a convex continuous operator and $B : M \to X$ is a linear continuous mapping which satisfy*

$(\mathcal{HL}_1)$ *for each* $x \in M$ *and* $y \in X$ *such that*

$$x = B(x) + y \Rightarrow y \in M,$$

and for every $y \in M$ *there exists unique* $x(y) \in M$ *such that*

$$x(y) = B(x(y)) + y.$$

Then there exists $y \in M$ *such that* $y = B(y) + A(y)$; *that is,* $A + B$ *has a fixed point in* M.

Proof. Using the first part of $(\mathcal{HL}_1)$, the operator $F : M \to M$, given by $F(x) = x - B(x)$, is well-defined. We show that the operator F is bijective. In fact, by the second part of $(\mathcal{HL}_1)$, F is surjective. That is, for every $y \in M$, there exists $x(y) \in M$ such that $F(x(y)) = y$. Next, let $x_1, x_2 \in M$, such that

$$F(x_1) = F(x_2) \Rightarrow -B(x_1) + x_1 = -B(x_2) + x_2.$$

Therefore

$$0 + B(x_2 - x_1) = x_2 - x_1.$$

Since M is symmetric, by $(\mathcal{HL}_1)$ we deduce that $x_1 = x_2$. Hence $(I - B)^{-1}$ exists. Now, we show that $(I - B)^{-1}$ is continuous. Let $\{x_\alpha\}_{\alpha \in J} \subset M$ be a net such that

$$\lim_{\alpha \in J} x_\alpha = x \text{ and } x_\alpha = y_\alpha - B(y_\alpha) \in M.$$

Since the set M is compact, there exists a convergent subnet $\{y_{\alpha_\beta}\}$ of the net $\{y_\alpha\}$. Thus

$$\lim_{\beta \in J} x_{\alpha_\beta} = \lim_{\beta \in J} y_{\alpha_\beta} - \lim_{\beta \in J} B(y_{\alpha_\beta}) \Rightarrow x = y - B(y).$$

Let us define $N : M \to M$ by

$$y \to N(y) = (I - B)^{-1} A(y).$$

Since A is a convex continuous map and $(I - B)^{-1}$ is a linear continuous map, from Lemma 6.58, N is a convex continuous operator. By Theorem 6.62 there exists $y \in M$ such that

$$y = B(y) + A(y).$$

$\square$

We shall prove, using Theorem 6.63, a fixed point Theorem for the multivalued mapping $B + G$, where B is a singlevalued mapping and G is a multivalued mapping.

Theorem 6.67. *Let* X *be a Hausdorff topological vector space with property (W), and* $M \subset X$ *be a symmetric convex compact subset. Assume that* $G : M \to \mathcal{P}_{cv,cp}(X)$ *is convex u.s.c. multivalued operator and* $B : X \to X$ *is a linear continuous mapping which satisfy*

$(\mathcal{HL}_2)$ *for each* $x \in X$ *and* $y \in M$ *such that*

$$x \in B(x) + G(y) \Rightarrow x \in M,$$

and for every $y \in M$ *there exists a unique* $x(y) \in M$ *such that*

$$x(y) = B(x(y)) + y.$$

Then there exists $y \in M$ *such that* $y \in B(y) + G(y)$.

Proof. Using the condition $(\mathcal{HL}_2)$, the operator $F : M \to \mathcal{P}_{cv,cp}(M)$ given by

$$F(y) = (I - B)^{-1} G(y)$$

is well-defined. Moreover, by Lemma 6.58, F is a convex *u.s.c.*, since M is convex compact; then from Theorem 6.63 there exists $y \in M$ such that

$$y \in F(y) \Rightarrow y \in B(y) + G(y).$$

$\square$

Chapter 7

Generalized Metric and Banach Spaces

In 1905, the French mathematician Maurice Fréchet [121, 122] introduced the concept of metric spaces, although the name "metric" is due to Hausdorff [102,157]. In 1934, the Serbian mathematician Duro Kurepa [196], a PhD student of Fréchet, introduced metric spaces in which an ordered vector space is used as the codomain of a metric instead of the set of real numbers. In the literature the metric spaces with vector valued metrics are known under various names such as pseudometric spaces, k-metric spaces, generalized metric spaces, cone-valued metric spaces, cone metric spaces, abstract metric spaces and vector valued metric spaces. Fixed point theory in K-metric spaces was developed by A.I.Perov in 1964 [237]. For more details on fixed point theory in K-metric and K-normed spaces, we refer the reader to [291].

7.1 Generalized metric space

In this section we define generalized metric space (or vector metric spaces) and prove some properties. If, $x, y \in \mathbb{R}^n$, $x = (x_1, \ldots, x_n)$, $y = (y_1, \ldots, y_n)$, by $x \leq y$ we mean $x_i \leq y_i$ for all $i = 1, \ldots, n$. Also $|x| = (|x_1|, \ldots, |x_n|)$ and $\max(x, y) = \max(\max(x_1, y_1), \ldots, \max(x_n, y_n))$. If $c \in \mathbb{R}$, then $x \leq c$ means $x_i \leq c$ for each $i = 1, \ldots, n$. For $x \in \mathbb{R}^n$, $(x)_i = x_i$, $i = 1, \ldots, n$.

Definition 7.1. *Let X be a nonempty set. By a generalized metric on X (or vector-valued metric) we mean a map $d : X \times X \to \mathbb{R}^n$ with the following properties:*

(i) $d(u, v) \geq 0$ for all $u, v \in X$; if $d(u, v) = 0$ then $u = v$.

(ii) $d(u, v) = d(v, u)$ for all $u, v \in X$.

(iii) $d(u, v) \leq d(u, w) + d(w, v)$ for all $u, v, w \in X$.

Note that for any $i \in \{1, \ldots, n\}$ $(d(u, v))_i = d_i(u, v)$ is a metric space in X.

We call the pair (X, d) a generalized metric space. For $r = (r_1, r_2, \ldots, r_n) \in \mathbb{R}^n_+$, we will denote by

$$B(x_0, r) = \{x \in X : d(x_0, x) < r\}$$

the open ball centered at x_0 with radius r and

$$\overline{B(x_0, r)} = \{x \in X : d(x_0, x) \leq r\}$$

the closed ball centered at x_0 with radius $r = (r_1, \ldots, r_n) > 0$, $r_i > 0$, $i = 1, \ldots, n$.

Definition 7.2. *Let (X, d) be a generalized metric space. A subset $A \subseteq X$ is called open if, for any $x_0 \in A$, there exists $r \in \mathbb{R}^n_+$ with $r > 0$ such that*

$$B(x_0, r) \subseteq A.$$

Any open ball is an open set and the collection of all open balls of X generates the generalized metric topology on X.

Definition 7.3. *Let (X, d) be a generalized metric space*

(a) *A sequence (x_p) in X converges (or $\mathbb{R}_+^n$-converges) to some $x \in X$, if for every $\epsilon \in \mathbb{R}_+^n$, $\epsilon > 0$ there exists $p_0(\epsilon) \in \mathbb{N}$ such that for each*

$$d(x_p, x) \leq \epsilon \quad \text{for all} \quad p \geq p_0(\epsilon).$$

(b) *A sequence (x_p) is called a Cauchy sequence if for every $\epsilon \in \mathbb{R}_+^n$, $\epsilon > 0$ there exists $p_0(\epsilon) \in \mathbb{N}$ such that for each*

$$d(x_p, x_q) \leq \epsilon \quad \text{for all} \quad p, q \geq p_0(\epsilon).$$

(c) *A generalized metric space X is called complete if each Cauchy sequence in X converges to a limit in X.*

(d) *A subset Y of a generalized metric space X is said to be closed whenever $(x_p) \subseteq Y$ and $x_p \to x$, as $p \to \infty$ imply $x \in Y$.*

Using the above definitions, we have the following properties: If $x_p \to x$ as $p \to \infty$, then

(i) The limit x is unique.

(ii) Every subsequence of (x_p) converges to x.

(iii) If also $x_p \to x$ as $p \to \infty$, then

$$d(x_p, y_p) \to d(x, y) \text{ as } p \to \infty.$$

Theorem 7.4. *For the generalized metric space (X, d) the following hold:*

(a) *Every convergent sequence is an Cauchy sequence,*

(b) *Every Cauchy sequence is bounded,*

(c) *If a Cauchy sequence (x_p) has a subsequence (x_{p_k}) such that*

$$x_{p_k} \to x \text{ as } p_k \to \infty,$$

then

$$x_p \to x \text{ as } p \to \infty.$$

Proof. (a) Let $(x_p)_{p \in \mathbb{N}}$ be a convergent sequence in X. The for every $\epsilon \in \mathbb{R}_+^n$ there exists $p_0(\epsilon) \in \mathbb{N}$ such that

$$d(x_p, x) \leq \frac{\epsilon}{2} \quad \text{for all } p \geq p_0(\epsilon).$$

Then for every $p, q \geq p_0(\epsilon)$ we have

$$d(x_p, x_q) \leq d(x_p, x) + d(x_q, x) \Rightarrow d(x_p, x_q) \leq \epsilon.$$

Hence $(x_p)_{p \in \mathbb{N}}$ is a Cauchy sequence in X.

(b) Let $(x_p)_{p \in \mathbb{N}}$ be a Cauchy sequence. Fix $\epsilon \in \mathbb{R}_+^n$. There exists $p_0(\epsilon) \in \mathbb{N}$ such that

$$d(x_p, x_q) \leq \epsilon, \quad \text{for all } p, q \geq p_0(\epsilon).$$

Hence for each $p \in \mathbb{N}$, we get

$$x_p \in B(x_{p_0(\epsilon)}, \epsilon + r), \quad r = \max_{1 \leq i,j \leq p_0(\epsilon)-1} d(x_i, x_j),$$

this implies that $(x_p)_{p \in \mathbb{N}}$ bounded in X.

(c) Let $(x_p)_{p \in \mathbb{N}}$ be a Cauchy sequence and let $(x_{p_k})_{p_k \in \mathbb{N}}$ be a subsequence of $(x_p)_{p \in \mathbb{N}}$ such that $\lim_{p_k \to \infty} x_{p_k} = x$. The for every $\epsilon \in \mathbb{R}_+^n$ there exist $p_*(\epsilon), q_*(\epsilon) \in \mathbb{N}$ such that

$$d(x_p, x_q) \leq \frac{\epsilon}{2} \quad \text{for all } p, q \geq p_*(\epsilon)$$

and

$$d(x_{p_k}, x) \leq \frac{\epsilon}{2} \quad \text{for all } p_k \geq q_*(\epsilon).$$

Then

$$d(x_p, x) \leq d(x_p, x_{p_k}) + d(x_{p_k}, x) \leq \epsilon \quad \text{for all } p \geq \max(q_*(\epsilon), p_*(\epsilon)).$$

Hence

$$x_p \to x \text{ as } p \to \infty.$$

$\square$

Definition 7.5. *Let (X, d) and (Y, ρ) be generalized metric spaces, and let $x \in X$.*

(a) *A function $f : X \to Y$ is said to be continuous (or topologically continuous) at x if for every $\epsilon \in \mathbb{R}_+^n, \epsilon > 0$ there exists some $\delta(\epsilon) \in \mathbb{R}_+^n, \delta(\epsilon) > 0$ such that*

$$\rho(f(x), f(y)) < \epsilon$$

whenever $x, y \in X$ and $d(x, y) < \delta(\epsilon)$.

The function f is said to be topologically continuous if it is topologically continuous at each point of X.

Definition 7.6. *Let (X, d) be a generalized metric space. We say that a subset $Y \subset X$ is closed if, $(x_p) \subset Y$ and $x_p \to x$ as $p \to \infty$ imply $x \in Y$.*

Definition 7.7. *Let (X, d) be a generalized metric space. A subset C of X is called compact if every open cover of C has a finite subcover. A subset C of X is sequentially compact if, every sequence in C contains a convergent subsequence with limit in C.*

Definition 7.8. *A subset C of X is totally bounded if, for each $\epsilon \mathbb{R}_+^n$ with $\epsilon > 0$, there exists a finite number of elements $x_1, x_2, \ldots, x_p \in X$ such that*

$$C \subseteq \cup_{i=1}^p B(x_i, \epsilon).$$

The set $x_1, \ldots, x_p$ is called a finite ϵ-net.

Theorem 7.9. *If C is a subset of X, then the following affirmations hold:*

i) *C is compact if and only if, C is sequentially compact if and only if, C is closed and totally bounded;*

ii) *C relatively compact, if and only if, C sequentially relatively compact, if and only if, C totally bounded.*

Definition 7.10. *Let (X, d) be a generalized metric space. If $A \subset X$ is a nonempty set, then the function*

$$\delta(A) = \sup\{d(x, y) : x; y \in A\}$$

is called the diameter of A. If $\delta(A) < \infty$, then A is called a bounded set.

Theorem 7.11. *Let (X, d) be a generalized metric space. For any compact set $A \subset X$ and for any closed set $B \subset X$ that is disjoint from A, there exist continuous functions $f : X \to [0, 1]$, $g : X \to [0, 1] \times [0, 1] \times \cdots \times [0, 1] := [0, 1]^n$ such that*

i) $f(x) = 0$ for all $x \in B$,

ii) $f(x) = 1$ for all $x \in A$,

iii) $g(x) = (1, \ldots, 1)$ for all $x \in B$,

iv) $g(x) = (0, \ldots, 0)$ for all $x \in A$.

Proof. Note that $d_i(x, B) = 0$ for any $x \in B$ and $d_i(x, A) = 0$ and $d_i(x, A) > 0$ for any $x \in A$. Thus we obtain i) and ii). Let $f : X \to [0, 1]$ be defined by

$$f(x) = \frac{\displaystyle\sum_{i=1}^{n} d_i(x, B)}{\displaystyle\sum_{i=1}^{n} d_i(x, A) + \sum_{i=1}^{n} d_i(x, B)}, \quad x \in X.$$

To prove that f is continuous, let $(x_m)_{m \in \mathbb{N}}$ be a sequence convergent to $x \in X$. Then

$|f(x_m) - f(x)|$

$$= \left| \frac{\displaystyle\sum_{i=1}^{n} d_i(x_m, B)}{\displaystyle\sum_{i=1}^{n} d_i(x_m, A) + \sum_{i=1}^{n} d_i(x_m, B)} - \frac{\displaystyle\sum_{i=1}^{n} d_i(x, B)}{\displaystyle\sum_{i=1}^{n} d_i(x, A) + \sum_{i=1}^{n} d_i(x, B)} \right|$$

$$= \left| \frac{\displaystyle\sum_{i=1}^{n} d_i(x_m, B) \sum_{i=1}^{n} d_i(x, A) - \sum_{i=1}^{n} d_i(x_m, A) \sum_{i=1}^{n} d_i(x, B)}{(\displaystyle\sum_{i=1}^{n} d_i(x, A) + \sum_{i=1}^{n} d_i(x, B))(\sum_{i=1}^{n} d_i(x_m, A) + \sum_{i=1}^{n} d_i(x_m, B))} \right|$$

$$\leq \frac{\displaystyle\sum_{i=1}^{n} d_i(x, A) \sum_{i=1}^{n} |d_i(x_m, B) - d_i(x, B)|}{(\displaystyle\sum_{i=1}^{n} d_i(x, A) + \sum_{i=1}^{n} d_i(x, B))(\sum_{i=1}^{n} d_i(x_m, A) + \sum_{i=1}^{n} d_i(x_m, B))}$$

$$+ \frac{\displaystyle\sum_{i=1}^{n} d_i(x, B) \sum_{i=1}^{n} |d_i(x_m, A) - d_i(x, A)|}{(\displaystyle\sum_{i=1}^{n} d_i(x, A) + \sum_{i=1}^{n} d_i(x, B))(\sum_{i=1}^{n} d_i(x_m, A) + \sum_{i=1}^{n} d_i(x_m, B))}.$$

Since for each $i = 1, \ldots, m$, we have

$$|d_i(x_m, B) - d_i(x, B)| \to 0, \ |d_i(x_m, A) - d_i(x, A)| \to 0 \text{ as } m \to \infty.$$

Therefore, as $m \to \infty$,

$$\frac{\sum_{i=1}^{n} d_i(x, A) \sum_{i=1}^{n} |d_i(x_m, B) - d_i(x, B)|}{(\sum_{i=1}^{n} d_i(x, A) + \sum_{i=1}^{n} d_i(x, B))(\sum_{i=1}^{n} d_i(x_m, A) + \sum_{i=1}^{n} d_i(x_m, B))} \to 0,$$

and

$$\frac{\sum_{i=1}^{n} d_i(x, A) \sum_{i=1}^{n} |d_i(x_m, A) - d_i(x, A)|}{(\sum_{i=1}^{n} d_i(x, A) + \sum_{i=1}^{n} d_i(x, B))(\sum_{i=1}^{n} d_i(x_m, A) + \sum_{i=1}^{n} d_i(x_m, B))} \to 0.$$

Thus, we get

$$|f(x_m) - f(x)| \to 0 \text{ as } m \to \infty.$$

We can easily prove that the following function $g : X \to [0, 1]^n$ defined by

$$g(x) = \begin{pmatrix} \frac{d_1(x, A)}{d_1(x, B) + d_1(x, B)} \\ \ldots \\ \frac{d_n(x, A)}{d_n(x, B) + d_n(x, B)} \end{pmatrix}, \quad x \in X$$

is continuous and satisfies iii) and iv). □

Let (X, d) be a generalized metric space. We define the following metric spaces: Let $X_i = X$, $i = 1, \ldots, n$. Consider $\prod_{i=1}^{n} X_i$ with $\bar{d}$ defined by

$$\bar{d}((x_1, \ldots, x_n), (y_1, \ldots, y_n)) = \sum_{i=1}^{n} d_i(x_i, y_i).$$

The diagonal space of $\prod_{i=1}^{n} X_i$ is defined by

$$\widetilde{X} = \{(x, \ldots, x) \in \prod_{i=1}^{n} X_i : x \in X, \ i = 1, \ldots, n\},$$

which is a metric space with the following distance

$$d_*((x, \ldots, x), (y, \ldots, y)) = \sum_{i=1}^{n} d_i(x, y), \text{ for each } x, y \in X.$$

It is clear that $\widetilde{X}$ is closed set in $\prod_{i=1}^{n} X_i$.

Intuitively, X and $\widetilde{X}$ are the same. This is shown in the following result.

Lemma 7.12. *Let (X, d) be a generalized metric space. Then there exists a homeomorphism map $h : X \to \widetilde{X}$.*

Proof. Consider $h : X \to \widetilde{X}$ defined by

$$h(x) = (x, \ldots, x) \quad \text{for all } x \in X.$$

Obviously h is bijective.

- To prove that h is a continuous map, let $x, y \in X$. Thus

$$d_*(h(x), h(y)) \leq \sum_{i=1}^{n} d_i(x, y).$$

For $\epsilon > 0$ we take $\delta = (\frac{\epsilon}{n}, \ldots, \frac{\epsilon}{n})$, let $x_0 \in X$ be fixed and $B(x_0, \delta) = \{x \in X : d(x_0, x) < \delta\}$. Then for every $x \in B(x_0, \delta)$ we have

$$d_*(h(x_0), h(x)) \leq \epsilon.$$

- Now, $h^{-1} : \widetilde{X} \to X$ is a map defined by

$$h^{-1}(x, \ldots, x) = x, \qquad (x, \ldots, x) \in \widetilde{X}.$$

To show that h^{-1} is continuous, let $(x, \ldots, x), (y, \ldots, y) \in \widetilde{X}$. Then

$$d(h^{-1}(x, \ldots, x), h^{-1}(y, \ldots, y)) = d(x, y).$$

Let $\epsilon = (\epsilon_1, \ldots, \epsilon_n) > 0$. We take $\delta = \frac{1}{n} \left(\min_{1 \leq i \leq n} \epsilon_i \right)$ and we fix $(x_0, \ldots, x_0) \in \widetilde{X}$. Set

$$B((x_0, \ldots, x_0), \delta) = \{(x, \ldots, x) \in \widetilde{X} : d_*((x_0, \ldots, x_0), (x, \ldots, x)) < \delta\}.$$

For $(x, \ldots, x) \in B((x_0, \ldots, x_0), \delta)$ we have

$$d_*((x_0, \ldots, x_0), (x, \ldots, x)) < \delta \Rightarrow \sum_{i=1}^{n} d_i(x_0, x) < \frac{1}{n} \left(\min_{1 \leq i \leq n} \epsilon_i \right).$$

Then

$$d_i(x_0, x) < \frac{1}{n} \left(\min_{1 \leq i \leq n} \epsilon_i \right), \quad i = 1, \ldots, n \Rightarrow d(x_0, x) < \epsilon.$$

Hence h^{-1} is continuous. $\qquad \square$

Theorem 7.13. *Every generalized metric space is paracompact.*

Proof. Let X be a generalized metric space. By Lemma 7.12 there exists $\widetilde{X}$, a metric space which is homeomorphic to X. Since every metric space is paracompact hence X is paracompact. $\qquad \square$

Theorem 7.14. *Let (X, d) be a generalized metric space. To any locally finite open covering $(U_i)_{i \in I}$ of X, we can associate a locally Lipschitzian partition of unity subordinated to it.*

Proof. From Theorem 7.13, X is paracompact, then there exists a family of locally finite open set, let us write,

$$\mathcal{V} = \{V_i | i \in I_*\},$$

covering of X such that

$$\overline{V}_i \subset U_i \quad \text{for every } i \in I_*.$$

Let us define for any $i \in I$ the function $f_i : X \to \mathbb{R}_+$ by

$$f_i(x) = \sum_{j=1}^{n} d_j(x, X \backslash V_i).$$

For each $x, y \in X$ we have

$$\left| \sum_{j=1}^{n} d_j(x, X \backslash V_i) - \sum_{j=1}^{n} d_j(y, X \backslash V_i) \right| \leq \sum_{j=1}^{n} d_j(x, y) \text{ for each } x, y \in X.$$

Hence

$$\left| \sum_{j=1}^{n} d_j(x, X \backslash V_i) - \sum_{j=1}^{n} d_j(y, X \backslash V_i) \right| \leq A d(x, y) \text{ for each } x, y \in X,$$

where $A = (1, \ldots, 1) \in \mathcal{M}_{1 \times n}(\mathbb{R}_+)$. Then for every $i \in I_*$, f_i is Lipschitzian and verifies

$$supp(f_i) = \overline{V}_i \subset U_i.$$

Let us introduce for any $i \in I_*$ the following function $\psi_i : X \to [0, 1]$ defined by

$$\psi_i(x) = \frac{f_i(x)}{\displaystyle\sum_{i \in I_*} f_i(x)} \quad \text{for all } x \in X.$$

a) Firstly, we prove that ψ_i is locally Lipschitz on X. Indeed, let $x \in X$, then there exists a neighborhood V_x of x which meets only a finite number of $\{\overline{V}_i | i \in I_*\}$. That is there is $\{i_1, \ldots, i_m\}$ such that

$$V_x \cap V_i = \emptyset \text{ for each } i \in I_* \backslash \{i_1, \ldots, i_p\} \Rightarrow \sum_{i \in I_*} f_i(y) = \sum_{k=1}^{p} f_{i_k}(y) > 0, \ y \in V_x.$$

By the continuity of $\displaystyle\sum_{k=1}^{p} f_{i_k}$ there exists a neighborhood $W_x \subset V_x$ of x and $m, \overline{M} > 0$ such that

$$m \leq \sum_{i \in I} f_i(y) = \sum_{k=1}^{p} f_{i_k}(y) \leq \overline{M} \quad \text{for any } y \in W_x.$$

Thus for $y, z \in W_x$, we get

$$
\begin{aligned}
|\psi_i(z) - \psi_i(y)| &= \left| \frac{f_i(y)}{\displaystyle\sum_{i \in I} f_i(y)} - \frac{f_i(z)}{\displaystyle\sum_{i \in I} f_i(z)} \right| \\
&= \left| \frac{\displaystyle\sum_{k=1}^{p} f_{i_k}(z) f_i(y) - \sum_{k=1}^{p} f_{i_k}(y) f_i(z)}{\displaystyle\sum_{k=1}^{p} f_{i_k}(z) \sum_{i=1}^{p} f_{i_k}(z)} \right| \\
&\leq \frac{1}{m^2} \left| \sum_{k=1}^{p} f_{i_k}(z) f_i(y) - \sum_{k=1}^{p} f_{i_k}(y) f_i(z) \right|
\end{aligned}
$$

$$\leq \frac{1}{m^2} \sum_{k=1}^{p} |f_{i_k}(z) f_i(y) - f_{i_k}(y) f_i(z)|$$

$$\leq \frac{1}{m^2} \sum_{k=1}^{p} |f_{i_k}(z) - f_{i_k}(y)||f_i(y)| + \sum_{k=1}^{p} |f_{i_k}(y)||f_i(y) - f_i(z)|.$$

Therefore

$$|\psi_i(z) - \psi_i(y)| \leq \frac{2\bar{M}p}{m^2} Ad(y, z) \text{ for any } y, z \in W_x.$$

b) Now, we show that ψ_i is continuous. Let $x_0 \in X$. Then there exists a neighborhood V_x of x which intersects only a finite number of elements of $\{\overline{V}_i | i \in I_*\}$. That is there is a set of indices $\{i_1, \ldots, i_p\}$ such that

$$V_{x_0} \cap V_i = \emptyset \text{ for each } i \in I_* \backslash \{i_1, \ldots, i_p\}.$$

This implies that, for every $i \in I_* \backslash \{i_1, \ldots, i_p\}$ we have

$$V_{x_0} \subset X \backslash V_i \Rightarrow f_i(V_{x_0}) = 0,$$

and

$$V_{x_0} \cap supp(f_i) = \emptyset \text{ for each } i \in I_* \backslash \{i_1, \ldots, i_p\}. \tag{7.1}$$

From a) we obtain

$$\sum_{i \in I_*} f_i(x) = \sum_{i=1}^{p} f_i(x) \text{ for each } x \in V_{x_0}.$$

Therefore,

$$\psi_i(x) = \frac{f_i(x)}{\sum_{k=1}^{p} f_{i_k}(x)} \text{ for every } x \in V_{x_0}.$$

It is clear that $\sum_{k=1}^{p} f_{i_k}(x_0) \neq 0$, since for each $i \in I_*$, f_i is a continuous function. Hence ψ_i is continuous on X.

$\square$

Definition 7.15. *Let (X, d) be a generalized metric space. A subset Y of X is called dense whenever every $B(x, r) \cap Y \neq \emptyset$ for each $x \in X$ and $r \in \mathbb{R}^n_+, r = (r_1, \ldots, r_n)$, $r_i > 0$, $i = 1, \ldots, n$.*

We already have the following result.

Corollary 7.16. *Let Y be a subset of a generalized metric space (X, d). Then, Y is dense if and only if for every $x \in X$ there exists a sequence $(x_p)_{p \in \mathbb{N}}$ in Y satisfying*

$$x_p \to x \text{ as } p \to \infty.$$

Theorem 7.17. *(Cantor's intersection theorem). Let (X, d) be a cmplete generalized metric space. Let $(F_p)_{p \in \mathbb{N}}$ be a decreasing sequence of nonempty closed subsets of X such that*

$$\delta_{p \to \infty}(F_p) = 0 \in \mathbb{R}^n_+.$$

Then $\cap_{p \in \mathbb{N}} F_p$ contains exactly one point.

Proof. For all $p \in \mathbb{N}$ choose $x_p \in F_p$. Since $\delta(F_p) \to 0$ as $p \to \infty$, this implies that $(x_p)_{p \in \mathbb{N}}$ is Cauchy. Hence there exists $x \in X$ such that

$$x_p \to x \quad \text{as } p \to \infty.$$

We show that $x \in F_p$ for every $p \in \mathbb{N}$. If $(x_p)_{p \in \mathbb{N}}$ is finite then $x_p = x$ for infinitely many p, so that $x \in F_p$ for infinitely many p. Since $F_{p+1} \subseteq F_p$ this implies $x \in F_p$ for each $p \in \mathbb{N}$. So suppose $(x_p)_{p \in \mathbb{N}}$ is infinite. For all $m \in \mathbb{N}$, $(x_m, x_{m+1}, \dots, x_{m+k}, \dots)$ is a sequence in F_m converging to x. Since $(x_p)_{p \geq m}$ is infinite, this implies that $x \in \overline{F_m}$. But F_m is closed, so $x \in F_m$. Therefore $x \in \cap_{p \in \mathbb{N}} F_p$. If $\cap_{p \in \mathbb{N}} F_p$ contains two points x and y then we have

$$d(x, y) \leq \delta(F_p) \to 0, \text{ as } p \to \infty \Rightarrow d(x, y) = 0.$$

Hence

$$\cap_{p \in \mathbb{N}} F_p = \{x\}.$$

$\square$

Theorem 7.18. *The following are equivalent for a generalized metric space (X, d)*

1) X is a complete space.

2) For any descending sequence $\{F_p\}$ of closed bounded subsets of X,

$$\lim_{p \to \infty} \delta(F_p) = 0 \in \mathbb{R}_+^n.$$

Theorem 7.19. *Every complete generalized metric space is a Baire space.*

7.2 Generalized Banach space

Definition 7.20. *Let E be a vector space on $\mathbb{K} = \mathbb{R}$ or $\mathbb{C}$. By a vector-valued norm on E we mean a map $\| \cdot \| : E \to \mathbb{R}_+^n$ with the following properties:*

(i) $\|x\| \geq 0$ for all $x \in E$; if $\|x\| = 0$ then $x = 0$

(ii) $\|\lambda x\| = |\lambda| \|x\|$ for all $x \in E$ and $\lambda \in \mathbb{K}$

(iii) $\|x + y\| \leq \|x\| + \|y\|$ for all $x, y \in E$.

The pair $(E, \|\cdot\|)$ is called a generalized normed space. If the generalized metric generated by $\| \cdot \|$ (i.e $d(x, y) = \|x - y\|$) is complete then the space $(E, \| \cdot \|)$ is called a generalized Banach space, where

$$\|x - y\| = \begin{pmatrix} \|x - y\|_1 \\ \dots \\ \|x - y\|_n \end{pmatrix}.$$

Notice that $\| \cdot \|$ is a generalized Banach space on E if and only if $\| \cdot \|_i$, $i = 1, \dots, n$ are norms on E.

Definition 7.21. *Let E and F be two generalized normed spaces, $K \subset E$ and let $N : K \to F$ be an operator. Then N is said to be:*

i) *compact, if for any bounded subset $A \subseteq K$ we have $N(A)$ is relatively compact, i.e. $\overline{N(A)}$ is compact;*

ii) *completely continuous, if N is continuous and compact;*

iii) *with relatively compact range, if N is continuous and $N(K)$ is relatively compact, i.e. $\overline{f(K)}$ is compact.*

Definition 7.22. *Let $(E, \| \cdot \|)$ be a generalized Banach space and $U \subset E$ an open subset such that $0 \in U$. The function $p_U : E \to \mathbb{R}_+$ defined by*

$$p_U(x) = \inf\{\alpha > 0 : x \in \alpha U\},$$

is called the Minkowski functional of U.

Lemma 7.23. *Let $(E, \| \cdot \|)$ be a generalized Banach space and $U \subset E$ an open subset such that $0 \in U$. Then*

i) *If $\lambda \geq 0$, then $p_U(\lambda x) = \lambda p_U(x)$.*

ii) *If U is convex we have*

    a)$p_U(x + y) \leq p_U(x) + p_U(y)$, for every $x, y \in U$.

    b)$\{x \in E : p_U(x) < 1\} \subset U \subset \{x \in E : p_U(x) \leq 1\}$.

    c) If U is symmetric; then $p_U(x) = p_U(-x)$.

iii) *p_U is continuous.*

Proof. i) Let $x \in E$ be arbitrary and $\lambda \geq 0$. We have

$$
\begin{aligned}
p_U(\lambda x) &= \inf\{\alpha > 0 : \lambda x \in \alpha U\} \\
&= \inf\{\alpha > 0 : x \in \lambda^{-1}\alpha U\} \\
&= \inf\{\lambda\beta > 0 : x \in \beta U\} \\
&= \lambda \inf\{\beta > 0 : x \in \beta U\} \\
&= \lambda p_U(x).
\end{aligned}
$$

$ii) - a)$ Let $\alpha_1 > 0$ and $\alpha_2 > 0$ such that

$$x \in \alpha_1 U \text{ and } y \in \alpha_2 U.$$

Then

$$x + y \in \alpha_1 U + \alpha_2 U \Rightarrow \frac{x + y}{\alpha_1 + \alpha_2} \in \frac{\alpha_1}{\alpha_1 + \alpha_2}U + \frac{\alpha_2}{\alpha_1 + \alpha_2}U.$$

Hence

$$x + y \in (\alpha_1 + \alpha_2)U. \tag{7.2}$$

For every $\epsilon > 0$ there exist $\alpha_\epsilon > 0$, $\beta_\epsilon > 0$ such that

$$\alpha_\epsilon \leq p_U(x) + \epsilon \text{ and } \beta_\epsilon \leq p_U(y) + \epsilon.$$

From (7.2) we have

$$p_U(x + y) \leq p_U(x) + p_U(y) + 2\epsilon \Rightarrow p_U(x + y) \leq p_U(x) + p_U(y) + 2\epsilon.$$

Letting $\epsilon \to 0$ we obtain

$$p_U(x + y) \leq p_U(x) + p_U(y) \text{ for every } x, y \in U.$$

b) Let $x \in E$ such that $p_U(x) < 1$, then there exists $\alpha \in (0,1)$ such that

$$p_U(x) \leq \alpha < 1 \quad \text{and} \quad x \in \alpha U \Rightarrow x = \alpha a + (1-\alpha)0 \in U.$$

Therefore

$$\{x \in E : p_U(x) < 1\} \subset U.$$

For $x \in U$ we have

$$x = \alpha x \in U, \ \alpha = 1 \Rightarrow p_U(x) \leq 1.$$

Then

$$\{x \in E : p_U(x) < 1\} \subset U \subset \{x \in E : p_U(x) \leq 1\}.$$

iii) Since $0 \in U$ then there exists $r > 0$ such that

$$B(0,r) = \{x \in E : \|x\| < r_*\} \subset U,$$

where

$$\|x\| = \begin{pmatrix} \|x\|_1 \\ \dots \\ \|x\|_n \end{pmatrix} \quad \text{and} \quad r_* = \begin{pmatrix} r \\ \dots \\ r \end{pmatrix}.$$

Given $\epsilon > 0$, then $x + \epsilon B(0, r_*)$ is a neighborhood of x. For every $y \in x + \epsilon B(0, r_*)$ we have

$$\frac{x-y}{\epsilon} \in B(0, r_*) \Rightarrow p_U\left(\frac{x-y}{\epsilon}\right) \leq 1.$$

It is clear that

$$|p_U(x) - p_U(y)| \leq p_U(x-y) = \epsilon p_U\left(\frac{x-y}{\epsilon}\right) \leq \epsilon.$$

Hence p_U is continuous. $\qquad\square$

Remark 7.24. *In generalized metric space in the sense of Perov, the notions of convergence sequence, Cauchy sequence, completeness, open subset and closed subset are similar to those for usual metric spaces.*

7.3 Matrix convergence

Definition 7.25. *A square matrix $M \in \mathcal{M}_{n \times n}(\mathbb{R})$ of real numbers is said to be convergent to zero if*

$$M^k \to 0, \quad \text{as } k \to \infty.$$

Lemma 7.26. *[277] Let M be a square matrix of nonnegative numbers. The following assertions are equivalent:*

(i) M is convergent to zero;

(ii) the matrix $I - M$ is non-singular and

$$(I - M)^{-1} = I + M + M^2 + \cdots + M^k + \cdots ;$$

(iii) $|\lambda| < 1$ *for every* $\lambda \in \mathbb{C}$ *with* $det(M - \lambda I) = 0$;

(iv) $(I - M)$ *is non-singular and* $(I - M)^{-1}$ *has nonnegative elements.*

Proof. Assume that M is convergent to zero. We show that $I - M$ is non-singular; it suffices to prove that the linear system

$$(I - M)x = 0 \tag{7.3}$$

has only the null solution. Let $x \in \mathbb{C}$ be a solution of the system (7.3), then

$$x = Mx = M^2 x = \cdots M^k x = \cdots$$

and letting $k \to \infty$ we deduce $x = 0$. Hence $I - M$ is non-singular. Furthermore, we have

$$I - (I - M)(I + M + M^2 + \cdots M^k) = M^{k+1} \to 0 \quad \text{as} \quad k \to \infty.$$

This implies that

$$(I - M)^{-1} = I + M + M^2 + \cdots M^k \cdots .$$

$\square$

Lemma 7.27. *A square matrix $M \in \mathcal{M}_{n \times n}(\mathbb{R})$ of real numbers is convergent to zero if and only if its spectral radius $\rho(M)$ is strictly less than 1. In other words, this means that all the eigenvalues of M are in the open unit disc.*

Lemma 7.28. *Let $M \in \mathcal{M}_{n \times n}(\mathbb{R}_+)$ be convergent to zero. Then*

$$z \le (I - M)^{-1} z \text{ for every } z \in \mathbb{R}_+^n.$$

Proof. Since $M \in \mathcal{M}_{n \times n}(\mathbb{R}_+)$ is convergent to zero, then from Lemma 7.26, $(I - M)^{-1} \in \mathcal{M}_{n \times n}(\mathbb{R}_+)$ and

$$(I - M)^{-1} = I + M + M^2 + \cdots .$$

Thus for every $z \in \mathbb{R}_+^n$ we have

$$(I - M)^{-1} z = \sum_{i=0}^{\infty} M^i z \Rightarrow z \le (I - M)^{-1} z.$$

$\square$

Lemma 7.29. *Let $M \in \mathcal{M}_{n \times n}(\mathbb{R}_+)$ be convergent to zero. Then*

$$P_M = \{z \in \mathbb{R}_+^n : (I - M)z > 0\}$$

is nonempty and coincides with the set

$$\{(I - M)^{-1} z_0 : z_0 \in \mathbb{R}^n, \ z_0 > 0\}.$$

Proof. It is clear that $I - M \in \mathcal{M}_{n \times n}(\mathbb{R}_+)$ and is a singular matrix, then for every $z \in \mathbb{R}_+^n, z = (z_1, \ldots, z_n)$ with $z_i > 0, \ i = 1, \ldots, n$, we get $(I - M)z > 0$. This implies that $P_m \ne \emptyset$. Now we show that

$$P_M = \{(I - M)^{-1} z_0 : z_0 \in \mathbb{R}^n, \ z_0 > 0\}.$$

Indeed, if $z_0 \in \mathbb{R}^n$ and $z_0 > 0$, then

$$z := (I - M)^{-1} z_0 \ge z_0 \Rightarrow z > 0.$$

Hence $(I - M)z > 0$ and so $z \in P_M$. Conversely, if $z \in P_M$, then $z_0 := (I - M)z > 0$ and $z = (I - M)^{-1} z$.

$\square$

Definition 7.30. *We say that a non-singular matrix $A = (a_{ij})_{1 \leq i,j \leq n} \in \mathcal{M}_{n \times n}(\mathbb{R})$ has the absolute value property if*

$$A^{-1}|A| \leq I,$$

where

$$|A| = (|a_{ij}|)_{1 \leq i,j \leq n} \in \mathcal{M}_{n \times n}(\mathbb{R}_+).$$

Some examples of matrices convergent to zero are the following:

1) $A = \begin{pmatrix} a & 0 \\ 0 & b \end{pmatrix}$, where $a, b \in \mathbb{R}_+$ and $\max(a, b) < 1$

2) $A = \begin{pmatrix} a & -c \\ 0 & b \end{pmatrix}$, where $a, b, c \in \mathbb{R}_+$ and $a + b < 1$, $c < 1$

3) $A = \begin{pmatrix} a & -a \\ b & -b \end{pmatrix}$, where $a, b, c \in \mathbb{R}_+$ and $|a - b| < 1$, $a > 1, b > 0$.

Lemma 7.31. *Let $M = (a_{ij})_{1 \leq i,j \leq n} \in \mathcal{M}_{n \times n}(\mathbb{R}_+)$ be a triangular matrix with*

$$\max\{|a_{ii}|, i = 1, \ldots, n\} < \frac{1}{2}.$$

Then the matrix $A = (I - M)^{-1}M$ is convergent to zero.

Proof. Suppose $M := \begin{pmatrix} a_{11} & \cdots & a_{1n} \\ \vdots & & \vdots \\ & \cdots & \\ 0 & \cdots & a_{nn} \end{pmatrix} \in \mathcal{M}_{n \times n}(\mathbb{R}_+)$. Then the eigenvalues of M are

$\lambda_i = \frac{a_{ii}}{1 - a_{ii}}$, for all $i = 1, \ldots, n$. Since all of the eigenvalues of M are in the open unit disc, the conclusion follows from Theorem 7.26. $\qquad\square$

Chapter 8

Fixed Point Theorems in Vector Metric and Banach Spaces

8.1 Banach principle theorem

The classical Banach contraction principle was extended for contractive maps on spaces endowed with vector-valued metric spaces by Perov in 1964 [237], Perov and Kibenko [238] and Precup [247]. For a version of Schauder's fixed point, see Viorel [279]. The purpose of this section is to present that version of Schaefer's fixed point theorem and the nonlinear alternative of Leary-Schauder type in generalized Banach spaces.

Theorem 8.1. *[237] Let (X, d) be a complete generalized metric space with $d : X \times X \longrightarrow \mathbb{R}^n$ and let $N : X \longrightarrow X$ be such that*

$$d(N(x), N(y)) \leq M d(x, y)$$

for all $x, y \in X$ and some square matrix M of nonnegative numbers. If the matrix M is convergent to zero, that is $M^k \longrightarrow 0$ as $k \longrightarrow \infty$, then N has a unique fixed point $x_* \in X$,

$$d(N^k(x_0), x_*) \leq M^k(I - M)^{-1} d(N(x_0), x_0),$$

for every $x_0 \in X$ and $k \geq 1$.

Proof. Let $x \in X$ and define the sequence $x_n = N^n(x)$, where $N^n = N \circ \cdots \circ N$. Using the fact that N is an M-contraction, we get

$$d(x_{k+1}, x_k) \leq M^k d(N(x), x)$$

and, as a consequence,

$$d(x_k, x_{k+m}) \leq (M^k + M^{k+1} + \cdots + M^{k+m-1}) d(N(x), x).$$

From Lemma 7.26 we deduce that

$$d(x_k, x_{k+m}) \leq M^k(I - M)^{-1} d(N(x), x).$$

Hence (x_k) is a Cauchy sequence with respect to d and thus converges to some limit $x_* \in X$. The continuity of N guarantees that

$$x_* = N(x_*).$$

For uniqueness, let y_1 and y_2 be two fixed points of N, then

$$d(y_1, y_2) = d(N^k(y_1), N^k(y_2)) \leq M^k d(N(y_1), N(y_2)).$$

Since $M^k \to 0$ as $k \to \infty$, this implies $d(y_1, y_2) = 0$, so $y_1 = y_2$. $\qquad \square$

The next result is an extension of Perov's Theorem.

Theorem 8.2. *Let (X, d) be a generalized complete metric space and let $N : X \to X$ be an (A, B, C, D, E)-contraction, $A, B, C, D, E \in \mathcal{M}_{n \times n}(\mathbb{R}_+)$ are such that the matrices E and $C + E$ or the matrices D and $B + D$ converge to zero and the matrix $M := (I - C - E)^{-1}(A + C + D)$ or the matrix $M_* := (I - B - D)^{-1}(A + B + E)$ converges to zero and*

$$
\begin{aligned}
d(N(x), N(y)) \;\leq\; & Ad(x, y) + Bd(y, N(x)) + Cd(x, N(y)) \\
& + Dd(x, N(x)) + Ed(y, N(y))
\end{aligned}
$$

for all $x, y \in X$. Then, the following conclusions hold:

1) *N has at least one fixed point and, for each $x_0 \in X$, the sequence $x_p := N^p(x_0)$ of successive approximations of N starting from x_0 converges to $x_*(x_0) \in Fix(N)$ as $n \to \infty$.*

2) *For each $x_0 \in X$ we have*

$$
d(x_p, x_*(x_0)) \leq M^p (I - M)^{-1} d(x_0, N(x_0)), \text{ for all } n \in \mathbb{N}
$$

or

$$
d(x_p, x_*(x_0)) \leq M_*^p (I - M_*)^{-1} d(x_0, N(x_0)), \text{ for all } n \in \mathbb{N}.
$$

3) *If, additionally, the matrix $A + B + C$ converges to zero, then N has a unique fixed point in X.*

With a stronger assumption on the space we get the following.

Theorem 8.3. *Let E be a Banach generalized space, $Y \subseteq E$ nonempty convex compact subset of E and $N : Y \to Y$ be a single valued map. Assume that*

$$
d(N(x), N(y)) \leq d(x, y) \text{ for all } x, y \in Y.
$$

Then N has a fixed point.

Proof. For every $m \in \mathbb{N}$, we have $\frac{I}{2^m} \in \mathcal{M}_{n \times n}(\mathbb{R}_+)$ and

$$
\frac{I}{2^{mk}} \to 0 \text{ as } k \to \infty.
$$

Thus, for some $x_0 \in Y$ the mapping $N_m : Y \to Y$ defined by

$$
N_m(x) = \left(1 - \frac{1}{2^m}\right) N(x) + \frac{1}{2^m} x_0 \in Y \text{ for all } x \in Y.
$$

Hence, we get

$$
d(N_m(x), N_m(y)) \leq \frac{I}{2^m} d(x, y) \text{ for all } x, y \in Y.
$$

From Theorem 8.1 there exists a unique $x_m \in Y$ such that

$$
x_m = N_m(x_m), \quad m \in \mathbb{N}.
$$

Since Y is compact, then there exists a subsequence of $(x_m)_{m \in \mathbb{N}}$ converging to $x \in Y$. Now we show that $x = N(x)$.

$$
d(x, N(x)) \;=\; \begin{pmatrix} d_1(x, N(x)) \\ \cdots \\ d_n(x, N(x)) \end{pmatrix}
$$

$$\leq \quad d(x,x_m) + d(x_m, N(x_m)) + d(N(x_m), N(x))$$
$$\leq \quad 2Id(x,x_m) + d(x_m, N(x_m))$$

and

$$d(x_m, N(x_m)) = \begin{pmatrix} d_1(x_m, N(x_m)) \\ \dots \\ d_n(x_m, N(x_m)) \end{pmatrix}$$
$$= \begin{pmatrix} \|x_m - N(x_m)\|_1 \\ \dots \\ \|x_m - N(x_m)\|_n \end{pmatrix}$$
$$= \begin{pmatrix} \|\frac{N(x_m)}{2^m} - \frac{1}{2^m}x_0\|_1 \\ \dots \\ \|\frac{N(x_m)}{2^m} - \frac{1}{2^m}x_0\|_n \end{pmatrix}$$
$$\leq \quad \frac{1}{2^m}d(x_m, x) + \frac{1}{2^m}d(N(x), x_0).$$

Hence

$$d(x, N(x)) \leq \left(2 + \frac{1}{2^m}\right) Id(x, x_m) + \frac{1}{2^m}d(N(x), x_0) \to 0 \text{ as } m \to \infty.$$

$\square$

Now, we examine a local version of the Perov fixed point theorem.

Theorem 8.4. *Let E be a generalized Banach space, $\overline{B(0,r)}$ be the closed ball of radius $r \in \mathbb{R}^n_+$ and $N : \overline{B(0,r)} \to E$ a contraction such that*

$$N(\partial\overline{B(0,r)}) \subset \overline{B(0,r)},$$

where

$$\partial\overline{B(0,r)} = \{x \in E : \sum_{i=1}^n \|x\|_i = \sum_{i=1}^n r_i \dots, n\}, \ r = (r_1, \dots, r_n).$$

Then N has a unique fixed point in $\overline{B(0,r)}$.

Proof. We defined $N_* : \overline{B(0,r)} \to E$ by

$$N_*(x) = \frac{x + N(x)}{2}, \quad x \in \overline{B(0,r)}.$$

First we show that

$$N_*(\overline{B(0,r)}) \subseteq \overline{B(0,r)}.$$

Let $x \in \overline{B(0,r)}$, and $x_* = \frac{x\sum_{i=1}^n r_i}{\sum_{i=1}^n \|x\|_i}$, then for each $x \in \overline{B(0,r)}$ such that $x \neq 0$, we have

$$\|N(x) - N(x_*)\| \leq M\|x - x_*\| \Rightarrow \|N(x) - N(x_*)\| \leq \left(1 - \frac{\sum_{i=1}^n r_i}{\sum_{i=1}^n \|x\|_i}\right) M\|x\|$$

and

$$\|N(x)\| \leq \|N(x) - N(x_*)\| + \|N(x_*)\|.$$

Therefore

$$\|N(x)\| \leq \left(1 - \frac{\displaystyle\sum_{i=1}^{n} r_i}{\displaystyle\sum_{i=1}^{n} \|x\|_i}\right) M\|x\| + r$$

$$\leq 2r - \frac{\displaystyle\sum_{i=1}^{n} r_i}{\displaystyle\sum_{i=1}^{n} \|x\|_i} M\|x\|.$$

Hence

$$\|N_*(x)\| \leq r - \frac{\displaystyle\sum_{i=1}^{n} r_i}{2\displaystyle\sum_{i=1}^{n} \|x\|_i} M\|x\|$$

$$\leq r.$$

By continuity we also have

$$\|N_*(0)\| \leq r,$$

and consequently

$$N_*(\overline{B(0,r)}) \subseteq \overline{B(0,r)}.$$

Moreover for every $x, y \in \overline{B(0,r)}$ we have

$$\|N_*(x) - N_*(y)\| \leq \frac{1}{2}(I + M)\|x - y\|.$$

Now, we show that $\frac{1}{2}(I + M)$ is convergent. Indeed let $z \in \mathbb{R}^n$ such that

$$z - \frac{1}{2}(I + M)z = 0 \Rightarrow Mz = z, \dots, z = M^k z \dots,$$

and letting $k \to \infty$ we get $z = 0$. This implies that $I - \frac{1}{2}(I + M)$ is nonsingular and $(I - \frac{1}{2}(I + M))^{-1} = \sum_{k=0}^{\infty} \frac{1}{2^k}(I + M)^k$. By Lemma 7.26, we deduce that $I - \frac{1}{2}(I + M)$ converges to zero. From Perov's fixed point theorem, Theorem 8.1, we deduce that N_* has a unique fixed point $x \in \overline{B(0,r)}$. Of course if $x = N_*(x)$ then $x = N(x)$. $\square$

As a consequence of Perov's fixed point theorem we have the following result.

Theorem 8.5. *Let E be a generalized Banach space and $N : E \to E$ be a contraction. The $I_E - N$ is a homeomorphism.*

Proof. The continuity of $I_E - N$ is obvious, since N is continuous. Now we show that $I_E - N$ is bijective, let us consider any $y \in E$ and the equation

$$(I_E - N)(x) = y, \ x \in E.$$

Consider the following operator $N_* : E \to E$ by

$$N_*(x) = N(x) + y, \quad x \in E.$$

Since N is an contraction, we get that N_* is an contraction too. Hence N_* has a unique fixed point $x_* \in E$. Thus $I_E - N$ is bijective. The continuity of $(I_E - N)^{-1}$ follows in the similar way for the case of usual Banach space. $\qquad \square$

8.2 Continuation methods for contractive maps

Let (X, d) be a complete generalized metric space and $U \subseteq X$ an open subset of X.

Definition 8.6. *Let $F : U \to X$ and $G : U \to X$ be two contractions; here $\overline{U}$ denotes the closure of U in X. We say that F and G are homotopic if there exists $H : U \times [0, 1] \to X$ with the following properties:*

a) $H(\cdot, 0) = G$ *and* $H(., 1) = F$;

b) $x \neq H(x, t)$ *for every* $x \in \partial U$ *and* $t \in [0, 1]$;

c) there exists $M \in \mathcal{M}_{n \times n}(\mathbb{R}_+)$ convergent to zero such that

$$d(H(x, t), H(y, t)) \leq M d(x, y) \text{ for every } x, y \in U \text{ and } t \in [0, 1];$$

d) there exists $M_* \in \mathcal{M}_{1 \times n}(\mathbb{R}_+)$, such that

$$d(H(x, t), H(x, s)) \leq M_* |t - s| \text{ for every } x \in U \text{ and } t, s \in [0, 1].$$

Theorem 8.7. *Let (X, d) be a generalized complete metric space and U an open subset of X. Suppose that $F : U \to X$ and $G : U \to X$ are two homotopic contractive maps and G has a fixed point in U. Then F has a fixed point in U.*

Proof. Let

$$\Lambda = \{\lambda \in [0, 1] : x = H(x, \lambda) \text{ for some } x \in U\},$$

where H is a homotopy between F and G. Then $H(\cdot, 0) = x$, so $0 \in \Lambda$. We will show that Λ is both open and closed in $[0, 1]$ and hence by connectedness we have that $\Lambda = [0, 1]$. As a result, F has a fixed point in U.

To prove that Λ is closed, let

$$\{\lambda_p\} \subseteq \Lambda \text{ with } \lambda_p \to \lambda \text{ as } p \to \infty.$$

Since $\lambda_p \in \Lambda$ for every $p \in \mathbb{N}$, there exists $x_p \in U$ with

$$x_p = H(x_p, \lambda_p).$$

Also for $p, q \in \mathbb{N}$ we have

$$
\begin{aligned}
d(x_p, x_q) &= d(H(x_p, \lambda_p), H(x_q, \lambda_q)) \\
&\leq d(H(x_p, \lambda_p), H(x_p, \lambda_q)) + d(H(x_p, \lambda_q), H(x_q, \lambda_q)) \\
&\leq M_* |\lambda_p - \lambda_q| + M d(x_p, x_q).
\end{aligned}
$$

From Lemma 7.29, we obtain

$$d(x_p, x_q) \;\leq\; (I - M)^{-1} M_* |\lambda_p - \lambda_q|.$$

Since $\{\lambda_p\}$ is a Cauchy sequence we have that $\{x_p\}$ is also a Cauchy sequence, and since X is complete there exists $x \in \overline{U}$ with $\lim\limits_{p \to \infty} x_p = x$. Clearly,

$$
\begin{aligned}
d(x_p, H(x, \lambda)) &= d(H(x_p, \lambda_p), H(x, \lambda)) \\
&\leq d(H(x_p, \lambda_p), H(x_p, \lambda)) + d(H(x_p, \lambda), H(x, \lambda)) \\
&\leq M_* |\lambda_p - \lambda| + M d(x_p, x) \to 0 \text{ as } p \to \infty.
\end{aligned}
$$

Hence $d(x, H(x, \lambda)) = 0$, this implies that $x = H(x, \lambda)$. By $b)$ in the above definition we get $x \in U$.

To prove that Λ is open in $[0, 1]$, let $\lambda \in [0, 1]$ such that

$$x = H(x, \lambda).$$

Since U is open from Lemma 7.29 there exists $r \in P_M$, $r = \begin{pmatrix} r_1 \\ \vdots \\ r_n \end{pmatrix}, r_i > 0, \ i = 1, \ldots, n$

such that

$$\overline{B(x, r)} \subset U.$$

By $d)$ of the above definition, there exists $\eta > 0$ such that

$$
\begin{aligned}
d(x, H(y, \mu)) &= d(H(x, \lambda), H(y, \mu)) \\
&\leq d(H(x, \lambda), H(x, \mu)) + d(H(x, \mu), H(y, \mu)) \\
&\leq M_* |\mu - \lambda| + M d(x, y) \\
&\leq (I - M)r + Mr \leq r,
\end{aligned}
$$

where $d(x, y) \leq r$ and $M_* |\mu - \lambda| \leq (I - M)r, M_* = \begin{pmatrix} a_1 \\ \vdots \\ a_n \end{pmatrix}, a_i > 0, \ i = 1, \ldots, n$. This

shows that $H(\cdot, \mu) : \overline{B(x, r)} \to \overline{B(x, r)}$. By Theorem 8.1 we deduce that $H(\cdot, \mu)$ has a unique fixed point in U. Thus Λ for any $\lambda \in (\mu - \min_{1 \leq i \leq n} \frac{\bar{r}_i}{a_i}, \mu + \min_{1 \leq i \leq n} \frac{\bar{r}_i}{a_i}) \cap [0, 1]$

where $\bar{r} = (I - M)r = \begin{pmatrix} \bar{r}_1 \\ \vdots \\ \bar{r}_n \end{pmatrix}$ and therefore Λ is open in $[0, 1]$. $\qquad\square$

For the next results we assume that X is a generalized Banach space. We now present a nonlinear alternative of Leray-Schauder type for contractive maps.

Theorem 8.8. *Let X be a generalized Banach space, U be an open subset of X and $N : \overline{U} \to X$ be a contraction with $F(\overline{U})$ bounded. Then either*

i) N has a fixed point in U, or

ii) there exist $\lambda \in (0, 1)$ and $x \in \overline{U} \backslash U$ with $x = \lambda N(x)$.

Proof. Assume that *ii*) does not hold. Hence

$$x \neq \lambda N(x) \quad \text{for each } x \in \overline{U} \backslash U \text{ and } \lambda \in [0, 1].$$

We define the following continuous operator $H : \overline{U} \times [0, 1] \to X$ by

$$H(x, \lambda) = \lambda N(x), \quad x \in \overline{U}, \ \lambda \in [0, 1].$$

Let G be the zero map. We can easily prove that H preserves homotopic, contractive mappings between G and N. Since G has a fixed point in U, from Theorem 8.7, we deduce that there exists $x \in U$ such that $x = N(x)$, that is, *i*) occurs. $\qquad \square$

8.3 Perov fixed point type for expansive mapping

Definition 8.9. *Let (X, d) be a generalized metric space and C be a subset of X. The mapping $B : C \to X$ is said to be expansive, if there exists a constant $k \in \mathbb{R}$, $k > 1$ such that*

$$d(B(x), B(y)) \geq kd(x, y) \quad \text{for all } x, y \in C.$$

Lemma 8.10. *Let X be a generalized metric space and $C \subseteq X$. Assume the mapping $B : C \to X$ is expansive with constant $k > 1$. Then the inverse of $B : C \to B(C)$ exists and*

$$d(B^{-1}(x), B^{-1}(y)) \leq \frac{I}{k} d(x, y), \ x, y \in B(C).$$

Proof. Let $x, y \in C$ and $B(x) = B(y)$, then

$$d(B(x), B(y)) \geq kd(x, y) \Rightarrow d(x, y) = 0 \Rightarrow x = y.$$

Thus $B : C \to B(C)$ is invertible. Let $x, y \in B(C)$, then there exist $a, b \in C$ such that

$$B(a) = x, \ B(b) = y.$$

Hence

$$d(a, b) = d(B^{-1}(x), B^{-1}(y)) \text{ and } d(x, y) = d(B(a), B(b)) \geq kd(a, b).$$

Therefore

$$d(B^{-1}(x), B^{-1}(y)) \leq \frac{I}{k} d(x, y) \text{ for all } x, y \in C.$$

$\qquad \square$

As a consequence of Perov's Theorem we have the following result.

Theorem 8.11. *Let X be a complete generalized metric space and C be a closed subset of X. Assume $B : C \to X$ is expansive and $C \subseteq B(C)$. Then there exists a unique point $x \in C$ such that $x = B(x)$.*

Proof. Since B is expansive, there exists $k > 1$ such that

$$d(B(x), B(y)) \geq kd(x, y) \quad \text{for all } x, y \in C.$$

From Lemma 8.10 the operator $B : C \to C$ is invertible and

$$d(B^{-1}(x), B^{-1}(y)) \leq \frac{I}{k} d(x, y), \quad x, y \in C.$$

Hence B^{-1} is contractive. By Theorem 8.1 there exists a unique $x \in C$ such that

$$B^{-1}(x) = x \Rightarrow x = B(x).$$

$\square$

Lemma 8.12. *Let $B : X \to X$ be a map such that B^m (mth power) is an expansive map for some $m \in \mathbb{N}$. Assume further that there exists a closed subset C of X such that C is contained $B(C)$. Then there exists a unique fixed point of B.*

Proof. Since B^m is an expansive map and $C \subseteq B^m(C)$, then from Theorem 8.11 there exists a unique fixed point of B^m. Let $x \in C$ be the unique fixed point of B^m. Using the fact that B^m is an expansive map, then there exists $k > 1$ such that

$$d(B^m(x), B^m(y)) \geq kd(x, y) \text{ for all } x, y \in C.$$

Hence
$$d(x, B(x)) = d(B^m(x), B^{m+1}(x)) \geq kd(x, B(x)) \Rightarrow d(x, B(x)) = 0.$$

Then B has the unique fixed point $x \in C$.

$\square$

8.4 Leray-Schauder type theorem

Theorem 8.13. *([279]) Let E be a generalized Banach space, $C \subset E$ be a nonempty closed convex subset of E and $N : C \to C$ be a continuous operator with relatively compact range. Then N has at least one fixed point in C.*

As a consequence of the Schauder fixed point theorem we present the version of Schaefer's fixed point theorem and nonlinear alternative Leary-Schauder type theorem in a generalized Banach space.

Theorem 8.14. *Let $(E, \|\cdot\|)$ be a generalized Banach space and $N : E \to E$ is a continuous compact mapping. Moreover assume that the set*

$$\mathcal{A} = \{x \in E : x = \lambda N(x) \quad \text{for some } \lambda \in (0, 1)\}$$

is bounded. Then N has a fixed point.

Proof. Let $K > 0$ such that

$$\sum_{i=1}^{n} \|x\|_i < nK \text{ for each } x \in \mathcal{A}.$$

Set $M_* = (nK, \dots, nK)$ and we, define $N_* : \overline{B(0, M_*)} \to \overline{B(0, M_*)}$ by

$$N_*(x) = \begin{cases} N(x) & \text{if } \sum_{i=1}^{n} \|N(x)\|_i \leq nK \\[2ex] \dfrac{KnN(x)}{\sum_{i=1}^{n} \|N(x)\|_i} & \text{if } \sum_{i=1}^{n} \|N(x)\|_i > nK. \end{cases}$$

We will show that N_* is continuous. Let $x \in \overline{B(0, M_*)}$ such that $\sum_{i=1}^{n} \|N(x)\|_i < nK$ then $N_*(x) = N(x)$.

If $(x_m)_{m \in \mathbb{N}} \in \overline{B(0, M_*)}$ and $\sum_{i=1}^{n} \|N(x_m)\|_i \le M_*$, then the continuity of N implies that $\|N(x_m) - N(x)\| \to 0$ as $m \to \infty \Rightarrow \|N_*(x_n) - N_*(x)\| \to 0$ as $m \to \infty$.

Now let $x \in \overline{B(0, M_*)}$ such that $\sum_{i=1}^{n} \|N(x)\|_i > nK$. Then $N_*(x) = \frac{KnN(x)}{\sum_{i=1}^{n} \|N(x)\|_i}$.

If $(x_m)_{m \in \mathbb{N}} \in \overline{B(0, M_*)}$ and $\sum_{i=1}^{n} \|N(x_m)\|_i > nK$, then $N_*(x_m) = \frac{KnN(x_m)}{\sum_{i=1}^{n} \|N(x_m)\|_i}$. By the continuity of N we have, for every $j = 1, \ldots, n$

$$
\begin{aligned}
\|N_*(x_m) - N_*(x)\|_j &= \left\| \frac{KnN(x_m)}{\sum_{i=1}^{n} \|N(x_m)\|_i} - \frac{KnN(x)}{\sum_{i=1}^{n} \|N(x)\|_i} \right\|_j \\
&= \left\| \frac{KnN(x_m) \sum_{i=1}^{n} \|N(x)\|_i - KnN(x) \sum_{i=1}^{n} \|N(x_m)\|_i}{\sum_{i=1}^{n} \|N(x_m)\|_i \sum_{i=1}^{n} \|N(x)\|_i} \right\|_j \\
&\le \frac{Kn\|N(x_m) - N(x)\|_j}{\sum_{i=1}^{n} \|N(x_m)\|_i} \\
&\quad + \frac{Kn\|N(x)\|_j \sum_{i=1}^{n} \|N(x_m) - N(x)\|_i}{\sum_{i=1}^{n} \|N(x_m)\|_i \sum_{i=1}^{n} \|N(x)\|_i}.
\end{aligned}
$$

Since $\sum_{i=1}^{n} \|N(x_m)\|_i > nM$, thus $\lim_{m \to \infty} \sum_{i=1}^{n} \|N(x_m)\|_i \ge nK$, and hence

$$\|N_*(x_m) - N_*(x)\| \to 0 \text{ as } m \to \infty.$$

Let $x \in \overline{B(0, M_*)}$ such that $\sum_{i=1}^{n} \|N(x)\|_i = nK$ then $N_*(x) = N(x)$.

If $(x_m)_{m \in \mathbb{N}} \in \overline{B(0, M_*)}$ and $\sum_{i=1}^{n} \|N(x_m)\|_i \le nK$ then the continuity of N implies that

$$\|N(x_m) - N(x)\| \to 0 \text{ as } m \to \infty \Rightarrow \|N_*(x_m) - N_*(x)\| \to 0 \text{ as } m \to \infty.$$

If $(x_m)_{m \in \mathbb{N}} \in \overline{B(0, M_*)}$ and $\sum_{i=1}^{n} \|N(x_m)\|_i > M_*$ then $N_*(x_m) = \dfrac{KnN(x_m)}{\sum_{i=1}^{n} \|N(x_m)\|_i}$, and hence

$$
\begin{aligned}
\|N_*(x_m) - N_*(x)\|_j &= \left\| \frac{KnN(x_m)}{\sum_{i=1}^{n} \|N(x_m)\|_i} - N(x) \right\|_j \\[2em]
&= \left\| \frac{KnN(x_m) - N(x) \sum_{i=1}^{n} \|N(x_m)\|_i}{\sum_{i=1}^{n} \|N(x_m)\|_i} \right\|_j \\[2em]
&\leq \frac{Kn\|N(x_m) - N(x)\|_j}{\sum_{i=1}^{n} \|N(x_m)\|_i} \\[2em]
&+ \frac{\|N(x)\|_j \left| Kn - \sum_{i=1}^{n} \|N(x_m)\|_i \right|}{\sum_{i=1}^{n} \|N(x_m)\|_i}.
\end{aligned}
$$

It is clear that

$$
\lim_{m \to \infty} \sum_{i=1}^{n} \|N(x_m)\|_i = \sum_{i=1}^{n} \|N(x)\|_i = Kn.
$$

Therefore

$$
\|N_*(x_m) - N_*(x)\|_j \to 0 \text{ as } m \to \infty.
$$

Thus, we conclude that N_* is continuous. Consider the following map

$$
\rho : \overline{B(0, M_*)} \to \overline{B(0, M_*)} \text{ defined by}
$$

$$
\rho(x) = \begin{cases} x & \text{if } \sum_{i=1}^{n} \|x\|_i \leq nK \\[2em] \frac{Knx}{\sum_{i=1}^{n} \|x\|_i} & \text{if } \sum_{i=1}^{n} \|x\|_i > nK. \end{cases}
$$

It is evident that ρ is continuous and $N_* = \rho \circ N$. The compactness of N implies that N_* is compact. By Theorem 8.13 there exists $x \in \overline{B(0, M_*)}$ such that $x = N_*(x)$.

Notice $\sum_{i=1}^{n} \|N(x)\|_i \leq Kn$ for otherwise,

$$
x = \lambda N(x), \ \lambda = \frac{Kn}{\sum_{i=1}^{n} \|N(x)\|_i} \text{ with } 0 < \lambda < 1 \Rightarrow x \in \mathcal{A}.
$$

This implies that

$$\sum_{i=1}^{n} \|x\|_i < Kn,$$

but

$$x = \frac{KnN(x)}{\sum_{i}^{n} \|N(x)\|_i} \Rightarrow \sum_{i=1}^{n} \|x\|_i = Kn.$$

This yields a contradiction with $x \in \mathcal{A}$. Hence we get that

$$x = N_*(x) = N(x).$$

$\square$

Next we state the nonlinear alternative of Leray-Schauder type.

Lemma 8.15. *Let X be a generalized Banach space, $U \subset E$ be a bounded, convex open neighborhood of zero and let $G : \overline{U} \to E$ be a continuous compact map. If G satisfies the boundary condition*

$$x \neq \lambda G(x)$$

for all $x \in \partial U$ and $0 \leq \lambda \leq 1$, then the set $Fix(G) = \{x \in U : x = G(x)\}$ is nonempty.

Proof. Let p be the Minkowski function of U and since $\overline{U}$ is bounded, then there exists $M > 0$ such that

$$G(\overline{U}) \subseteq \frac{1}{2} B(0, M_*), \quad M_* = (K, \dots, K).$$

Consider $G_* : \overline{B(0, M_*)} \to \overline{B(0, M_*)}$ defined by

$$G_*(x) = \begin{cases} G(x) & \text{if } x \in \overline{U} \\ \frac{1}{p(x)} G(\frac{x}{p(x)}) & \text{if } x \in E \backslash \overline{U}. \end{cases}$$

Clearly $\overline{B(0, M_*)}$ is a closed, convex, bounded subset of E and G_* is a continuous compact operator. Then from Theorem 8.13 there exists $x \in \overline{B(0, M_*)}$ such that $\overline{G}(x) = x$. If $x \in E \backslash \overline{U}$, and then

$$x = \frac{G\left(\frac{x}{p(x)}\right)}{p(x)} \Rightarrow \frac{x}{p(x)} = \frac{1}{p^2(x)} G\left(\frac{x}{p(x)}\right).$$

Since $x \in E \backslash \overline{U}$, then

$$p(x) = 1 \text{ or } p(x) > 1 \Rightarrow x \in \partial U, \quad \frac{x}{p(x)} \in \partial U.$$

This is a contradiction with

$$z \neq \lambda G(z), \quad \text{for each } \lambda \in [0, 1], \ z \in \partial U.$$

Consequently, there exists $x_* \in U$ such that $G(x_*) = x_*$. $\square$

Theorem 8.16. *Let $(E, \|\cdot\|)$ be a Banach space, $C \subset E$ a closed convex subset, $U \subset C$ a bounded set, open (with respect to the topology C) and such that $0 \in U$. Let $G : \overline{U} \to C$ be a compact continuous mapping. If the following assumption is satisfied:*

$$x \neq \lambda G(x), \text{ for all } x \in \partial_C U \text{ and all } \lambda \in (0, 1),$$

then G has a fixed point in U.

Proof. Let $C_* = \{x \in \overline{U} : x = \lambda G(x) \text{ for some } \lambda \in [0,1]\}$. Since $0 \in U$ then C_* is a nonempty set and by the continuity of G we conclude that C_* is closed. Clearly $\partial_C U \cap C_* = \emptyset$. From Theorem 7.11 there exists $f : \overline{U} \to [0,1]$ such that

$$f(x) = \begin{cases} 0 & \text{if } x \in \partial_C U \\ 1 & \text{if } x \in C_*. \end{cases}$$

Consider $G_* : C \to C$ defined by

$$G_*(x) = \begin{cases} f(x)G(x) & \text{if } x \in U \\ 0 & \text{if } x \in C \backslash U. \end{cases}$$

Since $G_*(x) = 0$, for each $x \in \partial_C U$, and G_* is continuous on U and $E \backslash U$, then G_* is continuous. Set $\Omega = \overline{co}(\{0\} \cup G(\overline{U}))$, which is convex and compact. We can easily prove that

$$G_*(\Omega) \subset \Omega.$$

Then from Theorem 8.13 there exists $x \in \Omega$ such that $G_*(x) = x$. From the definition of G_* we have $G(x) = x$. $\qquad\square$

From the above theorem we obtain the following:

Theorem 8.17. *Let $C \subset E$ be a closed convex subset and $U \subset C$ a bounded open neighborhood of zero(with respect to topology of C). If $G : \overline{U} \to E$ is compact continuous then*

i) *either G has a fixed point in $\overline{U}$, or*

ii) *there exists $x \in \partial U$ such that $x = \lambda G(x)$ or some $\lambda \in (0,1)$.*

Now, we state the nonlinear alternative of Schaefer fixed point theorem type.

Theorem 8.18. *Let X be a generalized Banach space and let $G : X \to X$ be completely continuous. Then, either*

(i) *the operator equation $x = Tx$ has a solution, or*

(ii) *the set*
$$\mathcal{E} = \{x \in X : x = \lambda N(x), \quad \lambda \in (0,1)\}$$
is unbounded.

8.5   Measure of noncompactness

In this section, by using Theorem 8.13 and the concept of a measure of noncompactness in vector-valued Banach spaces, we obtain a Sadovkii fixed point theorem.

Definition 8.19. *Let X be a generalized Banach space and $(\mathcal{A}, \leq)$ be a partially ordered set. A map $\beta : \mathcal{P}(X) \to \mathcal{A} \times \mathcal{A} \ldots \times \mathcal{A}$ is called a generalized measure of noncompactness (m.n.c.) on X, if*

$$\beta(\overline{co}\,\Omega) = \beta(\Omega) \text{ for every } \Omega \in \mathcal{P}(X),$$

where $\beta(\Omega) := \begin{pmatrix} \beta_1(\Omega) \\ \vdots \\ \beta_n(\Omega) \end{pmatrix}$.

Definition 8.20. *A measure of noncompactness β is called:*

(a) *Monotone if $\Omega_0, \Omega_1 \in \mathcal{P}(X)$, $\Omega_0 \subset \Omega_1$ implies $\beta(\Omega_0) \leq \beta(\Omega_1)$.*

(b) *Nonsingular if $\beta(\{a\} \cup \Omega) = \beta(\Omega)$ for every $a \in X$ and $\Omega \in \mathcal{P}(X)$.*

(c) *Invariant with respect to the union with compact sets if $\beta(K \cup \Omega) = \beta(\Omega)$ for every relatively compact set $K \subset X$ and $\Omega \in \mathcal{P}(X)$.*

(d) *Real if $\mathcal{A} = \overline{\mathbb{R}}_+$ and $\beta(\Omega) < \infty$ for every $i = 1, \ldots, n$ and every bounded Ω.*

(e) *Semi-additive if $\beta(\Omega_0 \cup \Omega_1) = \max(\beta(\Omega_0), \beta(\Omega_1))$ for every $\Omega_0, \Omega_1 \in \mathcal{P}(X)$.*

(f) *Lower-additive if β is real and $\beta(\Omega_0 + \Omega_1) \leq \beta(\Omega_0) + \beta(\Omega_1)$ for every $\Omega_0, \Omega_1 \in \mathcal{P}(X)$.*

(g) *Regular if the condition $\beta(\Omega) = 0$ is equivalent to the relative compactness of Ω.*

A typical example of an m.n.c. is the Hausdorff measure of noncompactness α defined for all $\Omega \subset X$ by

$$\alpha(\Omega) := \inf\{\epsilon \in \mathbb{R}_+^n : \text{there exists } n \in \mathbb{N} \text{ such that } \Omega \subseteq \cup_{i=1}^n B(x_i, \epsilon)\}.$$

Definition 8.21. *Let X, Y be two generalized normed spaces and $F : X \to \mathcal{P}(Y)$ be a multivalued map. Then F is called an M-contraction with (respect to β) if there exists $M \in M_{n \times n}(\mathbb{R})$ converging to zero such that for every $D \in \mathcal{P}(X)$, we have*

$$\beta(F(D)) \leq M\beta(D).$$

The next result is concerned with β−condensing maps.

Theorem 8.22. *Let $V \subset X$ be a bounded closed convex set and $N : V \to V$ be a generalized β−condensing continuous mapping, where β is a nonsingular measure of noncompactness defined on subsets of X. Then the set*

$$Fix(N) = \{x \in V : x \in N(x)\}$$

is nonempty.

Proof. Let $\mathcal{M}_1 = V$, $\mathcal{M}_{k+1} = \overline{co}N(\mathcal{M}_k)$, $k \in \mathbb{N}$. It is clear that the sequence $(\mathcal{M}_k)_{k \in \mathbb{N}}$ consists of a decreasing sequence of nonempty closed convex subsets of V. Since N is β-condensing,

$$\beta(\overline{co}N(\mathcal{M}_1)) = \beta(N(\mathcal{M}_1)) \leq M\beta(V).$$

Continuing this process, we obtain

$$\beta(\mathcal{M}_{k+1}) \leq M^{k+1}\beta(V).$$

Therefore,

$$\lim_{k \to \infty} \beta(\mathcal{M}_k) = 0.$$

Thus

$$C = \bigcap_{k=1}^{\infty} \mathcal{M}_k \neq \emptyset$$

is convex and compact. Furthermore, by the convexity of $\{\mathcal{M}_k\}_{k\in\mathbb{N}}$ and $N(\mathcal{M}_1) \subseteq \mathcal{M}_1$, we have

$$N(\mathcal{M}_2) \subseteq \mathcal{M}_2 \Rightarrow N(\mathcal{M}_3) \subseteq \mathcal{M}_3.$$

Proceeding by induction yields

$$N(\mathcal{M}_k) \subseteq \mathcal{M}_k \text{ for every } k \in \mathbb{N} \Rightarrow N(C) \subseteq C.$$

Hence, by Theorem 8.13, N has at least one fixed point. $\square$

As a consequence of Theorem 8.22, we present versions of Schaefer's fixed point theorem and the nonlinear alternative of Leray-Schauder type theorem for $\beta-$condensing operators in a generalized Banach space.

Theorem 8.23. *Let E be a generalized Banach space and $N : E \to E$ be a continuous and $\beta-$condensing operator. Assume that the set*

$$A = \{x \in E : x = \lambda N(x) \text{ for some } \lambda \in (0,1)\}$$

is bounded. Then N has a fixed point.

Theorem 8.24. *Let E be a generalized Banach space, $U \subset E$ be a bounded, convex, open neighborhood of zero, and let $G : \overline{U} \to E$ be a continuous and $\beta-$condensing mapping. If G satisfies the boundary condition*

$$x \neq \lambda G(x)$$

for all $x \in \partial U$ and $0 \leq \lambda \leq 1$, then the set $Fix(G) = \{x \in U : x = G(x)\}$ is nonempty.

8.6 Approximation method and Perov type fixed point theorem

For a fairly general class of multifunctions with compact and nonconvex values, and using a different method, approximate continuous selections have been constructed by Cellina [80], Górniewicz [141] and Górniewicz and Lassonde [142], and hence used to develop an index theory. In this section we use a technique of approximation proved recently for a version of a Krasnosel'skii-type fixed point result.

The following continuous approximation theorem is the key of this section.

Theorem 8.25. *Let X be a generalized normed space, Y be a generalized Banach space and $F : X \to \mathcal{P}_{cv}(Y)$ be an u.s.c. multivalued map. Then, for every $\epsilon \in \mathbb{R}_+^n$, there exists a locally Lipschitzian function $f_\epsilon : X \to Y$ such that*

$$f_\epsilon(X) \subseteq coF(X)$$

and

$$\Gamma(f_\epsilon) \subseteq \Gamma(F) + B(F(x), \epsilon),$$

where

$$B(F(x), \epsilon) = \{z \in X : d(z, F(x)) < \epsilon\}.$$

Proof. Fix $\epsilon = (\epsilon_1, \ldots, \epsilon_n) > 0$. For every $x \in X$ there exists $B(x, \delta(x)) \subset X$ such that

$$F(y) \subseteq F(B(x, \delta(x)) \subset F(x) + B(0, \epsilon) \quad \text{for each } y \in B(x, \delta(x)),$$

where $\delta(x) = (\delta_1(x), \ldots, \delta_n(x)) > 0$. We take $0 < \delta(x) \leq \frac{\epsilon}{2}$. The family $\{B(x, \delta(x))\}_{x \in X}$ covers X. From Theorem 7.13, X is paracompact. Let $\{U_i\}_{i \in I_*}$ be a local refinement and $\{f_i\}_{i \in I_*}$ be a locally Lipschitzean partition of unity subordinate to it. Choose for each $i \in I_*$ an $x_i \in U_i$ and define f_ϵ by

$$f_\epsilon(x) = \sum_{i \in I_*} f_i(x) z_i \quad \text{for each } x \in X,$$

where $z_i \in F(x_i)$. It is clear that f_ϵ is well defined, locally Lipschitzean and

$$f_\epsilon(X) \subseteq co(F(X)).$$

Now we show that f_ϵ is an approximate of F. Let $x \in X$ and $I_*(x)$ the subset of all $i \in I_*$ such that $f_i(x) \neq 0$; therefore $x \in \cap_{i \in I_*(x)} B(x_i, \delta_i(x))$. Let $i, j \in I_*(x)$ then

$$d(x_i, x_j) \leq d(x_i, x) + d(x, x_j) \leq \delta_i + \delta_j < \epsilon.$$

Let $k \in I_*(x)$ be such that

$$\delta_k = \max_{i \in I_*} \delta_i.$$

For every $i \in I_*(x)$ we have

$$F(x_i) \subset F(B(x_i, \delta_i)) \subset F(B(x_k, 2\delta_k)) \subset F(x_k) + B(0, \epsilon), \quad \text{for all } i \in I_*(x).$$

Using the fact that $F(x) + B(0, \epsilon)$ is convex, then

$$f_\epsilon(x) \in F(x) + B(0, \epsilon).$$

$\square$

We have the first result.

Theorem 8.26. *Let X be a generalized Banach space, C be a nonempty compact convex subset of X, and $G : C \to \mathcal{P}_{cp,cv}(C)$ be an u.s.c. multivalued map. Then the operator inclusion G has at least one fixed point, that is, there exists $x \in C$ such that $x \in G(x)$.*

Proof. From Theorem 8.25, given $\epsilon > 0$, there exists $f_\epsilon : C \to X$, a continuous map, such that

$$\Gamma(f_\epsilon) \subset \Gamma(G) + B(0, \epsilon) \tag{8.1}$$

and

$$f_\epsilon(C) \subset coG(C).$$

We consider $F_\epsilon : C \to C$ defined by

$$F_\epsilon(y) = f_\epsilon(x), \quad x \in C.$$

By Theorem 8.13, there exists $x_\epsilon \in C$ such that

$$x_\epsilon = f_\epsilon(x_\epsilon).$$

Let $\{\epsilon_n : n \in \mathbb{N}\}$ be such that

$$\epsilon_n \to 0 \text{ as } n \to \infty.$$

Since C is compact, there exists a subsequence x_{ϵ_n} converging to $x \in C$. So

$$f_{\epsilon_n}(x_{\epsilon_n}) \to x, \text{ as } n \to \infty. \tag{8.2}$$

From (8.2), we get

$$d((x_{\epsilon_n}, f_{\epsilon_n}(x_{\epsilon_n})), \Gamma(G)) \leq \epsilon_n, \text{ for every } n \in \mathbb{N}.$$

It is clear that G has a closed graph in $X \times X$ and consequently $x \in G(x)$. This implies that G has at least one fixed point. $\square$

As a consequence of the above result we present the multivalued version of Schaefer's fixed point theorem and the nonlinear alternative Leray-Schauder type theorem in generalized Banach spaces.

Theorem 8.27. *Let $(X, \|\cdot\|)$ be a generalized Banach space and $F : X \to \mathcal{P}_{cp,cv}(X)$ be a completely continuous multivalued mapping and u.s.c. Moreover assume that the set*

$$\mathcal{A} = \{x \in X : x \in \lambda F(x) \quad \text{for some } \lambda \in (0,1)\}$$

is bounded. Then F has a fixed point.

Proof. Let $K > 0$ be such that

$$\|x\|_i < K \text{ for each } x \in \mathcal{A}.$$

Set $M = (K, \ldots, K) \in \mathbb{R}^n_+$ and $U = B(0, M)$. Since F is completely continuous, there exists $M_* \in \mathbb{R}^n_+$ such that

$$F(\overline{U}) \subseteq \frac{1}{2} B(0, M_*).$$

We consider the following multivalued operator $F_* : X \to \mathcal{P}_{cp,cv}(X)$ defined by

$$F_*(x) = \begin{cases} F(x) & \text{if } x \in \overline{U} \\ \frac{1}{p_U(x)} F\left(\frac{x}{p_U(x)}\right) & \text{if } x \in X \setminus \overline{U}, \end{cases}$$

where p_U is the Minkowski function of U. Clearly, $F_*(\overline{B(0, M_*)}) \subseteq \overline{B(0, M_*)}$ and F_* is u.s.c. Hence by Theorem 8.26 there exists $x \in \overline{B(0, M_*)}$ such that $x \in F_*(x)$. If $x \in X \setminus \overline{U}$ then

$$x \in \frac{F\left(\frac{x}{p_U(x)}\right)}{p_U(x)} \Rightarrow \frac{x}{p_U(x)} \in \frac{1}{p_U^2(x)} F\left(\frac{x}{p_U(x)}\right).$$

Since $x \in X \setminus \overline{U}$, then for $p_U(x) > 1$ we have

$$\frac{x}{p_U(x)} \in \partial U \Rightarrow M < \|\frac{x}{p_U(x)}\| \leq M.$$

Consequently, there exists $x_* \in U$ such that $x_* \in F(x_*)$. $\square$

Next we state the nonlinear alternative of Leray-Schauder type.

Theorem 8.28. *Let X be a generalized Banach space, $U \subset X$ be a bounded, convex open neighborhood of zero and let $G : \overline{U} \to X$ be a compact u.s.c. multivalued map. If G satisfies the boundary condition*

$$x \notin \lambda G(x)$$

for all $x \in \partial U$ and $0 \leq \lambda < 1$, then the set $Fix(G) = \{x \in U : x \in G(x)\}$ is nonempty.

Proof. Let p be the Minkowski function of U, and since $\overline{U}$ is bounded, then there exists $M > 0$ such that

$$G(\overline{U}) \subseteq \frac{1}{2} B(0, M_*), \quad M_* = (K, \ldots, K).$$

Consider $G_* : \overline{B(0, M_*)} \to \mathcal{P}_{cp,cv}(\overline{B(0, M_*)})$ defined by

$$G_*(x) = \begin{cases} G(x) & \text{if } x \in \overline{U} \\ \frac{1}{p(x)} G(\frac{x}{p(x)}) & \text{if } x \in X \backslash \overline{U}. \end{cases}$$

It is clear that $\overline{B(0, M_*)}$ is a closed, convex, bounded subset of X and G_* is a compact and u.s.c. multivalued operator. Then from Theorem 8.26 there exists $x \in \overline{B(0, M_*)}$ such that $x \in G_*(x)$. If $x \in X \backslash \overline{U}$ then

$$x = \frac{G\left(\frac{x}{p(x)}\right)}{p(x)} \Rightarrow \frac{x}{p(x)} = \frac{1}{p^2(x)} G\left(\frac{x}{p(x)}\right).$$

Since $x \in X \backslash \overline{U}$, then

$$p(x) = 1 \text{ or } p(x) > 1 \Rightarrow x \in \partial U, \ \frac{x}{p(x)} \in \partial U.$$

This is a contradiction with

$$z \notin \lambda G(z), \quad \text{for each, } \lambda \in [0, 1], \ z \in \partial U.$$

Consequently, there exist $x_* \in U$ such that $x_* \in G(x_*)$. □

Theorems 8.26 and 8.1 immediately yield the following Krasnoel'skii fixed point for the sum of two operators. We need the next lemma in the proof.

Lemma 8.29. *[240] Let X be a generalized Banach space, $C \subset X$ be a closed set, and $B : C \to X$ be a contraction mapping. Then $I - B : C \to (I - B)(C)$ is a homeomorphism.*

Proof. Since B is continuous, $I - B$ is continuous. In order to prove that $I - B : C \to (I - B)(C)$ is bijective, let $y_1, y_2 \in X$ be such that

$$B(y_1) - y_1 = B(y_2) - y_2.$$

Then,

$$0 = \|B(y_1) - y_1 - B(y_2) + y_2\| \geq \|y_1 - y_2\| - \|B(y_1) - B(y_2)\| \quad \text{implies} \quad y_1 = y_2,$$

which shows that $I - B$ is one to one. Hence, the inverse of $I - B : C \to (I - B)(C)$ exists. The continuity of $(I - B)^{-1}$ follows in the usual way. □

Theorem 8.30. *Let X be a generalized Banach space, C be a nonempty compact convex subset of X, $G : C \to \mathcal{P}_{cp,cv}(X)$ be an u.s.c. multivalued map and $B : C \to C$ be a map. Assume that G and B satisfy the following hypotheses:*

$(\mathcal{H}_1)$ *B is a contraction mapping.*

$(\mathcal{H}_2)$ *$B(C) + G(C) \subset C$.*

Then the inclusion $x \in B(x) + G(x)$ has a solution, that is $B + G$ has a fixed point.

Proof. From Theorem 8.25, given $\epsilon > 0$, there exists $f_\epsilon : C \to X$, a continuous map, such that (8.2) and

$$f_\epsilon(C) \subset coG(C).$$

From $(\mathcal{H}_2)$, and since C is convex, we get

$$B(C) + f_\epsilon(C) \subseteq C.$$

For fixed $y \in C$, we consider $F_\epsilon^y : C \to C$ defined by

$$F_\epsilon^y(x) = B(x) + f_\epsilon(y), \quad x \in C.$$

By Theorem 8.1, there exists a unique $x_\epsilon(y) \in C$ such that

$$x_\epsilon(y) = B(x_\epsilon(y)) + f_\epsilon(y).$$

From [240] the mapping $I - B : C \to (I - B)(C)$ is an homeomorphism. We define the operator $N_\epsilon : C \to C$ by

$$N_\epsilon(x) = (I - B)^{-1} f_\epsilon(x).$$

It is easy to see that N_ϵ satisfies the conditions of Theorem 8.13, and so, there exists $x_\epsilon \in C$ such that

$$x_\epsilon = B(x_\epsilon) + f_\epsilon(x_\epsilon).$$

Let $\{\epsilon_n : n \in \mathbb{N}\}$ be such that

$$\epsilon_n \to 0 \text{ as } n \to \infty.$$

Since C is compact, there exists a subsequence x_{ϵ_n} converging to $x \in C$ such that (8.2). From (8.2), we get

$$d((x_{\epsilon_n}, f_{\epsilon_n}(x_{\epsilon_n})), Graph(G)) \leq \epsilon_n, \quad \text{for every } n \in \mathbb{N}.$$

It is clear that G has a closed graph in $X \times X$ and consequently $(I - B)(x) \in G(x)$. This implies that

$$x \in B(x) + G(x).$$

$\square$

As a consequence of Theorem 8.30, we can present a global multivalued version of Krasonsel'skii's fixed point theorem.

Theorem 8.31. *Let X be a generalized Banach and $G : X \to \mathcal{P}_{cp,cv}(X)$ be an u.s.c. multivalued operator. Assume condition $(\mathcal{H}_1)$ holds and G is compact. Then, either*

(a) $x \in \lambda B(\frac{x}{\lambda}) + \lambda G(x)$ has a solution for $\lambda = 1$,

or

(b) the set $\{x \in X : x \in \lambda B(\frac{x}{\lambda}) + \lambda G(x), \ \lambda \in (0, 1)\}$ is unbounded.

From the above result we can then easily prove an "equation" version.

Theorem 8.32. *Let X be a generalized Banach and $N : X \to X$ be a continuous operator. Assume condition $(\mathcal{H}_1)$ holds and N is compact. Then, either*

(c) $x = \lambda B(\frac{x}{\lambda}) + \lambda G(x)$ has a solution for $\lambda = 1$,

or

(d) the set $\{x \in X : x = \lambda B(\frac{x}{\lambda}) + \lambda G(x), \ \lambda \in (0, 1)\}$ is unbounded.

8.7 Covitz and Nadler type fixed point theorems

Theorem 8.33. *Let (X, d) be a generalized complete metric space, and let $F : X \to \mathcal{P}_{cl}(X)$ be a multivalued map. Assume that there exist $A, B, C \in \mathcal{M}_{n \times n}(\mathbb{R}_+)$ such that*

$$H_d(F(x), F(y)) \le Ad(x, y) + Bd(y, F(x)) + Cd(x, F(x)) \tag{8.3}$$

where $A + C$ converges to zero. Then there exists $x \in X$ such that $x \in F(x)$.

Proof. Let $x \in X$ and

$$D(x) = D(x, d(x, F(x)) := \{y \in X : d(x, y) \le d(x, F(x))\}.$$

Since $F(x)$ is closed, then

$$D(x) \cap F(x) \ne \emptyset.$$

So we can select $x_1 \in F(x)$ such that

$$d(x, x_1) \le d(x, F(x)) \le Ad(x, x_1) + Bd(x_1, F(x)) + Cd(x, F(x)),$$

thus

$$d(x, x_1) \le (A + C)d(x, F(x)). \tag{8.4}$$

For $x_2 \in F(x_1)$ we have

$$
\begin{aligned}
d(x_2, x_1) &\le d(x_1, F(x)) + H_d(F(x), F(x_1)) \\
&\le Ad(x, x_1) + Cd(x, F(x)) \\
&\le (A + C)d(x, x_1),
\end{aligned}
$$

then

$$d(x_2, x_1) \le (A + C)^2 d(x, F(x)). \tag{8.5}$$

Continuing this procedure we can find a sequence $(x_n)_{n \in \mathbb{N}}$ of X such that

$$d(x_n, x_{n+1}) \le (A + C)^{n+1} d(x, F(x)), \ n \in \mathbb{N}.$$

Let $p \in \mathbb{N}$. Since d is a metric we have

$$d(x_n, x_{n+p}) \le d(x_n, x_{n+1}) + \ldots + d(x_{n+p-1}, x_{n+p}).$$

Hence, for all $n, p \in \mathbb{N}$, the following estimation holds

$$d(x_n, x_{n+p}) \le (A + C)^{n+1}(I + (A + C) + (A + C)^2 + \cdots + (A + C)^{p-1})d(x, F(x))$$

Therefore

$$d(x_n, x_{n+p}) \to 0 \text{ as } n \to \infty,$$

so $(x_n)_{n \in \mathbb{N}}$ is a Cauchy sequence in the complete generalized metric space X. Then there exists $x_* \in X$ such that

$$d(x_n, x_*) \to 0 \text{ as } n \to \infty.$$

From (8.3) we obtain

$$
\begin{aligned}
d(x_*, F(x_*)) &\le d(x_*, x_n) + H_d(F(x_{n+1}), F(x_*)) \\
&\le d(x_n, x_*) + Ad(x_{n+1}, x_*) + Bd(x_*, F(x_{n+1}))
\end{aligned}
$$

$$+Cd(x_{n+1}, F(x_{n+1}))$$
$$\leq \quad d(x_n, x_*) + Ad(x_{n+1}, x_*) + Bd(x_*, F(x_{n+1}))$$
$$+Cd(x_{n+1}, F(x_{n+1}))$$
$$\leq \quad d(x_n, x_*) + Ad(x_{n+1}, x_*) + Bd(x_*, x_n)$$
$$+Cd(x_{n+1}, x_n) \to \text{ as } n \to \infty.$$

This implies that $x_* \in F(x_*)$. $\qquad\square$

Lemma 8.34. *Let (X, d) be a generalized Banach space and $F : X \to \mathcal{P}_{cl}(Y)$ be a multivalued map. Assume that there exist $p \in \mathbb{N}$ and $M \in \mathcal{M}_{n \times n}(\mathbb{R}_+)$ converging to zero such that*

$$H_d(F^p(x), F^p(y)) \leq Md(x, y), \text{ for each } x, y \in X$$

and

$$\sup_{a \in F^{p+1}(y)} d(a, F(x)) \leq d(y, F(x)).$$

Then there exists $x \in X$, such that $x \in F(x)$.

Proof. By Theorem 8.33, there exists $x \in X$ such that $x \in F^p(x)$. Now we show that $x \in F(x)$.

$$\begin{aligned} d(x, F(x)) &\leq \quad d(x, F^{p+1}(x)) + H_d(F^{p+1}(x), F(x)) \\ &\leq \quad H_d(F^p(x), F^{p+1}(x)) \\ &\leq \quad Md(x, F(x)). \end{aligned}$$

Hence

$$d(x, F(x)) \quad \leq \quad M^k d(x, F(x)) \to 0 \text{ as } k \to \infty \Rightarrow d(x, F(x)) = 0.$$

$\qquad\square$

Theorem 8.35. *Let (X, d) be a complete generalized metric space and $B(x_0, r_0) = \{x \in X : d(x, x_0) < r_0\}$ be the open ball in X with radius r_0 and centered at some point $x_0 \in X$. Assume that $F : B(x_0, r_0) \to \mathcal{P}_{cl}(X)$ is a contractive multivalued map such that*

$$H_d(x_0, F(x_0)) < (I - M)r_0,$$

where $M \in \mathcal{M}_{n \times n}(\mathbb{R}_+)$ is the matrix contraction for F. Then F has at least one fixed point.

Proof. Let $r_1 \in \mathbb{R}_+^n$ be such that

$$d(x_0, F(x_0)) \leq (I - M)r_1 < (I - M)r_0.$$

Set

$$K(x_0, r_1) = \{x \in X : d(x, x_0) \leq r_1\}.$$

It is clear that $K(x_0, r_1)$ is complete generalized metric space. Let us define a multivalued map

$$F_*(x) = F(x) \text{ for all } x \in K(x_0, r_1).$$

In view of Theorem 8.33, for the proof, it is sufficient to show that

$$F_*(K(x_0, r_1)) \subseteq K(x_0, r_1).$$

Let $x \in K(x_0, r_1)$; then we have:

$$d(x_0, y) \le \sup_{z \in F(x)} d(x_0, z) = H_d(x_0, F(x)), \quad \text{for all } y \in F(x).$$

Thus

$$
\begin{aligned}
d(x_0, y) &\le H_d(x_0, F(x_0)) + H_d(F(x_0), F(y)) \\
&\le (I - M)r_1 + Md(x_0, y) \le (I - M)r_1 + r_1 M = r_1,
\end{aligned}
$$

and the proof is completed. □

Lemma 8.36. *Let E be a generalized Banach space, $Y \subseteq E$ be a nonempty convex compact subset of E and $F : X \to \mathcal{P}_{cl}(Y)$ be a multivalued map such that*

$$H_d(F(x), F(y)) \le d(x, y), \quad \text{for each } x, y \in X.$$

Then there exists $x \in X$, such that $x \in F(x)$.

Proof. For every $m \in \mathbb{N}$, we have $\frac{I}{2^m} \in \mathcal{M}_{n \times n}(\mathbb{R}_+)$ and

$$\frac{I}{2^{mk}} \to 0 \text{ as } k \to \infty.$$

Thus, for some $x_0 \in Y$ the mapping $f_m : Y \to Y$ defined by

$$F_m(x) = (1 - \frac{1}{2^m})F(x) + \frac{1}{2^m}x_0 \in Y \quad \text{for all } x \in Y.$$

Then

$$H_d(F_m(x), F_m(y)) \le \frac{I}{2^m}d(x, y) \quad \text{for all } x, y \in Y.$$

From Theorem 8.33 there exists $x_m \in Y$ such that

$$x_m \in F_m(x_m), \quad m \in \mathbb{N}.$$

Since Y is compact, then there exists a subsequence of $(x_m)_{m \in \mathbb{N}}$ converging to $x \in Y$. Now we show that $x \in F(x)$.

$$
\begin{aligned}
d(x, F(x)) &= \begin{pmatrix} d_1(x, F(x)) \\ \cdots \\ d_n(x, F(x)) \end{pmatrix} \\
&\le d(x, x_m) + d(x_m, F(x_m)) + H_d(F(x_m), F(x)) \\
&\le 2Id(x, x_m) + d(x_m, F(x_m))
\end{aligned}
$$

and

$$
\begin{aligned}
d(x_m, F(x_m)) &= \begin{pmatrix} d_1(x_m, F(x_m)) \\ \cdots \\ d_n(x_m, F(x_m)) \end{pmatrix} \\
&= \begin{pmatrix} \|x_m - F(x_m)\|_1 \\ \cdots \\ \|x_m - F(x_m)\|_n \end{pmatrix} \\
&\le d(x_m, F(x)) + H_d(F(x), F(x_m)) \\
&\le d(x_m, z_m) + d(x, x_m)
\end{aligned}
$$

$$\leq \quad d(x_m, x) + \frac{1}{2^m} d(z_m, x_0)$$

where

$$z_m \in F(x) \text{ and } x_m = (1 - \frac{1}{2^m})z_m + \frac{1}{2^m}x_0.$$

Since $x_m \to x$ as $m \to \infty$, then

$$z_m \to x \text{ as } m \to \infty.$$

Hence

$$d(x, F(x)) \leq 3I d(x, x_m) + \frac{1}{2^m} d(z_m, x_0) \to 0 \text{ as } m \to \infty.$$

$\square$

We wish to prove a Kranosel'skii type fixed point theorem by using an expansive operator combined with a continuous operator.

Lemma 8.37. *Let E be a generalized normed space and $C \subseteq E$. Assume the mapping $B : C \to X$ is expansive with constant $k > 1$. Then the inverse of $I - B : C \to (I - B)(C)$ exists and*

$$d((I - B)^{-1}(x), (I - B)^{-1}(y)) \leq \frac{1}{k-1}d(x, y), \ x, y \in (I - B)(C).$$

Proof. Let $x, y \in C$ and $x - B(x) = y - B(y)$, then

$$
\begin{aligned}
0 = d(x - B(x), y - B(y)) &= \begin{pmatrix} \|x - B(x) - y + B(y)\|_1 \\ \cdots \\ \|x - B(x) - y + B(y)\|_n \end{pmatrix} \\
&\geq \begin{pmatrix} \|B(y) - B(x)\|_1 - \|x - y\|_1 \\ \cdots \\ \|B(y) - B(x)\|_n - \|x - y\|_n \end{pmatrix} \\
&\geq \begin{pmatrix} k\|y - x\|_1 - \|x - y\|_1 \\ \cdots \\ k\|y - x\|_n - \|x - y\|_n \end{pmatrix} \\
&= (k - 1)I d(x, y).
\end{aligned}
$$

Thus $I - B : C \to (I - B)(C)$ is invertible. Let $x, y \in (I - B)(C)$, then there exist $a, b \in C$ such that

$$a - B(a) = x, \ b - B(b) = y.$$

Hence

$$d(a, b) = d((I - B)^{-1}(x), (I - B)^{-1}(y)) \text{ and } d(x, y) \geq kd(a, b) - d(a, b).$$

Therefore

$$d(I - B)^{-1}(x), (I - B)^{-1}(y)) \leq \frac{I}{k-1}d(x, y) \text{ for all } x, y \in (I - B)(C).$$

$\square$

Theorem 8.38. *Let E be a generalized Banach space and C be a compact convex subset of E. Assume that $A : M \to X$ is continuous and $B : C \to E$ is a continuous expansive map satisfying*

$(\mathcal{H}_1)$ *for each* $x \in C$ *such that*

$$x = B(x) + A(y) \Rightarrow y \in C.$$

Then there exists $y \in C$ *such that* $y = By + A(y)$.

Proof. Let $y \in C$. Let $F_y : C \to X$ be an operator defined by

$$F_y(x) = B(x) + A(y), \quad x \in C.$$

From Theorem 8.11 there exists a unique $x(y) \in C$ such that

$$x(y) = B(x(y)) + A(y).$$

By Lemma 8.37 $I - B$ is invertible. Moreover, $(I - B)^{-1}$ is continuous. Let us define $N : C \to C$ by

$$y \to N(y) = (I - B)^{-1}A(y).$$

Let $x \in C$ and $N(x) = (I - B)^{-1}(A(x))$. Then

$$N(x) = (I - B)^{-1}(A(x)) \Rightarrow N(x) = B(N(x)) + A(x),$$

and thus $(\mathcal{H}_1)$ implies that $N(x) \in C$. Let $\{y_m : m \in \mathbb{N}\} \subseteq C$ be a sequence converging to y in C. We shall show that $N(y_m)$ converges to $N(y)$. Set $x_m = (I - B)^{-1}A(y_m)$, then

$$(I - B)(x_m) = A(y_m), \ m \in \mathbb{N}.$$

Since C is compact, there exists a subsequence of $\{x_m\}$ converging for some $x \in C$. Then

$$(I - B)(x_m) \to (I - B)(x) \text{ as } m \to \infty.$$

Hence

$$A(y_m) \to (I - B)(x) \ \text{ as } \ m \to \infty.$$

Therefore

$$N(y_m) \to N(y) \ \text{ as } \ m \to \infty.$$

Hence from Theorem 8.13, there exists $y \in C$ such that $y = (I - B)^{-1}A(y)$, and we deduce that $B + G$ has a fixed point in C. $\qquad \square$

8.8 Fixed point index

The classical idea of index theory is to assign to each continuous self map of an n-dimensional manifold (for example an open subset of $\mathbb{R}^n$ with a smooth boundary) a number that would count the number of fixed points.

Definition 8.39. *A subset* C *of generalized Banach space* E *is called a retract of* E *if there exists a continuous mapping* $r : E \to C$ *such that* $r(x) = x$ *for all* $x \in C$.

Lemma 8.40. *Let C be a retract subset of a generalized real Banach space E. Then $h(C)$ is a retract subset of the Banach space $(\widetilde{E}, \|\cdot\|_*)$ where*

$$\widetilde{E} = \{(x, \ldots, x) \in \Pi_{i=1}^n E_i : x \in E = E_i, \ i = 1, \ldots, n\},$$

$$\|(x, \ldots, x)\|_* = \sum_{i=1}^n \|x\|_i, \quad x \in E$$

and $h : E \to \widetilde{E}$ is defined by

$$h(x) = (x, \ldots, x), \ x \in E.$$

Proof. From the proof of Lemma 7.2, h is a homeomorphism. Set

$$h(C) = \{(x, \ldots, x) \in \widetilde{E} : x \in C\}.$$

Since C is a retract of E, there there exists a continuous map $r : E \to C$ such that $r(x) = x$ for $x \in C$. We define $r_* : \widetilde{E} \to h(C)$ by

$$r_*(x, \ldots, x) = (h \circ r \circ h^{-1})(x, \ldots, x), \quad x \in E.$$

It clear that r_* is continuous, and for all $x \in C$, we have $r_*(x, \ldots, x) = h(x)$. This implies that $h(C)$ is a retract of $\widetilde{E}$. $\qquad\square$

In what follows we establish the existence and uniqueness of the fixed point index (see Guo and Lakshmikantham [147]).

Theorem 8.41. *Let C_* be a retract of real Banach space X. Then for every bounded relatively open subset U of C_* and every completely continuous operator $A : \overline{U} \to C_*$ which has no fixed point on ∂U (relative to C_*), there exists an integer $i(A, U, C_*)$ satisfying the following conditions.*

C_1) *Normality: $i(A, U, C_*) = 1$ if $Ax = y_0$ for an $x \in \overline{U}$;*

C_2) *Additivity: $i(A, U, C_*) = i(A, U_1, C_*) + i(A, U_2, C_*)$ whenever U_1 and U_2 are disjoint open subsets of U such that A has no fixed points in $\overline{U} \backslash U_1 \cup U_2$;*

C_3) *Homotopy invariance: $i(H(t, .), U, C_*)$ is independent of $t \in [0, 1]$ where $H : [0, 1] \times \overline{U} \to C_*$ is completely continuous and $H(t, x) \neq x$ for any $(t, x) \in [0, 1] \times \partial U$;*

C_4) *Permanence: $i(A, U, C_*) = i(A, U \cap Y, Y)$ if Y is a retract of C_* and $A(\overline{U}) \subset Y$;*

C_5) *Excision: $i(A, U, C_*) = i(A, U_0, C_*)$ whenever U_0 is an open subset of U such that A has no fixed points in $\overline{U} \backslash U_0$;*

C_6) *Solution: if $i(A, U, C_*) \neq 0$, then A has at least one fixed point in U.*

From Lemma 8.40 and Theorem 8.41, we can easily prove the following result.

Theorem 8.42. *Let C be a retract of the real generalized Banach space E. Then for every bounded relatively open $U \subset C$ and every completely continuous operator $N : \overline{U} \to C$ which has no fixed point on ∂U (relative to C), there exists an integer $i(N, U, C) = i(N_*, h(U), h(C))$, where $N_* = h \circ N \circ h^{-1}$, satisfying the following conditions:*

$\overline{C}_1$) *Normality: $i(N_*, h(U), h(C)) = 1$ if $N_*(x, \ldots, x) = y_0$ for an $(x, \ldots, x) \in \overline{h(U)}$;*

$\overline{C}_2$) *Additivity:* $i(N_*, h(U), h(C)) = i(N_*, U_1, h(C)) + i(N_*, U_2, h(C))$ *where* U_1 *and* U_2 *are disjoint open subsets of* $h(U)$ *such that* N_* *has no fixed points in* $\overline{h(U)} \backslash U_1 \cup U_2$;

$\overline{C}_3$) *Homotopy invariance:* $i(H(t, .), h(U), h(C))$ *is independent of* $t \in [0, 1]$ *where* $H : [0, 1] \times \overline{h(U)} \to h(C)$ *is completely continuous and* $H(t, (x, \dots, x)) \neq (x, \dots, x)$ *for any* $(t, x) \in [0, 1] \times \partial h(U)$;

$\overline{C}_4$) *Permanence:* $i(N_*, h(U), h(C)) = i(N_*, h(U) \cap Y, Y)$ *if* Y *is a retract of* $h(C)$ *and* $N_*(\overline{h(U)}) \subset Y$;

$\overline{C}_5$) *Excision:* $i(N_*, h(U), h(C)) = i(N_*, U_0, h(C))$ *whenever* U_0 *is an open subset of* $h(U)$ *such that* N_* *has no fixed points in* $\overline{h(U)} \backslash U_0$;

$\overline{C}_6$) *Solution: if* $i(N_*, h(U), h(C)) \neq 0$, *then* N_* *has at least one fixed point in* $h(U)$.

8.9 Legggett-Williams type fixed point results

In this section we give some Legggett-Williams type fixed point type results via index theory.

Theorem 8.43. *[147] Let* C_* *be a retract of the real Banach space* X *and* C_1 *be a bounded convex retract of* C_*. *Let* U *be a nonempty relatively open subset of* C_* *with* $U \subset C_1$. *If* $A : C_1 \to C_*$ *is completely continuous,* $A(C_1) \subset C_1$, *and* A *has no fixed points in* $C_1 \backslash U$, *then* $i(A, U, C_*) = 1$.

Corollary 8.44. *[147] Let* C_* *be a nonempty closed convex set in a real Banach space* X. *If* $A : C_* \to C_*$ *is completely continuous, then* $i(A, U, C_*) = 1$.

The following result is vector version of the above theorem.

Theorem 8.45. *Let* C *be a retract of the real generalized Banach space* E *and* C_1 *be a bounded convex retract of* C. *Let* U *be a nonempty relatively open subset of* C *with* $U \subset C_1$. *Suppose that* $N : C_1 \to C_*$ *is completely continuous,* $N(C_1) \subset C_1$, *and* N *has no fixed points in* $C_1 \backslash U$. *Then* $i(N, U, C_*) = 1$.

Proof. From Theorem 8.42, we have $i(N, U, C_*) = i(N_*, h(U), h(C))$. By Theorem 8.43, $i(N, U, C_*) = 1$. $\square$

Corollary 8.46. *Let* C *be a nonempty closed convex subset of a real generalized Banach space* E. *If* $N : C \to C$ *is completely continuous, then* $i(N, U, C) = 1$.

8.10 Legggett-Williams type fixed point theorems in vector Banach spaces

In the last twenty years, there has been much interest focused on proving the existence of positive solutions of various types of nonlinear problems based on the Krasnosel'skii's [190] and Leggett-Williams [203] fixed point theorems. There have been a number of efforts to generalize these fixed point theorems; for example, see [6, 21–23, 294]. Recently, vector versions of Krasnosel'skii's fixed point theorem on cones were given by Budişan [65] and Precup [248]. Here we wish to give some vector versions of the Leggggett-Williams theorem.

Definition 8.47. *Let E be a real generalized Banach space. A nonempty closed convex set $P \subset E$ is a cone if it satisfies the following two conditions:*

(i) $x \in P$, and $\lambda \geq 0$, then $\lambda x \in P$;

(ii) $x \in P$ and $-x \in P$ implies $x = 0$.

Every cone $P \subset E$ induces an ordering in E given by

$$x \leq y \quad \text{if and only if } y - x \in P.$$

Definition 8.48. *A nonnegative continuous map $\Psi : P \longrightarrow \mathbb{R}^n_+$ defined on a cone P in a generalized Banach space E is concave if*

$$\Psi(tx + (1-t)y) \geq t\Psi(x) + (1-t)\Psi(y),$$

for all x, $y \in P$ and $t \in [0,1]$.

Definition 8.49. *A nonnegative continuous map $\Phi : P \longrightarrow \mathbb{R}^n_+$ defined on a cone P in a generalized Banach space E is convex if*

$$\Phi(tx + (1-t)y) \leq t\Phi(x) + (1-t)\Phi(y),$$

for all x, $y \in P$ and $t \in [0,1]$.

Let φ and θ be a nonnegative continuous convex functionals on P, ψ be a non-negative continuous functional, and ϕ be a nonnegative continuous concave functional on P. Then, for positive vectors $(a_1, \ldots, a_n)$, $(b_1, \ldots, b_n)$, and $(d_1, \ldots, d_n)$, we define the following sets:

- $P(\varphi, (d_1, \ldots, d_n)) = \{u \in P : \varphi(u) < d_i, \ i = 1, \ldots, n\}$;

- $P(\varphi, \phi, (b_1, \ldots, b_n), (d_1, \ldots, d_n)) = \{u \in P : b_i \leq \phi(u), \varphi(u) \leq d_i, \ i = 1, \ldots, n\}$;

- $P(\varphi, \theta, \psi, (b_1, \ldots, b_n), (c_1, \ldots, c_n), (d_1, \ldots, d_n))$
$$= \begin{cases} u \in P : b_i \leq \phi(u), \ b_i \leq \phi(u), \\ \theta(u) \leq c_i, \ \theta(u) \leq c_i, \qquad \text{for } i = 1, \ldots, n; \\ \varphi(u) \leq d_i, \ \varphi(u) \leq d_i, \end{cases}$$

- $R(\varphi, \psi, (a_1, \ldots, a_n), (d_1, \ldots, d_n)) = \begin{cases} u \in P : a_i \leq \psi(u), \ a_i \leq \psi(u), \\ \varphi(u) \leq d_i, \ \varphi(u) \leq d_i, \end{cases} \quad \text{for } i = 1, \ldots, n.$

8.11   Multiple fixed points

First we give some generalizations of the Leggett-Williams fixed point theorem in a Banach space.

Theorem 8.50. *[35] Let P be a cone in a real Banach space E and let c and L be positive numbers. Let α and Ψ be nonnegative continuous concave functionals on P, and let γ, β, and θ be nonnegative continuous convex functional on P with*

$$\alpha(x) \leq \beta(x) \quad and \quad \|x\| \leq L\gamma(x), \quad for \ all \ \ x \in \overline{P(\gamma, \alpha)}.$$

Assume that

$$A : \overline{P(\gamma, \alpha)} \to \overline{P(\gamma, \alpha)}$$

is completely continuous and there exist nonegative numbers h, d, a, and b with $0 < d < a$ such that:

(i) $\{x \in P(\gamma, \theta, \alpha, a, b, c) : \quad \alpha(x) > a\} \neq \emptyset$ *and* $\alpha(A(x)) > a$ *for* $x \in P(\gamma, \theta, \alpha, a, b, c)$;

(ii) $\{x \in P(\gamma, \beta, \psi, h, d, c) : \quad \beta(x) < d\} \neq \emptyset$ *and* $\beta(A(x)) \leq d$ *for* $x \in P(\gamma, \beta, h, d, c)$;

(iii) $\alpha(A(x)) > a$ *for* $x \in P(\gamma, \alpha, a, c)$ *with* $\theta(A(x)) > b$;

(iv) $\beta(A(x)) < d$ *for* $x \in P(\gamma, \beta, d, c)$ *with* $\psi(A(x)) < h$.

Then A has at least three fixed points x_1, x_2, $x_3 \in \overline{P(\gamma, \alpha)}$ with

$$\beta(x_1) < d, \quad a < \alpha(x_2)$$

and

$$d < \beta(x_3), \quad with \ \alpha(x_3) < a.$$

From Theorems 8.50 and 8.43, we can prove the following result.

Theorem 8.51. *Let P be a cone in a real generalized Banach space E, let $c \in \mathbb{R}_+^2$, L_1, $L_2 \in \mathbb{R}_+$, let α and Ψ be nonnegative continuous concave functionals on P, and let γ, β, and θ be nonegative continuous convex functionals on P with*

$$\alpha(x) \leq \beta(x) \quad and \ \|x\| \leq L\gamma(x), \quad for \ all \ x \in \overline{P(\gamma, \alpha)},$$

where

$$L = \begin{pmatrix} L_1 & 0 \\ 0 & L_2 \end{pmatrix}.$$

Assume that

$$A : \overline{P(\gamma, \alpha)} \to \overline{P(\gamma, \alpha)}$$

is completely continuous and there exist nonegative real vectors h, d, a, and b with $0_{\mathbb{R}_+} < d < a$ such that:

(i) $\{x \in P(\gamma, \theta, \alpha, a, b, c) : \quad \alpha(x) > a\} \neq \emptyset$ *and* $\alpha(A(x)) > a$ *for* $x \in P(\gamma, \theta, \alpha, a, b, c)$;

(ii) $\{x \in P(\gamma, \beta, \psi, h, d, c) : \quad \beta(x) < d\} \neq \emptyset$ *and* $\beta(A(x)) \leq d$ *for* $x \in P(\gamma, \beta, h, d, c)$;

(iii) $\alpha(A(x)) > a$ *for* $x \in P(\gamma, \alpha, a, c)$ *with* $\theta(A(x)) > b$;

(iv) $\beta(A(x)) < d$ *for* $x \in P(\gamma, \beta, d, c)$ *with* $\psi(A(x)) < h$.

Then A has at least three fixed points $x_1, x_2, x_3 \in \overline{P(\gamma, \alpha)}$ with

$$\beta(x_1) < d, \quad a < \alpha(x_2)$$

and

$$d < \beta(x_3), \quad with \ \alpha(x_3) < a.$$

Using the index Theorem 8.43 and an idea of Bai and Ge [39], we can obtain the following result.

Theorem 8.52. *Let P be a cone in a generalized Banach space E, φ and θ be non-negative, continuou,s and convex functionals on P, ψ be a non-negative, continuous, and concave functional on P, and ϕ be a non-negative continuous functional on P satisfying $\psi(kx) \leq kx$ for $(0,0) \leq (k,k) \leq (1,1)$, such that for some positive vectors (M_1, M_2) and (d_1, d_2),*

$$\phi(x) \leq \psi(x) \text{ and } \|x\| \leq (M_1\varphi(x), M_2\varphi(x))$$

for all $(x,y) \in \overline{P(\varphi, (d_1, d_2))}$. Suppose that

$$N : \overline{P(\varphi, (d_1, d_2))} \mapsto \overline{P(\varphi, (d_1, d_2))}$$

is completely continuous and there exist positive vectors (a_1, a_2), (b_1, b_2), (c_1, c_2) with $(a_1, a_2) < (b_1, b_2)$, and such that the following conditions are satisfied:

(S_1) $\{x \in P(\varphi, \theta, \Phi, (b_1, b_2)) : \psi(x) > (b_1, b_2)\} \neq \varnothing$ *and* $\phi(N(x)) > (b_1, b_2)$ *for* $x \in P(\varphi, \theta, \Phi, (b_1, b_2))$;

(S_2) $\phi(N(x)) > (b_1, b_2)$ *for* $x \in P(\varphi, \phi, (b_1, b_2), (d_1, d_2))$ *with* $\psi_1(x) = (a_1, a_2)$;

(S_2) $(0,0) \in R(\varphi, \phi, (a_1, a_2), (d_1, d_2)), \psi(N(x)) < (a_1, a_2)$ *for* $x \in R(\varphi, \phi, (a_1, a_2), (d_1, d_2))$ *with* $\psi(x) = (a_1, a_2)$.

Then N has at least three fixed points $x_1, x_2, x_3 \in \overline{P((\varphi_1, \varphi_2), (d_1, d_2))}$ with

$$\varphi(x_i) \leq (d_1, d_2), \text{ for } i = 1, 2, 3,$$

$$(b_1, b_2) < \phi(x_1), \quad (a_1, a_2) < \psi(x_1),$$

and

$$\psi(x_2) < (b_1, b_2), \quad \psi(x_3) < (a_1, a_2).$$

Chapter 9

Random Fixed Point Theorems

Probabilistic functional analysis is an important mathematical area of research due to its applications to probabilistic models in applied problems. Random operator theory is needed for the study of various classes of random equations. Indeed,in many cases, the mathematical models or equations used to describe phenomena in the biological, physical, engineering, and systems sciences contain certain parameters or coefficients which have specific interpretations, but whose values are unknown. Therefore, it is more realistic to consider such equations as random operator equations. These equations are much more difficult to handle mathematically than deterministic equations. Important contributions to the study of the mathematical aspects of such random equations have been undertaken in [50, 231, 260] among others. The problem of fixed points for random mappings was initiated by the Prague school of probabilities. The first results were studied in 1955-1956 by Špaček and Hanš in the context of Fredholm integral equations with random kernels. In a separable metric space, random fixed point theorems for contraction mappings were proved by Hanš [152, 153], Špaček [268], Hanš and, Špaček [154] and Mukherjee [219, 220]. Then random fixed point theorems of Schauder or Krasnosel'skii type were given by Mukherjea (cf. Bharucha-Reid [50], p. 110), Prakasa Rao [246] and Bharucha-Reid [51].

9.1 Principle expansive mapping

We begin with some auxiliary results.

Lemma 9.1. *Let X be a separable Banach space, M be a closed subset of X, and assume that the mapping $B : M \to X$ is expansive with constant $h > 1$. Then $(I - B)(M)$ is closed.*

Proof. Let $\{y_n : n \in \mathbb{N}\} \subset (I - B)(M)$ be a sequence converging to y. We will show that $y \in (I - B)(M)$. For every $n \in \mathbb{N}$ there exists $x_n \in M$ such that $y_n = (I - B)(x_n)$. By Lemma 6.27,

$$\|x_n - x_m\| = \|(I - B)^{-1}(y_n) - (I - B)^{-1}(y_m)\| \leq \frac{1}{h - 1}\|y_n - y_m\| \quad \text{for} \quad n, m \in \mathbb{N}.$$

Hence, $\{x_n\}$ is Cauchy sequence in X. Thus, there exists $x \in X$ such that $\lim_{n \to \infty} x_n = x \in M$. Since $(I - B)^{-1}$ is continuous, we concluded that

$$(I - B)^{-1}(y_n) \to (I - B)^{-1}(y), \quad \text{as } n \to \infty.$$

Therefore,

$$y = (I - B)(x) \in (I - B)(M).$$

$\square$

Theorem 9.2. *Let X be a Banach space, $M \subset X$ be a nonempty compact convex set and B and T be two maps from M into X such that*

(i) *B is expansive,*

(ii) *T is weakly-strongly continuous,*

(iii) *For every $y \in M$ we have*
$$M \subseteq B(M) + T(y).$$

Then the set $Fix(B+T) = \{x \in M : x = B(x) + T(x)\} \neq \emptyset$ and $Fix(B+T)$ is closed.

Proof. For fixed $y \in M$, we consider $F^y : M \to X$ defined by
$$F^y(x) = B(x) + T(y), \quad x \in M.$$

From Theorem 6.25, there exists a unique fixed point $x(y) \in M$ such that
$$x(y) = B(x(y)) + T(y).$$

Since B is expansive, there exists $h > 1$ such that
$$|B(x(y_1)) - B(x(y_2))| \geq h|x(y_1) - x(y_2)|.$$

Then
$$|x(y_1) - x(y_2)| \leq \frac{1}{h-1}|T(y_1) - T(y_2)|.$$

Using the fact that $T(\cdot)$ is weakly-strongly continuous, then $x(\cdot) : M \to M$ is continuous. Let $L : M \to M$ be defined by $L(z) = x(T(z))$. It is clear that L is continuous, and by Schauder's fixed point theorem, there exists $z \in M$ such that $x(T(z)) = z$. Hence
$$B(z) + T(z) = z.$$

$\square$

Definition 9.3. *The random operator $T : \Omega \times X \to X$ is said to be random expansive if there exists a positive real random variable $k(\omega) > 1$ such that*
$$\|T(\omega, x_1) - T(\omega, x_2)\| \geq k(\omega)\|x_1 - x_2\| \quad \text{for all } x_1, x_2 \in X, \ \omega \in \Omega. \tag{9.1}$$

Theorem 9.4. *Let (X, d) be a complete separable metric space, and let $N : \Omega \times X \to X$ be a random map. Assume that for every $\omega \in \Omega$ there exists $k(\omega) \in (0, 1)$ such that*
$$d(N(\omega, x), N(\omega, y)) \leq k(\omega)d(x, y), \quad \text{for each } x, y \in X, \omega \in \Omega. \tag{9.2}$$

Then there exists a unique random function $x : \Omega \to X$ such that
$$x(\omega) = N(\omega, x(\omega)), \quad \omega \in \Omega.$$

Proof. Let $y : \Omega \to X$ be any arbitrary measurable function. We define $(x_n(\omega))_{n \in \mathbb{N}}$ by $x_0 = y$ and $x_n(\omega) = N(\omega, x_{n-1}(\omega))$, $n \in \mathbb{N}$. By the triangle inequality, for all $a(\omega), b(\omega) \in X$, we have
$$d(a(\omega), b(\omega)) \leq \frac{1}{1 - k(\omega)} \left[d(a(\omega), N(\omega, a(\omega))) + d(N(\omega, b(\omega)), b(\omega))\right].$$

Taking $a(\omega) = x_n(\omega)$ and $b(\omega) = x_m(\omega)$, we see that

$$d(x_n(\omega), x_m(\omega)) \leq \frac{1}{1 - k(\omega)} [d(x_n(\omega), x_{n+1}(\omega,)) + d(x_m(\omega), x_{m+1}(\omega))],$$

so

$$d(x_n(\omega), x_{n+1}(\omega)) \leq (k(\omega))^n(\omega)d(x_0(\omega), x_1(\omega)), \ n \in \mathbb{N}.$$

Hence,

$$d(x_n(\omega), x_m(\omega)) \leq \frac{k^n(\omega) + k^m(\omega)}{1 - k(\omega)} d(x_0(\omega), x_1(\omega,)).$$

Therefore,

$$d(x_n(\omega), x_m(\omega)) \to 0 \text{ as } n, m \to \infty,$$

so $(x_n(\omega))_{n \in \mathbb{N}}$ is a Cauchy sequence in the complete metric space X. Therefore, there exists $x(\omega) \in X$ such that

$$d(x_n(\omega), x(\omega)) \to 0 \text{ as } n \to \infty.$$

From (9.2) we obtain

$$\begin{aligned}
d(x(\omega), N(\omega, x(\omega))) &\leq d(x(\omega), x_n(\omega)) + d(N(\omega, x_{n+1}(\omega)), N(\omega, x(\omega))) \\
&\leq d(x(\omega), x_n(\omega)) + k(\omega)d(x_{n+1}(\omega)n, x(\omega)) \to 0
\end{aligned}$$

as $n \to \infty$. This implies that

$$x(\omega) = N(\omega, x(\omega)), \text{ for every } \omega \in \Omega.$$

It is clear that x_0 is random variable, so for every $n \in \mathbb{N}$, $x_n : \Omega \to X$ is random variable. Since x is the limit of the sequence x_n, $x : \Omega \to X$ is a random variable. By the contraction condition, we can prove that x is the unique random fixed point of N. $\qquad \square$

We now give a second version of a random Banach fixed point theorem.

Theorem 9.5. *Let $(\Omega, \mathcal{U}, \mu)$ be a complete measurable space, X be a real separable Banach space, and $N : \Omega \times X \to X$ be a continuous random operator. Let $k : \Omega \to [0, 1)$ be a random variable such that*

$$d(N(\omega, x_1), N(\omega, x_2)) \leq k(\omega)d(x_1, x_2) \text{ for each } x_1, x_2 \in X, \ a.e.\omega \in \Omega.$$

Then there exists a random variable $x : \Omega \to X$ that is the unique random fixed point of F.

Theorem 9.6. *Let X be a real separable Banach space and M be a closed subset of X. Assume that the random operator $T : \Omega \times X \to X$ is continuous, random expansive, and*

$$M \subset T(\omega, M), \text{ for every } \omega \in \Omega.$$

Then there exists a random variable $y : \Omega \to M$ which is the unique fixed point of T.

Proof. Let $\omega \in \Omega$ and $T(\omega, \cdot) : M \to T_\omega(M)$. From (9.1), the inverse of T_ω, $T_\omega^{-1} : T_\omega(M) \to M$, exists. Let $x_1, x_2, y_1, y_2 \in M$ be such that

$$T(\omega, x_1) = y_1, \ T(\omega, x_2) = y_2.$$

Then,

$$\|y_1 - y_2\| \geq k(\omega)\|x_1 - x_2\| \quad \text{implies} \quad \|T_\omega^{-1}(y_1) - T_\omega^{-1}(y_2)\| \leq \frac{1}{k(\omega)}\|y_1 - y_2\|.$$

By Banach's fixed point theorem, there exists a unique $y_\omega \in M$ that is a fixed point of T_ω^{-1}. That is, $T_\omega^{-1}(y_\omega) = y_\omega$, and so $y_\omega = T_\omega(y_\omega)$, $\omega \in \Omega$. Define a multivalued mapping $G : \Omega \to \mathcal{P}(M)$ by

$$G(\omega) = \{y \in M : y = T(\omega, y)\}.$$

Since $T(\omega, \cdot)$ is a continuous mapping, for every $\omega \in \Omega$, we have $G(\omega) \in \mathcal{P}_{cp}(M)$. We define $G_n : \Omega \to \mathcal{P}_{cl}(M)$ by

$$
\begin{aligned}
G_n(\omega) &= \left\{ \omega \in \Omega : \|y - T(\omega, y)\| < \frac{1}{n} \right\} \\
&= \left\{ \omega \in \Omega : y - T(\omega, y) \in B\left(0, \frac{1}{n}\right) \right\}.
\end{aligned}
$$

Since $(\omega, y) \to y - T(\omega, y)$ is a Carathéodory function, from Theorem 3.21, the set multivalued map G_n is measurable, so

$$\overline{G_n(\omega)} = \overline{\left\{ \omega \in \Omega : y - T(\omega, y) \in B\left(0, \frac{1}{n}\right) \right\}}$$

is measurable. Moreover,

$$G(\omega) = \bigcap_{n=1}^{\infty} \overline{G_n(\omega)}, \quad \omega \in \Omega.$$

From Theorem 3.23, there exists a measurable selection $y : \Omega \to M$ of G that in turn is a unique random fixed point of T. $\qquad \square$

We can easily prove the following lemma.

Lemma 9.7. *Let X be a real separable Banach space, $(\Omega, \mathcal{U}, \mu)$ be a complete measurable space, and $T : \Omega \times X \to X$ be a continuous random operator, and let $k(\omega)$ be a nonnegative real valued random variable such that $k(\omega) > 1$ a.s. and for some integer $n \in \mathbb{N}$ we have*

$$\|T_\omega^n(x_1) - T_\omega^n(x_2)\| \geq k(\omega)\|x_1 - x_2\| \text{ for each } x_1, x_2 \in X,$$

Then there exists a random variable $y : \Omega \to X$ which is the unique fixed point of T.

Lemma 9.8. *Let X be a real separable Banach space and M be a closed subset of X. Assume that the random operator $T : \Omega \times M \to X$ is continuous and random expansive. Then the inverse of $T(\omega, \cdot) - I_X : M \to (T(\omega, \cdot) - I_X)(M)$, for every $\omega \in \Omega$, exists and*

$$
\begin{aligned}
&\|(T(\omega, \cdot) - I_X)^{-1}(y_1) - (T(\omega, \cdot) - I_X)^{-1}(y_2)\| \\
&\qquad \leq \frac{1}{k(\omega) - 1}\|y_1 - y_2\| \text{ for all } y_1, y_2 \in (T(\omega, \cdot) - I_X)^{-1}(M). \quad (9.3)
\end{aligned}
$$

Proof. Let $\omega \in \Omega$ and $x_1, x_2 \in M$ be such that

$$T(\omega, x_1) - x_1 = T(\omega, x_2) - x_2.$$

Then,

$$\|T(\omega, x_1) - x_1 - T(\omega, x_2) + x_2\| \geq (k(\omega) - 1)\|x_1 - x_2\| \quad \text{implies} \quad x_1 = x_2,$$

which shows that $T(\omega, \cdot) - I_X$ is one to one. Hence, the inverse of $T(\omega, \cdot) - I_X : M \to (T(\omega, \cdot) - I_X)(M)$ exists and (9.3) holds. $\qquad \square$

Theorem 9.9. *Let $M \subset X$ be a nonempty compact convex subset. Suppose that $T, B : \Omega \times M \to X$ are random operators such that*

$(\mathcal{A}_1)$ *T is a continuous random operator.*

$(\mathcal{A}_2)$ *B is a continuous random expansive operator.*

$(\mathcal{A}_3)$ *$T(\omega, M) \subseteq (I - B(\omega, \cdot))(M), \quad \omega \in \Omega$.*

Then $B + T$ has at least one random fixed point.

Proof. Let $\omega \in \Omega, y \in M$ and $F_{\omega, y} : M \to M$ be defined by

$$F_{\omega, y}(x) = B(\omega, x) + T(\omega, y).$$

From Theorem 8.11 there exists a unique fixed point of $F_{\omega, y}(\cdot)$, and by Lemma 9.8, $(I - B(\omega, \cdot))^{-1}$ exists. We define the operator $N_\omega : M \to M$ by $N_\omega(y) = (I - B(\omega, \cdot))^{-1} T(\omega, y)$. It is easy to see that by the Schauder fixed point theorem N_ω has at least one fixed point. Define a mapping $S : \Omega \to \mathcal{P}_{cp}(M)$ by

$$S(\omega) = \{y \in M : y = B(\omega, y) + T(\omega, y)\}.$$

Let K be a closed subset of M then

$$
\begin{aligned}
S_-^{-1}(K) &= \{\omega \in \Omega : S(\omega) \cap K \neq \emptyset\} \\
&= \{\omega \in \Omega : y = B(\omega, y) + T(\omega, y), \ y \in K\}.
\end{aligned}
$$

Since X is separable Banach space then there exist $\{y_i : i \in \mathbb{N}\} \subset M$ such that

$$\overline{\{y_i : i \in \mathbb{N}\}} = M.$$

Hence

$$S_-^{-1}(K) = \cap_{n=1}^\infty \cup_{x_i \in K_n} \left\{\omega \in \Omega : \|y_i - B(\omega, y_i) - T(\omega, y_i)\| < \frac{2}{n}\right\},$$

where

$$K_n = \left\{x \in C : d(x, K) < \frac{1}{n}\right\}.$$

Therefore $S^{-1}(K)$ is measurable. Since $B(\omega, \cdot) + T(\omega, \cdot)$ is a continuous mapping, by Theorem 3.23, there exists $y : \Omega \to M$ a measurable selection of S which is a random fixed point of $B + T$. $\qquad \square$

Theorem 9.10. *Let $M \subset X$ be a nonempty compact convex subset. Suppose that $T, B : \Omega \times M \to X$ are random operators such that*

$(\mathcal{A}_4)$ *T is a weakly-strongly continuous random operator.*

$(\mathcal{A}_5)$ *B is a continuous random expansive operator.*

$(\mathcal{A}_6)$ *For each $y \in M$; we have $M \subseteq B(\omega, \cdot)(M) + T(\omega, y), \quad \omega \in \Omega$.*

Then $B + T$ has at least one random fixed point.

9.2 Approximation method and Krasnosel'skii-type fixed point theorems

Theorem 9.11. *Let X be a separable Banach space, M be a compact convex subset of X, $G : \Omega \times M \to \mathcal{P}_{cp,cv}(X)$ be a random multivalued map and $B : \Omega \times M \to M$ be a continuous random operator. Assume that G and B satisfy the following hypotheses:*

$(\mathcal{H}_1)$ *$G(\omega, \cdot)$ is a u.s.c. multivalued mapping for every $\omega \in \Omega$ and $G(.,x)$ is measurable.*

$(\mathcal{H}_2)$ *$B(\omega, \cdot)$ is a contraction mapping for every $\omega \in \Omega$.*

$(\mathcal{H}_3)$ *$G(\omega, M) \subseteq (I - B(\omega, \cdot))(M), \quad \omega \in \Omega$.*

Then there exists $y : \Omega \to M$ a random operator such that

$$y(\omega) \in B(\omega, y(\omega)) + G(\omega, y(\omega)), \ \omega \in \Omega.$$

Proof. Let $\omega \in \Omega$, from Theorem 6.3, given $\epsilon > 0$, there exists $f_\epsilon(\omega, \cdot) : M \to X$, a continuous map, such that

$$\Gamma(f_\epsilon(\omega, \cdot)) \subset \Gamma(\omega, G) + \epsilon B_* \tag{9.4}$$

and

$$f_\epsilon(\omega, M) \subset coG(\omega, M).$$

From $(\mathcal{H}_3)$, and since M is convex, we get

$$B(\omega, M) + f_\epsilon(\omega, M) \subseteq M.$$

For fixed $y \in M$, we consider $F_\epsilon^y : M \to M$ defined by

$$F_\epsilon^y(\omega, x) = B(\omega, x) + f_\epsilon(\omega, y), \ x \in M.$$

By the Banach random fixed point theorem, there exists a unique $x_\epsilon(y) \in M$ such that

$$x_\epsilon(y) = B(x_\epsilon(y)) + f_\epsilon(y).$$

From $(\mathcal{H}_2)$ the mapping $I - B(\omega, \cdot) : M \to (I - B)(\omega, M)$ is a homeomorphism. We define the operator $N_\epsilon(\omega,) : M \to M$ by

$$N_\epsilon(\omega, x) = (I - B)^{-1} f_\epsilon(\omega, x).$$

It easy to see that N_ϵ satisfies the conditions of the Schauder's fixed point theorem, and so, there exists $x_\epsilon(\omega, .) \in M$ such that

$$x_\epsilon(\omega) = B(\omega, x_\epsilon) + f_\epsilon(\omega, x_\epsilon(\omega)).$$

Let $\{\epsilon_n : n \in \mathbb{N}\}$ be such that

$$\epsilon_n \to 0 \text{ as } n \to \infty.$$

Since M is compact, there exists a subsequence of $x_{\epsilon_n}(\omega)$ converging to $x(\omega) \in M$. So

$$f_{\epsilon_n}(\omega, x_{\epsilon_n}(\omega)) = (I - B)(\omega, x_{\epsilon_n}(\omega)) \to (I - B)(\omega, x(\omega)), \text{ as } n \to \infty. \tag{9.5}$$

From (9.4), we get

$$d(x_{\epsilon_n}(\omega), f_{\epsilon_n}(\omega, x_{\epsilon_n}(\omega)), \Gamma(G(\omega, .))) \leq \epsilon_n, \text{ for every } n \in \mathbb{N}.$$

It clear that $G(\omega, .)$ has a closed graph in $X \times X$ and consequently $(I - B)(\omega, x) \in G(\omega, x)$. This implies that

$$x(\omega) \in B(\omega, x(\omega)) + G(\omega, x(\omega)).$$

We define the multivalued operator $F : \Omega \to \mathcal{P}(X)$ by

$$F(\omega) = \{x \in X : x \in B(\omega, x) + G(\omega, x)\}.$$

Let $\phi : \Omega \times X \to X$ be defined by

$$\phi(\omega, x) = d(x, B(\omega, x) + G(\omega, x))$$

$$\begin{aligned}
\Gamma(F) &= \{(\omega, x) \in \Omega \times X : x \in F(\omega)\} \\
&= \{(\omega, x) \in \Omega \times X : d(x, \phi(\omega, x)) = 0\} \\
&= \phi^{-1}(\{0\}).
\end{aligned}$$

Since ϕ is measurable then $\Gamma(F)$ is measurable. By Theorem 3.31 F has a measurable selection $y : \Omega \to X$ which is a random fixed point of $B + G$. $\qquad\square$

Now, we can easily prove the following two results.

Theorem 9.12. *Let X be a separable Banach space and M be a closed bounded convex subset of X. Assume that $G : \Omega \times M \to \mathcal{P}_{cp,cv}(X)$ is a random multivalued map such that for each $\omega \in \Omega$, $G(\omega, \cdot)$ is u.s.c., B satisfies $(\mathcal{H}_2) - (\mathcal{H}_3)$, and for $\omega \in \Omega$,*

$(\mathcal{H}_4)$ $G(\omega, M)$ *is compact.*

Then the operator $B + G$ has at least one fixed point.

Theorem 9.13. *Let $(X, |\cdot|)$ be a Banach space and M be a closed bounded convex subset of X. Assume that $G : \Omega \times M \to \mathcal{P}_{cp,cv}(X)$ is a random multivalued map such that for each $\omega \in \Omega$, $G(\omega, \cdot)$ is u.s.c. and B satisfies $(\mathcal{H}_2)$ and the condition:*

$(\mathcal{H}_3)$ $G(\omega, M)$, *for $\omega \in \Omega$, is compact,*

$(\mathcal{H}_4)$ $B(\omega, M) + G(\omega, M) \subseteq coG(\omega, M)$, $\omega \in \Omega$.

Then the operator $B + G$ has at least one fixed point.

Proof. Fix $\omega \in \Omega$ and $\widetilde{M} = \overline{co}G(\omega, M)$. Then, $\widetilde{M}$ is compact convex. It suffices to prove only that $N_{\epsilon,\omega}(\widetilde{M}) \subseteq \widetilde{M}$, where N_ϵ is defined in the proof of Theorem 9.11. Indeed, let $x \in \widetilde{M}$ such that $y = N_{\epsilon,\omega}(x)$. Hence

$$y = (I - B(\omega, \cdot))^{-1} f_\epsilon(x) \Rightarrow x \in \overline{co}G(\omega, M).$$

Then

$$N_{\epsilon,\omega}(\widetilde{M}) \subseteq \widetilde{M}.$$

So, by Theorem 9.11, there exists a random function $y : \Omega \to M$ which is a fixed point of $B + G$. $\qquad\square$

Theorem 9.14. *Let X be a separable Banach space and M be a compact convex subset of X. Let $B : \Omega \times M \to X$ be a random continuous 1-coercive operator and $G : \Omega \times M \to \mathcal{P}_{cp,cv}(X)$ be a random multivalued map such that $G(\omega, \cdot)$ is u.s.c. Assume that*

$(\mathcal{H}_\alpha)$ $\alpha > 1$, $G(\omega, M) \subseteq (I - \alpha B(\omega, \cdot))(M)$.

Then $B + G$ has at least one random fixed point.

Proof. Fix $\omega \in \Omega$, and for each $n \in \mathbb{N}$, let $\alpha_n > 1$ with $\alpha_n \to 1$ as $n \to \infty$. Then $\alpha_n B(\omega, \cdot)$ is an expansive operator. From Theorem 9.11, for each $n \in \mathbb{N}$, there exists $x_n \in M$ such that

$$x_n \in \alpha_n B(\omega, x_n) + G(\omega, x_n) \quad \text{implies} \quad x_n \in B(\omega, \alpha_n x_n) + G(\omega, x_n).$$

Since M is compact, there exists a subsequence of $\{x_n\}$ converging to $x \in M$. Let $L_\omega : [0, 1] \times M \to \mathcal{P}_{cp}(X)$ be a multivalued operator defined by

$$(\alpha, x) \to L_\omega(\alpha, x) = \alpha B(\omega, x) + G(\omega, x).$$

Using the fact that $B(\omega, \cdot)$ is a continuous operator, $G(\omega, \cdot)$ is u.s.c., and M is compact, we know that L_ω has a closed graph, and so

$$x_n \in \alpha_n B(\omega, x_n) + G(\omega, x_n) \to x \in B(\omega, x) + G(\omega, x) \text{ as } n \to \infty.$$

Hence, $B + G$ has at least one random fixed point. $\square$

The final theorem in this section is our global multivalued version of the Krasonsel'skii fixed point theorem.

Theorem 9.15. *Let X be a separable Banach space and $G : \Omega \times X \to \mathcal{P}_{cp,cv}(X)$ be a multivalued map such that $\omega \to G(\omega, \cdot)$ is u.s.c. and $x \to G(\cdot, x)$ is measurable. Assume condition $(\mathcal{H}_2)$ holds and $G(\omega, \cdot)$ is compact. Then, either*

(a) $x(\omega) \in \lambda B(\omega, \frac{x}{\lambda}) + \lambda G(\omega, x)$ has a random solution for $\lambda = 1$,

or

(b) the set

$$\{x : \Omega \to X \text{ is a random variable} \mid x(\omega) \in \lambda B(\omega, \tfrac{x(\omega)}{\lambda}) + \lambda G(\omega, x(\omega)), \ \lambda \in (0, 1), \ \omega \in \Omega\}$$

is unbounded.

9.3   Random fixed point for a Cartesian product of operators

In this section we are concerned with solutions for the system

$$\begin{cases} x & = \ F(\omega, x, y) \\ y & = \ G(\omega, x, y). \end{cases} \tag{9.6}$$

Let $(X, \|\cdot\|_X)$, $(Y, \|\cdot\|_Y)$ be two Banach spaces and let $F : \Omega \times X \times Y \to X$, $G : \Omega \times X \times Y \to Y$.

Theorem 9.16. *Assume that:*

$(\mathcal{H}_5)$ *For every $x \in X$, the function $y \to F(x, y)$ is a continuous operator,*

$(\mathcal{H}_6)$ *There exists $k > 1$ such that*

$$\|F(x_1, y) - F(x_2, y)\|_X \geq k\|x_1 - x_2\|, \text{ for every, } x_1, x_2 \in X, \ y \in Y$$

and for each $y \in Y$ we have

$$X \subset F(X, y),$$

$(\mathcal{H}_7)$ $G(\cdot, \cdot)$ *is a continuous compact operator.*

Then the system

$$\begin{cases} x = F(x, y) \\ y = G(x, y) \end{cases} \tag{9.7}$$

admits a solution

Proof. Fix $y \in Y$ and define $T_y : X \to X$ by $T_y(x) = F(x, y)$. From $(\mathcal{H}_6)$ and Theorem 6.25, there exists a unique $x(y) \in X$ such that $x(y) = T_y(x(y))$. Therefore, for every $y \in Y$ there exists a unique $x(y) \in X$ such that

$$x(y) = F(x(y), y).$$

Define $\overline{T} : Y \to Y$ by

$$\overline{T}(y) = G(x(y), y), \text{ for every } y \in Y.$$

We show that $\overline{T}$ is a continuous compact operator. Let $(y_n)_n$ be a sequence converging to y in Y, thus

$$\begin{aligned} \|x(y_n) - x(y)\|_X &= \|F(x(y_n), y_n) - F(x(y), y)\|_X \\ &\geq \|F(x(y_n), y_n) - F(x(y), y_n)\|_X \\ &\quad -\|F(x(y), y_n) - F(x(y), y)\|_X \\ &\geq k\|x(y_n) - x(y))\|_X - \|F(x(y), y_n) - F(x(y), y)\|_X \end{aligned}$$

Hence

$$\|x(y_n) - x(y)\|_X \leq \frac{1}{k-1} \|F(x(y), y_n) - F(x(y), y)\|_X \to 0 \text{ as } n \to \infty.$$

Then $x(\cdot)$ is continuous, and by using the condition $(\mathcal{H}_7)$ we conclude that $\overline{T}$ is continuous. Hence $\overline{T} : Y \to Y$ is a continuous compact operator. So by the Schauder fixed point theorem, there exists $y \in Y$ such that

$$y = G(x(y), y).$$

Then $(x(y), y)$ is a solution of problem (9.7). $\square$

Theorem 9.17. *Let $(\Omega, \sum)$ be a measurable space, $(X, \|\cdot\|_X)$, $(Y, \|\cdot\|_Y)$ be two separable Banach spaces and let $F : \Omega \times X \times Y \to X$, $G : \Omega \times X \times Y \to Y$ are two random operators*

$(\mathcal{H}_8)$ *For every $\omega \in \Omega$ the function $(x, y) \to F(\omega, x, y)$ is a continuous operator,*

$(\mathcal{H}_9)$ *There exists $k(\omega) > 1$ such that*

$$\|F(x_1, y) - F(x_2, y)\|_X \geq k(\omega)\|x_1 - x_2\|, \text{ for } x_1, x_2 \in X, \ y \in Y, \omega \in \Omega$$

and for each $y \in Y, \omega \in \Omega$ we have

$$X \subset F(\omega, X, y),$$

$(\mathcal{H}_{10})$ *For every $\omega \in \Omega$ we have $G(\omega, \cdot, \cdot)$ is a continuous compact operator.*

Then the system (9.6) has at least one random solution.

Proof. Let $y \in Y$, $\omega \in \Omega$. Consider $T_{\omega,y} : X \to X$ by $T_{\omega,y}(x) = F(\omega, x, y)$. From Theorem 6.25 there exists unique $x(y)(\omega) \in X$ such that

$$x(y)(\omega) = F(\omega, x(y)(\omega), y).$$

Define $\mathcal{T}_y : X \to \mathcal{P}_{cl}(X)$ by

$$\mathcal{T}_y(\omega) = \{x \in X : x = F(\omega, x, y)\}.$$

Since $F(\omega, \cdot, y)$ is continuous, the function $\varphi : X \to \mathbb{R}$ defined by

$$\phi(x) = d(x, x - F(\omega, x, y))$$

is measurable. Hence $\mathcal{T}_y$ has measurable selection. Now, let $T : Y \to Y$ be defined by

$$T(y) = G(\omega, x(y)(\omega), y).$$

It is clear that T is continuous operator. Thus from the Schauder fixed point there exist $y(\omega) \in Y$ such that

$$y(\omega) = G(\omega, x(y)(\omega), y(\omega)), \quad \omega \in \Omega.$$

Using the fact that $G(\cdot, x, y)$ is measurable and $G(\omega, \cdot, \cdot)$ is continuous, then $y : \Omega \to Y$ is a measurable function. Then $(x(y)(\omega), y(\omega))$ is a solution of problem (9.6). $\qquad\square$

Theorem 9.18. *Let $(\Omega, \sum)$ be a measurable space, $(X, \| \cdot \|_X)$, $(Y, \| \cdot \|_Y)$ be two separable Banach spaces and let $F : \Omega \times X \times Y \to X$, $G : \Omega \times X \times Y \to Y$ be two random operators such that*

$(\overline{\mathcal{H}}_1)$ *For every $\omega \in \Omega$ the function $(x, y) \to F(\omega, x, y)$ is a continuous operator,*

$(\overline{\mathcal{H}}_2)$ *There exists $k(\omega) \in [0, 1)$ such that*

$$\|F(x_1, y) - F(x_2, y)\|_X \le k(\omega)\|x_1 - x_2\|, \text{for } x_1, x_2 \in X, \ y \in Y, \omega \in \Omega,$$

$(\overline{\mathcal{H}}_3)$ *For every $\omega \in \Omega$, we have $G(\omega, \cdot, \cdot)$ is a continuous compact operator.*

Then the system (9.6) has at least one random solution.

9.4   Measurable selection in vector metric space

As a consequence of Kuratowski-Ryll-Nardzewski and J.R. Aumann selection theorems we can conclude the following results.

Theorem 9.19. *Let $(\Omega, \sum)$ be a measurable space, Y be a separable generalized metric space and $F : \Omega \to \mathcal{P}_{cl}(Y)$ be a measurable multivalued function. Then F has a measurable selection.*

Proof. Consider $F_* : \Omega \to \mathcal{P}_{cl}(\widetilde{Y})$ defined by

$$F_*(\omega) = (h \circ F)(\omega), \quad \text{for all } \omega \in \Omega$$

where h is defined in Lemma 7.12. Let $C \subset \widetilde{X}$ be a open set, then

$$F_*^{-1}(C) \quad = \quad \{\omega \in \Omega : (h \circ F)(\omega) \cap C \ne \emptyset\}$$

$$= \{\omega \in \Omega : \ F(\omega) \cap h^{-1}(C) \neq \emptyset\}.$$

Since F is a measurable multifunction, hence $F_*^{-1}(C) \in \sum$. By Theorem 3.23 there exists $x : \Omega \to \widetilde{X}$ a measurable single function such that

$$x(\omega) \in (h \circ F)(\omega), \quad \text{for all } \omega \in \Omega \Rightarrow (h^{-1} \circ x)(\omega) \in F(\omega), \quad \text{for all } \omega \in \Omega.$$

Using the fact that h^{-1} is a continuous function, then $h^{-1} \circ x : \Omega \to X$ is a measurable selection of F. $\qquad\square$

Theorem 9.20. *Let $(\Omega, \sum)$ be a measurable space and Y be a separable generalized metric space. If $G : \Omega \to \mathcal{P}_{cp}(X)$ is a multivalued map such that the graph $\Gamma(G)$ of G is measurable, then G has a measurable selection.*

Proof. Let $G_* : \Omega \to \mathcal{P}_{cp}(\widetilde{X})$ defined by

$$G_*(\omega) = (h \circ G)(\omega), \quad \text{for all } \omega \in \Omega.$$

Then

$$
\begin{aligned}
\Gamma(G_*) &= \{(\omega, y) \in \Omega \times \widetilde{X} : \ y \in G_*(\omega)\} \\
&= \{(\omega, y) \in \Omega \times \widetilde{X} : \ y \in (h \circ G)(\omega)\} \\
&= \{(\omega, y) \in \Omega \times \widetilde{X} : \ h^{-1}(y) \in G(\omega)\} \\
&= \{(\omega, y) \in \Omega \times \widetilde{X} : \ h^{-1}(y) \in G(\omega)\} \\
&= \{(\omega, z) \in \Omega \times h^{-1}(\widetilde{X}) : \ z \in G(\omega)\}.
\end{aligned}
$$

Hence $\Gamma(G_*)$ is measurable. By Theorem 3.31 there exists $x : \Omega \to \widetilde{X}$ a measurable single function, such that

$$x(\omega) \in (h \circ G_*)(\omega), \quad \text{for all } \omega \in \Omega.$$

So, $h^{-1} \circ x$ is a measurable selection of the multivalued map G. $\qquad\square$

9.5 Perov random fixed point theorem

In this section we present the various random versions of the Perov fixed point theorem.

Theorem 9.21. *Let $(\Omega, \mathcal{F}, \mu)$ be a complete probability space, X be a real separable generalized Banach space and $F : \Omega \times X \to X$ be a continuous random operator, and let $M(\omega) \in \mathcal{M}_{n \times n}(\mathbb{R}_+)$ be a random variable matrix such that $M(\omega)$ converges to 0 a.s. and*

$$d(F(\omega, x_1), F(\omega, x_2)) \leq M(\omega) d(x_1, x_2) \text{ for each } x_1, x_2 \in X, \ \omega \in \Omega.$$

Then there exists a random variable $x : \Omega \to X$ which is the unique random fixed point of F.

Proof. Let

$$E = \{\omega \in \Omega : d(F(\omega, x_1), F(\omega, x_2)) \leq M(\omega) d(x_1, x_2) \text{ for each } x_1, x_2 \in X\},$$

then $\mu(E) = 1$. From Theorem 8.1, for every fixed $\omega \in E$, there exists a unique $x(\omega) \in X$ such that $F(\omega, x(\omega)) = x(\omega)$. Let $y : \Omega \to X$ be any arbitrary measurable function. We define $(x_n(\omega))_{n \in \mathbb{N}}$, $x_0(\omega) = y(\omega)$ by

$$x_n(\omega) = F(\omega, F^{n-1}(\omega, y(\omega))), \quad n \in \mathbb{N}.$$

Clearly, x_n is a random variable and for each $n, m \in \mathbb{N}$ we have

$$d(x_n(\omega), x_{n+k}(\omega)) \le (M^k(\omega) + \cdots + M^{n+k}(\omega))d(x_0(\omega), x_1(\omega)).$$

By Lemma 7.12, we get

$$d(x_n(\omega), x_{n+k}(\omega)) \le M^k(\omega)(I - M(\omega))^{-1}d(x_0(\omega), x_1(\omega)).$$

Hence $(x_n)_{n \in \mathbb{N}}$ is a Cauchy sequence. Then there exists a random variable $y_* : \Omega \to X$ such that

$$d(x_n(\omega), y_*(\omega)) \to 0 \quad \text{as } n \to \infty.$$

Then

$$d(y_*(\omega), F(\omega, y_*(\omega))) \le d(y_*(\omega), x_n(\omega)) + M(\omega)d(x_n(\omega), y_*(\omega)) \to 0 \quad \text{as } n \to \infty.$$

Thus

$$y_*(\omega) = F(\omega, y_*(\omega)) \text{ for each } \omega \in E,$$

so

$$y_*(\omega) = x(\omega), \quad \omega \in E.$$

$\square$

By simple modification we conclude the following result.

Theorem 9.22. *Let $(\Omega, \mathcal{F})$ be a measurable space, X be a real separable generalized Banach space and $F : \Omega \times X \to X$ be a continuous random operator, and let $M(\omega) \in \mathcal{M}_{n \times n}(\mathbb{R}_+)$ be a random variable matrix such that for every $\omega \in \Omega$ the matrix, $M(\omega)$ converges to 0 and*

$$d(F(\omega, x_1), F(\omega, x_2)) \le M(\omega)d(x_1, x_2) \text{ for each } x_1, x_2 \in X, \ \omega \in \Omega.$$

Then there exists a random variable $x : \Omega \to X$ which is the unique random fixed point of F.

Theorem 9.23. *Let $(\Omega, \mathcal{F}, \mu)$ be a complete probability space and X be a real separable generalized Banach space and $T : \Omega \times X \to X$ be a continuous random operator, and let $M(\omega) \in \mathcal{M}_{n \times n}(\mathbb{R}_+)$ be a nonnegative real matrix random variable such that $\rho(M(\omega)) < 1$ a.s. and*

$$\|T(\omega, x_1) - T(\omega, x_2)\| \le M(\omega)\|x_1 - x_2\| \text{ for each } x_1, x_2 \in X, \ \omega \in \Omega.$$

Then there exists a random variable $y : \Omega \to X$ which is the unique fixed point of T.

Proof. Let

$$E = \{\omega \in \Omega : \rho(M(\omega)) < 1\}, \ F = \{\omega \in \Omega : T(\omega, \cdot) \text{ is continuous}\}$$

and

$$G_{x,y} = \{\omega \in \Omega : \|T(\omega, x) - T(\omega, y)\| \le M(\omega)\|x - y\|\}.$$

Let D be a countable dense subset of X. We first show that

$$\left(\bigcap_{x,y \in X} G_{x,y} \cap E \cap F\right) = \left(\bigcap_{a,b \in D} G_{a,b} \cap E \cap F\right).$$

Let $\omega \in \left(\bigcap_{a,b \in D} G_{a,b} \cap E \cap F\right)$, then

$$\|T(\omega, a) - T(\omega, b)\| \leq M(\omega)\|a - b\|, \quad \text{for all } a, b \in D. \tag{9.8}$$

Let $x, y \in X$, thus there exist two sequences $(a_n)_{n \in \mathbb{N}}$, $(b_n)_{n \in \mathbb{N}} \in D$ such that

$$a_n \to x, \quad b_n \to y \text{ as } n \to \infty.$$

From (9.8), we get

$$\|T(\omega, a_n) - T(\omega, b_n)\| \leq M(\omega)\|a_n - b_n\|, \quad n \in \mathbb{N}.$$

By the continuity of $T(\omega, \cdot)$, as $n \to \infty$, we have

$$\|T(\omega, x) - T(\omega, y)\| \leq M(\omega)\|x - y\|.$$

This implies that

$$\left(\bigcap_{a,b \in D} G_{a,b} \cap E \cap F\right) \subseteq \left(\bigcap_{x,y \in X} G_{x,y} \cap E \cap F\right). \tag{9.9}$$

Also it is obvious that

$$\left(\bigcap_{x,y \in X} G_{x,y} \cap E \cap F\right) \subseteq \left(\bigcap_{a,b \in D} G_{a,b} \cap E \cap F\right). \tag{9.10}$$

From (9.9) and (9.10), we have

$$\left(\bigcap_{x,y \in X} G_{x,y} \cap E \cap F\right) = \left(\bigcap_{a,b \in D} G_{a,b} \cap E \cap F\right).$$

Hence

$$\left(\bigcap_{x,y \in X} G_{x,y} \cap E \cap F\right) \in \mathcal{F}.$$

Since $\mu(E) = 1$ and $\mu(F) = 1$ then

$$\mu(\Omega \backslash E) = 0, \quad \mu(\Omega \backslash F) = 0.$$

It is clear that

$$\bigcap_{x,y \in X} G_{x,y} = \{\omega \in \Omega : \|T(\omega, x) - T(\omega, y)\| \leq M(\omega)\|x - y\|, \text{ for all } x, y \in X\}.$$

Hence

$$\mu(\bigcap_{x,y \in X} G_{x,y}) = 1 \Rightarrow \mu(\Omega \backslash \bigcap_{x,y \in X} G_{x,y}) = 0.$$

Therefore,

$$\mu(E_*) = 1, \quad E_* = \bigcap_{x,y \in X} G_{x,y} \cap E \cap F.$$

Thus for for every $\omega \in E_*$, $T(\omega, \cdot)$ is a deterministic operator and hence has a unique fixed point in the deterministic case, which we denote by $\xi(\omega)$. Let $x : \Omega \to X$ be a function defined by

$$x(\omega) = \begin{cases} \xi(\omega), & \text{if } \omega \in E_*, \\ 0, & \text{if } \omega \in \Omega \backslash E_*. \end{cases}$$

By the same method used in Theorem 9.21 we can prove that x is the unique random fixed point of T. $\qquad \square$

Theorem 9.24. *Let (Ω, F, μ) be a complete measurable space and $T : \Omega \times X \to X$ be an almost surely continuous random operator. Assume that there exists $M(\omega) \in \mathcal{M}_{2 \times 2}(\mathbb{R}_+)$ a real matrix random variable such that*

$$\mu(\{\omega : \|T(\omega, x_1) - T(\omega, x_2)\| \le M(\omega)\|x - y\|\}) = 1.$$

Then for every real number $\lambda \ne 0$ such that $\rho(M(\omega)) < |\lambda|$ and

$$\mu\{\omega \in \Omega : \rho(M(\omega)) < |\lambda|\} = 1,$$

there exists a random operator S that is the inverse of the random operator $(T(\omega, \cdot) - \lambda I_X)$.

Proof. Let $\lambda \ne 0$ and $y \in Y$. We consider $T_y(\omega, \cdot) : X \to X$ by

$$T_y(\omega) = \frac{1}{|\lambda|} T(\omega, x) - y, \ x \in X.$$

Clearly T_y is a random contraction operator. Therefore by Theorem 9.21 there exists a unique random operator $x_y : \Omega \to X$ such that

$$x_y(\omega) = \frac{1}{|\lambda|} T(\omega, x_y(\omega)) - y, \ \text{a.s.}$$

Then there exists $S_1(\omega) : X \to X$ a random operator, such that

$$S_1(\omega) \left(\frac{1}{|\lambda|} T(\omega) - I \right) = I.$$

So, $S(\omega) = \frac{1}{|\lambda|} S_1(\omega)$ is the inverse of $T(\omega, \cdot) - \lambda I$. $\qquad \square$

Remark 9.25. *We can replace the constant λ of the above theorem by a real-valued random operator $\lambda : \Omega \to \mathbb{R}$ such that*

$$\rho(M(\omega)) < \lambda(\omega) \ a.s.$$

9.6 Schauder and Krasnosel'skii type random fixed point

In 1966 Mukherjea gave a random version of Schauder's fixed point theorem on an atomic measure space. Then Prakasa Rao extended this result and obtained a theorem of Krasnosel'skii type on the same measure spaces. Bharucha-Reid generalized results of Mukherjea and Prakasa Rao.

Theorem 9.26. *Let X be a generalized Banach space, C be a separable closed convex subset of X and $F : \Omega \times C \to C$ be a continuous random operator. Suppose that for every $\omega \in \Omega$, $F(\omega, C)$ is compact. Then there exists a random fixed point $x : \Omega \to C$ of F.*

Proof. Let $\omega \in \Omega$. We consider $F_\omega : C \to C$ by

$$F_\omega(x) = F(\omega, x), \ x \in X.$$

Therefore, by Theorem 8.13 there exists $x(\omega) \in C$ such that

$$x(\omega) = F(\omega, x(\omega)). \tag{9.11}$$

Now, we define $\widetilde{F}_* : \Omega \to \mathcal{P}(\widetilde{X})$ by

$$\widetilde{F}_*(\omega) = \{(x, x, \dots, x) : (x, \dots, x) = h \circ F(\omega, h^{-1}(x))\}$$

where $h : X \to \widetilde{X}$ is defined by

$$h(x) = (x, \dots, x), \quad x \in X,$$

and

$$\widetilde{X} = \{(x, \dots, x) \in \prod_{i=1}^{n} X : x \in X, \ i = 1, \dots, n\}.$$

Thus $\widetilde{X}$ is a Banach space with the following norm

$$\|(x, \dots, x)\| = \sum_{i=1}^{n} \|x\|, \text{ for each } x \in X.$$

From (9.11) and h is a homeomorphism, we get

$$\widetilde{F}_*(\omega) \in \mathcal{P}_{cl}(\widetilde{X}) \text{ for all } \omega \in \Omega.$$

Since for each $\omega \in \Omega$, $F(\omega, C)$ is a compact, then

$$\widetilde{F}_*(\omega) \in \mathcal{P}_{cp}(\widetilde{C}) \text{ for all } \omega \in \Omega,$$

where

$$\widetilde{C} = \{(x, \dots, x) : x \in C\}.$$

Let K be a nonempty closed set subset of $\widetilde{C}$, then

$$
\begin{aligned}
\widetilde{F}_*^{-1}(K) &= \{\omega \in \Omega : \widetilde{F}_*(\omega) \cap K \neq \emptyset\} \\
&= \cup_{x \in K} \{\omega \in \Omega : (x, \dots, x) = h \circ F(\omega, h^{-1}(x))\} \\
&= \cup_{x \in C} \{\omega \in \Omega : h^{-1}(x, \dots, x) = F(\omega, h^{-1}(x))\} \\
&= \cap_{m=1}^{\infty} \cup_{x_i \in h^{-1}(K_m)} \{\omega \in \Omega : \|h^{-1}(x_i, \dots, x_i) \\
&\quad - F(\omega, h^{-1}(x_i, \dots, x_i))\| < \epsilon_m\}
\end{aligned}
$$

where

$$\epsilon_m := \begin{pmatrix} \frac{1}{m} \\ \cdots \\ \frac{1}{m} \end{pmatrix} \text{ and } K_n = \left\{(x, \dots, x) \in \widetilde{C} : d((x, \dots, x), K) < \frac{1}{m}\right\}.$$

Then

$$\widetilde{F}_*^{-1}(K) \;=\; \bigcap_{m=1}^{\infty} \bigcup_{x_i \in h^{-1}(K_m)} F_-^{-1}(B(x_i, \epsilon_m), x_i) \in \mathcal{F}.$$

Then from Theorem 9.19 there exists a measurable function $x : \Omega \to C$ such that

$$x(\omega) = F(\omega, x(\omega)), \quad \omega \in \Omega.$$

$\square$

By the above result we present the following random nonlinear alternative.

Theorem 9.27. *Let X be a separable generalized Banach space and let $F : \Omega \times X \to X$ be a completely continuous random operator. Then, either*

(i) *the random equation $F(\omega, x) = x$ has a random solution, i.e., there is a measurable function $x : \Omega \to X$ such that $F(\omega, x(\omega)) = x(\omega)$ for all $\omega \in \Omega$, or*

(ii) *the set $\mathcal{M} = \{ x : \Omega \to X$ is measurable $\mid \lambda(\omega) F(\omega, x) = x \}$ is unbounded for some measurable $\lambda : \Omega \to X$ with $0 < \lambda(\omega) < 1$ on Ω.*

Finally, we prove a random Krasnosel'skii fixed point theorem type.

Theorem 9.28. *Let $C \subset X$ be a nonempty compact convex subset of a separable generalized Banach space X. Suppose that $T, B : \Omega \times C \to X$ are random operators such that*

$(\mathcal{A}_1)$ *T is a continuous random operator.*

$(\mathcal{A}_2)$ *B is a continuous, random, and $M(\omega)-$contraction operator.*

$(\mathcal{A}_3)$ *The matrix $I - M(\omega)$ has the absolute value property.*

$(\mathcal{A}_4)$ *$B(\omega, C) + T(\omega, C) \subset C, \quad \omega \in \Omega.$*

Then $B + T$ has at least one random fixed point.

Proof. Let $\omega \in \Omega, y \in C$ and $F_{w,y} : C \to C$ be defined by

$$F_{w,y}(x) = B(\omega, x) + T(\omega, y).$$

From Theorem 9.21 there exists a unique fixed point of $F_{\omega,y}(\cdot)$ and by Theorem 9.23 $(I - B(\omega, \cdot))^{-1}$ exists. We define the operator $N_\omega : C \to C$ by $N_\omega(y) = (I - B(\omega, \cdot))^{-1} T(\omega, y)$. It is easy to see that by Schauder's fixed point theorem, N_ω has at least one fixed point. Defining a mapping $S : \Omega \to \mathcal{P}_{cp}(C)$ by

$$S(\omega) = \{ y \in C : y = B(\omega, y) + T(\omega, y) \}.$$

Let K be closed subset of C. Then

$$\begin{aligned} S_-^{-1}(K) &= \{ \omega \in \Omega : S(\omega) \cap K \neq \emptyset \} \\ &= \{ \omega \in \Omega : y = B(\omega, y) + T(\omega, y), \ y \in K \}. \end{aligned}$$

Since X is a separable generalized Banach space, then there exists $\{ y_i : i \in \mathbb{N} \} \subset K$ such that

$$\overline{\{ y_i : i \in \mathbb{N} \}} = K.$$

Hence,

$$S_-^{-1}(K) = \cap_{n=1}^{\infty} \cup_{x_i \in K_n} \{\omega \in \Omega : \|y_i - B(\omega, y_i) - T(\omega, y_i)\| < 2\epsilon_n\}$$

where

$$K_n = \{x \in C : d(x, K) < \epsilon_n\}.$$

and

$$\epsilon_n = \begin{pmatrix} \frac{1}{n} \\ \vdots \\ \frac{1}{n} \end{pmatrix}, \quad (x, y) \in X \times X.$$

Therefore, $S^{-1}(K)$ is measurable. Since $B(\omega, \cdot) + T(\omega, \cdot)$ is a continuous mapping, from Theorem 9.19, there exists $y : \Omega \to C$, a measurable selection of S, which is a random fixed point of $B + T$. $\qquad\square$

An immediate corollary to the above theorem in applicable form is:

Corollary 9.29. *Let X be a separable generalized Banach space X. Suppose that $T, B : \Omega \times X \to X$ are two random operators, such that*

$(\bar{A}_1)$ *T is a completely continuous random operator.*

$(\bar{A}_2)$ *B is a continuous random and $M(\omega)-$contraction operator.*

$(\bar{A}_3)$ *The matrix $I - M$ has the absolute value property.*

If

$$\mathcal{M} = \{x : \Omega \to X \text{ is measurable} \mid \lambda(\omega)T(\omega, x) + \lambda(\omega)B(\frac{x}{\lambda(\omega)}, \omega) = x\}$$

is bounded for all measurable $\lambda : \Omega \to \mathbb{R}$ with $0 < \lambda(\omega)) < 1$ on Ω, then the random equation

$$x = T(\omega, x) + B(\omega, x), \quad x \in X,$$

has at least one solution.

Chapter 10

Semigroups

10.1 C_0-semigroups

Let E be a Banach space and $B(E)$ be the Banach space of linear bounded operators.

Definition 10.1. *A semigroup of class (C_0) is a one parameter family $\{S(t) \mid t \geq 0\} \subset B(E)$ satisfying the conditions:*

(i) $S(t) \circ S(s) = S(t + s)$, for $t, s \geq 0$.

(ii) $S(0) = I$.

(iii) the map $t \to S(t)(x)$ is strongly continuous, for each $x \in E$, i.e;

$$\lim_{t \to 0} S(t)x = x, \ \forall x \in E.$$

A semigroup of bounded linear operators $S(t)$, is uniformly continuous if

$$\lim_{t \to 0} \|S(t) - I\| = 0.$$

Here I denotes the identity operator in E.

We note that if a semigroup $\{S(t)\}$ is of class (C_0) then it satisfies a growth condition.

Proposition 10.2. *Let $\{S(t)\}_{t \geq 0}$ be a semigroup of bounded linear operators. Then there exists some constant $M \geq 0$ and $\omega \in \mathbb{R}$ such that*

$$\|S(t)\|_{B(E)} \leq M e^{\omega t}, \quad for \ t \geq 0.$$

If, in particular $M = 1$ and $\beta = 0$, i.e; $\|S(t)\|_{B(E)} \leq 1$, for $t \geq 0$, then the semigroup $S(t)$ is called a *contraction semigroup* (C_0).

Definition 10.3. *Let $S(t)$ be a semigroup of class (C_0) defined on E. The infinitesimal generator A of $S(t)$ is the linear operator defined by*

$$A(x) = \lim_{h \to 0} \frac{S(h)(x) - x}{h}, \quad for \ x \in D(A),$$

where $D(A) = \{x \in E \mid \lim_{h \to 0} \frac{S(h)(x) - x}{h} \ exists \ in \ E\}$.

Let us recall the following property:

Proposition 10.4. *The infinitesimal generator A is closed linear and densely defined operator in E. If $x \in D(A)$, then $S(t)(x)$ is a C^1-map and*

$$\frac{d}{dt} S(t)(x) = A(S(t)(x)) = S(t)(A(x)) \quad on \ [0, \infty).$$

Theorem 10.5. (Hille and Yosida) *[236]. Let A be a densely defined linear operator with domain and range in a Banach space E. Then A is the infinitesimal generator of uniquely determined semigroup S(t) of class (C_0) satisfying*

$$\|S(t)\|_{B(E)} \le M \exp(\omega t), \quad t \ge 0,$$

where $M > 0$ and $\omega \in \mathbb{R}$ if and only if $(\lambda I - A)^{-1} \in B(E)$ and $\|(\lambda I - A)^{-n}\| \le M/(\lambda - \omega)^n$, $n = 1, 2, \ldots$, for all $\lambda \in \mathbb{R}$.

For more details on strongly continuous operators, we refer the reader to the books of Goldstein [135], Pazy [236].

10.1.1 Analytic semigroups

Definition 10.6. *Let $\Delta = \{z : \varphi_1 < arg z < \varphi_2, \varphi_1 < 0 < \varphi_2\}$ and for $z \in \Delta$, let $S(z)$ be a bounded linear operator. The family $S(z)$, $z \in \Delta$ is an analytic semigroup in Δ if*

(i) *$z \to S(z)$ is analytic in Δ.*

(ii) *$S(0) = I$ and $\lim_{z \to 0} S(z)x = x$ for every $x \in E$.*

(iii) *$S(z_1 + z_2) = S(z_1)S(z_2)$ for $z_1, z_2 \in \Delta$.*

A semigroup $S(t)$ will be called analytic if it is analytic in some sector Δ containing the nonnegative real axis.

Clearly, the restriction of an analytic semigroup to the real axis is a C_0 semigroup. We will be interested below in the possibility of extending a given C_0 semigroup to an analytic semigroup in some sector Δ around the nonnegative real axis.

Theorem 10.7. *[236] Let S(t) be a uniformly bounded C_0 semigroup. Let A be the infinitesimal generator of S(t) and assume $0 \in \rho(A)$. The following statements are equivalent:*

(a) *$S(t)$ can be extended to an analytic semigroup in a sector $\Delta_\delta = \{z : |arg z| < \delta\}$ and $\|S(z)\|$ is uniformly bounded in every closed subsector $\Delta_{\delta'}$, $\delta' < \delta$, of Δ_δ.*

(b) *There exists a constant C such that for every $\sigma > 0$, $\tau \ne 0$*

$$\|R(\sigma + it : A)\| \le \frac{C}{\tau}.$$

(c) *There exist $0 < \delta < \pi/2$ and $M > 0$ such that*

$$\rho(A) \supset \Sigma = \{\lambda : |arg\lambda| < \frac{\pi}{2} + \delta\} \cup \{0\}$$

and

$$\|R(\lambda : A)\| \le \frac{M}{|\lambda|} \text{ for } \lambda \in \Sigma, \lambda \ne 0.$$

(d) *$S(t)$ is differentiable for $t > 0$ and there is a constant C such that*

$$\|AS(t)\| \le \frac{C}{t}, \quad t > 0.$$

10.2 Fractional powers of closed operators

For our definition we will make the following assumption.

(DDC) Let A be a densely defined closed linear operator for which

$$\rho(A) \supset \Sigma^+ = \{\lambda : 0 < \omega < |arg\lambda| \leq \pi\} \cup V$$

where V is a neighborhood of zero, and

$$\|R(\lambda : A)\| \leq \frac{M}{1 + |\lambda|} \text{ for } \lambda \in \Sigma^+.$$

If $M = 1$ and $w = \frac{\pi}{2}$, then $-A$ is the infinitesimal generator of a C_0 semigroup. If $w < \frac{\pi}{2}$ then, by Theorem 10.7, $-A$ is the infinitesimal generator of an analytic semigroup. The assumption that $0 \in \rho(A)$, and therefore a whole neighborhood V of zero is in $\rho(A)$, was made mainly for convenience. Most of the results on fractional powers that we will obtain in this section remain true even if $0 \in \rho(A)$.

Definition 10.8. *Let A satisfy Assumption (DDC) with $w < \frac{\pi}{2}$. For every $\alpha > 0$ we define*

$$A^\alpha = (A^{-\alpha})^{-1}.$$

For $\alpha = 0$, $A^\alpha = I$.

Theorem 10.9. *[236] Let A^α be defined by Definition 10.8. Then*

(a) *A^α is a closed operator with domain $D(A^\alpha) = R(A^{-\alpha}) =$ the range of $A^{-\alpha}$.*

(b) *$\alpha \geq \beta > 0$ implies $D(A^\alpha) \subset D(A^\beta)$.*

(c) *$\overline{D(A^\alpha)} = E$ for every $\alpha \geq 0$.*

(d) *If α, β are real then*

$$A^{\alpha+\beta}x = A^\alpha \cdot A^\beta x$$

for every $x \in D(A^\gamma)$ where $\gamma = \max(\alpha, \beta, \alpha + \beta)$.

Theorem 10.10. *[236] Let $-A$ be the infinitesimal generator of an analytic semigroup $S(t)$. If $0 \in \rho(A)$, then*

(a) *$S(t) : E \to D(A^\alpha)$ for every $t > 0$ and $\alpha \geq 0$.*

(b) *For every $x \in D(A^\alpha)$ we have $S(t)A^\alpha x = A^\alpha S(t)x$.*

(c) *For every $t > 0$ the operator $A^\alpha S(t)$ is bounded and*

$$\|A^\alpha S(t)\| \leq M_\alpha t^{-\alpha} e^{-\delta t}.$$

(d) *Let $0 < \alpha \leq 1$ and $x \in D(A^\alpha)$ then*

$$\|S(t)x - x\| \leq C_\alpha t^\alpha \|A^\alpha x\|.$$

Chapter 11

Systems of Impulsive Differential Equations on Half-lines

In this chapter we study the existences, uniqueness, continuous dependence on initial conditions, and boundedness of solutions for a system of impulsive differential equations using a fixed point approach in vector Banach spaces. In addition, the compactness of the solution space and the *u.s.c.* of solutions are investigated. More precisely we consider the system of impulsive differential equations

$$
\begin{cases}
x'(t) & = & f(t, x, y), t \in J := [0, \infty), t \neq t_k, \ k = 1, \dots, \\
y'(t) & = & g(t, x, y), t \in J, \quad t \neq t_k, \ k = 1, \dots, \\
x(t_k^+) - x(t_k^-) & = & I_k(x(t_k), y(t_k)), \quad k = 1, \dots, \\
y(t_k^+) - y(t_k^-) & = & \overline{I}_k(x(t_k), y(t_k)), \quad k = 1, \dots, \\
x(0) & = & x_0, \\
y(0) & = & y_0,
\end{cases}
\tag{11.1}
$$

where $x_0,\ y_0 \in \mathbb{R}$, f, $g : J \times \mathbb{R} \times \mathbb{R} \to \mathbb{R}$ are given functions, and I_k, $\overline{I}_k \in C(\mathbb{R} \times \mathbb{R}, \mathbb{R})$. The notations $x(t_k^+) = \lim_{h \to 0^+} x(t_k + h)$ and $x(t_k^-) = \lim_{h \to 0^+} x(t_k - h)$ stand for the right and left hand limits of the function y at $t = t_k$, respectively. For all the results in this chapter see [49].

11.1 Uniqueness and continuous dependence on initial data

In order to define a solution for problem (11.1), consider the space of piecewise continuous functions

$$
PC_b = \{y \in PC([0, \infty), \mathbb{R}) : y \text{ is bounded}\},
$$

where

$$
PC([0, \infty), \mathbb{R}) = \{y : [0, \infty) \to \mathbb{R}, y_k \in C((t_k, t_{k+1}], \mathbb{R}), \ k = 0, \dots, y(t_k^-)
$$

$$
\text{and } y(t_k^+) \text{ exist and satisfy } y(t_k) = y(t_k^-) \text{ for } k = 1, \dots\}.
$$

Note that PC_b is a Banach space with the norm

$$
\|y\|_b = \sup\{|y(t)| : \ t \in [0, \infty)\}.
$$

Definition 11.1. *A function* $(x, y) \in PC(J, \mathbb{R}) \times PC(J, \mathbb{R})$ *is said to be a solution of (11.1) if and only if*

$$
\begin{cases}
x(t) & = & x_0 + \int_0^t f(s, x(s), y(s))ds + \displaystyle\sum_{0 < t_k < t} I_k(x(t_k), y(t_k)), t \in J, \\
y(t) & = & y_0 + \int_0^t g(s, x(s), y(s))ds + \displaystyle\sum_{0 < t_k < t} \overline{I}_k(x(t_k), y(t_k)), t \in J.
\end{cases}
$$

In this section we assume the following conditions:

$(\mathcal{D}_1)$ There exist functions $l_i \in L^1(J, \mathbb{R}^+)$, $i = 1, \ldots, 4$, such that

$$|f(t, x, y) - f(s, \overline{x}, \overline{y})| \leq l_1(t)|x - \overline{x}| + l_2(t)|y - \overline{y}|, \text{ for all } x, \overline{x}, y, \overline{y} \in \mathbb{R}$$

and

$$|g(t, x, y) - g(s, \overline{x}, \overline{y})| \leq l_3(t)|x - \overline{x}| + l_4(t)|y - \overline{y}|, \text{ for all } x, \overline{x}, y, \overline{y} \in \mathbb{R}.$$

$(\mathcal{D}_2)$ There exist constants $a_{1k}, a_{2k} \geq 0$, $k = 1, \ldots,$ such that

$$|I_k(x, y) - I_k(\overline{x}, \overline{y})| \leq a_{1k}|x - \overline{x}| + a_{2k}|y - \overline{y}|, \text{ for all } x, \overline{x}, y, \overline{y} \in \mathbb{R}$$

and

$$\sum_{k=1}^{\infty} |I_k(0, 0)| < \infty.$$

$(\mathcal{D}_3)$ There exist constants $b_{1k}, b_{2k} \geq 0$, $k = 1, \ldots,$ such that

$$|\overline{I}_k(x, y) - \overline{I}_k(\overline{x}, \overline{y})| \leq b_{1k}|x - \overline{x}| + b_{2k}|y - \overline{y}|, \text{ for all } x, \overline{x}, y, \overline{y} \in \mathbb{R}$$

and

$$\sum_{k=1}^{\infty} |\overline{I}_k(0, 0)| < \infty.$$

We will use the Perov fixed point theorem to prove that a solution of problem (11.1) is bounded and tends to zero as $t \to \infty$.

Theorem 11.2. *Assume that $(\mathcal{D}_1) - (\mathcal{D}_3)$ are satisfied, $f(\cdot, 0, 0), g(\cdot, 0, 0) \in L^1(J, \mathbb{R})$,*

$$\sum_{k=1}^{\infty} a_{ik} < \infty \text{ and } \sum_{k=1}^{\infty} b_{ik} < \infty, i = 1, 2,$$

and the matrix

$$M = \begin{pmatrix} \|l_1\|_{L^1} + \sum_{k=1}^{\infty} a_{1k} & \|l_2\|_{L^1} + \sum_{k=1}^{\infty} a_{2k} \\ \|l_3\|_{L^1} + \sum_{k=1}^{\infty} b_{1k} & \|l_4\|_{L^1} + \sum_{k=1}^{\infty} b_{2k} \end{pmatrix} \in \mathcal{M}_{2 \times 2}(\mathbb{R}^+) \tag{11.2}$$

converges to zero. Then the problem (11.1) has unique a solution. If, in addition,

$$\sum_{k=1}^{\infty} a_{1k} + \sum_{k=1}^{\infty} a_{2k} + \sum_{k=1}^{\infty} b_{1k} + \sum_{k=1}^{\infty} b_{2k} < 1,$$

the unique solution of (11.1) is bounded.

Proof. Consider the operator $N : PC \times PC \to PC \times PC$ defined by

$$N(x, y) = (N_1(x, y), N_2(x, y))$$

where

$$N_1(x,y)(t) = x_0 + \int_0^t f(s, x(s), y(s))ds + \sum_{0 < t_k < t} I_k(x(t_k), y(t_k)), t \in [0, \infty)$$

and

$$N_2(x,y)(t) = y_0 + \int_0^t g(s, x(s), y(s))ds + \sum_{0 < t_k < t} \bar{I}_k(x(t_k), y(t_k)), t \in [0, \infty).$$

We begin by showing that the operator N is well defined.

Let $(x, y) \in PC_b \times PC_b, t \in [0, \infty)$; then

$$\|N_1(x,y)\|_b \leq |x_0| + \int_0^t |f(s, x(s), y(s))|ds + \sum_{0 < t_k < t} |I_k(x(t_k), y(t_k))|$$

$$\leq \|l_1\|_{L^1}\|x\|_b + \|l_2\|_{L^1}\|y\|_b + \sum_{0 < t_k < t} (a_{1k}\|x\|_b + a_{2k}\|y\|_b)$$

$$+ \|f(\cdot, 0, 0)\|_{L^1} + \sum_{0 < t_k < t} (|I_k(0,0)| + |\bar{I}_k(0,0)|).$$

Similarly, we have

$$\|N_2(x,y)\|_b \leq \|l_3\|_{L^1}\|x\|_b + \|l_4\|_{L^1}\|y\|_b + \sum_{0 < t_k < t} (b_{1k}\|x\|_b + b_{2k}\|y\|_b)$$

$$+ \|g(\cdot, 0, 0)\|_{L^1} + \sum_{0 < t_k < t} (|I_k(0,0)| + |\bar{I}_k(0,0)|).$$

Thus,

$$\left(\begin{array}{c} \|N_1(x,y)\|_b \\ \|N_1(x,y)\|_b \end{array} \right) \leq \left(\begin{array}{cc} \|l_1\|_{L^1} + \sum_{k=1}^{\infty} a_{1k} & \|l_2\|_{L^1} + \sum_{k=1}^{\infty} a_{2k} \\ \|l_3\|_{L^1} + \sum_{k=1}^{\infty} b_{1k} & \|l_4\|_{L^1} + \sum_{k=1}^{\infty} b_{1k}\|x\| + b_{2k} \end{array} \right) \left(\begin{array}{c} \|x\|_b \\ \|y\|_b \end{array} \right)$$

$$+ \left(\begin{array}{c} \|f(\cdot, 0, 0)\|_{L^1} + \sum_{k=1}^{\infty} (\|I_k(0,0)\|_b + \|\bar{I}_k(0,0)\|_b) \\ \|g(\cdot, 0, 0)\|_{L^1} + \sum_{k=1}^{\infty} (|I_k(0,0)| + |\bar{I}_k(0,0)|) \end{array} \right).$$

Hence, the operator N is well defined.

Clearly, the fixed points of N are solutions of the problem (11.1). To show that N is a contraction, let $(x, y), (\bar{x}, \bar{y}) \in PC_b \times PC_b$. Then $(\mathcal{D}_1)$ and $(\mathcal{D}_2)$ imply

$$|N_1(x,y)(t) - N_1(\bar{x}, \bar{y})(t)| \leq \int_0^t |f(s, x(s), y(s)) - f(s, \bar{x}(s), \bar{y}(s))|ds$$

$$+ \sum_{0 < t_k < t} |I_k(x(t_k), y(t_k)) - I_k(\bar{x}(t_k), \bar{y}(t_k))|$$

$$\leq \int_0^t (l_1(s)|x(s) - \overline{x}(s)| + l_2(s)|y(s) - \overline{y}(s)|)\, ds$$
$$+ \sum_{0 < t_k < t} (a_{1k}|x(t_k) - \overline{x}(t_k)| + a_{2k}|y(t_k) - \overline{y}(t_k)|).$$

Thus,

$$\|N_1(x, y) - N_1(\overline{x}, \overline{y})\|_b \leq \left(\|l_1\|_{L^1} + \sum_{k=1}^{\infty} a_{1k}\right)\|x - \overline{x}\|_b$$
$$+ \left(\|l_2\|_{L^1} + \sum_{k=1}^{\infty} a_{2k}\right)\|y - \overline{y}\|_b.$$

Similarly, we have

$$\|N_2(x, y) - N_2(\overline{x}, \overline{y})\|_b \leq \left(\|l_3\|_{L^1} + \sum_{k=1}^{\infty} b_{1k}\right)\|x - \overline{x}\|_b$$
$$+ \left(\|l_4\|_{L^1} + \sum_{k=1}^{\infty} b_{2k}\right)\|y - \overline{y}\|_b.$$

Therefore,

$$\|N(x, y) - N(\overline{x}, \overline{y})\|_b \leq M \begin{pmatrix} \|x - \overline{x}\|_b \\ \|y - \overline{y}\|_b \end{pmatrix}, \quad \text{for all } (x, y), (\overline{x}, \overline{y}) \in PC_b \times PC_b.$$

Hence, by Theorem 8.1, the operator N has at least one fixed point that is a solution of (11.1).

Next, we show that the solution (x, y) is bounded. Let $t \in [0, \infty)$; then

$$|x(t)| \leq |x_0| + \int_0^t |f(s, x(s), y(s))|ds + \sum_{0 < t_k < t} |I_k(x(t_k), y(t_k))|$$
$$\leq |x_0| + \int_0^t (l_1(s)|x| + l_2(s)|y|)ds + \sum_{k=1}^{\infty} a_{1k}|x(t_k)| + \sum_{k=1}^{\infty} a_{2k}|y(t_k)|$$
$$+ \|f(\cdot, 0, 0)\|_{L^1} + \|g(\cdot, 0, 0)\|_{L^1} + \sum_{k=1}^{\infty} |I_k(0, 0)| + \sum_{k=1}^{\infty} |\overline{I}_k(0, 0)|$$

and

$$|y(t)| \leq |y_0| + \int_0^t (l_3(s)|x(s)| + l_4(s)|y(s)|)ds + \sum_{k=1}^{\infty} b_{1k}|x(t_k)| + \sum_{k=1}^{\infty} b_{2k}|y(t_k)|$$
$$+ \|f(\cdot, 0, 0)\|_{L^1} + \|g(\cdot, 0, 0)\|_{L^1} + \sum_{k=1}^{\infty} |I_k(0, 0)| + \sum_{k=1}^{\infty} |\overline{I}_k(0, 0)|.$$

Thus,

$$|x(t)| + |y(t)|$$
$$\leq |x_0| + |y_0| + \int_0^t ((l_1(s) + l_3(s))|x(s)| + (l_2(s) + l_4(s))|y(s)|)ds$$

$$+ \left(\sum_{k=1}^{\infty} a_{1k} + \sum_{k=1}^{\infty} a_{2k} + \sum_{k=1}^{\infty} b_{1k} + \sum_{k=1}^{\infty} b_{2k} \right)(|x(t_k)| + |y(t_k)|)$$

$$+ 2\|f(\cdot,0,0)\|_{L^1} + 2\|g(\cdot,0,0)\|_{L^1} + 2\sum_{k=1}^{\infty} |I_k(0,0)| + 2\sum_{k=1}^{\infty} |\bar{I}_k(0,0)|.$$

Hence,

$$\sup_{s \in (0,t)} (|x(s)| + |y(s)|)$$

$$\leq |x_0| + |y_0| + \int_0^t (l_1(s) + l_3(s) + l_2(s) + l_4(s))$$

$$\times \sup_{s \in [0,t]} (|x(s)| + |y(s)|) ds$$

$$+ \left(\sum_{k=1}^{\infty} a_{1k} + \sum_{k=1}^{\infty} a_{2k} + \sum_{k=1}^{\infty} b_{1k} + \sum_{k=1}^{\infty} b_{2k} \right) \sup_{s \in [0,t]} (|x(t_k)| + |y(t_k)|)$$

$$+ 2\|f(\cdot,0,0)\|_{L^1} + 2\|g(\cdot,0,0)\|_{L^1} + 2\sum_{k=1}^{\infty} |I_k(0,0)| + 2\sum_{k=1}^{\infty} |\bar{I}_k(0,0)|.$$

This implies that

$$\sup_{s \in (0,t)} (|x(s)| + |y(s)|) \leq \alpha + \int_0^t l(s) \sup_{s \in [0,t]} (|x(s)| + |y(s)|) ds$$

where

$$\alpha = \frac{|x_0| + |y_0| + 2\|f(\cdot,0,0)\|_{L^1} + 2\|g(\cdot,0,0)\|_{L^1} + 2\sum_{k=1}^{\infty} |I_k(0,0)| + 2\sum_{k=1}^{\infty} |\bar{I}_k(0,0)|}{1 - \left(\sum_{k=1}^{\infty} a_{1k} + \sum_{k=1}^{\infty} a_{2k} + \sum_{k=1}^{\infty} b_{1k} + \sum_{k=1}^{\infty} b_{2k} \right)}$$

and

$$l(s) = \frac{l_1(s) + l_2(s) + l_3(s) + l_4(s)}{1 - \left(\sum_{k=1}^{\infty} a_{1k} + \sum_{k=1}^{\infty} a_{2k} + \sum_{k=1}^{\infty} b_{1k} + \sum_{k=1}^{\infty} b_{2k} \right)}.$$

Applying the Gronwall-Bellman lemma, we obtain

$$\sup_{s \in [0,t]} (|x(s)| + |y(s)|) \leq \alpha \exp \left(\int_0^t l(s) ds \right),$$

so

$$\|x\|_b + \|y\|_b \leq \alpha \exp \left(\int_0^{\infty} l(s) ds \right).$$

This implies that the solution (x, y) is bounded. $\qquad \square$

The next result establishes the continuous dependence of solutions on initial conditions.

Theorem 11.3. *Assume that conditions $(\mathcal{D}_1) - (\mathcal{D}_3)$ hold, the matrix M defined in (11.2) converges to zero, $I_k(0,0) = \bar{I}_k(0,0)$, $k = 1, \ldots$, and $f(t,0,0) = g(t,0,0) = 0$ for $t \in J$. For every $(x_0, y_0) \in \mathbb{R} \times \mathbb{R}$ we denote by $(x(t,x_0), y(t,y_0))$ the solution of (11.1). Then the map $(x_0, y_0) \to (x(\cdot, x_0), y(\cdot, y_0))$ is continuous.*

Proof. Let $(x_0, y_0), (\bar{x}_0, \bar{y}_0) \in \mathbb{R} \times \mathbb{R}$. Then from Theorem 11.2, there exist $(x(\cdot, x_0), y(\cdot, y_0))$, $(\bar{x}(\cdot, \bar{x}_0), \bar{y}(\cdot, \bar{y}_0)) \in PC_b \times PC_b$ such that

$$x(t, x_0) = x_0 + \int_0^t f(s, x(s, x_0), y(s, y_0)) ds + \sum_{0 < t_k < t} I_k(x(t_k, x_0), y(t_k, y_0)),$$

$$y(t, y_0) = y_0 + \int_0^t g(s, x(s, x_0), y(s, y_0)) ds + \sum_{0 < t_k < t} \bar{I}_k(x(t_k, x_0), y(t_k, y_0)),$$

$$x(t, \bar{x}_0) = \bar{x}_0 + \int_0^t f(s, x(s, \bar{x}_0), y(s, \bar{y}_0)) ds + \sum_{0 < t_k < t} I_k(x(t_k, \bar{x}_0), y(t_k, \bar{y}_0)),$$

and

$$\bar{y}(t, \bar{y}_0) = \bar{y}_0 + \int_0^t g(s, x(s, \bar{x}_0), y(s, \bar{y}_0)) ds + \sum_{0 < t_k < t} \bar{I}_k(x(t_k, \bar{x}_0), y(t_k, \bar{y}_0))$$

for $t \in [0, \infty)$. Hence from the proof of Theorem 11.2 we deduce that

$$\|x(\cdot, x_0) - \bar{x}(\cdot, \bar{x}_0)\|_b + \|y(\cdot, y_0) - \bar{y}(\cdot, \bar{y}_0)\|_b$$

$$\leq \frac{|x_0 - \bar{x}_0| + |y_0 - \bar{y}_0|}{1 - \left(\sum_{k=1}^{\infty} a_{1k} + \sum_{k=1}^{\infty} a_{2k} + \sum_{k=1}^{\infty} b_{1k} + \sum_{k=1}^{\infty} b_{2k} \right)} \times \exp \left(\int_0^{\infty} l(s) ds \right).$$

Then,

$$\|x(\cdot, x_0) - \bar{x}(\cdot, \bar{x}_0)\|_b + \|y(\cdot, y_0) - \bar{y}(\cdot, \bar{y}_0)\|_b \to 0, \text{ as } (x_0, y_0) \to (\bar{x}_0, \bar{y}_0).$$

$\square$

11.2   Existence and compactness of solution sets

In this section we present an application of a Krasnosel'skii type fixed point theorem to problem (11.1). The following compactness criterion on unbounded domains is a simple extension of a compactness criterion in $C_b(\mathbb{R}_+, \mathbb{R})$ (see [85, p. 62]).

Lemma 11.4. *Let $C \subset PC_b$. Then C is relatively compact if it satisfies the following conditions:*

(a) *C is uniformly bounded in $PC_\ell(\mathbb{R}^+, \mathbb{R}^n)$;*

(b) *The functions belonging to C are almost equicontinuous on $\mathbb{R}^+$, i.e., equicontinuous on every compact subinterval of $\mathbb{R}^+$;*

(c) *The functions in C are equiconvergent, that is, given $\varepsilon > 0$ there corresponds $T(\varepsilon) > 0$ such that $|x(\tau_1) - x(\tau_2)| < \varepsilon$ for any $\tau_1, \tau_2 \geq T(\varepsilon)$ and $x \in C$.*

Theorem 11.5. *In addition to $(\mathcal{D}_1)$ assume that:*

$(\mathcal{D}_4)$ *There exist $\alpha_k, \beta_k \geq 0$, $k = 1, \ldots$, such that*

$$|I_k(x, y)| \leq \alpha_k |x| + \beta_k |y| + c_k, \text{ for all } (x, y) \in \mathbb{R} \times \mathbb{R};$$

$(\mathcal{D}_5)$ *There exist* $\overline{\alpha_k}$, $\overline{\beta_k} \geq 0$, $k = 1, \ldots$, *such that*

$$|\overline{I_k}(x, y)| \leq \overline{\alpha_k}|x| + \overline{\beta_k}|y| + \overline{c_k}, \text{ for all } (x, y) \in \mathbb{R} \times \mathbb{R}.$$

If

$$M_* = \begin{pmatrix} \|l_1\|_{L^1} & \|l_2\|_{L^1} \\ \|l_3\|_{L^1} & \|l_4\|_{L^1} \end{pmatrix} \in \mathcal{M}_{2\times2}(\mathbb{R}^+) \tag{11.3}$$

converges to zero,

$$\sum_{k=1}^{\infty} \alpha_k + \sum_{k=1}^{\infty} \overline{\alpha_k} + \sum_{k=1}^{\infty} \beta_k + \sum_{k=1}^{\infty} \overline{\beta_k} < 1,$$

$$\sum_{k=1}^{\infty} c_k < \infty, \text{ and } \sum_{k=1}^{\infty} \overline{c_k} < \infty,$$

then problem (11.1) has at least one bounded solution.

Proof. Let $N : PC_b \times PC_b \to PC_b \times PC_b$ be the operator defined in the proof of Theorem 11.2, and let $N = A + B$ where A, $B : PC_b \times PC_b \to PC_b \times PC_b$ are given by

$$A(x(t), y(t)) = (A_1(x(t), y(t)), A_2(x(t), y(t)), \quad t \in J,$$

and

$$B(x(t), y(t)) = (B_1(x(t), y(t)), B_2(x(t), y(t)), \quad t \in J,$$

where

$$\begin{cases} A_1(x, y) &= \displaystyle\sum_{0 < t_k < t}^{\infty} I_k(x(t_k), y(t_k)) \\ A_2(x, y)) &= \displaystyle\sum_{0 < t_k < t}^{\infty} \overline{I_k}(x(t_k), y(t_k)). \end{cases}$$

and

$$\begin{cases} B_1(x, y) &= x_0 + \int_0^t f(s, x(s), y(s))ds \\ B_2(x, y)) &= y_0 + \int_0^t g(s, x(s), y(s))ds \end{cases}$$

Step 1 *B is a contraction.* Let (x, y), $(\overline{x}, \overline{y}) \in PC_b \times PC_b$. Then

$$|B_1(x(t), y(t)) - B_1(\overline{x}t), \overline{y}(t))| \leq \int_0^t |f(s, x(s), y(s)) - f(s, \overline{x}(s), \overline{y}(s))|ds$$

$$\leq \int_0^t (l_1(s)|x(s) - \overline{x}(s)| + l_2(s)|y(s) - \overline{y}(s)|) \, ds.$$

Hence,

$$\|B_1(x, y) - B_1(\overline{x}, \overline{y})\|_b \leq \|l_1\|_{L^1}\|x - \overline{x}\|_b + \|l_2\|_{L^1}\|y - \overline{y}\|_b.$$

Similarly,

$$\|B_2(x, y) - B_2(\overline{x}, \overline{y})\|_b \leq \|l_3\|_{L^1}\|x - \overline{x}\|_b + \|l_4\|_{L^1}\|y - \overline{y}\|_b.$$

Therefore,

$$\|B(x, y) - B(\overline{x}, \overline{y})\|_b \leq \begin{pmatrix} \|l_1\|_{L^1} & \|l_2\|_{L^1} \\ \|l_3\|_{L^1} & \|l_4\|_{L^1} \end{pmatrix} \begin{pmatrix} \|x - \overline{x}\|_b \\ \|y - \overline{y}\|_b \end{pmatrix}.$$

Step 2. *A is continuous.* Given $(x_n, y_n) \to (x, y)$ in $PC_b \times PC_b$, there exists $M, M' > 0$ such that

$$\|x_n\|_b \leq M \text{ and } \|y_n\|_b \leq M' \text{ for every } n \in \mathbb{N},$$

and

$$|(A_1 x_n)(t) - (A_1 x)(t)| \leq \sum_{0 \leq t_k < t} |I_k(x_n, y_n) - I_k(x, y)|.$$

Since

$$\sum_{k=1}^{\infty} \alpha_k < \infty \quad \text{and} \quad \sum_{k=1}^{\infty} \beta_k < \infty,$$

for every $\epsilon > 0$, there exists $k_0 \in \mathbb{N}$ such that

$$\sum_{k=k_0}^{\infty} \alpha_k < \frac{\epsilon}{6M} \quad \text{and} \quad \sum_{k=k_0}^{\infty} \beta_k < \frac{\epsilon}{6M'}.$$

Using the fact $\lim_{k \to \infty} t_k = \infty$, there exists $n_0 \in \mathbb{N}$ such that for each $k \geq n_0$ we have $t_k \geq k_0$. From $(\mathcal{D}_4)$, we obtain

$$
\begin{aligned}
\| A_1(x_n, y_n) - A_1(x, y) \|_{PC_b} &\leq \sum_{0 \leq t_k \leq t_{n_0 - 1}} |I_k(x_n, y_n) - I_k(x, y)| \\
&+ \sum_{k=k_0}^{\infty} (2M\alpha_k + 2M'\beta_k) \\
&\leq \sum_{k=1}^{k_0 - 1} |I_k(x_n, y_n) - I_k(x, y)| + \frac{2\epsilon}{3}.
\end{aligned}
$$

Using the fact that the I_k are continuous functions, we have

$$\sum_{k=0}^{n_0 - 1} |I_k(x_n, y_n) - I_k(x, y)| \to 0 \text{ as } n \to \infty.$$

Hence,

$$\|A_1(x_n, y_n) - A_1(x, y)\|_b \to 0 \text{ as } n \to \infty.$$

Similarly, we have

$$\|A_2(x_n, y_n) - A_2(x, y)\|_b \to 0 \text{ as } n \to \infty.$$

Thus,

$$\|A(x_n, y_n) - A(x, y)\|_b \to 0 \text{ as } n \to \infty.$$

Step 3 From $(\mathcal{D}_4)$, we can easily prove that A maps bounded sets into bounded sets in $PC \times PC$.

We next will show that $A(M)$ is contained in a compact set.

Step 4. *A maps bounded sets in $PC_b \times PC_b$ into almost equicontinuous sets of $PC_b \times PC_b$.* Let $r = (r_1, r_2) > 0$, $B_r := \{(x, y) \in PC_b \times PC_b : \|(x, y)\|_\infty \leq r\}$ be a bounded set in $PC \times PC$, $\tau_1, \tau_2 \in [0, \infty)$, $\tau_1 < \tau_2$, and $\phi \in B_r$. We have

$$A\phi(\tau_1) = (A_1\phi(\tau_1), A_2\phi(\tau_1))$$

where

$$
\begin{cases}
A_1\phi(\tau_1) = \sum_{0 \leq t_k \leq \tau_1} I_k(\phi_1(t_k), \phi_2(t_k)), \\
A_2\phi(\tau_1). = \sum_{0 \leq t_k \leq \tau_2} I_k(\phi_1(t_k), \phi_2(t_k)).
\end{cases}
$$

Then

$$|A_1\phi(\tau_2) - A_1\phi(\tau_1)| \leq \sum_{\tau_1 \leq t_k \leq \tau_2} I_k(\phi_1(t_k), \phi_2(t_k)),$$

so

$$|A_1\phi(\tau_2) - A_1\phi(\tau_1)| \to 0 \text{ as } \tau_1 \to \tau_2.$$

Similarly we have

$$|A_2\phi(\tau_2) - A_2\phi(\tau_1)| \to 0 \text{ as } \tau_1 \to \tau_2.$$

Thus,

$$|A\phi(\tau_1) - A\phi(\tau_2)| \to 0 \text{ as } \tau_1 \to \tau_2.$$

Step 5. We now show that the set $A(\overline{B}(0,r))$ is equiconvergent, i.e. for every $\varepsilon > 0$, there exists $T(\varepsilon) > 0$ such that $\|A(\phi(t)) - A(\phi(s))\| \leq \varepsilon$ for every $t, s \geq T(\epsilon)$ and each $\phi \in \overline{B}(0,r)$. Letting $\phi \in \overline{B}(0,r)$, for every $\epsilon > 0$, there exists $k_0 \in \mathbb{N}$ such that

$$\sum_{k=k_0}^{\infty} \alpha_k < \frac{\epsilon}{2r_1}, \ \sum_{k=k_0}^{\infty} \beta_k < \frac{\epsilon}{2r_2},$$

and

$$
\begin{aligned}
|A_1\phi(t) - A_1\phi(s)| &\leq \sum_{s \leq t_k \leq t} I_k(\phi_1(t_k), \phi_2(t_k)) \\
&\leq \sum_{s \leq t_k \leq t} (\alpha_k r_1 + \beta_k r_2).
\end{aligned}
$$

Then for every $s, t > k_0$, we obtain

$$|A_1\phi(t) - A_1\phi(s)| \leq r_1 \sum_{k=k_0}^{\infty} \alpha_k + r_2 \sum_{k=k_0}^{\infty} \beta_k.$$

Therefore for all $\phi \in B(0,r)$ and $s, t > k_0$ we have

$$|A_1\phi(t) - A_1\phi(s)| \leq \epsilon.$$

Similarly, we can prove that there exists $\bar{k}_0 > 0$ such that for all $\phi \in B(0,r)$ and $s, t > \bar{k}_0$ we have

$$|A_2\phi(t) - A_2\phi(s)| \leq \epsilon.$$

Thus, for every $(\epsilon, \epsilon) > 0$ there exists $(k_0, \bar{k}_0) > 0$ such that for all $s, t > k_0$ and $s, t > \bar{k}_0$ we have

$$|A\phi(\tau_1) - A\phi(\tau_2)| \leq (\epsilon, \epsilon), \text{ for every } \phi \in B(0,r).$$

Step 6 Now, we show that the set

$$\mathcal{M} = \{(x,y) \in PC_b \times PC_b : (x,y) = \lambda B(\frac{x}{\lambda}, \frac{y}{\lambda}) + \lambda A(x,y), \lambda \in (0,1)\}$$

is bounded. Let $(x,y) \in \mathcal{M}$; then

$$
\begin{aligned}
|x(t)| &\leq |x_0| + \int_0^t \lambda \left| f\left(s, \frac{x(s)}{\lambda}, \frac{y(s)}{\lambda}\right) \right| ds + \sum_{0 < t_k < t} |I_k(x(t_k), y(t_k))| \\
&\leq |x_0| + \int_0^t (l_1(s)|x(s)| + l_2(s)|y(s)|)ds + \sum_{k=1}^{\infty} \alpha_k |x(t_k)|
\end{aligned}
$$

$$+ \sum_{k=1}^{\infty} \beta_k |y(t_k)| + \sum_{k=1}^{\infty} c_k$$

and

$$|y(t)| \leq |y_0| + \int_0^t (l_3(s)|x(s)| + l_4(s)|y(s)|)ds + \sum_{k=1}^{\infty} \overline{\alpha}_k |x(t_k)|$$

$$+ \sum_{k=1}^{\infty} \overline{\beta}_k |y(t_k)| + \sum_{k=1}^{\infty} \overline{c}_k.$$

Thus,

$$|x(t)| + |y(t)| \leq |x_0| + |y_0| + \int_0^t ((l_1(s) + l_3(s))|x(s)| + (l(s)_2 + l(s)_4)|y(s)|)ds$$

$$+ (\sum_{k=1}^{\infty} \alpha_k + \sum_{k=1}^{\infty} \overline{\alpha}_k)|x(t_k| + (\sum_{k=1}^{\infty} \beta_k + \sum_{k=1}^{\infty} \overline{\beta}_k)|y(t_k|$$

$$+ \sum_{k=1}^{\infty} c_k + \sum_{k=1}^{\infty} \overline{c}_k.$$

Hence,

$$\sup_{s \in [0,t]} (|x(s)| + |y(s)|)$$

$$\leq |x_0| + |y_0| + \int_0^t (l_1(s) + l_3(s) + l_2(s) + l_4(s)) \times \sup_{s \in [0,t]} (|x(s)| + |y(s)|)ds$$

$$+ (\sum_{k=1}^{\infty} \alpha_k + \sum_{k=1}^{\infty} \overline{\alpha}_k) \sup_{s \in (0,t)} |x(t_k| + (\sum_{k=1}^{\infty} \beta_k + \sum_{k=1}^{\infty} \overline{\beta}_k) \sup_{s \in (0,t)} |y(t_k|$$

$$+ \sum_{k=1}^{\infty} c_k + \sum_{k=1}^{\infty} \overline{c}_k.$$

This implies that

$$\sup_{s \in (0,t)} (|x(s)| + |y(s)|) \leq \beta + \int_0^t l_*(s) \sup_{s \in [0,t]} (|x(s)| + |y(s)|)ds$$

where

$$\beta = \frac{|x_0| + |y_0| + \sum_{k=1}^{\infty} c_k + \sum_{k=1}^{\infty} \overline{c}_k}{1 - (\sum_{k=1}^{\infty} \alpha_k + \sum_{k=1}^{\infty} \overline{\alpha}_k + \sum_{k=1}^{\infty} \beta_k + \sum_{k=1}^{\infty} \overline{\beta}_k)}$$

and

$$l_*(s) = \frac{l_1(s) + l_3(s) + l_2(s) + l_4(s)}{1 - (\sum_{k=1}^{\infty} \alpha_k + \sum_{k=1}^{\infty} \overline{\alpha}_k + \sum_{k=1}^{\infty} \beta_k + \sum_{k=1}^{\infty} \overline{\beta}_k)}.$$

By the Gronwall-Bellman lemma, we have

$$\sup_{s \in (0,t)} (|x(s)| + |y(s)|) \leq \beta e^{\|l_*\|_{L^1}},$$

so

$$\|x\|_b \le \beta e^{\|l_*\|_{L^1}}, \quad \text{and} \quad \|y\|_b \le \beta e^{\|l_*\|_{L^1}}.$$

Hence, from Theorem 8.32, the problem (11.1) has at least one solution. $\square$

By a simple modification in the proof we can obtain the following result.

Theorem 11.6. *In addition to* $(\mathcal{D}_2)$ *assume that*

$(\mathcal{D}_6)$ *There exists* $p \in L^1(J, \mathbb{R}_+)$, *and a continuous nondecreasing function* $\psi : \mathbb{R}_+ \to (0, \infty)$
such that

$$|f(t, x, y)| \le p(t)\psi(|x| + |y|), \quad \text{for all } (x, y) \in \mathbb{R} \times \mathbb{R}$$

and

$$|g(t, x, y)| \le p(t)\psi(|x| + |y|), \quad \text{for all } (x, y) \in \mathbb{R} \times \mathbb{R}.$$

If

$$\bar{M}_* = \begin{pmatrix} \displaystyle\sum_{k=1}^{\infty} a_{1k} & \displaystyle\sum_{k=1}^{\infty} a_{2k} \\ \displaystyle\sum_{k=1}^{\infty} b_{1k} & \displaystyle\sum_{k=1}^{\infty} b_{2k} \end{pmatrix} \in \mathcal{M}_{2\times2}(\mathbb{R}^+) \tag{11.4}$$

converges to zero, then the problem (11.1) has at least one bounded solution.

By the nonlinear alternative in generalized Banach spaces we can also prove the following result.

Theorem 11.7. *Assume that* $(\mathcal{D}_2) - (\mathcal{D}_6)$ *hold. If*

$$\sum_{k=1}^{\infty} \alpha_k + \sum_{k=1}^{\infty} \overline{\alpha}_k + \sum_{k=1}^{\infty} \beta_k + \sum_{k=1}^{\infty} \overline{\beta}_k < \infty,$$

$$\sum_{k=1}^{\infty} c_k < \infty, \quad \text{and} \sum_{k=1}^{\infty} \overline{c}_k < \infty,$$

then the problem (11.1) has at least one solution. Moreover, the solution set

$$S(x_0, y_0) = \{(x, y) \in PC_b \times PC_b : (x, y) \text{ is a solution of } (11.1)\}$$

is compact and the multivalued map $S : (x_0, y_0) \to S(x_0, y_0)$ *is u.s.c.*

Chapter 12

Differential Inclusions

12.1 Filippov's theorem on a bounded intervals

In this section we present the systems version of Filippov's theorem and relaxation result for linear and semilinear differential inclusions.

$$
\begin{cases}
x'(t) & \in & A_1 x(t) + F(t, x(t), y(t)), \ \ a.e. \ t \in J := [0, b], \\[2mm]
y'(t) & \in & A_2 y(t) + G(t, x(t), y(t)), \ \ a.e. \ t \in J := [0, b], \\
x(0) & = & a, \\
y(0) & = & \bar{a},
\end{cases}
\tag{12.1}
$$

where $a \in L^1([0, b], \mathbb{R})$ and $A_i : D(A_i) \subset E \to E, i = 1, 2$ is the generator of an integral resolvent family defined on a complex Banach space E, and $F, G : [0, b] \times E \to \mathcal{P}(E)$ are multi-valued maps.

Theorem 12.1. *Assume the conditions:*

(12.1.1) The function $F \colon J \times E \to \mathcal{P}_{cl}(E)$ is such that
(a) for all $x, y \in E$, the maps $t \mapsto F(t, x, y)$ and $t \mapsto G(t, x, y)$ are measurable,
(b) the maps

$$
\gamma : t \mapsto d(g(t), F(t, x(t), y(t))) \ \ and \ \ \bar{\gamma} : t \mapsto d(\bar{g}(t), G(t, x, y))
$$

are integrable.

(12.1.2) There exists a function $p \in L^1(J, \mathbb{R}^+)$ such that, for all $x_1, x_2, y_1, y_2 \in E$, we have

$$
H(F(t, x_1, y_1), F(t, x_2, y_2)) \leq \frac{p(t)}{2} (\|x_1 - x_2\| + \|y_1 - y_2\|),
$$

and

$$
H(G(t, x_1, y_1), G(t, x_2, y_2)) \leq \frac{p(t)}{2} (\|x_1 - x_2\| + \|y_1 - y_2\|).
$$

Then, for every $\epsilon > 0$, Problem (12.1) has at least one solution (x_ϵ, y_ϵ) satisfying, for a.e. $t \in [0, b]$, the estimates

$$
\|x_\epsilon(t) - x(t)\| \ \leq \ M \int_0^t [\gamma_*(u) + \epsilon] \exp\left(2Me^{P(t) - P(s)}\right) ds, \ \ t \in [0, b],
$$

$$
\|y_\epsilon(t) - y(t)\| \ \leq \ M \int_0^t [\gamma_*(u) + \epsilon] \exp\left(2Me^{P(t) - P(s)}\right) ds, \ \ t \in [0, b],
$$

where

$$
P(t) = \int_0^t p(s)ds, \ \ \gamma_*(t) = \gamma(t) + \bar{\gamma}(t), \ \ t \in J.
$$

Proof. We construct a sequence of functions $(z_n)_{n \in \mathbb{N}}$ which will be shown to converge to some solution of Problem (12.1) on the interval $[0, b]$, namely to

$$x(t) \in S(t)a + \int_0^t S(t-s)F(s, x(s), y(s))ds, \quad t \in [0, b]. \tag{12.2}$$

Let $f_0 = g$ on $[0, b]$ and $z_0(t) = x(t)$, $t \in [0, b)$, i.e.

$$z_0(t) = S(t)a + \int_0^t S(t-s)f_0(s)ds, \quad t \in [0, b],$$

and

$$y(t) \in S(t)\bar{a} + \int_0^t S(t-s)G(s, x(s), y(s))ds, \quad t \in [0, b]. \tag{12.3}$$

Let $\bar{f}_0 = \bar{g}$ on $[0, b]$ and $\bar{z}_0(t) = y(t)$, $t \in [0, b)$, i.e.

$$\bar{z}_0(t) = S(t)a + \int_0^t S(t-s)\bar{f}_0(s)ds, \quad t \in [0, b].$$

Then define the multi-valued maps $U_1, \bar{U}_1 \colon [0, b] \to \mathcal{P}(E)$ by

$$U_1(t) = F(t, z_0(t), \bar{z}_0(t)) \cap B(g(t), \gamma(t) + \epsilon).$$

and

$$\bar{U}_1(t) = G(t, z_0(t), \bar{z}_0(t)) \cap B(\bar{g}(t), \gamma(t) + \epsilon)$$

Since $g, \bar{g}, \gamma$ and $\bar{\gamma}$ are measurable, Theorem III.4.1 in [79] tells us that the balls $B(g(t), \gamma(t) + \epsilon)$ and $B(\bar{g}(t), \bar{\gamma}(t) + \epsilon)$ are measurable. Moreover $F(t, z_0(t), \bar{z}_0(t))$ and $G(t, z_0(t), \bar{z}_0(t))$ are measurable (see [34]), and U_1 and $\bar{U}_1$ are nonempty. Indeed, since $v = \frac{\epsilon}{2} > 0$ is a measurable function, from Lemma 3.48, there exist functions u and $\bar{u}$ which are measurable selections of $F(t, z_0(t), \bar{z}_0(t))$ and $G(t, z_0(t), \bar{z}_0(t))$, respectively, and such that

$$|u(t) - g(t)| \leq d(g(t), F(t, z_0(t), \bar{z}_0(t))) + \frac{\epsilon}{2} = \gamma(t) + \frac{\epsilon}{2}$$

and

$$|\bar{u}(t) - \bar{g}(t)| \leq d(\bar{g}(t), G(t, z_0(t), \bar{z}_0(t))) + \epsilon = \bar{\gamma}(t) + \frac{\epsilon}{2}.$$

Then $u \in U_1(t)$ and $\bar{u} \in \bar{U}_1(t)$, proving our claim. We deduce that the intersection multivalued operators $U_1(t)$ and $\bar{U}_1(t)$ are measurable (see [34, 79, 138]). By Lemma 3.23 (Kuratowski-Ryll-Nardzewski selection theorem), there exist functions $t \to f_1(t)$, $\bar{f}_1(t)$ which are measurable selections for U_1 and $\bar{U}_1(t)$, respectively. Consider

$$z_1(t) = S(t)a + \int_0^t S(t-s)f_1(s)ds, \quad t \in [0, b],$$

and

$$\bar{z}_1(t) = S(t)\bar{a} + \int_0^t S(t-s)\bar{f}_1(s)ds, \quad t \in [0, b].$$

For each $t \in [0, b]$, we have

$$\|z_1(t) - z_0(t)\| \leq M \int_0^t \|f_0(s) - f_1(s)\|ds, \tag{12.4}$$

and

$$\|\bar{z}_1(t) - \bar{z}_0(t)\| \le M \int_0^t \|\bar{f}_0(s) - \bar{f}_1(s)\| ds. \tag{12.5}$$

Hence

$$\|z_1(t) - z_0(t)\| \le M \int_0^t \gamma(s) ds + Mt\frac{\epsilon}{2}$$

and

$$\|\bar{z}_1(t) - \bar{z}_0(t)\| \le M \int_0^t \bar{\gamma}(s) ds + Mt\frac{\epsilon}{2}.$$

Then

$$\|z_1(t) - z_0(t)\| + \|\bar{z}_1(t) - \bar{z}_0(t)\| \le M \int_0^t (\gamma(s) + \bar{\gamma}(s)) ds + Mt\epsilon.$$

Using the fact that $F(t, z_1(t), \bar{z}_1(t))$ and $G(t, z_1(t), \bar{z}_1(t))$ are measurable, the balls

$$B(f_1(t), \frac{p(t)}{2}(\|z_1(t) - z_0(t)\| + \|\bar{z}_1(t) - \bar{z}_0(t)\|))$$

and

$$B(\bar{f}_1(t), \frac{p(t)}{2}(\|\bar{z}_1(t) - \bar{z}_0(t)\| + \|\bar{z}_1(t) - \bar{z}_0(t)\|))$$

are also measurable by [79, Theorem III.4.1]. From (12.1.2) we have

$$H_d(F(t, z_1, \bar{z}_1(t)), F(t, z_0(t), \bar{z}_0(t))) \le \frac{p(t)}{2}(\|z_1(t) - z_0(t)\| + \|\bar{z}_1(t) - \bar{z}_0(t)\|),$$

and

$$H_d(G(t, z_1, \bar{z}_1(t)), G(t, z_0(t), \bar{z}_0(t))) \le \frac{p(t)}{2}(\|z_1(t) - z_0(t)\| + \|\bar{z}_1(t) - \bar{z}_0(t)\|).$$

Hence there exist $w \in F(t, z_1(t), \bar{z}_1(t))$ and $\bar{w} \in G(t, z_1(t), \bar{z}_1(t))$ such that

$$\|f_1(t) - w\| \le \frac{p(t)}{2}(\|z_1(t) - z_0(t)\| + \|\bar{z}_1(t) - \bar{z}_0(t)\|)$$

and

$$\|\bar{f}_1(t) - \bar{w}\| \le p(t)(\|z_1(t) - z_0(t)\| + \|\bar{z}_1(t) - \bar{z}_0(t)\|).$$

We consider the following multivalued maps

$$U_2(t) = F(t, z_1(t), \bar{z}_1(t)) \cap B(f_1(t), p(t)(\|z_1(t) - z_0(t)\| + \|\bar{z}_1(t) - \bar{z}_0(t)\|))$$

and

$$\bar{U}_2(t) = G(t, z_1(t), \bar{z}_1(t)) \cap B(f_1(t), p(t)(\|z_1(t) - z_0(t)\| + \|\bar{z}_1(t) - \bar{z}_0(t)\|))$$

which are nonempty. Therefore the intersection multi-valued operator U_2 is measurable with nonempty, closed values (see [34, 79, 138]). By Lemma 3.23, there exist a functions $f_2, \bar{f}_2$, which are a measurable selections for U_2, $\bar{U}_2$ respectively. Thus $f_2(t) \in F(t, z_1(t), \bar{z}_1(t))$, $\bar{f}_2(t) \in G(t, z_1(t), \bar{z}_1(t))$ and

$$\|f_1(t) - f_2(t)\| \le \frac{p(t)}{2}(\|z_1(t) - z_0(t)\| + \|\bar{z}_1(t) - \bar{z}_0(t)\|) \tag{12.6}$$

and

$$\|\bar{f}_1(t) - \bar{f}_2(t)\| \le \frac{p(t)}{2}(\|z_1(t) - z_0(t)\| + \|\bar{z}_1(t) - \bar{z}_0(t)\|). \qquad (12.7)$$

Define

$$z_2(t) = S(t)a + \int_0^t S(t-s)f_2(s)ds, \quad t \in [0, b]$$

and

$$\bar{z}_2(t) = S(t)\bar{a} + \int_0^t S(t-s)\bar{f}_2(s)ds, \quad t \in [0, b].$$

Using (12.4), (12.6), (12.5) and (12.7), a simple integration by parts yields the following estimates, valid for every $t \in [0, b]$,

$$
\begin{aligned}
\|z_2(t) - z_1(t)\| &\le \int_0^t \|S(t-s)\|_{B(E)}\|f_2(s) - f_1(s)\|ds \\
&\le \int_0^t M\frac{p(s)}{2}(\|z_1(s) - z_0(s)\| + \|\bar{z}_1(s) - \bar{z}_0(s)\|))ds
\end{aligned}
$$

and

$$
\begin{aligned}
\|\bar{z}_2(t) - \bar{z}_1(t)\| &\le \int_0^t \|S(t-s)\|_{B(E)}\|\bar{f}_2(s) - \bar{f}_1(s)\|ds \\
&\le \int_0^t M\frac{p(s)}{2}(\|z_1(s) - z_0(t)\| + \|\bar{z}_1(s) - \bar{z}_0(s)\|))ds, \ t \in [0, b].
\end{aligned}
$$

Then

$$
\begin{aligned}
\|z_2(t) - z_1(t)\| &\le \int_0^t \|S(t-s)\|_{B(E)}\|f_2(s) - f_1(s)\|ds \\
&\le \int_0^t Mp(s)\left(M\int_0^s \gamma_*(u)du + M\epsilon s\right)ds \\
&\le M^2 \int_0^t [\gamma_*(s) + \epsilon]e^{P(t)-P(s)}ds, \quad t \in [0, b],
\end{aligned}
$$

and

$$
\begin{aligned}
\|\bar{z}_2(t) - \bar{z}_1(t)\| &\le \int_0^t \|S(t-s)\|_{B(E)}\|\bar{f}_2(s) - \bar{f}_1(s)\|ds \\
&\le \int_0^t Mp(s)\left(M\int_0^s \gamma(u)du + M\epsilon s\right)ds \\
&\le M^2 \int_0^t [\gamma_*(s) + \epsilon]e^{P(t)-P(s)}ds, \quad t \in [0, b].
\end{aligned}
$$

Let

$$U_3(t) = F(t, z_2(t), \bar{z}_2(t)) \cap B(f_2(t), \frac{p(t)}{2}(\|z_2(t) - z_1(t)\| + \|\bar{z}_2(t) - \bar{z}_1(t)\|))$$

and

$$\bar{U}_3(t) = G(t, z_2(t), \bar{z}_2(t)) \cap B(f_2(t), \frac{p(t)}{2}(\|z_2(t) - z_1(t)\| + \|\bar{z}_2(t) - \bar{z}_1(t)\|)).$$

Arguing as for U_2, $\bar{U}_2$ we can prove that U_3, $\bar{U}_3$ are measurable multi-valued maps with nonempty values; so there exist measurable selections $f_3(t) \in U_3(t)$ and $\bar{f}_3(t) \in \bar{U}_3(t)$. This allows us to define

$$z_3(t) = S(t)a + \int_0^t S(t-s)f_3(s)ds, \quad t \in [0,b],$$

and

$$\bar{z}_3(t) = S(t)\bar{a} + \int_0^t S(t-s)\bar{f}_3(s)ds, \quad t \in [0,b].$$

For $t \in [0,b]$, we have

$$
\begin{aligned}
\|z_3(t) - z_2(t)\| &\leq M \int_0^t \|f_2(s) - f_3(s)\| \, ds \\
&\leq M \int_0^t p(s)(\|z_2(s) - z_1(s)\| + \|\bar{z}_2(s) - \bar{z}_1(s)\|) \, ds,
\end{aligned}
$$

and

$$
\begin{aligned}
\|\bar{z}_3(t) - \bar{z}_2(t)\| &\leq M \int_0^t \|\bar{f}_2(s) - \bar{f}_3(s)\| \, ds \\
&\leq M \int_0^t \frac{p(s)}{2} (\|z_2(s) - z_1(s)\| + \|\bar{z}_2(s) - \bar{z}_1(s)\|) \, ds.
\end{aligned}
$$

Then

$$\|z_2(t) - z_3(t)\| \leq M \left(\int_0^t M^2 \int_0^s (\gamma(r) + \epsilon)e^{P(s)-P(r)} ds dr \right).$$

Performing an integration by parts, we obtain, since P is a nondecreasing function, the following estimates

$$
\begin{aligned}
\|z_3(t) - z_2(t)\| &\leq \frac{M^3}{2} \int_0^t 2p(s) \left(\int_0^s [\gamma_*(u) + \epsilon]e^{P(s)-P(u)} du \right) ds \\
&\leq \frac{M^3}{2} \left(\int_0^t 2p(s)ds \int_0^s [\gamma_*(u) + \epsilon]e^{2(P(s)-P(u))} du \right) \\
&\leq \frac{M^3}{2} \left(\int_0^t \left(e^{2P(s)} \right)' ds \int_0^s [\gamma(u) + \epsilon]e^{-2P(u)} du \right) \\
&\leq \frac{M^3}{2} \left(e^{2P(t)} \int_0^t [\gamma_*(s) + \epsilon]e^{-2P(s)} ds - \int_0^t [\gamma(s) + \epsilon] ds \right) \\
&\leq \frac{M^3}{2} \left(\int_0^t [\gamma_*(s) + \epsilon]e^{2(P(t)-P(s))} ds \right),
\end{aligned}
$$

and

$$
\begin{aligned}
\|\bar{z}_3(t) - \bar{z}_2(t)\| &\leq \frac{M^3}{2} \int_0^t 2p(s) \left(\int_0^s [\gamma_*(u) + \epsilon]e^{P(s)-P(u)} du \right) ds \\
&\leq \frac{M^3}{2} \left(\int_0^t 2p(s)ds \int_0^s [\gamma_*(u) + \epsilon]e^{2(P(s)-P(u))} du \right)
\end{aligned}
$$

$$\leq \frac{M^3}{2}\left(\int_0^t \left(e^{2P(s)}\right)' ds \int_0^s [\gamma_*(u) + \epsilon]e^{-2P(u)} du\right)$$

$$\leq \frac{M^3}{2}\left(e^{2P(t)}\int_0^t [\gamma_*(s) + \epsilon]e^{-2P(s)} ds - \int_0^t [\gamma_*(s) + \epsilon] ds\right)$$

$$\leq \frac{M^3}{2}\left(\int_0^t [\gamma_*(s) + \epsilon]e^{2(P(t)-P(s))} ds\right).$$

Let

$$U_4(t) = F(t, z_3(t), \bar{z}_3(t)) \cap B(f_3(t), \frac{p(t)}{2}(\|z_3(t) - z_2(t)\| + \|\bar{z}_3(t) - \bar{z}_2(t)\|))$$

and

$$\bar{U}_4(t) = G(t, z_3(t), \bar{z}_3(t)) \cap B(f_3(t), \frac{p(t)}{2}(\|z_3(t) - z_2(t)\| + \|\bar{z}_3(t) - \bar{z}_2(t)\|)).$$

Then, arguing again as for $U_1, U_2, U_3, \bar{U}_1, \bar{U}_2, \bar{U}_3,$, we show that $U_4, \bar{U}_4$ are measurable multi-valued maps with nonempty values and that there exist measurable selections $f_4(t) \in U_4(t)$ and $\bar{f}_4(t) \in \bar{U}_4(t)$. Define

$$z_4(t) = S(t)a + \int_0^t S(t-s)f_4(s)ds, \quad t \in [0, b],$$

and

$$\bar{z}_4(t) = S(t)\bar{a} + \int_0^t S(t-s)\bar{f}_4(s)ds, \quad t \in [0, b].$$

For $t \in [0, b]$, we have

$$\|z_4(t) - z_3(t)\| \leq M\int_0^t \|f_4(s) - f_3(s)\| ds$$

$$\leq M\int_0^t \frac{p(t)}{2}(\|z_3(t) - z_2(t)\| + \|\bar{z}_3(t) - \bar{z}_2(t)\|) ds$$

$$\leq \frac{M^4}{2}\int_0^t p(s)\left(\int_0^s [\gamma_*(s) + \epsilon]e^{2(P(s)-P(u))} du\right) ds$$

$$\leq \frac{M^4}{6}\int_0^t 3p(s)e^{3P(s)} ds \int_0^s [\gamma_*(s) + \epsilon]e^{-3P(u)} du$$

$$\leq \frac{M^4}{6}\left(\int_0^t [\gamma_*(s) + \epsilon]e^{3(P(t)-P(s))} ds\right).$$

Repeating the process for $n = 0, 1, 2, 3, \ldots,$ we arrive at the following bound

$$\|z_n(t) - z_{n-1}(t)\| \leq \frac{M^n}{(n-1)!}\int_0^t [\gamma_*(s) + \epsilon]e^{(n-1)(P(t)-P(s))} ds, \quad t \in J \qquad (12.8)$$

and

$$\|\bar{z}_n(t) - \bar{z}_{n-1}(t)\| \leq \frac{M^n}{(n-1)!}\int_0^t [\gamma_*(s) + \epsilon]e^{(n-1)(P(t)-P(s))} ds, \quad t \in J. \qquad (12.9)$$

By induction, suppose that (12.8) holds for some n and check (12.8), (12.9) for $n + 1$. Let

$$U_{n+1}(t) = F(t, z_n(t), \bar{z}_n(t)) \cap B(f_n, \frac{p(t)}{2}(\|z_n(t) - z_{n-1}(t)\| + \|\bar{z}_n(t) - \bar{z}_{n-1}(t)\|))$$

and

$$\bar{U}_{n+1}(t) = G(t, z_n(t), \bar{z}_n(t)) \cap B(f_n, \frac{p(t)}{2}(\|z_n(t) - z_{n-1}(t)\| + \|\bar{z}_n(t) - \bar{z}_{n-1}(t)\|).$$

Since U_{n+1} and $\bar{U}_{n+1}$ are nonempty measurable sets, there exist measurable selections $f_{n+1}(t) \in U_{n+1}(t)$ and $\bar{f}_{n+1} \in \bar{U}_{n+1}$, which allows us to define for $n \in \mathbb{N}$

$$z_{n+1}(t) = S(t)a + \int_0^t S(t-s)f_{n+1}(s)ds, \quad t \in [0, b]. \tag{12.10}$$

and

$$\bar{z}_{n+1}(t) = S(t)\bar{a} + \int_0^t S(t-s)\bar{f}_{n+1}(s)ds, \quad t \in [0, b]. \tag{12.11}$$

Therefore, for a.e. $t \in [0, b]$, we have

$$\begin{aligned}
\|z_{n+1}(t) - z_n(t)\| &\leq M \int_0^t \|f_{n+1}(s) - f_n(s)\| ds \\
&\leq \frac{M^{n+1}}{(n-1)!} \int_0^t p(s)ds \left(\int_0^s [\gamma(u) + \epsilon]e^{(n-1)(P(s)-P(u))}du \right) \\
&\leq \frac{M^{n+1}}{n!} \int_0^t np(s)e^{nP(s)}ds \int_0^s [\gamma(u) + \epsilon]e^{-nP(u)}du.
\end{aligned}$$

Again, an integration by parts leads to

$$\|z_{n+1}(t) - z_n(t)\| \leq \frac{M^{(n+1)}}{n!} \int_0^t [\gamma(s) + \epsilon]e^{n(P(t)-P(s))}ds,$$

and

$$\|\bar{z}_{n+1}(t) - \bar{z}_n(t)\| \leq \frac{M^{(n+1)}}{n!} \int_0^t [\gamma(s) + \epsilon]e^{n(P(t)-P(s))}ds.$$

Consequently, (12.8) and (12.9) hold true for all $n \in \mathbb{N}$. We infer that $\{y_n\}$ is a Cauchy sequence in $C([0, b], E)$, converging uniformly to limit functions $x, y \in C([0, b], E)$. Moreover, from the definition of $\{U_n\}$ and $\{\bar{U}_n\}$ we have

$$\|f_{n+1}(t) - f_n(t)\| \leq p(t)(\|z_n(t) - z_{n-1}(t)\| + \|\bar{z}_n(t) - \bar{z}_{n-1}(t)\|), \text{ for a.e. } t \in [0, b].$$

Hence, for almost every $t \in [0, b]$, $\{f_n(t)\}$ and $\{\bar{f}_n(t)\}$ are also Cauchy sequences in E and converge almost everywhere to some measurable function $f(\cdot)$ in E. In addition, since $f_0 = g$, and $\bar{f}_0 = \bar{g}$ we have for a.e. $t \in [0, b]$

$$\begin{aligned}
\|f_n(t)\| &\leq \sum_{k=1}^n \|f_k(t) - f_{k-1}(t)\| + \|f_0(t)\| \\
&\leq \sum_{k=1}^n p(t)(\|z_{k-1}(t) - z_{k-2}(t)\| + \|\bar{z}_{k-1}(t) - \bar{z}_{k-2}(t)\|) \\
&\quad + \gamma(t) + \bar{\gamma}(t) + \|g(t)\| + \|\bar{g}(t)\| + \epsilon \\
&\leq \sum_{k=1}^\infty p(t)(\|z_{k-1}(t) - z_{k-2}(t)\| + \|\bar{z}_{k-1}(t) - \bar{z}_{k-2}(t)\|) \\
&\quad + \gamma_*(t) + \|g(t)\| + \|\bar{g}(t)\| + \epsilon.
\end{aligned}$$

Hence

$$\|f_n(t)\| \le MH(t)p(t) + \gamma_*(t) + \|g(t)\| + \|\bar{g}(t)\| + \epsilon,$$

and

$$\|\bar{f}_n(t)\| \le MH(t)p(t) + \gamma_*(t) + \|g(t)\| + \|\bar{g}(t)\| + \epsilon,$$

where

$$H(t) := M \int_0^t [\gamma_*(s) + \epsilon] \exp\left(Me^{P(t)-P(s)}\right) ds. \tag{12.12}$$

From the Lebesgue dominated convergence theorem, we deduce that $\{f_n\}$ and $\{\bar{f}_n\}$ converge to f and $\bar{f}$, respectively, in $L^1([0, b], E)$. Passing to the limit in (12.10) and (12.11), we find that the pair

$$x_*(t) = S(t)a + \int_0^t S(t-s)f(s)ds, \quad t \in [0, b]$$

and

$$y_*(t) = S(t)\bar{a} + \int_0^t S(t-s)\bar{f}(s)ds, \quad t \in [0, b],$$

is a solution to Problem (12.1) on $[0, b]$. Moreover, for a.e. $t \in [0, b]$, we have

$$
\begin{aligned}
\|x(t) - x_*(t)\| &= \left\| \int_0^t S(t-s)g(s)ds - \int_0^t S(t-s)f(s)ds \right\| \\
&\le M \int_0^t \|f(s) - f_0(s)\| ds \\
&\le \int_0^t \|f(s) - f_n(s)\| ds + M \int_0^t \|f_n(s) - f_0(s)\| ds \\
&\le M \int_0^t \|f(s) - f_n(s)\| ds + M \int_0^t p(s)H(s)ds \\
&\le M \int_0^t \|f(s) - f_n(s)\| ds + M \int_0^t p(s)[M \int_0^s [\gamma_*(u) \\
&\quad + \epsilon] \exp\left(Me^{P(s)-P(u)}\right) dsdu \\
&\le M \int_0^t \|f(s) - f_n(s)\| ds \\
&\quad + M \int_0^t (e^{P(s)})' \Big[M \int_0^s e^{-P(u)} [\gamma_*(u) + \epsilon] \times \\
&\quad \exp\left(Me^{P(s)-P(u)}\right) dsdu \Big].
\end{aligned}
$$

and

$$
\begin{aligned}
\|y(t) - y_*(t)\| &\le M \int_0^t \|f(s) - f_n(s)\| ds \\
&\quad + M \int_0^t (e^{P(s)})' \Big[M \int_0^s e^{-P(u)} [\gamma_*(u) + \epsilon] \times \\
&\quad \exp\left(Me^{P(s)-P(u)}\right) dsdu \Big].
\end{aligned}
$$

Passing to the limit as $n \to \infty$, we get

$$\|x(t) - x_*(t)\| \le \eta(t), \quad \text{a.e. } t \in [0, b] \tag{12.13}$$

and

$$\|y(t) - y_*(t)\| \le \eta(t), \quad \text{a.e. } t \in [0, b] \tag{12.14}$$

with

$$\eta(t) := M \int_0^t [\gamma_*(u) + \epsilon] \exp\left(2Me^{P(t)-P(s)}\right) ds.$$

$\square$

12.2 Impulsive semilinear differential inclusions

Differential equations with impulses were considered for the first time by Milman and Myshkis [217] and then followed by a period of active research which culminated with the monograph by Halanay and Wexler [150].

The dynamics of many processes in physics, population dynamics, biology, medicine may be subject to abrupt changes such as shocks and perturbations (see for instance [14, 191] and the references therein). These perturbations may be seen as impulses. For instance, in the periodic treatment of some diseases, impulses correspond to the administration of a drug treatment. In environmental sciences, impulses correspond to seasonal changes of the water level of artificial reservoirs. Their models are described by impulsive differential equations and inclusions.

Important contributions to the study of the mathematical aspects of such equations have been undertaken in [105, 145, 198, 257] among others.

In this section we consider the system with impulse effects

$$x'(t) - A_1 x(t) \in F_1(t, x(t), y(t)), \quad y'(t) - A_2 y(t) \in F_2(t, x(t), y(t)), \quad t \in [0, b], \quad (12.15)$$

$$x(t_k^+) - x(t_k^-) \in I_k(x(t_k)), \quad y(t_k^+) - y(t_k^-) \in \overline{I}_k(y(t_k)), \quad k = 1, \ldots, m, \quad (12.16)$$

$$x(0) = x_0, \quad y(0) = y_0, \quad (12.17)$$

where $J := [0, b]$, E is a Banach space, $F_1, F_2 \colon J \times E \times E \to \mathcal{P}(E)$ are multifunctions, $x_0, y_0 \in E$, $0 = t_0 < t_1 < \ldots < t_m < t_{m+1} = b$. The operators A_i, $i = 1, 2$ are infinitesimal generators of C_0-semigroups $\{T_i(t)\}_{t \geq 0}$ on a Banach space $(E, |\cdot|)$ respectively, $I_k, \overline{I}_k \colon E \to \mathcal{P}(E)$ $(k = 1, \ldots, m)$, and $\Delta y|_{t=t_k} = y(t_k^+) - y(t_k^-)$. The notations $y(t_k^+) = \lim\limits_{h \to 0^+} y(t_k + h)$ and $y(t_k^-) = \lim\limits_{h \to 0^+} y(t_k - h)$ stand for the right and the left limits of the function y at $t = t_k$, respectively.

In the absence of impulses and multimap framework the above system was used to study the initial value problems and boundary value problems for nonlinear competitive or cooperative differential systems from mathematical biology [222] and mathematical economics [179] where the model can be set in the operator form (12.15)-(12.17).

In [56, 251, 252] the authors present existence and uniqueness results for systems of semilinear differential equations without impulses. Recently Precup [251] proved the role of matrix convergence and vector metric in the study of semilinear operator systems.

In order to define mild solutions for problem $(12.15) - (12.17)$, we consider the space

$$PC = \{z \colon [0, b] \to E, \ z_k \in C(J_k, E), \ k = 0, \ldots, m, \text{ such that }$$
$$z(t_k^-) \text{ and } z(t_k^+) \text{ exist and satisfy } z(t_k^-) = z(t_k) \text{ for } k = 1, \ldots, m\}.$$

Endowed with the norm

$$\|z\|_{PC} = \max\{\|z_k\|_\infty, \ k = 0, \ldots, m\},$$

PC is a Banach space. Throughout this section, A is an infinitesimal generator of a C_0-semigroup $\{T_i(t)\}_{t \geq 0}, i = 1, 2$ and there exists $M > 0$ such that

$$\|T_i(t)\| \leq M \quad t \in \mathbb{R}_+.$$

Definition 12.2. *A function* $(x, y) \in PC \times PC$ *is said to be a mild solution of problem (12.15)-(12.17) if there exists* $v_1, v_2 \in L^1(J, E)$ *such that* $v_i(t) \in F_i(t, x(t), y(t))$ $i = 1, 2$ *a.e. on* J*, and* $\mathcal{I}_k(x(t_k)) \in I_k(x(t_k))$*,* $\overline{\mathcal{I}}_k(x(t_k)) \in \overline{I}_k(x(t_k))$*,* $k = 1, \ldots, m$

$$x(t) = T_1(t)x_0 + \int_0^t T_1(t-s)v_i(s)ds + \sum_{0 < t_k < t} T_1(t - t_k)\mathcal{I}_k(x(t_k^-)),$$

and

$$y(t) = T_2(t)y_0 + \int_0^t T_2(t-s)v_i(s)ds + \sum_{0 < t_k < t} T_2(t - t_k)\overline{\mathcal{I}}_k(x(t_k^-)).$$

12.2.1   Existence results

The following result is known as the Gronwall-Bihari Theorem.

Lemma 12.3. *[53] Let* $u, g \colon J \to \mathbb{R}$ *be positive real continuous functions. Assume there exist* $c > 0$ *and a continuous nondecreasing function* $h \colon \mathbb{R} \to (0, +\infty)$ *such that*

$$u(t) \leq c + \int_a^t g(s)h(u(s))\, ds, \quad \forall\, t \in J.$$

Then

$$u(t) \leq H^{-1}\left(\int_a^t g(s)\, ds \right), \quad \forall\, t \in J$$

provided

$$\int_c^{+\infty} \frac{dy}{h(y)} > \int_a^b g(s)\, ds.$$

Here H^{-1} *refers to the inverse of the function* $H(u) = \int_c^u \frac{dy}{h(y)}$ *for* $u \geq c$.

Let $(E, |\cdot|)$ be a separable Banach space and $F_i \colon J \times E \times E \to \mathcal{P}_{cl,b,cv}(E), i = 1, 2$ be Carathéodory multimaps which satisfy some of the following assumptions:

$(\mathcal{G}_1)$ There exist a functions $p_i \in L^1(J, \mathbb{R}^+)$ and a continuous nondecreasing functions $\psi_i \colon [0, \infty) \to [0, \infty), i = 1, 2$ such that

$$\|F_i(t, x, y)\|_{\mathcal{P}} \leq p(t)\psi_i(|x| + |y|) \text{ for a.e. } t \in J \text{ and each } x, y \in E,$$

with

$$\int_0^b p_i(s)ds < \int_1^\infty \frac{du}{\psi(u)}.$$

$(\mathcal{G}_2)$ $I_k, \overline{I}_k \colon E \to \mathcal{P}_{cv}(E), k = 1, \ldots, m$ are closed and there exist constants $c_k, \overline{c}_k > 0$ and continuous functions $\phi_k, \overline{\phi}_k \colon \mathbb{R}^+ \to \mathbb{R}^+$ such that

$$\|I_k(z)\|_{\mathcal{P}} \leq c_k\phi_k(|z|) \text{ for each } z \in E, k = 1, \ldots, m$$

and

$$\|\overline{I}_k(z)\|_{\mathcal{P}} \leq \overline{c}_k\overline{\phi}_k(|z|) \text{ for each } z \in E, k = 1, \ldots, m.$$

$(\mathcal{G}_3)$ The semigroup $T_i(\cdot)$, $i = 1, 2$ is compact for $t > 0$.

We need the following lemma.

Lemma 12.4. *[200]. Given a Banach space X, let $F : [a,b] \times X \longrightarrow \mathcal{P}_{cp,cv}(X)$ be an L^1-Carathéodory multi-valued map such that, for each $y \in C([a,b], X)$, $S_{F,y} \neq \emptyset$, and let Γ be a linear continuous mapping from $L^1([a,b], X)$ into $C([a,b], X)$. Then the operator*

$$\Gamma \circ S_F : C([a,b], X) \longrightarrow \mathcal{P}_{cp,cv}(C([a,b], X)),$$
$$y \longmapsto (\Gamma \circ S_F)(y) := \Gamma(S_{F,y})$$

has a closed graph in $C([a,b], X) \times C([a,b], X)$.

Theorem 12.5. *Assume that F satisfies $(\mathcal{G}_1)$, $(\mathcal{G}_2)$ and $(\mathcal{G}_3)$. Then the set of solutions for problem (12.15)-(12.17) is nonempty and compact.*

Proof. Consider the operator $N : PC \times PC \to \mathcal{P}(PC \times PC)$ defined for $(x,y) \in PC \times PC$ by

$$N(x,y) = \{(h_1, h_2) \in PC \times PC : (h_1(t), h_2(t)) = \{(N_1(x,y), N_2(x,y))\}\},$$

where

$$N_1(x,y) = \left\{ h_1 \in PC : h_1(t) = \left\{ \begin{array}{l} T_1(t)x_0 + \displaystyle\int_0^t T_1(t-s)v_1(s)ds \\ + \displaystyle\sum_{0 < t_k < t} T_1(t-t_k)\mathcal{I}_k(y(t_k)), \ t \in J \end{array} \right. \right\}$$

and

$$N_2(x,y) = \left\{ h_2 \in PC : h_2(t) = \left\{ \begin{array}{l} T_2(t)y_0 + \displaystyle\int_0^t T_2(t-s)v_2(s)ds \\ + \displaystyle\sum_{0 < t_k < t} T_2(t-t_k)\overline{\mathcal{I}}_k(y(t_k)), \ t \in J \end{array} \right. \right\}$$

and where $v_i \in S_{F_i,x,y} = \{v \in L^1(J,E) : f(t) \in F_i(t, x(t), y(t)), \text{ a.e. } t \in J\}$ and $\mathcal{I}_k(x(t_k)) \in I_k(x(t_k)), \overline{\mathcal{I}}_k(y(t_k)) \in I_k(y(t_k))$, $k = 1, \ldots, m$. That is,

$$N(x,y) = (N_1(x,y), N_2(x,y)) \quad \text{for every } (x,y) \in PC \times PC,$$

and clearly, fixed points of the operator N are solutions of Problem (12.15)-(12.17).

Since, for each $(x,y) \in PC \times PC$, the nonlinearity F takes convex values, the selection set $S_{F,x,y}$ is convex and by $(\mathcal{G}_2)$, then N has convex values. From $(\mathcal{G}_1)$ and $(\mathcal{G}_2)$, we can prove that N maps bounded sets.

Step 1. N maps bounded sets into equicontinuous sets of $PC \times PC$ into bounded sets. It suffices to prove that $N(\mathcal{B}_q \times \mathcal{B}_q)$ is relatively compact in $PC \times PC$, where $\mathcal{B}_q = \{z \in PC : \|z\|_{PC} \leq q\}$. First, $N(\mathcal{B}_q \times \mathcal{B}_q)$ is an equicontinuous set of $PC \times PC$. To see this, let $0 < \tau_1 < \tau_2 \leq b$, $(x,y) \in \mathcal{B}_q \times \mathcal{B}_q$, and $(h_1, h_2) \in N(x,y)$. Then there exists $v_i \in S_{F_i,x,y}, i = 1, 2$, and $\mathcal{I}_k(x(t_k)) \in I_k(x(t_k)), \overline{\mathcal{I}}_k(y(t_k)) \in I_k(y(t_k))$, $k = 1, \ldots, m$, such that

$$h_1(t) = T_1(t)x_0 + \int_0^t T_1(t-s)v_1(s)ds + \sum_{0 < t_k < t} T_1(t-t_k)\mathcal{I}_k(x(t_k)), \ t \in J,$$

and

$$h_2(t) = T_2(t)y_0 + \int_0^t T_2(t-s)v_2(s)ds + \sum_{0 < t_k < t} T_2(t-t_k)\overline{\mathcal{I}}_k(y(t_k)), \ t \in J.$$

Letting $d_k = \sup\limits_{|r|\leq q} \phi_k(r)$ and $\bar{d}_k = \sup\limits_{|r|\leq q} \bar{\phi}_k(r)$, we obtain the estimates

$$
\begin{aligned}
|h_1(\tau_2) - h_1(\tau_1)| \;\leq\; & |T(\tau_2)x_0 - T(\tau_1)x_0| \\
& + \int_0^{\tau_1} \|T_1(\tau_2 - s) - T_1(\tau_1 - s)\|_{B(E)} p_1(s)\psi_1(q)ds \\
& + \int_{\tau_1}^{\tau_2} \|T_1(\tau_2 - s)\|_{B(E)} p_1(s)\psi_1(q)ds \\
& + \sum_{k=1}^{m} T_1(\tau_2 - \tau_1) I_k(y_k) \\
& + \sum_{0 < t_k < \tau_1} d_k \|T_1(\tau_1 - t_k) - T_1(\tau_2 - t_k)\|_{B(E)}.
\end{aligned}
$$

Hence

$$
\begin{aligned}
|h_1(\tau_2) - h_1(\tau_1)| \;\leq\; & \|T_1(\tau_2 - \tau_1) - Id\|_{B(E)} |x_0| \\
& + M_1 \psi_1(q)\|T(\tau_2 - \tau_1) - Id\|_{B(E)} \int_0^{\tau_1} p_1(s)ds \\
& + M_1 \psi_1(q) \int_{\tau_1}^{\tau_2} p_1(s)ds + M_1 \sum_{\tau_1 < t_k < \tau_2} d_k \\
& + \|T_1(\tau_2 - \tau_1) - Id\|_{B(E)} \sum_{0 < t_k < \tau_1} d_k.
\end{aligned}
$$

The terms in the right-hand side tend to zero as $\tau_1 - \tau_2 \to 0$. Now we show that $H_1(t) = \{N_1(x(t), y(t)) : t \in J, \ (x,y) \in \mathcal{B}_q \times \mathcal{B}_q\}$ is a precompact set in E. Let $0 < t \leq b$ and $0 < \epsilon < t$. Then for $(x,y) \in \mathcal{B}_q \times \mathcal{B}_q$ we have

$$
\begin{aligned}
(N_\epsilon(x(t), y(t)))(t) \;=\; & \Big\{T_1(t)x_0 + \int_0^{t-\epsilon} T_1(t - s)v_1(s)ds \\
& + \sum_{0 < t_k < t} T_1(t - t_k)I_k(x(t_k))\Big\} \\
\;=\; & T_1(\epsilon)\Big\{T(t - \epsilon)x_0 + \int_0^{t-\epsilon} T_1(t - \epsilon - s)v_1(s)ds \\
& + \sum_{0 < t_k < t} T_1(t - \epsilon + t_k)I_k(x(t_k))\Big\}.
\end{aligned}
$$

Since $T_1(\epsilon)$ is compact by $(\mathcal{G}_3)$, then the set

$$
\begin{aligned}
\widetilde{H}_\epsilon(t) \;=\; & \{(N_\epsilon(x(t), y(t)) : (x,y) \in \mathcal{B}_q \times \mathcal{B}_q\} \\
\;=\; & T_1(\epsilon)\Big\{T_1(t - \epsilon)x_0 + \int_0^{t-\epsilon} T_1(t - \epsilon - s)v_1(s)ds \\
& + \sum_{0 < t_k < t} T_1(t - \epsilon + t_k)\mathcal{I}_k(x(t_k)), \\
& (x,y) \in \mathcal{B}_q \times \mathcal{B}_q, \ v_1 \in S_{F,x,y}, \ \mathcal{I}_k(x(t_k)) \in I_k(x(t_k))\},
\end{aligned}
$$

is precompact in E. Moreover for every $(h_1(t), h_\epsilon(t)) \in N_1(x(t), y(t)) \times N_\epsilon(x(t), y(t))$ such that

$$
h_1(t) \;=\; T_1(t)x_0 + \int_0^t T_1(t - s)v_1(s)ds + \sum_{0 < t_k < t} T_1(t - t_k)\mathcal{I}_k(x(t_k))\}
$$

and

$$
h_\epsilon(t) \;=\; T_1(t)x_0 + \int_0^{t-\epsilon} T_1(t - s)v_1(s)ds + \sum_{0 < t_k < t} T_1(t - t_k)\mathcal{I}_k(x(t_k)),
$$

we have

$$
|h_1(t) - h_\epsilon(t)| \;\leq\; M_1 \int_{t-\epsilon}^t p(s)\psi(r)ds
$$

which tends to 0 as $\epsilon \to 0$. Therefore, there are precompact sets arbitrarily closed to the set $H(t)$. Then $H(t)$ is precompact in X. It is clear that $H(0) = \{u_0\}$ is precompact in X. Hence for each $t \in [0, b]$ the set $H(t)$ is precompact in E.

The Arzelá-Ascoli theorem implies that $N_i \colon PC \times PC \to \mathcal{P}_{cp,cv}(PC)$ are completely continuous operators.

Step 2. N has a closed graph.

Let $(x_n, y_n) \to (x, y), h_n \in N(x_n, y_n)$ and $h_n := (h_n^1, h_n^2) \to h := (h_1, h_2)$. We shall prove that $h_* \in N(x, y)$. Now $h_n \in N(x_n, y_n)$ means there exist $v_n^i \in S_{F_i, x, y_n}, i = 1, 2$ and $\mathcal{I}_k^n(x(t_k)) \in I_k(x_n(t_k)), \overline{\mathcal{I}}_k^n(y_n(t_k)) \in I_k(y_n(t_k)), \ k = 1, \ldots, m$, such that

$$h_n^1(t) = T_1(t)x_0 + \int_0^t T_1(t - s)v_n^1(s)ds + \sum_{0 < t_k < t} T_1(t - t_k)\mathcal{I}_k^n(x_n(t_k)), \ t \in J,$$

and

$$h_n^2(t) = T_2(t)y_0 + \int_0^t T_2(t - s)v_n^2(s)ds + \sum_{0 < t_k < t} T_2(t - t_k)\overline{\mathcal{I}}_k^n(y_n(t_k)), \ t \in J.$$

Writing h_n^1 and h_n^2 in the form

$$h_n^1(t) = \begin{cases} L_0(v_n^1)(t), & \text{if } t \in [0, t_1], \\ L_1(v_n^1)(t), & \text{if } t \in (t_1, t_2], \\ \ldots & \ldots \\ L_{m-1}(v_n^1)(t), & \text{if } t \in (t_{m-1}, t_m], \end{cases} \tag{12.18}$$

where

$$
\begin{aligned}
L_0(v_n^1)(t) &= T_1(t)x_0 + \int_0^t T_1(t - s)v_n^1(s)ds, \ t \in [0, t_1] \\
L_1(v_n^1)(t) &= T_1(t - t_1)[L_0(v_n^1)(t_1) + \mathcal{I}_1^n(L_0(v_n^1)(t_1))] \\
&\quad + \int_{t_1}^t v_n^1(s)ds, \ t \in (t_1, t_2] \\
L_2(v_n^1)(t) &= T_1(t - t_2)[L_1(v_n^1)(t_2) + \mathcal{I}_2^n(L_1(v_n^1)(t_2))] \\
&\quad + \int_{t_2}^t v_n^1(s)ds, \ t \in (t_2, t_3] \\
\ldots \ \ldots \ \ldots \\
L_{m-1}(v_n^1)(t) &= T_1(t - t_{m-1})[L_{m-2}(v_n^1)(t_{m-1}) + \mathcal{I}_{m-1}^n(L_{m-2}(v_n^1)(t_{m-1}))] \\
&\quad + \int_{t_{m-1}}^t v_n^1(s)ds, \ t \in (t_{m-1}, t_m],
\end{aligned}
$$

and

$$h_n^2(t) = \begin{cases} \bar{L}_0(v_n^2)(t), & \text{if } t \in [0, t_1], \\ \bar{L}_1(v_n^2)(t), & \text{if } t \in (t_1, t_2], \\ \ldots & \ldots \\ \bar{L}_{m-1}(v_n^2)(t), & \text{if } t \in (t_{m-1}, t_m], \end{cases} \tag{12.19}$$

where

$$
\begin{aligned}
\bar{L}_0(v_n^2)(t) &= T_2(t)y_0 + \int_0^t T_2(t - s)v_n^1(s)ds, \ t \in [0, t_1] \\
\bar{L}_1(v_n^2)(t) &= T_2(t - t_1)[L_0(v_n^2)(t_1) + \mathcal{I}_1^n(L_0(v_n^2)(t_1))]
\end{aligned}
$$

$$+ \int_{t_1}^{t} v_n^2(s)ds, \ t \in (t_1, t_2]$$

$$\bar{L}_2(v_n^2)(t) = T_2(t - t_2)[L_1(v_n^2)(t_2) + \mathcal{I}_2^n(L_1(v_n^2)(t_2))]$$

$$+ \int_{t_2}^{t} v_n^2(s)ds, \ t \in (t_2, t_3]$$

$$\cdots \quad \cdots \quad \cdots$$

$$\bar{L}_{m-1}(v_n^2)(t) = T_2(t - t_{m-1})[L_{m-2}(v_n^2)(t_{m-1}) + \mathcal{I}_{m-1}^n(L_{m-2}(v_n^2)(t_{m-1}))]$$

$$+ \int_{t_{m-1}}^{t} v_n^1(s)ds, \ t \in (t_{m-1}, t_m].$$

Consider the linear continuous operator

$$\Gamma_1 : L^1([0, t_1], E) \to C([0, t_1], E)$$

defined by

$$v \to \Gamma_1(v)(t) = \int_0^t T_1(t - s)v(s)ds, \ t \in [0, t_1],$$

and

$$\Gamma_1 : L^1([t_k, t_{k+1}], E) \to C_*([t_k, t_{k+1}], E)$$

defined by

$$v \to \Gamma_{k+1}(v)(t) = \int_{t_k}^t T_1(t - s)v(s)ds, \ t \in [t_k, t_{k+1}], \ k = 1, \ldots, m$$

where

$$C_*([t_k, t_{k+1}], E) = \{y \in C((t_k, t_{k+1}), E) : y(t_k^+), \ \text{exists}\}$$

From Lemma 12.4, it follows that $\Gamma_k \circ S_{F_1^k}$ has closed graph, where

$$S_{F_1^k} = \{v \in L^1([t_k, t_{k+1}], E): \ v(t) \in F_1(t, x(t), y(t)),$$

$$\text{a.e.} \ t \in [t_k, t_{k+1}], \ k = 0, 1, \ldots, m\}.$$

Moreover, we have that

$$L_0(v_n^1)(t) - T_1(t)x_0 \in \Gamma_1 \circ S_{F_1^0, x(t), y(t)},$$

$$L_1(v_n^1)(t) - T_1(t - t_1)[L_0(v_n^1)(t_1) + \mathcal{I}_1^n(L_0(v_n^1)(t_1))] \in \Gamma_1 \circ S_{F_1^1, x(t), y(t)},$$

$$L_2(v_n^1)(t) - T_1(t - t_2)[L_1(v_n^1)(t_2) + \mathcal{I}_2^n(L_1(v_n^1)(t_2))] \in \Gamma_1 \circ S_{F_1^2, x(t), y(t)},$$

$$\cdots \quad \cdots \quad \cdots$$

$$L_{m-1}(v_n^1)(t) - T_1(t - t_{m-1})[L_{m-2}(v_n^1)(t_{m-1})$$

$$+ \mathcal{I}_{m-1}^n(L_{m-2}(v_n^1)(t_{m-1}))] \in \Gamma_m \circ S_{F_1^m, x(t), y(t)}.$$

Then there exists $v_0 \in S_{F_1^0, x, y}$ such that

$$h_1(t) = T_1(t)x_0 + \int_0^t T_1(t - s)v_0(s)ds, \quad t \in [0, t_1].$$

Using the fact that $\mathcal{I}_1$ has a closed graph, then there exist $I_1(x) \in \mathcal{I}_1(x)$ and $v_1 \in S_{F_1^1, x, y}$ such that

$$h_1(t) = T_1(t - t_1)[L_0(v_0)(t_1) + I_1(L_0(v_n^1)(t_1)) + \int_{t_1}^t T(t - s)v_1(s)ds, \quad t \in (t_1, t_2].$$

We continue this process, and we get

$$h_1(t) = T_1(t - t_{m-1})[L_{m-2}(v_{m-1})(t_{m-1}) + \mathcal{I}_{m-1}(L_{m-2}(v_{m-1})(t_{m-1}))]$$
$$+ \int_{t_m}^t T(t - s)v_m(s)ds, \quad t \in (t_m, b].$$

Set

$$v(t) = \begin{cases} v_0(t) & t \in [0, t_1] \\ v_1(t) & t \in (t_1, t_2] \\ \dots & \\ v_m(t) & t \in (t_m, b]. \end{cases}$$

Hence

$$h_1(t) = T_1(t)x_0 + \int_0^t v(s)ds + \sum_{0 < t_k < t} T_1(t - t_k)I_k(x(t_k)), \quad t \in J.$$

Similarly we can prove that there exist $\bar{v} \in S_{F_2, x, y}$ and $\bar{I}_k(y) \in \mathcal{I}_k(x)$, $k = 1, \dots, m$ such that

$$h_2(t) = T_2(t)y_0 + \int_0^t \bar{v}(s)ds + \sum_{0 < t_k < t} T_2(t - t_k)\bar{I}_k(y(t_k)), \quad t \in J.$$

Then we have $(x, y, h) \in Graph(N)$.

Step 3. A priori bounds on solutions.

Let $(x, y) \in PC \times PC$ be such that $(x, y) \in N(x, y)$. Then there exist $(v_1, v_2) \in S_{F_1, x, y} \times S_{F_2, x, y}$ and $\mathcal{I}_k(x(t_k)) \in I_k(x(t_k)), \overline{\mathcal{I}}_k(y(t_k)) \in I_k(y(t_k))$, $k = 1, \dots, m$, such that

$$x(t) = T_1(t)x_0 + \int_0^t T_1(t - s)v_1(s)ds + \sum_{0 < t_k < t} T_1(t - t_k)\mathcal{I}_k(x(t_k)), \quad t \in J$$

and

$$y(t) = T_2(t)y_0 + \int_0^t T_2(t - s)v_2(s)ds + \sum_{0 < t_k < t} T_2(t - t_k)\overline{\mathcal{I}}_k(y(t_k)), \quad t \in J.$$

• For $t \in [0, t_1]$, we have

$$E|x(t)| \leq M|x_0| + M \int_0^t |v_1(s)|ds.$$

Hence

$$|x(t)| \leq M|x_0| + M \int_0^t p_1(s)\psi_1(|x(s)| + |y(s)|)ds,$$

and

$$|y(t)| \leq M|y_0| + M \int_0^t p_2(s)\psi_2(|x(s)| + |y(s)|)ds.$$

Therefore

$$|x(t)| + |y(t)| \leq M|x_0| + M|y_0| + \int_0^t p(s)\phi(|x(s)| + |y(s)|)ds,$$

where

$$\gamma_0 = M(|x_0| + |y_0|), \ p(t) = p_1(t) + p_2(t), \ t \in [0, t_1].$$

By Lemma 12.3, we have

$$|x(t)| + |y(t)| \leq \Psi_0^{-1}\left(\int_0^{t_1} p(s)ds\right) := K_0, \text{ for each } t \in [0, t_1],$$

where

$$\Psi_0(z) = \int_{\gamma_0}^z \frac{du}{\psi(u)}.$$

Consequently

$$\|x\|_\infty \leq K_0 \text{ and } \|y\|_\infty \leq K_0.$$

For $t \in (t_1, t_2]$, we have

$$|x(t)| \leq M\|I_1(x(t_1))\|_\mathcal{P} + M\int_{t_1}^t |v_1(s)|ds.$$

Hence

$$|x(t)| \leq M\phi_1(K_0) + M\int_{t_1}^t p_1(s)\psi_1(|x(s)| + |y(s)|)ds,$$

and

$$|y(t)| \leq M\bar\phi(K_0) + M\int_{t_1}^t p_2(s)\psi_2(|x(s)| + |y(s)|)ds.$$

Therefore

$$|x(t)| + |y(t)| \leq \gamma_1 + \int_{t_1}^t p(s)\psi(|x(s)| + |y(s)|)ds,$$

where

$$\gamma_1 = M(\phi_1(K_0) + \bar\phi_1(K_0)), \ p(t) = p_1(t) + p_2(t), \ t \in [t_1, t_2].$$

By Lemma 12.3, we have

$$|x(t)| + |y(t)| \leq \Psi_1^{-1}\left(\int_{t_1}^{t_2} p(s)ds\right) := K_1, \text{ for each } t \in [t_1, t_2],$$

where

$$\Psi_1(z) = \int_{\gamma_1}^z \frac{du}{\psi(u)}.$$

This implies that

$$\|x\|_\infty \leq K_1 \text{ and } \|y\|_\infty \leq K_1.$$

We continue this process, and we get

$$|x(t)| + |y(t)| \leq \Psi_m^{-1}\left(\int_{t_m}^b p(s)ds\right) := K_1, \text{ for each } t \in [t_m, b],$$

where

$$\Psi_m(z) = \int_{\gamma_m}^z \frac{du}{\psi(u)}, \ \gamma_m = M(\phi_m(K_{m-1}) + \bar\phi_m(K_{m-1})).$$

Then

$$\|x\|_\infty \leq K_m \text{ and } \|y\|_\infty \leq K_m.$$

This shows that

$$\mathcal{E} = \{(x, y) \in PC \times PC : (x, y) \in \lambda N(x, y), \ \lambda \in (0, 1)\}$$

is bounded. As a consequence of Theorem 8.27 we deduce that N has a fixed point $(x, y) \in PC \times PC$ which is a solution to the problem (12.15)-(12.17). $\square$

12.3 Impulsive Stokes differential inclusions

In this section we use the abstract result proved in the above section to study the existence of mild solution for impulsive Stokes differential inclusions.

Let $\Omega \subset \mathbb{R}^3$ be a bounded open domain with the smooth boundary $\partial\Omega$ and and let $n(x)$ be the outward normal to Ω at the point $x \in \partial\Omega$. Let

$$Y = \{u \in (C_c^\infty(\Omega))^3 : \nabla u = 0 \text{ in } \Omega \text{ and } n \cdot u = 0 \text{ on } \partial\Omega\}$$

and $E = \overline{Y}^{(L^2(\Omega))^3}$ be the closure of Y in $(L^2(\Omega))^3$. It is clear that, endowed with the standard inner product of the space $(L^2(\Omega))^3$, defined by

$$\langle u, v \rangle = \sum_{i=1}^{3} \langle u_i, v_i \rangle_{L^2(\Omega)},$$

E is a Hilbert space. Let $P : (L^2(\Omega))^3 \to E$ denote the orthogonal projection of $(L^2(\Omega))^3$ in E.

Consider the following system of impulsive Stokes type partial differential inclusions:

$$
\begin{cases}
u_t - P(\Delta u) & \in & F(t, u(t,x), v(t,x)), & \text{a.e. } \in [0,b], \ x \in \Omega, \\
v_t - P(\Delta v) & \in & G(t, u(t,x), v(t,x)), & \text{a.e.} t \in [0,b], \ x \in \Omega, \\
u(t_k^+, x) - u(t_k^-, x) & \in & I_k(u(t_k, x)), \\
v(t_k^+, x) - v(t_k, x) & \in & \overline{I}_k(v(t_k, x)), & k = 1, \ldots, m \\
\nabla u = \nabla v = 0, & & & (t,x) \in [0,b] \times \partial\Omega \\
u = v = 0, & & & (t,x) \in [0,b] \times \partial\Omega \\
u(0,x) = u_0(x), & v(0,x) = v_0(x) & & x \in \Omega,
\end{cases}
\tag{12.20}
$$

where $P(\Delta)$ is the Stokes operator. Let $A : D(A) \subset E \to E$ be defined by

$$
\begin{cases}
D(A) & = & (H^2(\Omega) \cap H_0^1(\Omega))^3 \cap E \\
Au & = & -P(\Delta u), \ u \in D(A).
\end{cases}
$$

Lemma 12.6. *(Fujita-Kato) (Theorem 7.3.4, [281]) The operator A, defined as above, is the generator of a compact and analytic C_0-semigroup of contractions in E.*

Let us assume that

$(\mathcal{K}_1)$ $f_i, g_i : [0,b] \times \Omega \times \mathbb{R} \times \mathbb{R} \to \mathbb{R}$, $i = 1, 2$ are functions such that

$$f_1(t, x, u, v) \le f_2(t, x, u, v), \ g_1(t, x, u, v) \le g_2(t, x, u, v),$$
$$\text{for all } (t, x, u, v) \in [0,b] \times \Omega \times \mathbb{R} \times \mathbb{R}.$$

$(\mathcal{K}_2)$ there exist $\phi_1, \phi_2 \in L^1([0,b], \mathbb{R}_+) \cap L^\infty([0,b], \mathbb{R}_+)$ such that

$$|f_i(t, x, u, v)| \le \phi_1(t)(|u| + |v|) + \phi_2(t) \text{ and}$$
$$|g_i(t, x, u, v)| \le \phi_1(t)(|u| + |v|) + \phi_2(t),$$
$$i = 1, 2, \text{ for each } (t, x, u, v) \in [0,b] \times \Omega \times \mathbb{R} \times \mathbb{R}.$$

$(\mathcal{K}_3)$ f_1 and g_1 are l.s.c. and f_2 and g_2 are u.s.c.

Lemma 12.7. *[281] Let $f_i, g_i : [0, b] \times \Omega \times \mathbb{R} \times \mathbb{R} \to \mathbb{R}$, $i = 1, 2$ are functions satisfying $(\mathcal{K}_2) - (\mathcal{K}_3)$. Let $F : G : [0, b] \times L^2(\Omega) \times L^2(\Omega) \to \mathcal{P}(L^2(\Omega))$ be multivalued maps defined by*

$$F(t, u, v) = \{f \in L^2(\Omega) : f(x) \in [f_1(t, x, u, v), f_2(t, x, u, v]\}$$

and

$$G(t, u, v) = \{g \in L^2(\Omega) : g(x) \in [g_1(t, x, u, v), g_2(t, x, u, v)])\}.$$

Then F and G are nonempty, u.s.c. with weakly compact and convex values. Moreover $F(\cdot, \cdot, \cdot), G(\cdot, \cdot, \cdot) \in \mathcal{P}_{cl,b,cv}(L^2(\Omega))$.

Let

$$x(t)(\xi) = u(t, \xi) \quad t \in J, \quad \xi \in \Omega,$$
$$I_k(x(t_k)) = K_k u(t_k^-, \xi), \quad \xi \in \Omega, \quad k = 1, \cdots, m,$$
$$I_k(y(t_k)) = \bar{K}_k v(t_k^-, \xi), \quad \xi \in \Omega, \quad k = 1, \cdots, m,$$
$$x(0)(\xi) = u_0(0, \xi), \quad y(0)(\xi) = v_0(0, \xi) \quad \xi \in \Omega,$$

where $K_k, \bar{K}_k \in \mathcal{P}_{cp,cv}(L^2(\Omega))$. Assume that $(\mathcal{K}_1) - (\mathcal{K}_3)$ are satisfied. Thus the problem (12.20) can be written in the abstract form

$$\begin{cases} x'(t) - A_1 x(t) & \in & F_1(t, x(t), y(t)), \quad t \in [0, b] \\ y'(t) - A_2 y(t) & \in & F_2(t, x(t), y(t)), \quad t \in [0, b], \\ x(t_k^+) - x(t_k^-) & \in & I_k(x(t_k)), \\ y(t_k^+) - y(t_k^-) & \in & \bar{I}_k(y(t_k)), \quad k = 1, \ldots, m \\ x(0) = x_0, & & y(0) = y_0, \end{cases} \quad (12.21)$$

where $A_1 = A_2 = A$.

Thanks to those assumptions, it is straightforward to check that from Lemmas 12.6, 12.7, $(\mathcal{G}_1) - (\mathcal{G}_3)$ hold. Then assumptions in Theorem 12.5 are fulfilled, and we can conclude that the system (12.21) has at least one mild solution.

12.4 Differential inclusions in Almgren sense

Multiple-valued functions in the sense of Almgren [17] have applications in the framework of geometric measure theory. They give a useful tool to approximate abstract objects arising from geometric measure theory. For example, Almgren [17] used multiple-valued functions to approximate mass minimizing rectifiable currents, and hence successfully obtained their partial interior regularity. Solomon [263] gave proofs of the closure theorem without using the structure theorem; his proofs rely on various facts about multiple-valued functions. There are also other objects similar to these functions, such as the union of Sobolev's functions graphs introduced by Ambrosio, Gobbino, and Pallara (see [19]).

In complex function theory we often consider the two valued function $f(z) = \sqrt{z}$. This can be viewed as a function from $\mathbb{C} \to \mathcal{A}_2(\mathbb{C})$. Almgren [18] introduced $\mathcal{A}_Q(\mathbb{R}^n)$-valued functions to tackle the problem of estimating the size of the singular set of mass-minimizing integral currents (see [17] for a summary). Almgren's multiple-valued functions are a fundamental tool for understanding geometric variational problems in codimensions higher than 1. The success of Almgren's regularity theory raises the need for further studying multiple-valued functions. For additional information concerning multiple-valued functions, we suggest the references [100, 101, 131–134].

For some existence results on differential inclusions in the sense of Almgren, we suggest the paper by Golbet [131].

12.4.1 Multiple-valued function in Almgren sense

In this section, we recall some notations, definitions, and auxiliary results that will be used throughout this part of the book.

We denote by $[[p_i]]$ the Dirac mass in $p_i \in \mathbb{R}^n$.

Definition 12.8. *For every* T_1, $T_2 \in \mathcal{A}_Q(\mathbb{R}^n)$, *with* $T_1 = \sum_i [[p_i]]$ *and* $T_2 = \sum_i [[s_i]]$, *we define*

$$d_\mathcal{A}(T_1, T_2) := \min_{\sigma \in \mathcal{P}_Q} \sqrt{\sum_{i=1}^{Q} |p_i - s_{\sigma(i)}|^2},$$

$$d_\mathcal{A}(T_1, T_2) := \min_{\sigma \in \mathcal{P}_Q} \sum_{i=1}^{Q} |p_i - s_{\sigma(i)}|,$$

or

$$d_\mathcal{A}(T_1, T_2) := \min_{\sigma \in \mathcal{P}_Q} \{\max |p_i - s_{\sigma(i)}| : i = 1, \ldots, Q\},$$

where $\mathcal{P}_Q$ *denotes the group of permutations of* $\{1, \ldots, Q\}$.

Remark 12.9. In the above definition, we designated each expression with the same symbol because the results in this paper are independent of which definition of $d_\mathcal{A}$ is used.

Definition 12.10. *A multiple-valued function in the sense of Almgren is a map* $T : \Omega \to \mathcal{A}_k(\mathbb{R}^n)$, *where* $\Omega \subset \mathbb{R}^n$ *and*

$$\mathcal{A}_k(\mathbb{R}^n) = \left\{ \sum_{i=1}^{k} [[p_i]] : p_i \in \mathbb{R}^n \text{ for every } i = 1, \ldots, k \right\}$$

are equipped with the metric $d_\mathcal{A}$.

Definition 12.11. *Let* $\Omega \subset \mathbb{R}^m$ *and* $f : \Omega \to \mathcal{A}_Q(\mathbb{R}^n)$ *be a* Q-*valued function. If there exist single valued maps* $g_i : \Omega \to \mathbb{R}^m$, $i = 1, \ldots, Q$, *such that*

$$f(x) = \sum_{i=1}^{Q} [[g_i(x)]] \quad \text{for each } x \in \mathbb{R}^m,$$

then we say that the vector $(g_1, \ldots, g_Q)$ *is a selection for* f.

Theorem 12.12. ([18, 100]) *Let* $f : [0, b] \to \mathcal{A}_Q(\mathbb{R}^n)$ *be a continuous multiple-valued function. Then there are continuous functions* $f_1, \ldots, f_Q : [0, b] \to \mathbb{R}^n$ *such that*

$$f = \sum_{i=1}^{Q} f_i.$$

Remark 12.13. *If for each* $i \in \{1, \ldots, Q\}$, g_i *is continuous, then* f *has a continuous selection.*

Lemma 12.14. ([131]) *Let* $f : \mathbb{R} \to \mathcal{A}_Q(\mathbb{R}^n)$ *be a continuous multiple-valued function and* $g : \mathbb{R} \to \mathbb{R}^n$ *be a continuous function. If* $h : [0, b] \times \mathbb{R} \to \mathcal{A}_{Q-1}(\mathbb{R}^n)$ *satisfies*

$$f = [[g]] + h,$$

then h *is a continuous function.*

Remark 12.15. *An* $\mathcal{A}_Q(\mathbb{R}^n)$-*valued function is essentially a rule assigning* Q *unordered and not necessarily distinct elements of* $\mathbb{R}^n$ *to each element of its domain.*

Lemma 12.16. ([131]) *Let $(f_i) : [0, b] \to \mathcal{A}_Q(\mathbb{R}^n)$ be a sequence of multiple-valued functions pointwise converging to f, and let $(g_i) : [0, b] \to \mathbb{R}^n$ be a sequence of functions pointwise converging to g such that g_i is a selection for f_i for each $i \in \mathbb{N}$. Then g is a selection for f.*

Theorem 12.17. ([18]) *Suppose $f_1, \ldots, f_Q : [0, b] \to \mathbb{R}^n$ are continuous functions and $f = \sum_{i=1}^{Q} [[f_i]] : [0, b] \to \mathcal{A}_Q(\mathbb{R}^n)$. Then there exists a constant $C_{n,Q} > 0$, depending only on n and Q, such that*

$$\omega_{f_i} \leq C_{n,Q} \omega_f, \quad \text{for each } i = 1, \ldots, Q,$$

where ω_f is the modulus of continuity of f, i.e.,

$$\omega_f(\delta) = \sup\{d_{\mathcal{A}}(f(s_1), f(s_2)) : s_1, s_2 \in [0, b] \text{ and } |s_1 - s_2| \leq \delta\},$$

and

$$\omega_{f_i}(\delta) = \sup\{|f_i(s_1) - f(s_2)| : s_1, s_2 \in [0, b] \text{ and } |s_1 - s_2| \leq \delta\}.$$

12.4.2   Existence result on unbounded domains

We consider the following problem,

$$\begin{cases} y'(t) & \in & \{f_1(t, y(t)), \ldots f_Q(t, y(t))\}, \quad t \in [0, \infty), \\ y(0) & = & a, \end{cases} \tag{12.22}$$

where $f_i : [0, \infty) \times \mathbb{R}^N \to \mathbb{R}^N, 1 \leq i \leq Q$, are single-valued functions. Now we present our main result in this section; see Henderson and Ouahab [164].

Theorem 12.18. *Let $f_i : [0, b] \times \mathbb{R}^N \to \mathbb{R}^N$, $i = 1, \ldots, Q$, be single-valued functions for which we associate the continuous multiple-valued function in the sense of Almgren*

$$f = \sum_{i=1}^{Q} [[f_i]] : [0, \infty) \times \mathbb{R} \to \mathcal{A}_Q(\mathbb{R}^N).$$

Assume that there exist $M_1, M_2 > 0$ such that

$$d_{\mathcal{A}}(f(t, x), Q(0)) \leq M_1 + M_2 |x|, \quad \text{for all } x \in \mathbb{R}^N, \; t \in [0, \infty). \tag{12.23}$$

Then Problem (12.22) has at least one solution in $C([0, \infty), \mathbb{R}^N)$.

Proof. The proof involves several steps.

Step 1: First, we construct two sequences $\{y_m\}_{m=0}^{\infty}$ and $\{g_m\}_{m=0}^{\infty}$ by

$$y_0(t) = a, \text{ for all } t \in [0, n],$$

$$y_1(t) = \begin{cases} y_0(t), & \text{if } t \in [0, \frac{n}{2}], \\ a + \displaystyle\int_0^{t - \frac{n}{2}} g_{2,1}(s) ds, & \text{if } t \in (\frac{n}{2}, n], \end{cases}$$

where $g_{2,1} : [0, \frac{n}{2}] \to \mathbb{R}^N$ is a continuous selection of $f(\cdot, y_1(\cdot)) : [0, \frac{n}{2}] \times \mathbb{R}^N \to \mathcal{A}_Q(\mathbb{R}^N)$. From Theorem 12.12 we can find a continuous selection $g_1 : [0, n] \to \mathbb{R}^N$ of $f(\cdot, y_1(\cdot)) : [0, n] \times \mathbb{R}^N \to \mathcal{A}_Q(\mathbb{R}^N)$ such that $g_1(\cdot) = g_{2,1}(\cdot)$ on $[0, \frac{n}{2}]$. We define

$$y_2(t) = \begin{cases} y_0(t), & \text{if } t \in [0, \frac{n}{3}], \\ a + \displaystyle\int_0^{t - \frac{n}{3}} g_{3,1}(s) ds, & \text{if } t \in [\frac{n}{3}, \frac{2n}{3}], \\ a + \displaystyle\int_0^{t - \frac{n}{3}} g_{3,2}(s) ds, & \text{if } t \in (\frac{n}{3}, n], \end{cases}$$

where $g_{3,1} : [0, \frac{n}{3}] \to \mathbb{R}^N$ is a continuous selection of $f(\cdot, y_2(\cdot)) : [0, \frac{n}{3}] \to \mathcal{A}_Q(\mathbb{R}^N)$ and $g_{3,2} :$ $[0, \frac{2n}{3}] \to \mathbb{R}^N$ is a continuous selection of $f(\cdot, y_2(\cdot)) : [0, \frac{2n}{3}] \to \mathcal{A}_Q(\mathbb{R}^N)$ such that $g_{3,1}(\cdot) =$ $g_{3,2}(\cdot)$ on $[0, \frac{n}{3}]$. By Theorem 12.12 we can choose a continuous selection of $f(\cdot, y_3(\cdot)) :$ $[0, n] \to \mathcal{A}_Q(\mathbb{R}^N)$ such that $g_2(\cdot) = g_{3,2}(\cdot)$ on $[0, \frac{2n}{3}]$. Finally we can conclude by induction the following sequence,

$$y_m(t) = \begin{cases} y_0(t), & \text{if } t \in [0, \frac{n}{m}], \\ a + \int_0^{t - \frac{n}{m}} g_m(s)ds, & \text{if } t \in (\frac{n}{m}, n], \end{cases}$$

where $g_m : [0, n] \to \mathbb{R}^N$ is a continuous selection of $f(\cdot, y_m(\cdot)) : [0, n] \to \mathcal{A}_Q(\mathbb{R}^N)$.

Now, we show that $\{y_m : m \in \mathbb{N} \cup \{0\}\}$ is relatively compact. First, $\{y_m\}_{m=0}^\infty$ is bounded. By

$$\begin{aligned} |y_m(t)| &\leq |a| + \int_0^t |g_m(s)|ds \\ &\leq |a| + \int_0^t (M_1 + M_2|y_m(s)|)ds, \end{aligned}$$

and from Gronwall's Lemma, there exists $M > 0$ such that

$$\|y_m\|_\infty \leq M \text{ for each } m \in \mathbb{N} \cup \{0\}.$$

Next, we show that $\{y_m\}_{m=0}^\infty$ is equicontinuous. Let $t_1, t_2 \in [0, \frac{n}{m}]$. Then

$$|y_m(t_1) - y_m(t_2)| = 0 \text{ if } t_1, t_2 \in [0, \frac{n}{m}],$$

for $0 < t_1 \leq \frac{b}{m} < t_2 < b$, and we have

$$\begin{aligned} |y_m(t_1) - y_m(t_2)| &\leq \int_0^{t_2 - \frac{b}{m}} |g_m(s)|ds \\ &\leq M|t_2 - \frac{b}{m}| \\ &\leq M|t_2 - t_1|, \end{aligned}$$

and

$$\begin{aligned} |y_m(t_1) - y_m(t_2)| &\leq \int_{t_1 - \frac{n}{m}}^{t_2 - \frac{b}{m}} |g_m(s)|ds \\ &\leq M|t_1 - t_2| \quad t_1, t_2 \in (\frac{b}{m}, n]. \end{aligned}$$

Consequently, $\{y_m\}_{m=0}^\infty$ is bounded and equicontinuous. By Ascoli-Arzela, there exists a subsequence of $\{y_m\}_{m=0}^\infty$ converging to some y in $C([0, b], \mathbb{R}^N)$. Let $K = [0, b] \times B(0, M)$, and

$$\begin{aligned} \omega|_{f|_K}(\delta) = \ &\sup\{d_{\mathcal{A}}(f(t_1, x_1), f(t_2, x_2)) : |(t_1, x_1) - (t_2, x_2)| \leq \delta \\ &\text{where } (t_1, x_1), (t_2, x_2) \in K\} \end{aligned}$$

be a modulus of continuity of f restricted to K. Hence for each $m \in \mathbb{N} \cup \{0\}$, we have

$$\omega|_{f(\cdot, y_m(\cdot))}(\delta_2) = \sup\{d_{\mathcal{A}}(f(t_1, y_m(t_1)), f(t_2, y_m(t_2))) : |t_1 - t_2| \leq \delta,$$

$$\text{and } t_1, t_2 \in [0, n]\}$$
$$\leq \quad \sup\{d_{\mathcal{A}}(f(t_1, x_1), f(t_2, x_2)) : |t_1 - t_2| \leq \delta_2,$$
$$|x_1 - x_2| \leq \psi(M)\delta, \text{ and } (t_1, x_1), (t_2, x_2) \in K\}$$
$$\leq \quad \omega|_{f_K}(\delta\sqrt{1 + M^2}).$$

It is clear that $f(\cdot, y_m(\cdot)) - [[g_m(\cdot)]] : [0, n] \to \mathcal{A}_{Q-1}(\mathbb{R}^N)$ is a continuous multiple-function. Then there exist $h_1^m, \ldots, h_{Q-1}^m : [0, n] \to \mathbb{R}^N$ continuous functions such that

$$f(\cdot, y_m(\cdot)) = [[g_m(\cdot)]] + \sum_{i=1}^{Q-1} [[h_i^m(\cdot)]].$$

Then

$$\|g_m\|_\infty \leq L_1 \text{ for each } m \in \mathbb{N} \cup \{0\}$$

and

$$\omega|_{g_m} \leq \omega|_{f_K}(\delta_2), \text{ for every } m \in \mathbb{N} \cup \{0\}.$$

Consequently, $\{g_m\}_{m=0}^\infty$ is bounded and equicontinuous. From the Ascoli-Arzela theorem, we can conclude that $\{g_m\}_{m=0}^\infty$ is compact in $C([0, n], \mathbb{R}^N)$. Hence there exists a subsequence, denoted $\{g_m\}_{m=0}^\infty$, converging uniformly to g_n. Hence

$$\|y_m - z\|_\infty \leq n\|g_m - g\|_\infty \to 0 \text{ as } m \to \infty,$$

where

$$z(t) = a + \int_0^t g(s)ds := y^n(t), \quad t \in [0, n].$$

By Lemma 12.16 we conclude that g_n is a continuous selection of $f(\cdot, y_n(\cdot))$ on $[0, n]$. Then

$$y^n(t) = a + \int_0^t g_n(s)ds, \quad t \in [0, n],$$

is a solution of Problem (12.22) on $[0, n]$.

Step 2: By the same methods used in Step 1, we construct two sequences $(y_m)_{m=0}^\infty$ and $(g_m)_{m=0}^\infty$ by

$$y_0(t) = y^n(n), \text{ for all } t \in [n, n+1],$$

$$y_1(t) = \begin{cases} y_0(t), & \text{if } t \in [n, \dfrac{n+1}{2}], \\ y^n(n) + \displaystyle\int_0^{t-\frac{n+1}{2}} g_{2,1}(s)ds, & \text{if } t \in (\dfrac{n+1}{2}, n+1], \end{cases}$$

where $g_{2,1} : [n, \frac{n+1}{2}] \to \mathbb{R}^N$ is a continuous selection of $f(., y_1(.)) : [n, \frac{n+1}{2}] \times \mathbb{R}^N \to \mathcal{A}_Q(\mathbb{R}^N)$. From Theorem 12.12 we can find a continuous selection $g_1 : [n, n+1] \to \mathbb{R}^N$ of $f(\cdot, y_1(\cdot)) : [n, n+1] \times \mathbb{R}^N \to \mathcal{A}_Q(\mathbb{R}^N)$ such that $g_1(\cdot) = g_{2,1}(\cdot)$ on $[n, \frac{n+1}{2}]$. We define

$$y_2(t) = \begin{cases} y_0(t), & \text{if } t \in [n, \frac{n+1}{3}], \\ y^n(n) + \displaystyle\int_n^{t-\frac{n+1}{3}} g_{3,1}(s)ds, & \text{if } t \in [\frac{n+1}{3}, \frac{2(n+1)}{3}], \\ y^n(n) + \displaystyle\int_n^{t-\frac{n+1}{3}} g_{3,2}(s)ds, & \text{if } t \in (\frac{n+1}{3}, n+1], \end{cases}$$

where $g_{3,1} : [n, \frac{n+1}{3}] \to \mathbb{R}^N$ is a continuous selection of $f(\cdot, y_2(\cdot)) : [n, \frac{n+1}{3}] \to \mathcal{A}_Q(\mathbb{R}^N)$ and

$g_{3,2} : [n, \frac{2(n+1)}{3}] \to \mathbb{R}^N$ is a continuous selection of $f(\cdot, y_2(\cdot)) : [n, \frac{2(n+1)}{3}] \to \mathcal{A}_Q(\mathbb{R}^N)$ such that $g_{3,1}(\cdot) = g_{3,2}(\cdot)$ on $[n, \frac{n+1}{3}]$. By Theorem 12.12 we can choose a continuous selection of $f(\cdot, y_3(\cdot)) : [n, n+1] \to \mathcal{A}_Q(\mathbb{R}^N)$ such that $g_2(\cdot) = g_{3,2}(\cdot)$ on $[n, \frac{2(n+1)}{3}]$. Finally we can conclude by induction the following sequence,

$$
y_m(t) = \begin{cases} y_0(t), \text{ if } t \in [n, \frac{n+1}{m}], \\ y^n(n) + \int_0^{t - \frac{n+1}{m}} g_m(s)ds, \text{ if } t \in (\frac{n+1}{m}, n+1], \end{cases}
$$

where $g_m : [n, n+1] \to \mathbb{R}^N$ is a continuous selection of $f(\cdot, y_m(\cdot)) : [n, n+1] \to \mathcal{A}_Q(\mathbb{R}^N)$. From Step 1, we can show that there exist $y^{n+1} \in C([n, n+1], \mathbb{R}^N)$ and $g_{n+1} \in C([n, n+1], \mathbb{R}^N)$, $g_{n+1}(n) = g_n(n)$ continuous selection of $f(\cdot, y^{n+1}(\cdot)) : [n, n+1] \to \mathcal{A}_Q(\mathbb{R}^N)$ such that

$$
y^{n+1}(t) = y^n(n) + \int_n^t g_{n+1}(s)ds, \ t \in [n, n+1],
$$

which is a solution of problem (12.22) on $[n, n+1]$, with the initial condition $y(n) = y^n(n)$.

For the last part of the proof, we now employ the following diagonalization process. For $k \in \mathbb{N}$, let

$$
u_k(t) = \begin{cases} \widetilde{y}_k(t) & t \in [0, n_k], \\ \widetilde{y}_k(n_k) & t \in [n_k, \infty). \end{cases}
$$

Next, let $\{n_k\}_{k \in \mathbb{N}}$ be a sequence of numbers satisfying

$$
0 < n_1 < n_2 \cdots < n_k < \cdots \uparrow \infty.
$$

Notice

$$
\widetilde{y}_2(t) = \begin{cases} y_1(t) & t \in [0, n_1], \\ y_2(t) & t \in [n_1, n_2], \end{cases}
$$

where

$$
y_1(t) = \begin{cases} y^1(t) & t \in [0, 1], \\ y^2(t) & t \in [1, 2], \\ \vdots \\ y^{n_1}(t) & t \in [n_1 - 1, n_1], \end{cases}
$$

and

$$
y_2(t) = \begin{cases} y^{n_1+1}(t) & t \in [n_1, n_1 + 1], \\ y^{n_1+2}(t) & t \in [n_1 + 1, n_1 + 2], \\ \vdots \\ y^{n_2}(t) & t \in [n_2 - 1, n_2]. \end{cases}
$$

Set $S = \{u_{n_k}\}_{k=1}^\infty$. It is clear that there exists $M_* > 0$ such that, for every solution y of problem (12.22), we have

$$
\|y\|_* = \sup\{e^{-M_2 t} |y(t)| : t \in [0, \infty)\} \leq M_*.
$$

Notice

$$
|u_{n_k}(t)| \leq e^{n_k M_2} M_* \text{ for each } t \in [0, n_k], \ k \in \mathbb{N},
$$

and

$$
u_{n_k}(t) = a + \int_0^t g_{n_k}(t)dt \text{ for every } t \in [0, n_k].
$$

Then, for each $t, \tau \in [0, n_1]$ and $k \in \mathbb{N}$, we have

$$
\begin{aligned}
|u_{n_k}(t) - u_{n_k}(\tau)| &= \left| \int_0^t g_{n_1}(s)ds - \int_0^\tau g_{n_1}(s)ds \right| \\
&\leq \int_\tau^t |g_{n_1}(s)|ds \\
&\leq e^{n_1 M_2} M_* |t - \tau|.
\end{aligned}
$$

The Arzelà-Ascoli Theorem guarantees that there is a subsequence N_1 of $\mathbb{N}$ and a function $z_1 \in C([0, n_1], \mathbb{R}^N)$ such that $u_{n_k} \to z_1$ in $C([0, n_1], \mathbb{R}^N)$ as $k \to \infty$ through N_1. Let $N_1^* = N_1 \backslash \{1\}$. Notice that

$$
|u_{n_k}(t)| \leq M \quad \text{for every } t \in [0, n_2], \ k \in \mathbb{N}.
$$

Also for $k \in \mathbb{N}$, and $t, \tau \in [0, n_2]$, we have

$$
\begin{aligned}
|u_{n_k}(t) - u_{n_k}(\tau)| &= \left| \int_0^t g_{n_2}(s)ds - \int_0^\tau g_{n_2}(s)ds \right| \\
&\leq \int_\tau^t |g_{n_2}(s)|ds \\
&\leq M_* e^{n_2 M_2} |t - \tau|.
\end{aligned}
$$

The Arzelà-Ascoli Theorem guarantees that there is a subsequence N_2 of N_1^* and a function $z_2 \in C([0, n_2], \mathbb{R}^N)$ such that $u_{n_k} \to z_2$ in $C([0, n_2], \mathbb{R}^N)$ as $k \to \infty$ through N_2. Note $z_1 = z_2$ on $[0, n_1]$ since $N_2 \subset N_1^*$. Let $N_2^* = N_2 \backslash \{2\}$. Proceed inductively to obtain from $m \in \{2, 3, \ldots\}$ a subsequence N_m of N_{m-1}^* and a function $z_m \in C([0, n_m], \mathbb{R}^N)$ with $u_{n_k} \to z_m$ in $C([0, n_m], \mathbb{R}^N)$ as $k \to \infty$ through N_m. Let $N_m^* \backslash \{m\}$.

Define a function as follows: for $t \in [0, \infty)$ and let $n \in \mathbb{N}$ with $t \leq n_m$, define $y(t) = z_m(t)$. Then $y \in C^1([0, \infty), \mathbb{R}^N)$, $y(0) = a$ and $|y(t)| \leq M$ for each $t \in [0, \infty)$. Fix $t \in [0, \infty)$ and let $m \in \mathbb{N}$ with $t \leq n_m$. Then for each $n \in N_m^*$

$$
u_{n_k}(t) = a + \int_0^t g_{n_k}(s)ds.
$$

Let $n_k \to \infty$ through N_m^* to obtain

$$
z_m(t) = a + \int_0^t g_m(s)ds,
$$

where is g_m is a continuous selection of $f(\cdot, z_m(\cdot))$. Thus

$$
y(t) = a + \int_0^t g(s)ds, \ t \in [0, \infty)
$$

where is g is a continuous selection of $f(\cdot, y(\cdot))$. $\square$

12.5   Fractional differential inclusions in Almgren sense with Riemann-Liouville derivatives

Differential equations of fractional order recently have proved to be valuable tools in the modeling of many physical phenomena [104, 129, 130, 208, 213]. There has also been a significant theoretical development in fractional differential equations in recent years; see the monographs of Kilbas *et al.* [184], Miller and Ross [216], Oustaloup [230], Podlubny [244], and Samko *et al.* [256].

Applied problems requiring definitions of fractional derivatives are those that are physically interpretable for initial conditions containing $y(0)$, $y'(0)$, etc. The same requirements are true for boundary conditions. Caputo's fractional derivative satisfies those demands. For more details on geometric and physical interpretations for fractional derivatives of the Riemann-Liouville type, see [244].

In this section we consider the following problem

$$\begin{cases} D^\alpha y(t) \in \{f_1(t, t^{1-\alpha}y), \dots f_k(t, t^{1-\alpha}y)\}, & t \in (0, b], \\ \lim_{t \to 0^+} t^{1-\alpha} y(0) = c, \end{cases} \tag{12.24}$$

where each $f_i \colon [0, b] \times \mathbb{R}^N \to \mathbb{R}^N$ is a single valued function, and D^α denotes the usual Riemann-Liouville fractional derivative of order $\alpha \in (0, 1]$.

12.5.1   Fractional calculus

According to the Riemann-Liouville approach to fractional calculus, the notation of fractional integral of order α ($\alpha > 0$) is a natural consequence of the well known formula (usually attributed to Cauchy), that reduces the calculation of the $n-$fold primitive of a function $f(t)$ to a single integral of convolution type. The Cauchy formula reads

$$I^n f(t) := \frac{1}{(n-1)!} \int_0^t (t-s)^{n-1} f(s) ds, \quad t > 0, \ n \in \mathbb{N}.$$

Definition 12.19. *The fractional integral of order $\alpha > 0$ of a function $f \in L^1([a, b], \mathbb{R})$ is defined by*

$$I_{a+}^\alpha f(t) = \int_a^t \frac{(t-s)^{\alpha-1}}{\Gamma(\alpha)} f(s) ds,$$

where Γ is the gamma function. When $a = 0$, we write $I^\alpha f(t) = f(t) * \phi_\alpha(t)$ with $\phi_\alpha(t) :$ $\mathbb{R} \to \mathbb{R}$ defined by

$$\phi_\alpha(t) = \begin{cases} \frac{t^{\alpha-1}}{\Gamma(\alpha)}, & \text{if } t > 0, \\ 0, & \text{if } t \leq 0, \end{cases}$$

and $\phi_\alpha(t) \to \delta(t)$ as $\alpha \to 0$, where δ is the delta function and Γ is the Euler gamma function defined by

$$\Gamma(\alpha) = \int_0^\infty t^{\alpha-1} e^{-t} dt, \quad \alpha > 0.$$

For consistency, I^0 will denote the identity operator, i.e., $I^0 f(t) = f(t)$. Furthermore, by $I^\alpha f(0^+)$ we mean the limit (if it exists) of $I^\alpha f(t)$ as $t \to 0^+$ (this limit may be infinite).

After the notion of fractional integral, that of fractional derivative of order α ($\alpha > 0$) becomes a natural requirement, and it is tempting to replace α with $-\alpha$ in the above formulas. However, this generalization needs some care in order to guarantee the convergence of the integral and preserve the well known properties of the ordinary derivative of integer order. Denoting by D^n, with $n \in \mathbb{N}$, the operator of the derivative of order n, we first note that

$$D^n I^n = Id, \quad I^n D^n \neq Id, \quad n \in \mathbb{N},$$

i.e., D^n is the left-inverse (and not the right-inverse) of the corresponding integral operator J^n. We can easily prove that

$$I^n D^n f(t) = f(t) - \sum_{k=0}^{n-1} f^{(k)}(a^+) \frac{(t-a)^k}{k!}, \quad t > 0.$$

As consequence we expect that D^α is defined as the left-inverse to I^α. For this purpose, introducing the positive integer n such that $n - 1 < \alpha \leq n$, we can define the fractional derivative of order $\alpha > 0$ as follows.

Definition 12.20. *For a function f given on interval $[a, b]$, the αth Riemann-Liouville fractional order derivative of f is defined by*

$$D^\alpha f(t) \;=\; \frac{1}{\Gamma(n-\alpha)} \left(\frac{d}{dt}\right)^n \int_a^t (t-s)^{-\alpha+n-1} f(s)ds,$$

where $n = [\alpha] + 1$ and $[\alpha]$ is the integer part of α.

Defining, for consistency, $D^0 = I^0 = Id$, then we easily recognize that

$$D^\alpha I^\alpha = I^0, \quad \alpha \geq 0, \tag{12.25}$$

and

$$D^\alpha t^\gamma = \frac{\Gamma(\gamma+1)}{\Gamma(\gamma+1-\alpha)} t^{\gamma-\alpha}, \quad \alpha > 0, \ \gamma > -1, \ t > 0. \tag{12.26}$$

Of course, the properties (12.25) and (12.26) are a natural generalization of those known when the order is a positive integer.

Note the remarkable fact that the fractional derivative $D^\alpha f$ is not zero for the constant function $f(t) = 1$, if $\alpha \notin \mathbb{N}$. In fact, (12.26) with $\gamma = 0$ illustrates that

$$D^\alpha 1 = \frac{(t-a)^{-\alpha}}{\Gamma(1-\alpha)}, \quad \alpha > 0, \ t > 0. \tag{12.27}$$

It is clear that $D^\alpha 1 = 0$, for $\alpha \in \mathbb{N}$, due to the poles of the gamma function at the points $0, -1, -2, \ldots$.

We now observe an alternative definition of fractional derivative, originally introduced by Caputo [73,74] in the late 1960s and adopted by Caputo and Mainardi [75] in the framework of the theory of linear viscoelasticity (see a review in [208]).

Definition 12.21. *Let $f \in AC^n([a, b])$. The Caputo fractional-order derivative of f is defined by*

$$({}^c D^\alpha f)(t) := \frac{1}{\Gamma(n-\alpha)} \int_a^t (t-s)^{n-\alpha-1} f^{(n)}(s)ds.$$

This definition is of course more restrictive than the Riemann-Liouville one in that it requires the absolute integrability of the derivative of order m. Whenever we use the operator ${}^c D^\alpha$ we (tacitly) assume that this condition is met. We easily recognize that in general

$$D^\alpha f(t) := D^m I^{m-\alpha} f(t) \neq J^{m-\alpha} D^m f(t) := {}^c D^\alpha f(t), \tag{12.28}$$

unless the function $f(t)$, along with its first $n - 1$ derivatives, vanishes at $t = a^+$. In fact, assuming that the passage of the m-th derivative under the integral is legitimate, we recognize that, for $m - 1 < \alpha < m$ and $t > 0$,

$$D^\alpha f(t) = {}^c D^\alpha f(t) + \sum_{k=0}^{m-1} \frac{(t-a)^{k-\alpha}}{\Gamma(k-\alpha+1)} f^{(k)}(a^+), \tag{12.29}$$

and therefore, recalling the fractional derivative of the power function (12.26)

$$D^\alpha \left(f(t) - \sum_{k=0}^{m-1} \frac{(t-a)^{k-\alpha}}{\Gamma(k-\alpha+1)} f^{(k)}(a^+) \right) = {}^c D^\alpha f(t). \tag{12.30}$$

The alternative definition, that is, Definition 12.21, for the fractional derivative thus incorporates the initial values of the function and its lower orders. The subtraction of the Taylor polynomial of degree $n-1$ at $t = a^+$ from $f(t)$ means a sort of regularization of the fractional derivative. In particular, according to this definition, the important property that the fractional derivative of a constant is zero still holds, i.e.,

$$^c D^\alpha 1 = 0, \quad \alpha > 0. \tag{12.31}$$

For further readings and details on fractional calculus, we refer the reader to the monographs by Kilbas et al. [184], Podlubny [244], Samko et al. [256].

12.5.2   Existence result

We consider the following Banach space of continuous function.

$$C_*([0,b],\mathbb{R}^n) = \{y \in C((0,b],\mathbb{R}^n) : \lim_{t\to 0^+} t^{1-\alpha} y(t) \text{ exists}\}$$

A norm in this space is given by

$$\|y\|_* = \sup\{|t^{1-\alpha} y(t)| : t \in (0,b]\}.$$

The following theorem is a simple variant of the classical Arzelà-Ascoli Theorem. For $\mathcal{A}$ a subset of the space $C_*([0,b],\mathbb{R})$, define $\mathcal{A}_\alpha$ by

$$\mathcal{A}_\alpha = \{y_\alpha : y \in \mathcal{A}\},$$

where

$$y_\alpha(t) = \begin{cases} t^{1-\alpha} y(t), & \text{if} \quad t \in (0,b] \\ \lim_{t\to 0} t^{1-\alpha} y(t), & \text{if} \quad t = 0. \end{cases}$$

Theorem 12.22. *[81] Let $\mathcal{A}$ be a bounded set in $C_*([0,b],\mathbb{R})$. Assume that $\mathcal{A}_\alpha$ is equicontinuous on $[0,b]$. Then $\mathcal{A}$ is relatively compact in $C_*([0,b],\mathbb{R})$.*

We state the following generalization of Gronwall's lemma for singular kernels, whose proof can be found in [168, Lemma 7.1.1].

Lemma 12.23. *Let $v : [0,b] \to [0,\infty)$ be a real function and $w(\cdot)$ be a nonnegative, locally integrable function on $[0,b]$, and suppose there are constants $a > 0$ and $0 < \alpha < 1$ such that*

$$v(t) \le w(t) + a \int_0^t \frac{v(s)}{(t-s)^\alpha} ds.$$

Then, there exists a constant $K = K(\alpha)$ such that

$$v(t) \le w(t) + Ka \int_0^t \frac{w(s)}{(t-s)^\alpha} ds,$$

for every $t \in [0,b]$.

Now, we present our main results.

Theorem 12.24. *Let $f_i : \mathbb{R} \times \mathbb{R}^n \to \mathbb{R}^n$, $i = 1, \dots, k$, be single-valued functions such that we assume the multiple-valued function in the sense of Almgren*

$$f = \sum_{i=1}^{k} [[f_i]] : [0, b] \times \mathbb{R} \to \mathcal{A}_k(\mathbb{R}^n),$$

is continuous. Assume that there exist $M > 0$ such that

$$d_{\mathcal{A}}(f(t, x), k[[0]]) \le M, \quad \text{for all} \quad x \in \mathbb{R}, \ t \in (0, b]. \tag{12.32}$$

Then Problem (12.24) has at least one solution. Moreover the solution set of problem (12.24) is compact in $C_*([0, b], \mathbb{R}^n)$.

Proof. The proof involves several steps.

Step 1: We shall define the solution on the interval $(0, b]$. We construct two sequences $\{y_i\}_{i=0}^{i=\infty}$ and $\{g_i\}_{i=0}^{i=\infty}$ in the space $C_*([0, b], \mathbb{R}^n)$, respectively, by

$$y_\alpha^0(t) = \lim_{s \to t^+} t^{\alpha-1} y_0(s) = c, \quad t \in [0, b],$$

where

$$y_0(t) = ct^{\alpha-1}, \text{ if } \in (0, b].$$

$$y_\alpha^1(t) = \begin{cases} c, \text{ if } t \in [0, \frac{b}{2}], \\ \lim_{s \to t^+} y_1(s), \text{ if } t \in (\frac{b}{2}, b], \end{cases}$$

where

$$y_1(t) = \begin{cases} y_0(t), \text{ if } t \in (0, \frac{b}{2}], \\ ct^{\alpha-1} + \frac{1}{\Gamma(\alpha)} \int_0^{t-\frac{b}{2}} (t-s)^{\alpha-1} g_{2,1}(s) ds, \text{ if } t \in (\frac{b}{2}, b], \end{cases}$$

where $g_{2,1} : [0, \frac{b}{2}] \to \mathbb{R}^n$ is a continuous selection of $f(\cdot, y_\alpha^1(\cdot)) : [0, \frac{b}{2}] \times \mathbb{R}^n \to \mathcal{A}_k(\mathbb{R}^n)$. From Theorem 12.12 we can find a continuous selection $g_1 : [0, b] \to \mathbb{R}^n$ of $f(\cdot, y_\alpha^1(\cdot)) : [0, b] \times \mathbb{R}^n \to \mathcal{A}_k(\mathbb{R}^n)$ such that $g_1(\cdot) = g_{2,1}(\cdot)$ on $[0, \frac{b}{2}]$. We define

$$y_\alpha^2(t) = \begin{cases} c, & \text{if } t \in [0, \frac{b}{3}], \\ \lim_{s \to t^+} s^{1-\alpha} y_2(s), & \text{if } t \in (\frac{b}{3}, b], \end{cases}$$

where

$$y_2(t) = \begin{cases} y_0(t), & \text{if } t \in (0, \frac{b}{3}], \\ ct^{\alpha-1} + \frac{1}{\Gamma(\alpha)} \int_0^{t-\frac{b}{3}} (t-s)^{\alpha-1} g_{3,1}(s) ds, & \text{if } t \in (\frac{b}{3}, \frac{2b}{3}], \\ ct^{\alpha-1} + \frac{1}{\Gamma(\alpha)} \int_0^{t-\frac{b}{3}} (t-s)^{\alpha-1} g_{3,2}(s) ds, & \text{if } t \in (\frac{2b}{3}, b], \end{cases}$$

where $g_{3,1} : [0, \frac{b}{3}] \to \mathbb{R}^n$ is a continuous selection of $f(\cdot, y_\alpha^2(\cdot)) : [0, \frac{b}{3}] \to \mathcal{A}_k(\mathbb{R}^n)$ and $g_{3,2} : [0, \frac{2b}{3}] \to \mathbb{R}^n$ is a continuous selection of $f(\cdot, y_\alpha^2(\cdot)) : [0, \frac{2b}{3}] \to \mathcal{A}_k(\mathbb{R}^n)$ such that $g_{3,1}(\cdot) = g_{3,2}(\cdot)$ on $[0, \frac{b}{3}]$. By Theorem 12.12 we can choose a continuous selection of $f(\cdot, y_\alpha^3(\cdot))$:

$[0, b] \to \mathcal{A}_k(\mathbb{R}^n)$ such that $g_2(\cdot) = g_{3,2}(\cdot)$ on $[0, \frac{2b}{3}]$. Finally we can conclude by induction the following sequence,

$$y_\alpha^i(t) = \begin{cases} c, & \text{if } t \in [0, \frac{b}{i}], \\ \lim_{s \to t^+} y_i(s), & \text{if } t \in (\frac{b}{i}, b], \end{cases}$$

where

$$y_i(t) = \begin{cases} ct^{\alpha-1}, & \text{if } t \in (0, \frac{b}{i}], \\ ct^{\alpha-1} + \frac{1}{\Gamma(\alpha)} \int_0^{t-\frac{b}{i}} (t-s)^{\alpha-1} g_i(s) ds, & \text{if } t \in (\frac{b}{i}, b], \end{cases}$$

where $g_i : [0, b] \to \mathbb{R}^n$ is a continuous selection of $f(\cdot, y_\alpha^i(\cdot)) : [0, b] \times \mathbb{R} \to \mathcal{A}_k(\mathbb{R}^n)$. We have $(g_i) \in C([0, b], \mathbb{R}^n)$, and there exists a constant $M_* > 0$ such that

$$|g_i(t)| \leq M \qquad t \in [0, b].$$

We successively observe that, if $\tau_2, \tau_1 \in (0, \frac{b}{i}]$, then

$$|y_\alpha^i(\tau_2) - y_\alpha^i(\tau_2)| = 0,$$

if $0 \leq \tau_1 \leq \frac{b}{i} \leq \tau_2 \leq b$, then

$$\begin{aligned} |y_\alpha^i(\tau_2) - y_\alpha^i(\tau_1)| &\leq \frac{\tau_2^{1-\alpha}}{\Gamma(\alpha)} \int_0^{\tau_2 - \frac{b}{i}} |(\tau_2 - s)^{\alpha-1} g_i(s)| ds \\ &\leq \frac{\tau_2^{1-\alpha}}{\Gamma(\alpha)} \int_0^{\tau_2 - \tau_1} (\tau_2 - s)^{\alpha-1} ||g_i(s)| ds \\ &\leq \frac{M \tau_2^{1-\alpha}}{\Gamma(\alpha+1)} (\tau_2^\alpha - \tau_1^\alpha) \end{aligned}$$

and if $\frac{b}{i} < \tau_1 < \tau_2 < b$, then

$$\begin{aligned} |y_\alpha^i(\tau_2) - y_\alpha^i(\tau_1)| &\leq \frac{|\tau_2^{1-\alpha} - \tau_1^{1-\alpha}|}{\Gamma(\alpha)} \int_0^{\tau_1} (\tau_2 - s)^{\alpha-1} |g_i(s)| ds \\ &+ \frac{\tau_1^{1-\alpha}}{\Gamma(\alpha)} \int_0^{\tau_1} [(\tau_1 - s)^{\alpha-1} - (\tau_2 - s)^{\alpha-1}] |g_i(s)| ds \\ &+ \frac{\tau_2^{1-\alpha}}{\Gamma(\alpha)} \int_{\tau_1 - \frac{b}{i}}^{\tau_2 - \frac{b}{i}} (\tau_2 - s)^{\alpha-1} |g_i(s)| ds \\ &\leq \frac{M|\tau_2^{1-\alpha} - \tau_1^{1-\alpha}|}{\Gamma(\alpha)} \int_0^{\tau_1} (\tau_2 - s)^{\alpha-1} ds \\ &+ \frac{M \tau_1^{1-\alpha}}{\Gamma(\alpha)} \int_0^{\tau_1} [(\tau_1 - s)^{\alpha-1} - (\tau_2 - s)^{\alpha-1}] ds \\ &+ \frac{M \tau_2^{1-\alpha}}{\Gamma(\alpha)} \int_{\tau_1 - \frac{b}{i}}^{\tau_2 - \frac{b}{i}} \left(\frac{\tau_2 - s - \frac{b}{i}}{\tau_2 - s} \right)^{1-\alpha} \left(\tau_2 - s - \frac{b}{i} \right)^{\alpha-1} ds \\ &\leq \frac{M|\tau_2^{1-\alpha} - \tau_1^{1-\alpha}|}{\Gamma(\alpha)} \int_0^{\tau_1} (\tau_2 - s)^{\alpha-1} ds \\ &+ \frac{M \tau_1^{1-\alpha}}{\Gamma(\alpha)} \int_0^{\tau_1} [(\tau_1 - s)^{\alpha-1} - (\tau_2 - s)^{\alpha-1}] ds \\ &+ \frac{M \tau_2^{1-\alpha}}{\Gamma(\alpha)} \int_{\tau_1 - \frac{b}{i}}^{\tau_2 - \frac{b}{i}} \left(\tau_2 - s - \frac{b}{i} \right)^{\alpha-1} ds. \end{aligned}$$

Then

$$
\begin{aligned}
|y_\alpha^i(\tau_2) - y_\alpha^i(\tau_1)| \;\leq\; & \frac{M|\tau_2^{1-\alpha} - \tau_1^{1-\alpha}|}{\Gamma(\alpha)} \int_0^{\tau_2} (\tau_2 - s)^{\alpha-1} ds \\
& + \frac{M\tau_1^{1-\alpha}}{\Gamma(\alpha)} \int_0^{\tau_1} [(\tau_1 - s)^{\alpha-1} - (\tau_2 - s)^{\alpha-1}] ds \\
& + \frac{M\tau_2^{1-\alpha}}{\Gamma(\alpha)} \int_{\tau_1-\frac{b}{i}}^{\tau_2-\frac{b}{i}} \left(\tau_2 - s - \frac{b}{i}\right)^{\alpha-1} ds \\
\leq\; & \frac{M|\tau_2^{1-\alpha} - \tau_1^{1-\alpha}|}{\Gamma(\alpha+1)} \tau_2^\alpha + \frac{M\tau_1^{1-\alpha}}{\Gamma(\alpha+1)} |\tau_2 - \tau_1|^\alpha \\
& + \frac{M\tau_1^{1-\alpha}}{\Gamma(\alpha+1)} [\tau_1^\alpha - \tau_2^\alpha] + \frac{M\tau_2^{1-\alpha}}{\Gamma(\alpha+1)} |\tau_2 - \tau_1|^{\alpha-1} \\
\leq\; & \frac{M|\tau_2^{1-\alpha} - \tau_1^{1-\alpha}|}{\Gamma(\alpha+1)} \tau_2^\alpha + \frac{M\tau_1^{1-\alpha}}{\Gamma(\alpha+1)} |\tau_2 - \tau_1|^\alpha \\
& + \frac{M\tau_2^{1-\alpha}}{\Gamma(\alpha+1)} |\tau_2 - \tau_1|^{\alpha-1}.
\end{aligned}
$$

As $\tau_2 \longrightarrow \tau_1$ the right-hand side of the above inequality tends to zero. Consequently the sequence (y_α^i) is equicontinous. Also,

$$
\begin{aligned}
|t^{1-\alpha} y_i(t)| \;\leq\; & |c| + \frac{t^{1-\alpha}}{\Gamma(\alpha)} \int_0^{t-\frac{b}{i}} (t-s)^{\alpha-1} |g_i(s)| ds \\
\leq\; & |c| + \frac{b^{1-\alpha}}{\Gamma(\alpha)} \int_0^{t-\frac{b}{i}} (t-s)^{\alpha-1} |g_i(s)| ds \\
\leq\; & |c| + \frac{Mb^{1-\alpha}}{\Gamma(\alpha)} \left[-\left(\frac{b}{i}\right)^\alpha + t^\alpha \right] \\
\leq\; & |c| + \frac{Mb}{\Gamma(\alpha+1)} := M_*,
\end{aligned}
$$

and then, the sequence (y_α^i) is uniformly bounded and equicontinuous. By Ascoli's theorem there exists a subsequence still denoted $\{y_\alpha^i\}_{i=0}^{i=\infty}$ converging uniformly to some function y in $C_*(([0,b], \mathbb{R}^n)$.

Let $K = [0,b] \times B(c, M)$, and

$$
\begin{aligned}
\omega|_{f|_K}(\delta) \;=\; & \sup \{ d_{\mathcal{A}}(f(\tau_2, y_2), f(\tau_1, y_1)) : |(\tau_2, y_2) - (\tau_1, y_1)| \leq \delta \\
& \text{where } (\tau_1, y_1),\ (\tau_2, y_2) \in K \}
\end{aligned}
$$

be a modulus of continuity of f restricted to K. For each $\epsilon > 0$ there exists $\delta_1 > 0$ such that for every $|\tau_2 - \tau_1| \leq \delta_1$

$$
\frac{M|\tau_2^{1-\alpha} - \tau_1^{1-\alpha}|}{\Gamma(\alpha+1)} \tau_2^{\alpha-1} + \frac{M\left(\tau_1^{1-\alpha} + \tau_2^{1-\alpha}\right)}{\Gamma(\alpha+1)} |\tau_2 - \tau_1|^\alpha \leq \epsilon,
$$

and

$$
\frac{M\tau_2^{1-\alpha}}{\Gamma(\alpha+1)} (\tau_2^\alpha - \tau_1^\alpha) \leq \epsilon.
$$

Hence for each $i \in \mathbb{N} \cup \{0\}$, we have

$$
|\tau_2^{1-\alpha} y_i(\tau_2) - \tau_1^{1-\alpha} y_i(\tau_1)| \leq \epsilon, \text{ for all } \tau_1, \tau_2 \in [0,b] \text{ and } |\tau_2 - \tau_1| \leq \delta_1.
$$

This implies that

$$
\begin{aligned}
\omega|_{f(\cdot, y^i_\alpha(\cdot))}(\delta_2) &= \sup \left\{ d_{\mathcal{A}}(f(\tau_2, y^i_\alpha(\tau_2)), f(\tau_1, y^i_\alpha(\tau_1))) : |\tau_2 - \tau_1| \le \delta_1, \right. \\
&\qquad \left. \text{and } \tau_1, \tau_2 \in [0, b] \right\} \\
&\le \sup \left\{ d_{\mathcal{A}}(f(\tau_2, y_1), f(\tau_1, y_2)) : |\tau_2 - \tau_1| \le \delta_2, |y_1 - y_2| \le \delta_2, \right. \\
&\qquad \left. \text{and } (\tau_2, y_2), \ (\tau_1, y_1) \in K \right\} \\
&\le \omega|_{f_K}(\delta_2), \text{ where } \delta_2 = \max(\delta_1, \epsilon).
\end{aligned}
$$

It is clear that $f(\cdot, y^i_\alpha(\cdot)) - [[g_i(\cdot)]] : [0, b] \to \mathcal{A}_{k-1}(\mathbb{R}^n)$ is a continuous multiple-function. Then there exist $h^i_1, \ldots, h^i_{k-1} : [0, b] \to \mathbb{R}^n$ continuous functions such that

$$
f(\cdot, y^i_\alpha(\cdot)) = [[g_i(\cdot)]] + \sum_{j=1}^{k-1} [[h^i_j(\cdot)]].
$$

Then

$$
|g_i|_* \le M \text{ for each } n \in \mathbb{N} \cup \{0\}
$$

and

$$
\omega|_{g_i} \le \omega|_{f_K}(\delta_2), \text{ for every } n \in \mathbb{N} \cup \{0\}.
$$

Consequently, $\{g_i\}_{i=0}^{i=\infty}$ is bounded and equicontinuous. From the Ascoli-Arzelá theorem, we can conclude that $\{g_i\}_{i=0}^{i=\infty}$ is compact in $C([0, b], \mathbb{R}^n)$. Hence there exists a subsequence, denoted $\{g_i\}_{i=0}^{i=\infty}$, converging uniformly to g. Hence

$$
\begin{aligned}
|t^{1-\alpha} y_i(t) - t^{1-\alpha} z(t)| &\le \frac{t^{1-\alpha}}{\Gamma(\alpha)} \int_0^t (t - s)^{\alpha-1} |g_i(s) - g(s)| ds \\
&\le \frac{t}{\Gamma(\alpha + 1)} \|g_i - g\|_\infty \\
&\le \frac{b}{\Gamma(\alpha + 1)} \|g_i - g\|_\infty \to 0 \text{ as } i \to \infty.
\end{aligned}
$$

Hence

$$
z(t) = ct^{1-\alpha} + \frac{1}{\Gamma(\alpha)} \int_0^t (t - s)^{\alpha-1} g(s) ds := y(t), \quad t \in (0, b].
$$

Set

$$
y_\alpha(t) = \begin{cases} c, & \text{if } \ t = 0, \\ \lim_{s \to t^+} s^{1-\alpha} y(s), & \text{if } \ t \in (0, b], \end{cases}
$$

By Lemma 12.16 we conclude that g is a continuous selection of $f(\cdot, y_\alpha(\cdot))$ on $[0, b]$.
Step 2: Now we show that the set

$$
S = \{y \in C_*([0, b], \mathbb{R}^n) \mid y \text{ is a solution of } (12.24)\}
$$

is compact.

Let $\{y_i\}_{i=1}^{i=\infty}$ be a sequence in S. Then

$$
y_i(t) = ct^{1-\alpha} + \frac{1}{\Gamma(\alpha)} \int_0^t (t - s)^{\alpha-1} g_i(s) ds, \ t \in [0, b],
$$

where $g_i(\cdot)$ is a continuous selection of $f(\cdot, y_i(\cdot))$. As in Step 1, we conclude there exists a subsequence, still denoted $\{y_i\}_{i=1}^{i=\infty}$, which converges to a continuous function $y : [0, b] \to \mathbb{R}^n$.

It is clear that $f(\cdot, y_\alpha^i(\cdot)) - [[g_i(\cdot)]] : [0, b] \to \mathcal{A}_{k-1}(\mathbb{R}^n)$ is a continuous multiple-function. Then there exist $h_1^j, \ldots, h_{k-1}^j : [0, b] \to \mathbb{R}^n$ continuous functions such that

$$f(\cdot, y_\alpha^i(\cdot)) = [[g_i(\cdot)]] + \sum_{j=1}^{k-1} [[h_j^i(\cdot)]].$$

As in Step 1 we can prove that (g_i) is bounded and equicontinuous. Then from the Ascoli-Arzelá theorem, we can conclude that $\{g_i : i \in \mathbb{N}\}$ is compact in $C_*([0, b], \mathbb{R}^n)$, hence there exists a subsequence, denoted $(g_i)_{i \in \mathbb{N}}$, converging uniformly to g. Hence

$$|y_i - z|_* \leq |g_i - g|_* \to 0 \text{ as } i \to \infty,$$

where

$$z(t) = ct^{\alpha-1} + \frac{1}{\Gamma(\alpha)} \int_0^t (t-s)^{\alpha-1} g(s)ds := y(t), \quad t \in (0, b].$$

By Lemma 12.16, we conclude that g is a continuous selection of $f(\cdot, y(\cdot))$ on $(0, b]$. $\quad\square$

Remark 12.25. *We can replace the condition (12.32) by*

$$d_\mathcal{A}(f(t, x), k[[0]]) \leq M_1|x| + M_2, \quad \text{for all } x \in \mathbb{R}, \ t \in [0, b]. \tag{12.33}$$

Then the sequence $(y_i)_{i \in \mathbb{N} \cup \{0\}}$ defined in the proof of Theorem 12.24 is bounded in $C_*([0, b], \mathbb{R}^n)$.

Proof. For each $i \in \mathbb{N}$ we have

$$y_i(t) = \begin{cases} ct^{\alpha-1}, & \text{if} \quad t \in (0, \frac{b}{i}], \\ ct^{\alpha-1} + \frac{1}{\Gamma(\alpha)} \int_0^{t-\frac{b}{i}} (t-s)^{\alpha-1} g_i(s)ds, & \text{if} \quad t \in (\frac{b}{i}, b], \end{cases}$$

where $g_i : [0, b] \to \mathbb{R}^n$ is a continuous selection of $f(., y_\alpha^i(.)) : [0, b] \times \mathbb{R} \to \mathcal{A}_k(\mathbb{R}^n)$. We have $(g_i) \in C([0, b], \mathbb{R}^n)$ and from (12.33) we have,

$$|g_i(t)| \leq M_1 t^{1-\alpha}|y(t)| + M_2 \text{ for each } i \in \mathbb{N}.$$

Then

$$\begin{aligned}
|t^{1-\alpha} y_i(t)| &\leq |c| + \frac{t^{1-\alpha}}{\Gamma(\alpha)} \int_0^t (t-s)^{\alpha-1}|g_m(s)|ds \\
&\leq |a| + \frac{b^{1-\alpha}}{\Gamma(\alpha)} \int_0^t (t-s)^{\alpha-1}(M_1 s^{1-\alpha}|y_m(s)| + M_2)ds,
\end{aligned}$$

from Gronwall's Lemma 12.23, there exists $M > 0$ such that

$$\|y_i\|_* \leq M \text{ for each } i \in \mathbb{N} \cup \{0\}.$$

$\quad\square$

Theorem 12.26. *(Peano theorem) Let $\Omega \subset \mathbb{R} \times \mathbb{R}^n$ be an open set and $f : \Omega \to \mathcal{A}_k(\mathbb{R}^n)$ be a continuous function and $(0, c) \in \Omega$. Then there exist $h > 0$, $\eta > 0$ and $y : (0, h] \to \mathbb{R}^n$ such that*

$$t^{1-\alpha}y(t) \in \overline{B}(c, \eta), \quad \text{for all } t \in [0, h],$$

which is solution of Problem (12.24).

Proof. Let $\epsilon > 0$. Then by continuity of f there exists $\eta > 0$ such that

$$|t| \leq \eta, \quad \|y - c\| \leq \eta.$$

This implies that

$$d_{\mathcal{A}}(f(t,y), f(0,c)) \leq \epsilon.$$

Since the set $G = [-\eta, \eta] \times \overline{B}(c, \eta) \subset \Omega$ is compact. By the continuity of f on the compact set G, there exists a constant $M_1 > 0$ such that

$$d_{\mathcal{A}}(f(t,x), k[[0]]) < M_1, \quad \text{for all} \quad (t,x) \in G.$$

Consider the Cauchy problem

$$
\begin{cases}
{}^c D^\alpha y(t) & \in \quad \{f_1(t, t^{1-\alpha} y), \ldots f_k(t, t^{1-\alpha} y)\}, \quad t \in [0, h], \\
\lim\limits_{t \to 0+} t^{1-\alpha} y(t) & = \quad c,
\end{cases}
\tag{12.34}
$$

where $h \leq \min\left(\eta, \left(\frac{\eta \Gamma(\alpha+1)}{M_1}\right)^{\frac{1}{\alpha}}\right)$. By the same method used in the proof of Theorem 12.12 we define the following sequences

$$
y_i(t) =
\begin{cases}
ct^{\alpha-1}, \text{ if } t \in (0, \frac{h}{i}], \\
ct^{\alpha-1} + \frac{1}{\Gamma(\alpha)} \int_0^{t - \frac{h}{i}} (t-s)^{\alpha-1} g_i(s) ds, \text{ if } t \in (\frac{h}{i}, h],
\end{cases}
$$

where $i \in \mathbb{N}$, $g_i : [0, h] \to \mathbb{R}^n$ is a continuous selection of $f(\cdot, y^i_\alpha(\cdot)) : [0, h] \times \mathbb{R} \to \mathcal{A}_k(\mathbb{R}^n)$. If

$$
\begin{aligned}
|t^{1-\alpha} y_i(t) - c| &\leq \frac{t^{1-\alpha}}{\Gamma(\alpha)} \int_0^t (t-s)^{\alpha-1} |g_i(s)| ds \\
&\leq \frac{M_1 h^\alpha}{\Gamma(\alpha+1)} \leq \eta,
\end{aligned}
$$

then

$$t^{1-\alpha} y_i(t) \in \overline{B}(c, \eta), \quad \text{for all } t \in [0, h].$$

As in Theorem 12.24 we can prove that $(y_i)_{i \in \mathbb{N}}$ and $(g_i)_{i \in \mathbb{N}}$ are relatively compact in $C_*([0, h], \overline{B}(c, \eta))$ and $C([0, h], \overline{B}(c, \eta))$ respectively. Then there exist subsequences of $(y_i)_{i \in \mathbb{N}}$ and $(g_i)_{i \in \mathbb{N}}$ converging uniformly to y and g, respectively, and from Lemma 12.16 we conclude that

$$y(t) = ct^{\alpha-1} + \frac{1}{\Gamma(\alpha)} \int_0^{t - \frac{h}{i}} (t-s)^{\alpha-1} g(s) ds, \quad t \in (0, h],$$

where g is the continuous selection of $f(\cdot, y_\alpha(\cdot))$ and y is the solution of problem (12.24). $\square$

12.6 Differential inclusions via Caputo fractional derivative

We consider the following fractional differential inclusions in the Almgren sense via a Caputo derivative

$$
\begin{cases}
{}^c D^\alpha y(t) & \in \quad \{f_1(t, y(t)), \ldots f_Q(t, y(t))\}, \quad t \in [0, b], \\
y(0) & = \quad a \in \mathbb{R}^n,
\end{cases}
\tag{12.35}
$$

where each $f_i \colon [0, b] \times \mathbb{R}^n \to \mathbb{R}^n$, $i = 1, 2, \ldots Q$, is a single valued function, and $^cD^\alpha$ denotes the Caputo fractional derivative of order $\alpha \in (0, 1]$. The goal of this section is to present an existence result and prove the compactness of the solution set for the problem (12.35) by using an iteration method combined with a compactness argument. The results of this section are due to Graef *et al.* [146].

12.6.1 Existence and compactness result

In this subsection, we present an existence result and prove the compactness of the solution set for the problem (12.35) by using an iteration method combined with a compactness argument. First, we define the concept of a solution for Problem (12.35).

Definition 12.27. *A function $y \in C([0, b], \mathbb{R}^n)$ is called a solution of Problem (12.35), if there exists a continuous selection $g \in C([0, b], \mathbb{R}^n)$ of $f(\cdot, y(\cdot))$ such that*

$$y(t) = a + \frac{1}{\Gamma(\alpha)} \int_0^t (t - s)^{\alpha - 1} g(s) ds, \quad t \in [0, b].$$

Theorem 12.28. *Let $f_i \colon \mathbb{R} \times \mathbb{R}^n \to \mathbb{R}^n$, $i = 1, \ldots, Q$, be single valued functions such that the associated multiple-valued function in the sense of Almgren*

$$f = \sum_{i=1}^{Q} [\![f_i]\!] \colon [0, b] \times \mathbb{R} \to \mathcal{A}_Q(\mathbb{R}^n),$$

is continuous. Assume that there exist $M_1, M_2 > 0$ such that

$$d_{\mathcal{A}}(f(t, x), Q(0)) \le M_1 \|x\| + M_2, \quad \text{for all} \ \ x \in \mathbb{R}, \ t \in [0, b]. \tag{12.36}$$

Then Problem (12.35) has at least one solution. Moreover the solution set of problem (12.35) is compact in $C([0, b], \mathbb{R}^n)$.

Proof. The proof involves several steps.

Step 1: First, we construct two sequence $\{y_m\}_{m=0}^{m=\infty}$ and $\{g_m\}_{m=0}^{m=\infty}$ by

$$y_0(t) = a, \ \text{for all} \ t \in [0, b],$$

$$y_1(t) = \begin{cases} y_0(t), & if \ \ t \in [0, b/2], \\ a + \frac{1}{\Gamma(\alpha)} \displaystyle\int_0^{t - \frac{b}{2}} (t - s)^{\alpha - 1} g_{2,1}(s) ds, & if \ \ t \in (b/2, b], \end{cases}$$

where $g_{2,1} \colon [0, \frac{b}{2}] \to \mathbb{R}^n$ is a continuous selection of $f(\cdot, y_1(\cdot)) \colon [0, \frac{b}{2}] \times \mathbb{R}^n \to \mathcal{A}_Q(\mathbb{R}^n)$. From Theorem 12.12, we can find a continuous selection $g_1 \colon [0, b] \to \mathbb{R}^n$ of $f(\cdot, y_1(\cdot)) \colon [0, b] \times \mathbb{R}^n \to \mathcal{A}_Q(\mathbb{R}^n)$ such that $g_1(\cdot) = g_{2.1}(\cdot)$ on $[0, \frac{b}{2}]$. We define

$$y_2(t) = \begin{cases} y_0(t), & if \ \ t \in [0, \frac{b}{3}], \\ a + \frac{1}{\Gamma(\alpha)} \displaystyle\int_0^{t - \frac{b}{3}} (t - s)^{\alpha - 1} g_{3,1}(s) ds, & if \ \ t \in [\frac{b}{3}, \frac{2b}{3}], \\ a + \frac{1}{\Gamma(\alpha)} \displaystyle\int_0^{t - \frac{b}{3}} (t - s)^{\alpha - 1} g_{3,2}(s) ds, & if \ \ t \in (\frac{b}{3}, b], \end{cases}$$

where $g_{3,1} \colon [0, \frac{b}{3}] \to \mathbb{R}^n$ is a continuous selection of $f(\cdot, y_2(\cdot)) \colon [0, \frac{b}{3}] \to \mathcal{A}_Q(\mathbb{R}^n)$ and $g_{3,2} \colon [0, \frac{2b}{3}] \to \mathbb{R}^n$ is a continuous selection of $f(\cdot, y_2(\cdot)) \colon [0, \frac{2b}{3}] \to \mathcal{A}_Q(\mathbb{R}^n)$ such that $g_{3,1}(\cdot) = g_{3,2}(\cdot)$ on $[0, \frac{b}{3}]$. By Theorem 12.12 we can choose a continuous selection of $f(\cdot, y_3(\cdot)) \colon$

$[0, b] \to \mathcal{A}_Q(\mathbb{R}^n)$ such that $g_2(\cdot) = g_{3,2}(\cdot)$ on $[0, \frac{2b}{3}]$. Finally, we can conclude by induction the following sequence,

$$y_m(t) = \begin{cases} y_0(t), & \text{if } t \in [0, b/m], \\ a + \frac{1}{\Gamma(\alpha)} \int_0^{t - \frac{b}{m}} (t - s)^{\alpha-1} g_m(s) ds, & \text{if } t \in (b/m, b], \end{cases}$$

where $g_m : [0, b] \to \mathbb{R}^n$ is a continuous selection of $f(\cdot, y_m(\cdot)) : [0, b] \times \mathbb{R} \to \mathcal{A}_Q(\mathbb{R}^n)$.

Now, we show that $\{y_m : m \in \mathbb{N} \cup \{0\}\}$ is relatively compact. First, we show that $\{y_m\}_{m=0}^{m=\infty}$ is bounded. Since

$$\begin{aligned} |y_m(t)| &\leq |a| + \frac{1}{\Gamma(\alpha)} \int_0^t (t - s)^{\alpha-1} |g_m(s)| ds \\ &\leq |a| + \frac{1}{\Gamma(\alpha)} \int_0^t (t - s)^{\alpha-1} (M_1 |y_m(s)| + M_2) ds, \end{aligned}$$

by Gronwall's Lemma 12.23, there exists $M > 0$ such that

$$\|y_m\|_\infty \leq M \text{ for each } m \in \mathbb{N} \cup \{0\}.$$

Next, we show that $\{y_m\}_{m=0}^{m=\infty}$ is equicontinuous. Let $t_1, t_2 \in [0, \frac{b}{m}]$. Then

$$|y_m(t_1) - y_m(t_2)| = 0 \text{ if } t_1, t_2 \in [0, \frac{b}{m}].$$

For $0 < t_1 \leq \frac{b}{m} < t_2 < b$, we have

$$\begin{aligned} |y_m(t_1) - y_m(t_2)| &\leq \frac{1}{\Gamma(\alpha)} \int_0^{t_2 - \frac{b}{m}} (t - s)^{\alpha-1} |g_m(s)| ds \\ &\leq \frac{M_*}{\Gamma(\alpha+1)} \left| t_2^\alpha - \left(\frac{b}{m}\right)^\alpha \right| \leq \frac{M_*}{\Gamma(\alpha+1)} |t_2^\alpha - t_1^\alpha|, \end{aligned}$$

and for $t_1, t_2 \in (\frac{b}{m}, b]$,

$$\begin{aligned} |y_m(t_1) - y_m(t_2)| &\leq \frac{1}{\Gamma(\alpha)} \int_{t_1 - \frac{b}{m}}^{t_2 - \frac{b}{m}} (t_2 - s)^{\alpha-1} |g_m(s)| ds \\ &+ \frac{1}{\Gamma(\alpha)} \int_0^{t_1 - \frac{b}{m}} |(t_2 - s)^{\alpha-1} - (t_1 - s)^{\alpha-1}| |g_m(s)| ds \\ &\leq \frac{M_*}{\Gamma(\alpha+1)} \left[-\left(\frac{b}{m}\right)^\alpha + \left(t_2 - t_1 + \frac{b}{m}\right)^\alpha \right] \\ &+ \frac{M_*}{\Gamma(\alpha+1)} \left[\left(t_2 - t_1 + \frac{b}{m}\right)^\alpha - t_2^\alpha - \left(\frac{b}{m}\right)^\alpha + t_1^\alpha \right] \\ &\leq \frac{M_*}{\Gamma(\alpha+1)} [2(2t_2 - t_1)^\alpha - t_2^\alpha - t_1^\alpha], \end{aligned}$$

where

$$M_* = M_1 M + M_2.$$

Consequently, $\{y_m\}_{m=0}^{m=\infty}$ is bounded and equicontinuous. By the Ascoli-Arzelà theorem, there exists a subsequence of $\{y_m\}_{m=0}^{m=\infty}$ converging to some y in $C([0, b], \mathbb{R}^n)$. Let $K = [0, b] \times B(0, M)$, and

$$\omega|_{f|_K}(\delta) = \sup \{d_\mathcal{A}(f(t_1, x_1), f(t_2, x_2)) : |(t_1, x_1) - (t_2, x_2)| \leq \delta,$$

where $(t_1, x_1), (t_2, x_2) \in K\}$

be a modulus of continuity of f restricted to K. For each $\epsilon > 0$ there exists $\delta_1 > 0$ such that for every $|t_1 - t_2| \le \delta_1$, we have

$$\frac{M_*}{\Gamma(\alpha + 1)}|t_2^\alpha - t_1^\alpha| \le \epsilon \quad \text{and} \quad \frac{M_*}{\Gamma(\alpha + 1)}[2(2t_2 - t_1)^\alpha - t_2^\alpha - t_1^\alpha] \le \epsilon.$$

Hence, for each $m \in \mathbb{N} \cup \{0\}$, we have

$$|y_m(t_1) - y_m(t_2)| \le \epsilon, \text{ for all } t_1, t_2 \in [0, b] \text{ and } |t_1 - t_2| \le \delta_1.$$

This implies that

$$\begin{aligned}
\omega|_{f(\cdot, y_m(\cdot))}(\delta_2) &= \sup \{d_{\mathcal{A}}(f(t_1, y_m(t_1)), f(t_2, y_m(t_2))) : |t_1 - t_2| \le \delta_1, \\
&\quad \text{and } \tau_1, \tau_2 \in [0, b]\} \\
&\le \sup \{d_{\mathcal{A}}(f(t_1, x_1), f(t_2, x_2)) : |t_1 - t_2| \le \delta_2, |x_1 - x_2| \le \delta_2, \\
&\quad \text{and } (t_1, x_1), (t_2, x_2) \in K\} \\
&\le \omega|_{f_K}(\delta_2), \text{ where } \delta_2 = \max(\delta_1, \epsilon).
\end{aligned}$$

It is clear that $f(\cdot, y_m(\cdot)) - [[g_m(\cdot)]] : [0, b] \to \mathcal{A}_{Q-1}(\mathbb{R}^n)$ is a continuous multiple-function. Then there exist $h_1^m, \ldots, h_{Q-1}^m : [0, b] \to \mathbb{R}^n$ continuous functions such that

$$f(\cdot, y_m(\cdot)) = [[g_m(\cdot)]] + \sum_{i=1}^{Q-1} [[h_i^m(\cdot)]].$$

Thus,

$$\|g_m\|_\infty \le L_1 \text{ for each } m \in \mathbb{N} \cup \{0\}$$

and

$$\omega|_{g_m} \le \omega|_{f_K}(\delta_2), \text{ for every } m \in \mathbb{N} \cup \{0\}.$$

Consequently, $\{g_m\}_{m=0}^{m=\infty}$ is bounded and equicontinuous. From the Ascoli-Arzelà theorem, we can conclude that $\{g_m\}_{m=0}^{m=\infty}$ is compact in $C([0, b], \mathbb{R}^n)$. Hence, there exists a subsequence, again denoted by $\{g_m\}_{m=0}^{m=\infty}$, that converges uniformly to g. Hence,

$$\|y_m - z\|_\infty \le \frac{b^\alpha}{\Gamma(\alpha + 1)}\|g_m - g\|_\infty \to 0 \quad \text{as} \quad m \to \infty,$$

where

$$z(t) = a + \frac{1}{\Gamma(\alpha)}\int_0^t (t - s)^{\alpha-1}g(s)ds := y(t), \quad t \in [0, b].$$

By Lemma 12.16, we conclude that g is a continuous selection for $f(\cdot, y(\cdot))$ on $[0, b]$.

Step 2: Now we show that the set

$$S = \{y \in C([0, b], \mathbb{R}^n) \mid y \text{ is a solution of } (12.35)\}$$

is compact.

Let $\{y_m\}_{m=1}^{m=\infty}$ be a sequence in S. Then

$$y_m(t) = a + \frac{1}{\Gamma(\alpha)}\int_0^t (t - s)^{\alpha-1}g_m(s)ds, \quad t \in [0, b],$$

where $g_m(\cdot)$ is a continuous selection for $f(\cdot, y_m(\cdot))$. As in Step 1, we conclude there exists a subsequence, still denoted $\{y_m\}_{m=1}^{m=\infty}$, which converges to a continuous function $y : [0, b] \to \mathbb{R}^n$. It clear that $f(\cdot, y_m(\cdot)) - [[g_m(\cdot)]] : [0, b] \to \mathcal{A}_{Q-1}(\mathbb{R}^n)$ is a continuous multiple-valued function. Then there exist continuous functions $h_1^m, \ldots, h_{Q-1}^m : [0, b] \to \mathbb{R}^n$ continuous functions such that

$$f(\cdot, y_m(\cdot)) = [[g_m(\cdot)]] + \sum_{i=1}^{Q-1} [[h_i^m(\cdot)]].$$

As in Step 1 we can prove that (g_n) is bounded and equicontinuous. Then from the Ascoli-Arzelà theorem, we can conclude that $\{g_m : m \in \mathbb{N}\}$ is compact in $C([0, b], \mathbb{R}^n)$ and so there is a subsequence, denoted by $(g_m)_{m \in \mathbb{N}}$ again, that converges uniformly to g. Therefore,

$$\|y_m - z\|_\infty \le b\|g_m - g\|_\infty \to 0 \text{ as } n \to \infty,$$

where

$$z(t) = a + \frac{1}{\Gamma(\alpha)} \int_0^t (t - s)^{\alpha-1} g(s) ds := y(t), \quad t \in [0, b].$$

By Lemma 12.16, we conclude that g is a continuous selection of $f(\cdot, y(\cdot))$ on $[0, b]$. $\qquad\square$

Chapter 13

Random Systems of Differential Equations

Random differential equations and random integral equations have been studied systematically by Ladde and Lakshmikantham [197] and Bharucha-Reid [50], respectively. Such equations are good models in various branches of science and engineering when random factors and uncertainties have been taken into consideration. Hence, the study of the fractional differential equations with random parameters seems to be a natural one. We refer the reader to the monographs [267, 269], and the references therein. In this chapter we are interested in applying some of the results developed in Chapter 9 to random systems of differential equations with initial and boundary conditions.

13.1 Random Cauchy problem

In this section we shall use a random version of Perov type and study nonlinear initial value problems of random differential equations of first order for different aspects of the solutions under suitable conditions. In particular, in this section we study the following systems

$$\begin{cases} x'(t,\omega) = f(t, x(t,\omega), y(t,\omega), \omega) \\ y'(t,\omega) = g(t, x(t,\omega), y(t,\omega), \omega) \\ x(0,\omega) = x_0(\omega), \ \omega \in \Omega \\ y(0,\omega) = y_0(\omega), \ \omega \in \Omega, \end{cases} \tag{13.1}$$

where $f, g : [0,b] \times \mathbb{R} \times \mathbb{R} \times \Omega \to \mathbb{R}$, $(\Omega, \mathcal{A})$ is a measurable space and $x_0, y_0 : \Omega \to \mathbb{R}$ are random variables.

Definition 13.1. *A function $f : [0,b] \times \mathbb{R} \times \Omega \to \mathbb{R}$ is called random Carathéodory if the following conditions are satisfied:*

(i) the map $(t,\omega) \to f(t,x,\omega)$ is jointly measurable for all $x \in \mathbb{R}$,

(ii) the map $x \to f(t,x,\omega)$ is continuous for all $t \in [0,b]$ and $\omega \in \Omega$.

Definition 13.2. *A Carathéodory function $f : [0,b] \times \mathbb{R} \times \Omega \to \mathbb{R}$ is called random L^1-Carathéodory if for each real number $r > 0$ there is a measurable and bounded function $h_r \in L^1([0,b], \mathbb{R})$ such that*

$$|f(t,x,\omega)| \le h_r(t,\omega), \quad a.e. \ \ t \in [0,b]$$

for all $\omega \in \Omega$ and $x \in \mathbb{R}$ with $|x| \le r$.

Theorem 13.3. *(Carathéodory) Let X be a separable metric space and $G : \Omega \times X \to X$ be a mapping such that $G(\cdot, x)$ is measurable for all $x \in X$ and $G(\omega, \cdot)$ is continuous for all $\omega \in \Omega$. Then the map $(\omega, x) \to G(\omega, x)$ is jointly measurable.*

As a consequence of the above theorem we can easily prove the following result.

Lemma 13.4. *Let X be a separable generalized metric space and $G : \Omega \times X \to X$ be a mapping such that $G(\cdot, x)$ is measurable for all $x \in X$ and $G(\omega, \cdot)$ is continuous for all $\omega \in \Omega$. Then the map $(\omega, x) \to G(\omega, x)$ is jointly measurable.*

Proof. Let $C \subset X$ be a closed set of X, then

$$
\begin{aligned}
G^{-1}(C) &= \{(\omega, x) \in \Omega \times X : \ G(\omega, x) \in C\} \\
&= \{(\omega, x) \in \Omega \times X : \ d(G(\omega, x), C) = 0\}
\end{aligned}
$$

where

$$
d(x, y) = \begin{pmatrix} d_1(x, y) \\ \vdots \\ d_n(x, y) \end{pmatrix}, \quad (x, y) \in X \times X.
$$

Since X is a separable generalized metric space then there is a countable dense subset D of X. Let

$$
C_m = \{x \in X : d(x, C) < \epsilon_m\},
$$

where

$$
\epsilon_m := \begin{pmatrix} \frac{1}{m} \\ \vdots \\ \frac{1}{m} \end{pmatrix}.
$$

Hence $G(\omega, x) \in C$ if and only if for every $m \geq 1$ there exists $a \in D$ such that

$$
d(x, a) < \epsilon_m \text{ and } G(\omega, x) \in C_m,
$$

and therefore

$$
\begin{aligned}
G^{-1}(C) &= \cap_{m \geq 1} \cup_{a \in D} \{\omega \in \Omega : G(\omega, a) \in C_m\} \times \{x \in X : d(x, a) < \epsilon_m\} \\
&\in \mathcal{A} \otimes B(X).
\end{aligned}
$$

This implies that $G(\cdot, \cdot)$ is jointly measurable. $\qquad \square$

Theorem 13.5. *Let $f, g : [0, b] \times \mathbb{R} \times \mathbb{R} \times \Omega \to \mathbb{R}$ be two Carathédory functions. Assume that the following condition*

$(\mathcal{RL}_1)$ *There exist random variables $p_1, p_2, p_3, p_4 : \Omega \to \mathbb{R}$ such that*

$$
|f(t, x, y, \omega) - f(t, \widetilde{x}, \widetilde{y}, \omega)| \leq p_1(\omega)|x - \widetilde{x}| + p_2(\omega)|y - \widetilde{y}|
$$

and

$$
|g(t, x, y, \omega) - g(t, \widetilde{x}, \widetilde{y}, \omega)| \leq p_3(\omega)|x - \widetilde{x}| + p_4(\omega)|y - \widetilde{y}|
$$

where

$$
M(\omega) = \begin{pmatrix} bp_1(\omega) & bp_2(\omega) \\ bp_3(\omega) & bp_4(\omega) \end{pmatrix}.
$$

If $M(\omega)$ converges to 0, then problem (13.1) has a unique random solution.

Proof. Consider the operator $N : C([0, b], \mathbb{R}) \times C([0, b], \mathbb{R}) \times \Omega \to C([0, b], \mathbb{R}) \times C([0, b], \mathbb{R})$, defined by

$$N((x, y, \omega) = (A_1(t, x, y, \omega), A_2(t, x, y, \omega)),$$

where

$$A_1(x, y, \omega) = \int_0^t f(s, x(s, \omega), y(s, \omega), \omega) ds + x_0(\omega)$$

and

$$A_2(x, y, \omega) = \int_0^t g(s, x(s, \omega), y(s, \omega), \omega) ds + y_0(\omega).$$

First we show that N is a random operator on $C([0, b], \mathbb{R}) \times C([0, b], \mathbb{R}) \times \Omega$. Since f and g are Carathéodory functions, then $\omega \to f(t, x, y, \omega)$ and $\omega \to g(t, x, y, \omega)$ are measurable maps in view of Lemma 13.4. Further, the integral is a limit of a finite sum of measurable functions, therefore, the maps

$$\omega \to A_1(x(t, \omega), y(t, \omega), \omega), \quad \omega \to A_2(x(t, \omega), y(t, \omega), \omega)$$

are measurable. As a result, N is a random operator on $C([0, b], \mathbb{R}) \times C([0, b], \mathbb{R}) \times \Omega$ into $C([0, b], \mathbb{R}) \times C([0, b], \mathbb{R})$. We show that N satisfies all the conditions of Theorem 9.21 on $C([0, b], \mathbb{R}) \times C([0, b], \mathbb{R})$. Let $(x, y), (\widetilde{x}, \widetilde{y}) \in C([0, b], \mathbb{R}) \times C([0, b], \mathbb{R})$ then

$$
\begin{aligned}
|A_1(x(t), y(t), \omega) - A_1(\widetilde{x}(t), \widetilde{y}(t), \omega)| &= \left| \int_0^t f(s, x(s, \omega), y(s, \omega), \omega) ds \right. \\
&\quad \left. - \int_0^t f(s, \widetilde{x}(s, \omega), \widetilde{y}(s, \omega), \omega) ds \right| \\
&\leq \int_0^t |f(s, x(s, x), y(s, \omega), \omega) \\
&\quad - f(s, \widetilde{x}(s, \omega), \widetilde{y}(s, \omega), \omega)| ds \\
&\leq \int_0^t p_1(\omega) |x(s, \omega) - \widetilde{x}(s, \omega)| ds \\
&\quad + \int_0^t p_2(\omega) |y(s, \omega) - \widetilde{y}(s, \omega)| ds.
\end{aligned}
$$

Then

$$\|A_1(x, y, \omega) - A_1(\widetilde{x}, \widetilde{y}, \omega))\|_\infty \leq \|x - \widetilde{x}\|_\infty p_1(\omega) b + \|y - \widetilde{y}\|_\infty p_2(\omega) b,$$

and similarly, we obtain

$$\|A_2(x, y, \omega) - A_2(\widetilde{x}, \widetilde{y}, \omega)\|_\infty \leq \|x - \widetilde{x}\|_\infty p_3(\omega) b + \|y - \widetilde{y}\|_\infty p_4(\omega) b.$$

Hence

$$d(N(x, y, \omega), N(\widetilde{x}, \widetilde{y}, \omega)) \leq \widetilde{M}(\omega) d((x, y), (\widetilde{x}, \widetilde{y}))$$

where

$$d(x, y) = \left(\begin{array}{c} \|x - y\|_\infty \\ \|x - y\|_\infty \end{array} \right)$$

and

$$\widetilde{M}(\omega) = \left(\begin{array}{cc} b p_1(\omega) & b p_2(\omega) \\ b p_3(\omega) & b p_4(\omega) \end{array} \right) = b M(\omega).$$

From Theorem 9.22 there exists a unique random solution of problem (13.1). $\qquad \square$

We consider the following set of hypotheses in what follows:

(RH_1) The functions f and g are random Carathéodory on $[0, b] \times \mathbb{R} \times \Omega$.

(RH_2) There exist measurable and bounded functions $\gamma_1, \gamma_2 : \Omega \to L^1([0, b], \mathbb{R})$ and continuous and nondecreasing functions $\psi_1, \psi_2 : \mathbb{R}_+ \to (0, \infty)$ such that,

$$|f(t, x, y)| \leq \gamma_1(t, \omega)\psi(|x| + |y|), \quad |g(t, x, y)| \leq \gamma_2(t, \omega)\psi_2(|x| + |y|),$$

a.e. $t \in [0, b]$, and for all $\omega \in \Omega$ and $x, y \in \mathbb{R}$.

Now, we give an existence result for problem (13.1) by using the Leray-Schauder random fixed point theorem type in generalized Banach space.

Theorem 13.6. *Assume that the hypotheses (RH_1) and (RH_2) hold. If*

$$\int_0^b |\gamma_1(\omega)| + |\gamma_2(\omega)| < \int_{|x_0(\omega)| + |y_0(\omega)|}^{\infty} \frac{du}{\psi_1(u) + \psi_2(u)}, \quad \text{for all } \omega \in \Omega,$$

then the problem (13.1) has a random solution defined on $[0, b]$.

Proof. Let $N : C([0, b], \mathbb{R}) \times C([0, b], \mathbb{R}) \times \Omega \to C([0, b], \mathbb{R}) \times C([0, b], \mathbb{R})$ be the random operator defined in Theorem 13.5.

Clearly, the random fixed points of N are solutions to (13.1). In order to apply theorem 9.27, we first show that N is completely continuous. The proof will be given in several steps.

Step 1. $N(\cdot, \cdot, \omega) = (A_1(\cdot, \cdot, \omega), A_2(\cdot, \cdot, \omega))$ is continuous.
Let (x_n, y_n) be a sequence such that $(x_n, y_n) \to (x, y) \in C([0, b], \mathbb{R}) \times C([0, b], \mathbb{R})$ as $n \to \infty$. Then

$$|A_1(x_n(t, \omega), y_n(t, \omega), \omega) - A_1(x(t, \omega), y(t, \omega), \omega)|$$
$$\leq \int_0^t |f(s, x_n(s, \omega), y_n(s, \omega), \omega) - f(s, x(s, \omega), y(s, \omega), \omega)|ds.$$

Then

$$\|A_1(x_n(\cdot, \omega), y_n(\cdot, \omega), \omega) - A_1(x(\cdot, \omega), y(\cdot, \omega), \omega)\|_{\infty}$$
$$\leq \int_0^b |f(s, x_n(s, \omega), y_n(s, \omega), \omega) - f(s, x(s, \omega), y(s, \omega), \omega)|ds.$$

Since f is a Carathéodory function, by the Lebesgue dominated convergence theorem, we get

$$\|A_1(x_n(\cdot, \omega), y_n(\cdot, \omega), \omega) - A_1(x(\cdot, \omega), y(\cdot, \omega), \omega)\|_{\infty} \to 0 \text{ as } n \to \infty.$$

Similarly

$$\|A_2(x_n(\cdot, \omega), y_n(\cdot, \omega), \omega) - A_2(x(\cdot, \omega), y(\cdot, \omega), \omega)\|_{\infty} \to 0 \text{ as } n \to \infty.$$

Thus N is continuous.

Step 2. N maps bounded sets into bounded sets in $C([0, b], \mathbb{R}) \times C([0, b], \mathbb{R})$. Indeed, it is enough to show that for any $q > 0$ there exists a positive constant l such that for each $(x, y) \in B_q = \{(x, y) \in C([0, b], \mathbb{R}) \times C([0, b], \mathbb{R}) : \|x\|_{\infty} \leq q, \|y\|_{\infty} \leq q\}$, we have

$$\|N(x, y, \omega)\|_{\infty} \leq l = (l_1, l_2).$$

Then for each $t \in [0, b]$, we get

$$|A_1(x(t, \omega), y(t, \omega), \omega)| = \left| x_0(\omega) + \int_0^t f(s, x(s, \omega), y(s, \omega), \omega) ds \right|$$

$$\leq |x_0(\omega)| + \int_0^b |f_1(s, x(s, \omega), y(s, \omega))| ds.$$

From (RH_2), we get

$$\|A_1(x(\cdot, \omega), y(\cdot, \omega), \omega)\|_\infty \leq |x_0(\omega)| + \psi_1(2q) \int_0^b \gamma_1(s, \omega) ds := l_1.$$

Similarly, we have

$$\|A_2(x(\cdot, \omega), y(\cdot, \omega), \omega)\|_\infty \leq |y_0(\omega)| + \psi_2(2q) \int_0^b \gamma_2(s, \omega) ds := l_2.$$

Step 3. N maps bounded sets into equicontinuous sets of $C([0, b], \mathbb{R}) \times C([0, b], \mathbb{R})$. Let B_q be a bounded set in $C([0, b], \mathbb{R}) \times C([0, b], \mathbb{R})$ as in Step 2. Let $r_1, r_2 \in J, r_1 < r_2$ and $u \in B_q$. Thus we have

$$|A_1(x(r_2, \omega), y(r_2, \omega), \omega) - A_1(x(r_1, \omega), y(r_1, \omega), \omega)|$$
$$\leq \int_{r_1}^{r_2} |f^1(s, x(s, \omega), y(s, \omega), \omega)| ds.$$

Hence

$$|A_1(x(r_2, \omega), y(r_2, \omega), \omega) - A_1(x(r_1, \omega), y(r_1, \omega), \omega)|$$
$$\leq \psi_1(2q) \int_{r_1}^{r_2} \gamma_1(s, \omega) ds$$

and

$$|A_2(x(r_2, \omega), y(r_2, \omega), \omega) - A_2(x(r_1, \omega), y(r_1, \omega), \omega)|$$
$$\leq \psi_2(2q) \int_{r_1}^{r_2} \gamma_2(s, \omega) ds.$$

The right-hand term tends to zero as $|r_2 - r_1| \to 0$.

As a consequence of Steps 1 to 3 together with the Arzela-Ascoli, we conclude that N maps B_q into a precompact set in $C([0, b], \mathbb{R}) \times C([0, b], \mathbb{R})$.

Step 4. It remains to show that

$$\mathcal{A} = \{(x, y) \in C([0, b], \mathbb{R}) \times C([0, b], \mathbb{R}) : (x, y) = \lambda(\omega) N(x, y), \lambda(\omega) \in (0, 1)\}$$

is bounded.

Let $(x, y) \in \mathcal{A}$. Then $x = \lambda(\omega) A_1(x, y)$ and $y = \lambda(\omega) A_2(x, y)$ for some $0 < \lambda < 1$. Thus, for $t \in [0, b]$, we have

$$|x(t, \omega)| \leq |x_0(\omega)| + \int_0^t |f_1(s, x(s, \omega), y(s, \omega), \omega)| ds$$

$$\leq |x_0(\omega)| + \int_0^t \gamma_1(s, \omega) \psi_1(|x(s, \omega)| + |y(s, \omega)|) ds.$$

Hence

$$|x(t,\omega)| \leq |x_0(\omega)| + \int_0^t \gamma_1(s,\omega)\psi_1(|x(s,\omega)| + |y(s,\omega)|)ds$$

and

$$|y(t,\omega)| \leq |y_0(\omega)| + \int_0^t \gamma_2(s,\omega)\psi_2(|x(s,\omega)| + |y(s,\omega)|)ds.$$

Therefore

$$|x(t,\omega)| + |y(t,\omega)| \leq c|x_0(\omega)| + |y_0(\omega)|$$
$$+ \int_0^t p(s)\phi(|x(s,\omega)| + |y(s,\omega)|)ds,$$

where

$$c = |x_0(\omega)| + |y_0(\omega)|, \phi = \psi_1 + \psi_2, \text{ and } p = \gamma_1 + \gamma_2.$$

By Lemma 12.3, we have

$$|x(t,\omega)| + |y(t,\omega)| \leq \Gamma^{-1}\left(\int_c^b p(s)ds\right) := K_*, \text{ for each } t \in [0,b],$$

where

$$\Gamma(z) = \int_c^z \frac{du}{\phi(u)}.$$

Consequently

$$\|x\|_\infty \leq K_* \text{ and } \|y\|_\infty \leq K_*.$$

This shows that $\mathcal{E}$ is bounded. As a consequence of Theorem 9.27 we deduce that N has a random fixed point (x,y) which is a solution to the problem (13.1). □

13.2   Boundary value problems

Nonlocal boundary problems arise in many applied sciences. For example, the vibrations of a guy wire of a uniform cross-section and composed of N parts of different densities, and some problems in the theory of elastic stability, can be modelled by multi-point boundary value problems (see [218, 232]). The existence of solutions for systems of local and nonlocal boundary value problems (BVPs) has received an increased attention by researchers, see for example the papers of Agarwal, O'Regan and Wong [8–10], Henderson, Ntouyas and Purnaras [163], Precup [249,250] and references therein. Very recently the coupled systems of BVPs with local and nonlocal conditions was studied by Bolojan-Nica *et al* [56], Precup [251] and references therein.

In this section we are interested in random solutions of the following coupled systems:

$$\begin{cases} x''(t,\omega) = f(t,x(t,\omega),y(t,\omega),\omega), & t \in (0,1) \\ y''(t,\omega) = g(t,x(t,\omega),y(t,\omega),\omega), & t \in (0,1) \\ x(0,\omega) = 0, \quad x(1,\omega) = L_1(\int_0^1 x(t,\omega)d\alpha(t)), & \omega \in \Omega \\ y(0,\omega) = 0, \quad y(1,\omega) = L_2(\int_0^1 y(t,\omega)d\beta(t)), & \omega \in \Omega, \end{cases}$$

where $L_1, L_2 \in C(\mathbb{R}, \mathbb{R})$, and $\int_0^1 x(t, \omega)d\alpha(t)$, $\int_0^1 x(t, \omega)d\alpha(t)$ denote Riemann-Stieltjes integrals.

Lemma 13.7. *Let $f, g : [0, 1] \times \mathbb{R} \times \mathbb{R} \to \mathbb{R}$ be two continuous functions for the system,*

$$\begin{cases} x''(t) = f(t, x(t), y(t)), & t \in (0, 1) \\ y''(t) = g(t, x(t), y(t)), & t \in (0, 1) \\ x(0) = 0, \quad x(1) = L_1\left(\int_0^1 x(t)d\alpha(t)\right) \\ y(0) = 0, \quad y(1) = L_2\left(\int_0^1 y(t)d\beta(t)\right). \end{cases} \tag{13.2}$$

where $L_1, L_2 \in C(\mathbb{R}, \mathbb{R})$, and $\int_0^1 x(t)d\alpha(t)$, $\int_0^1 x(t)d\alpha(t)$ denote Riemann-Stieltjes integrals. The couple $(x, y) \in C([0, 1], \mathbb{R}) \times C([0, 1], \mathbb{R})$ is a solution of problem (13.2) if and only if

$$\begin{cases} x(t) = \int_0^1 \bar{K}(t, s)f(s, x(s), y(s))ds + L_1\left(\int_0^1 x(s)d\alpha(s)\right)t, & t \in (0, 1) \\ y(t) = \int_0^1 \bar{K}(t, s)g(s, x(s), y(s))ds + L_2\left(\int_0^1 y(s)d\beta(s)\right)t, & t \in (0, 1) \end{cases}$$

where

$$\bar{K}(t, s) = \begin{cases} t(1 - s), & 0 \leq t \leq s \leq 1 \\ s(1 - t), & 0 \leq s \leq t \leq 1. \end{cases}$$

Theorem 13.8. *Assume $(\mathcal{RL}_1)$ and the following condition is satisfied.*

(RH_3) there exist $0 < K_1 < 1, 0 < K_2 < 1$, two positive real constants, such that

$$|L_1(x) - L_1(y)| \leq K_1|x - y|, \quad |L_2(x) - L_2(y)| \leq K_1|x - y|,$$

for each $x, y \in \mathbb{R}$.

If

$$M_*(\omega) = \begin{pmatrix} p_1(\omega) + K_1 & p_2(\omega) + K_2 \\ p_3(\omega) + K_1 & p_4(\omega) + K_2 \end{pmatrix}$$

converges to 0, then problem (13.2) has unique random solution.

Proof. Consider the operator $N_* : C([0, b], \mathbb{R}) \times C([0, b], \mathbb{R}) \times \Omega \to C([0, b], \mathbb{R}) \times C([0, b], \mathbb{R})$,

$$(x, y, \omega) \mapsto (\bar{A}_1(t, x, y, \omega), \bar{A}_2(x(t, \omega), y(t, \omega), \omega))$$

where

$$A_1(, x(t, \omega), y(t, \omega), \omega) = \int_0^1 \bar{K}(t, s)f(s, x(s, \omega), y(s, \omega), \omega)ds$$

$$+ L_1\left(\int_0^1 x(s, \omega)d\alpha(s)\right)t$$

and

$$\bar{A}_2(t, x, y, \omega) = \int_0^1 \bar{K}(t, s)g(s, x(s, \omega), y(s, \omega), \omega)ds$$

$$+ L_1\left(\int_0^1 y(s, \omega)d\alpha(s)\right)t.$$

Clearly from Lemma 13.7, the random fixed points of N_* are solutions of problem (13.2).

As in Theorem 13.5, we can prove that N_* has a unique random fixed point which is the solution of problem (13.2) $\qquad\square$

To prove our next result we need the following lemma.

Lemma 13.9. *([287]) Let*

$$Q = \begin{pmatrix} a & -b \\ -c & d \end{pmatrix}$$

where a, b, c, $d \geq 0$ and $\det Q > 0$. Then Q^{-1} is order preserving.

Theorem 13.10. *Assume that (RH_1), and (RH_3), with $C_1 + K_1, \bar{C}_2 + K_2 \in [0, 1)$, and the following condition holds:*

(RH_4) *There exist $C_1, C_2, C_3, \bar{C}_1, \bar{C}_2, \bar{C}_3 > 0$ positive real constants such that*

$$|f(t, x, y, \omega)| \leq C_1 |x| + C_2 |y| + C_3, \ |g(t, x, y, \omega)| \leq \bar{C}_1 |x| + \bar{C}_2 |y| + \bar{C}_3,$$

a.e. $t \in [0, b]$, and for all $\omega \in \Omega$ and $x, y \in \mathbb{R}$.

If

$$\bar{M} = \begin{pmatrix} 1 - C_1 - K_1 & -C_2 \\ -\bar{C}_1 & 1 - \bar{C}_2 - K_2 \end{pmatrix}$$

and if $\det \bar{M} > 0$, then problem (13.2) has at least one random solution.

Proof. Let $N : C([0, 1], \mathbb{R}) \times C([0, 1], \mathbb{R}) \times \Omega \to C([0, 1], \mathbb{R}) \times C([0, 1], \mathbb{R})$ be the operator defined by

$$N(x, y, \omega) = F(x, y, \omega) + B(x, y, \omega), \quad (x, y, \omega) \in C([0, 1], \mathbb{R}) \times C([0, 1], \mathbb{R}) \times \Omega,$$

where

$$F(x, y, \omega) = (F_1(x, y, \omega), F_2(x, y, \omega)), \ B(x, y, \omega) = (B_1(x, y, \omega), B_2(x, y, \omega))$$

and

$$F_1(t, x, y, \omega) = \int_0^1 \bar{K}(t, s) f(s, x(s, \omega), y(s, \omega), \omega) ds,$$

$$B_1(x, y, \omega) = L_1 \left(\int_0^1 x(s, \omega) d\alpha(s) \right) t,$$

$$F_2(t, x, y, \omega) = \int_0^1 \bar{K}(t, s) g(s, x(s, \omega), y(s, \omega), \omega) ds,$$

$$B_2(x, y, \omega) = L_2 \left(\int_0^1 y(s, \omega) d\beta(s) \right) t.$$

Since $K_1, K_2 \in [0, 1)$ then

$$\bar{M} = \begin{pmatrix} K_1 & 0 \\ 0 & K_2 \end{pmatrix}$$

converges to zero. This implies that B is a contraction operator. By the same method used in Theorem 9.22 we can prove that F is a completely continuous operator.

Now, we show that the following set

$$\begin{aligned}
\mathcal{M} = \ & \{(x, y) : \Omega \to C([0, 1], \mathbb{R}) \times C([0, 1], \mathbb{R}) \\
& \text{is measurable} \mid \lambda(\omega) F(x, y, \omega) \\
& + B(\frac{x}{\lambda(\omega)}, \frac{y}{\lambda(\omega)}, \omega) = (x, y)\}
\end{aligned}$$

is bounded for some measurable $\lambda : \Omega \to \mathbb{R}$ with $0 < \lambda(\omega)) < 1$ on Ω. Let $(x, y) \in \mathcal{M}$, then define

$$
\begin{aligned}
x(t, \omega) &= \lambda(\omega) \int_0^1 \bar{K}(t, s) f(s, x(s, \omega), y(s, \omega), \omega) ds \\
&\quad + \lambda(\omega) L_1 \left(\int_0^1 \frac{x(s, \omega)}{\lambda(\omega)} d\alpha(s) \right) t
\end{aligned}
$$

and

$$
\begin{aligned}
y(t, \omega) &= \lambda(\omega) \int_0^1 \bar{K}(t, s) g(s, x(s, \omega), y(s, \omega), \omega) ds \\
&\quad + \lambda(\omega) L_2 \left(\int_0^1 \frac{y(s, \omega)}{\lambda(\omega)} d\beta(s) \right) t.
\end{aligned}
$$

Thus

$$
|x(t, \omega)| \leq C_1 |x(t, \omega)| + C_2 |y(t, \omega)| + C_3 + K_1 |x(t, \omega)| + |L_1(0)|
$$

and

$$
|y(t, \omega)| \leq \bar{C}_1 |x(t, \omega)| + \bar{C}_2 |y(t, \omega)| + \bar{C}_3 + K_2 |y(t, \omega)| + |L_2(0)|.
$$

This implies that

$$
\begin{pmatrix} 1 - C_1 - K_1 & -C_2 \\ -\bar{C}_1 & 1 - \bar{C}_2 - K_2 \end{pmatrix} \begin{pmatrix} x(t, \omega) \\ y(t, \omega) \end{pmatrix} \leq \begin{pmatrix} C_3 + |L_1(0)| \\ \bar{C}_3 + |L_2(0)| \end{pmatrix}.
$$

Therefore

$$
\bar{M} \begin{pmatrix} x(t, \omega) \\ y(t, \omega) \end{pmatrix} \leq \begin{pmatrix} C_3 + |L_1(0)| \\ \bar{C}_3 + |L_2(0)| \end{pmatrix}. \tag{13.3}
$$

Since $\bar{M}$ satisfies the hypotheses of Lemma 13.9, so $(\bar{M})^{-1}$ is order preserving. We apply $(\bar{M})^{-1}$ to both sides of the inequality (13.3) we obtain

$$
\begin{pmatrix} x(t, \omega) \\ y(t, \omega) \end{pmatrix} \leq (\bar{M})^{-1} \begin{pmatrix} C_3 + |L_1(0)| \\ \bar{C}_3 + |L_2(0)| \end{pmatrix}.
$$

Hence, from Lemma 9.29, the operator N has at least one random fixed point which is solution of problem (13.2).   □

13.3　An example

Let $\Omega = \mathbb{R}$ be equipped with the usual $\sigma-$ algebra consisting of Lebesgue measurable subsets of $(-\infty, 0)$ and let $J := [0, 1]$.

Consider the following random differential equation system, for $t \in J$,

$$
\begin{cases}
x'(t, \omega) = \frac{t\omega^2 x^2(t, \omega)}{(1+\omega^2)(1+x^2(t, \omega)+y^2(t, \omega))} \\
y'(t, \omega) = \frac{t\omega^2 y^2(t, \omega)}{(1+\omega^2)(1+x^2(t, \omega)+y^2(t, \omega))} \\
x(0, \omega) = \sin \omega, \ \omega \in \Omega \\
y(0, \omega) = \cos \omega, \ \omega \in \Omega.
\end{cases} \tag{13.4}
$$

Here

$$f(t, x, y, \omega) = \frac{t\omega^2 x^2}{2(1 + \omega^2)(1 + x^2 + y^2)}$$

and

$$g(t, x, y, \omega) = \frac{t\omega^2 y^2}{2(1 + \omega^2)(1 + x^2 + y^2)}.$$

Clearly, the map $(t, \omega) \mapsto f(t, x, y, \omega)$ is jointly continuous for all $x, y \in [1, \infty)$. The same is true for the map g. Also the maps $x \mapsto f(t, x, y, \omega)$ and $y \mapsto f(t, x, y, \omega)$ are continuous for all $t \in J$ and $\omega \in \Omega$. Similarly for the maps corresponding to function g. Thus the functions f and g are Carathéodory on $J \times [1, \infty) \times [1, \infty) \times \Omega$. Firstly, we show that f and g are Lipschitz functions. Indeed, let $x, y \in \mathbb{R}$, then

$$
\begin{aligned}
|f(t, x, y, \omega) - f(t, \widetilde{x}, \widetilde{y}, \omega)| &= \left| \frac{t\omega^2 x^2}{2(1 + \omega^2)(1 + x^2 + y^2)} \right. \\
&\quad \left. - \frac{t\omega^2 x^2}{2(1 + \omega^2)(1 + \widetilde{x}^2 + \widetilde{y}^2)} \right| \\
&= \left| \frac{t\omega^2[(1 + \widetilde{x}^2 + \widetilde{y}^2)x^2 - (1 + x^2 + y^2)\widetilde{x}^2]}{2(1 + \omega^2)(1 + x^2 + y^2)(1 + \widetilde{x}^2 + \widetilde{y}^2)} \right| \\
&= \frac{t\omega^2}{2(1 + \omega^2)(1 + x^2 + y^2)(1 + \widetilde{x}^2 + \widetilde{y}^2)} \times \\
&\quad |x^2 + \widetilde{y}^2 x^2 - \widetilde{x}^2 - y^2\widetilde{x}^2| \\
&\leq \frac{\omega^2}{2(1 + \omega^2)}|x - \widetilde{x}| + \frac{\omega^2}{2(1 + \omega^2)}|y - \widetilde{y}|.
\end{aligned}
$$

Then

$$|f(t, x, y, \omega) - f(t, \widetilde{x}, \widetilde{y}, \omega)| \leq \frac{\omega^2}{2(1 + \omega^2)}|x - \widetilde{x}| + \frac{\omega^2}{2(1 + \omega^2)}|y - \widetilde{y}|.$$

Analogously for the function g, we get

$$|g(t, x, y, \omega) - g(t, \widetilde{x}, \widetilde{y}, \omega)| \leq \frac{\omega^2}{2(1 + \omega^2)}|x - \widetilde{x}| + \frac{\omega^2}{2(1 + \omega^2)}|y - \widetilde{y}|.$$

We take,

$$p_1(\omega) = p_2(\omega) = p_3(\omega) = p_4(\omega) = \frac{\omega^2}{2(1 + \omega^2)}$$

and

$$M(\omega) = \begin{pmatrix} \frac{\omega^2}{2(1+\omega^2)} & \frac{\omega^2}{2(1+\omega^2)} \\ \frac{\omega^2}{2(1+\omega^2)} & \frac{\omega^2}{2(1+\omega^2)} \end{pmatrix}.$$

We remark that

$$|\rho(M(\omega))| = \frac{\omega^2}{2(1 + \omega^2)} < 1,$$

and then

$$M(\omega) \text{ converges to } 0.$$

Therefore, all the conditions of Theorem 9.22 are satisfied. Hence the problem (13.4) has a unique random solution.

Chapter 14

Random Fractional Differential Equations

Random Fractional Differential Equations via Hadamard Fractional Derivatives

Another kind of fractional derivative that appears side by side with the Riemann-Liouville and Caputo derivatives in the literature is the fractional derivative due to Hadamard that was introduced in 1892 [148]. This derivative differs from the aforementioned derivatives in the sense that the kernel of the integral in the definition of the Hadamard derivative contains a logarithmic function with an arbitrary exponent. Details and properties of the Hadamard fractional derivative and integral can be found in [69–71, 182–184].

14.1 Hadamard fractional derivative

In this section we introduce some notations and definitions from the fractional calculus for Hadamard derivatives and integrals. We refer the reader to [125, 177] for the lemmas below and additional details.

Definition 14.1. *The Hadamard fractional integral of order $\alpha \in \mathbb{R}_+$ of a function $f : [a, b] \to \mathbb{R}^m$, $0 < a < b \leq \infty$, is defined by*

$$J^\alpha f(t) = \frac{1}{\Gamma(\alpha)} \int_a^t \left(\ln \frac{t}{s} \right)^{\alpha-1} f(s) \frac{ds}{s},$$

where $\Gamma(\cdot)$ is the Euler-Gamma function.

Definition 14.2. *The Hadamard derivative of order $\alpha \in [n-1, n)$, of the function $f : [a, b] \to \mathbb{R}^m$, $0 < a < b \leq \infty$, is given by*

$$^H Df(t) = \delta^n (J^{n-\alpha} f)(t) = \frac{1}{\Gamma(n-\alpha)} \left(t \frac{d}{dt} \right)^n \int_a^t \left(\ln \frac{t}{s} \right)^{n-\alpha-1} f(s) \frac{ds}{s},$$

where $\delta := t \frac{d}{dt}$, $\delta^0 f(t) = f(t)$, and $n = [\alpha] + 1$ with $[\alpha]$ denoting the smallest integer greater than or equal to α.

Definition 14.3 ([177]). *For an $n-$times differentiable function and $c > 0$, the Caputo type Hadamard fractional derivative of order $\alpha > 0$ of a function $f : [a, \infty) \to \mathbb{R}^m$ is*

$$^{CH} D_{c+}^\alpha g(t) = \frac{1}{\Gamma(n-\alpha)} \int_c^t \left(\ln \frac{t}{s} \right)^{n-\alpha-1} \delta^n g(s) \frac{ds}{s} = J_a^{n-\alpha} \delta^n f(t),$$

where $\alpha < n \leq \alpha + 1$, i.e., $n = [\alpha] + 1$, provided that the right-hand side exists.

The Hadamard fractional derivative is the left-inverse operator to the Hadamard fractional integral in the space $L^p[a, b]$, $1 \leq p \leq \infty$, that is $^H D^\alpha J^\alpha f = f$.

We take
$$AC_\delta^n([a,b]) = \{f : [a,b] \to \mathbb{R}^m : \delta^n f \in AC^n([a,b])\}.$$

The Caputo-type modifications of the left-sided and right-sided Hadamard fractional derivatives are defined respectively by

$$^{CH}D^\alpha f(t) = \ ^H D^\alpha \left[f(t) - \sum_{k=0}^{n-1} \frac{\delta^k(a)}{k!} \left(\ln \frac{t}{a} \right)^k \right]$$

and

$$^{CH}D^\alpha f(t) = \ ^H D^\alpha \left[f(t) - \sum_{k=0}^{n-1} \frac{\delta^k(b)}{k!} \left(\ln \frac{b}{t} \right)^k \right].$$

In particular, if $0 < \alpha < 1$, then

$$^{CH}D^\alpha f(t) = \ ^H D^\alpha \left[f(t) - f(a) \right],$$

and

$$^{CH}D^\alpha f(t) = \ ^H D^\alpha \left[f(t) - f(b) \right].$$

Lemma 14.4. *Let $\alpha > 0$ and $\beta > 0$. Then, given $0 < a < b < \infty$ and $1 \le p < \infty$, for every $f \in L^p(a,b)$,*
$$D^\beta J^\alpha f = J^{\alpha-\beta} f \quad \text{and} \quad J^\alpha J^\beta f = J^{\alpha+\beta} f.$$

Lemma 14.5. *Let $\alpha > 0$, $n = [\alpha] + 1$, and $f \in C[a,b]$. Then*
$$^{CH}D^\alpha J^\alpha f(t) = f(t) \quad t \in [a,b].$$

Lemma 14.6. *Let $\alpha > 0$, $n = [\alpha] + 1$, and $f \in AC_\delta^n[a,b]$ or $f \in C^n([a,b])$. Then,*
$$J^\alpha \ ^{CH}Df(t) = f(t) - \sum_{k=0}^{n-1} \frac{\delta^k(a)}{k!} \left(\ln \frac{t}{a} \right)^k.$$

14.2 Random fractional derivative

Let $(\Omega, \mathcal{A})$ be a measurable space. We equip the metric space X with a σ-algebra $\mathcal{B}(X)$ of Borel subsets of X so that $(X, \mathcal{B}(X))$ becomes a measurable space. Let X and Y be two locally compact metric spaces. By $C(X, Y)$ we denote the space of continuous functions from X into Y endowed with the compact-open topology.

Lemma 14.7. *([233]) The function $f : \Omega \times X \to Y$ is a Carathéodory function if and only if $\omega \to f(\omega, \cdot)$ is a measurable function from $\Omega \to C(X, Y)$.*

We say that $x(\cdot, \cdot) : [1, b] \times \Omega \to \mathbb{R}^m$ is sample path Lebesgue integrable on $[1, b]$ if $x(\cdot, \omega) : [1, b] \longrightarrow \mathbb{R}^m$ is Lebesgue integrable on $[1, b]$ for a.e. $\omega \in \Omega$.

Let $\alpha > 0$. If $x : [1, b] \times \Omega \longrightarrow \mathbb{R}^m$ is sample path Lebesgue integrable on $[1, b]$, then we can consider the fractional integral

$$J^\alpha x(t, \omega) = \frac{1}{\Gamma(\alpha)} \int_1^t \left(\ln \frac{t}{s} \right)^{\alpha-1} x(s, \omega) \frac{ds}{s}. \tag{14.1}$$

which will be called the sample path fractional integral of x.

Remark 14.8. *If $x(\cdot, \omega) : [1, b] \longrightarrow \mathbb{R}^m$ is Lebesgue integrable on $[1, b]$ for each $\omega \in \Omega$, then $t \longmapsto J^\alpha x(t, \omega)$ is also Lebesgue integrable on $[1, b]$ for each $\omega \in \Omega$.*

Recall that $x : [1, b] \times \Omega \longrightarrow \mathbb{R}^m$ is a Carathéodory function if $t \longmapsto x(t, \omega)$ is continuous for a.e. $\omega \in \Omega$ and $\omega \longmapsto x(t, \omega)$ is measurable for each $t \in [1, b]$. Also, a Carathéodory function is a product measurable function.

Proposition 14.9. *If $x : [1, b] \times \Omega \longrightarrow \mathbb{R}^m$ is a Carathéodory function, then the function $(t, \omega) \longmapsto J^\alpha x(t, \omega)$ is also a Carathéodory function.*

Proof. It is clear that $J^\alpha : C([1, b], \mathbb{R}^m) \to \mathbb{R}^m$ is a continuous operator. Let $L : \Omega \to C([1, b], \mathbb{R}^m)$ be defined by $L(\omega)(\cdot) = x(\cdot, \omega)$. From Lemma 14.7, $L(\cdot)$ is measurable, so the operator $\omega \to (J^\alpha \circ L)(\omega)(\cdot)$ is measurable. Since the function $t \to I^\alpha x(t, \omega)$ is a continuous function, $(t, \omega) \to J^\alpha x(t, \omega)$ is a Carathéodory function. $\square$

Definition 14.10. *A function $x : [1, b] \times \Omega \longrightarrow \mathbb{R}^m$ is said to have a sample path derivative at $t \in [1, b]$ if the function $t \longmapsto x(t, \omega)$ has a derivative at t for a.e. $x \in \Omega$. We will denote by $\frac{d}{dt} x(t, \omega)$ or by $x'(t, \omega)$ the sample path derivative of $x(\cdot, \omega)$ at t. We say that $x : [1, b] \times \Omega \longrightarrow \mathbb{R}^m$ is sample path differentiable on $[1, b]$ if $x(\cdot, \cdot)$ has a sample path derivative for each $t \in [1, b]$ and possesses a one-sided sample path derivative at the end points 1 and b.*

Proposition 14.11. *If $x : [1, b] \times \Omega \longrightarrow \mathbb{R}^m$ is sample path absolutely continuous on $[1, b]$ (that is, $t \longmapsto x(t, \omega)$ is absolutely continuous on $[1, b]$ for a.e. $w \in \Omega$), then x has a the sample path derivative $x'(t, \omega)$ for a.e. $t \in [1, b]$.*

At the end of this section we present some properties of two special functions. Denote by $E_{\alpha, \beta}$ the generalized Mittag-Leffler special function defined by

$$E_{\alpha, \beta}(z) := \sum_{k=0}^{\infty} \frac{z^k}{\Gamma(k\alpha + \beta)} = \frac{1}{2\pi i} \int_\Upsilon \frac{\lambda^{\alpha - \beta} e^\lambda}{\lambda^\alpha - z} d\lambda$$

where Υ is a contour that starts and ends at $-\infty$ and encircles the disc $|\lambda| \leq |z|^{\frac{1}{\alpha}}$ counterclockwise. If $0 < \lambda < 1$ and $\beta > 0$, then the asymptotic expansion of $E_{\alpha, \beta}$ as $z \to \infty$ is given by

$$E_{\alpha, \beta}(z) = \begin{cases} \frac{1}{\alpha} z^{1 - \beta} \exp\left(z^{\frac{1}{\alpha}}\right) + \mathcal{E}_{\alpha, \beta}(z), & \text{for } |arg(z)| \leq \frac{1}{2}\alpha\pi, \\ \mathcal{E}_{\alpha, \beta}(z), & \text{for } |arg(-z)| \leq (1 - \frac{1}{2})\alpha\pi, \end{cases}$$

where

$$\mathcal{E}_{\alpha, \beta}(z) = -\sum_{k=1}^{n-1} \frac{z^{-k}}{\Gamma(\beta - \alpha n)} + O(|z|^{-n}) \text{ as } z \to \infty.$$

Set

$$E_\alpha = E_{\alpha, 1} \text{ and } e_\alpha = E_{\alpha, \alpha}.$$

14.3 Existence of solution

Fractional differential equations with random parameters via Caputo derivative, have been studied by Lupulescu *et al.* [206], Lupulescu and Ntouyas [205], and Vu *et al.* [282].

We consider the system of Hadamard-type fractional differential equations

$$\begin{cases} {}^{CH}D^\alpha x(t,\omega) = f(t, x(t,\omega), y(t,\omega), \omega), \ 0 < \alpha < 1, \ t \in [0,b], \\ {}^{CH}D^\beta y(t,\omega) = g(t, x(t,\omega), y(t,\omega), \omega), \ 0 < \beta < 1, \ t \in [0,b], \\ x(1,\omega) = x_0(\omega), \ \omega \in \Omega, \\ y(1,\omega) = y_0(\omega), \ \omega \in \Omega, \end{cases} \qquad (14.2)$$

where $f, g : [1,b] \times \mathbb{R}^m \times \mathbb{R}^m \times \Omega \to \mathbb{R}^m$, $(\Omega, \mathcal{F})$ is a measurable space, and x_0, $y_0 : \Omega \to \mathbb{R}^m$ are random variables. We let ${}^{CH}D^\alpha x$ be the Caputo-modification of the Hadamard fractional derivative. This problem has been recently studied by Henderson *et al.* [165].

Definition 14.12. *A function $f : [1,b] \times \mathbb{R}^m \times \Omega \to \mathbb{R}^m$ is called random Carathéodory if the following conditions are satisfied:*

   (*i*) *the map $(t,\omega) \to f(t, x, \omega)$ is jointly measurable for all $x \in \mathbb{R}^m$;*

   (*ii*) *the map $x \to f(t, x, \omega)$ is continuous for all $t \in [1,b]$ and $\omega \in \Omega$.*

Lemma 14.13. *For the problem*

$$\begin{cases} D^\alpha x(t,\omega) = f(t, x(t,\omega), y(t,\omega), \omega), \ 0 < \alpha < 1, \\ x(1,\omega) = x_0(\omega), \ \omega \in \Omega, \end{cases} \qquad (14.3)$$

if $(t,\omega) \mapsto f(t, x(t,\omega), y(t,\omega), \omega)$ is product measurable and $t \mapsto f(t, x(t,\omega), y(t,\omega), \omega)$ is Lebesgue integrable on $[1,b]$ for a.e. $\omega \in \Omega$, then the function $x : [1,b] \times \Omega \longrightarrow \mathbb{R}^m$ is a solution of (14.3) if and only if

$$x(t,\omega) = x_0(\omega) + \frac{1}{\Gamma(\alpha)} \int_1^t \left(\ln \frac{t}{s} \right)^{\alpha-1} f(s, x(s,\omega), y(s,\omega), \omega) \frac{ds}{s}$$

for all $t \in [1,b]$ and for a.e. $\omega \in \Omega$.

Proof. We have:

$$ {}^{CH}D^\alpha x(t,\omega) = f(t, x(t,\omega), y(t,\omega), \omega). $$

Then

$$ J^\alpha D^\alpha x(t,\omega) = J^\alpha f(t, x(t,\omega), y(t,\omega), \omega). $$

By Lemma 14.6,

$$ J^{\alpha\,CH}D^\alpha x(t,\omega) = x(t,\omega) - x(1,\omega). $$

Thus

$$ x(t,\omega) - x(1,\omega) = J^\alpha f(t, x(t,\omega), y(t,\omega), \omega). $$

Next, from the definition of J^α, we have:

$$ x(t,\omega) - x(1,\omega) = \frac{1}{\Gamma(\alpha)} \int_1^t \left(\ln \frac{t}{s} \right)^{\alpha-1} f(s, x(s,\omega), y(s,\omega), \omega) \frac{ds}{s}. $$

Hence

$$ x(t,\omega) = x_0(\omega) + \frac{1}{\Gamma(\alpha)} \int_1^t \left(\ln \frac{t}{s} \right)^{\alpha-1} f(s, x(s,\omega), y(s,\omega), \omega) \frac{ds}{s}. $$

The converse is straightforward. $\qquad\square$

   Our first main result is the existence and uniqueness of a random solution to the problem (14.2).

Theorem 14.14. *Let f, $g : [1, b] \times \mathbb{R}^m \times \mathbb{R}^m \times \Omega \to \mathbb{R}^m$ be two Carathéodory functions. Assume that*

($\mathcal{JH}_1$) There exist random variables p_1, p_2, p_3, $p_4 : \Omega \to \mathbb{R}_+$ such that for all x, y, $\widetilde{x}$, $\widetilde{y} \in \mathbb{R}^m$ and $\omega \in \Omega$,

$$\|f(t, x, y, \omega) - f(t, \widetilde{x}, \widetilde{y}, \omega)\| \leq p_1(\omega)\|x - \widetilde{x}\| + p_2(\omega)\|y - \widetilde{y}\|,$$

and

$$\|g(t, x, y, \omega) - g(t, \widetilde{x}, \widetilde{y}, \omega)\| \leq p_3(\omega)\|x - \widetilde{x}\| + p_4(\omega)\|y - \widetilde{y}\|.$$

If for every $\omega \in \Omega$,

$$\widetilde{M}(\omega) = \begin{pmatrix} \frac{(\ln b)^\alpha p_1(\omega)}{\Gamma(\alpha+1)} & \frac{(\ln b)^\alpha p_2(\omega)}{\Gamma(\alpha+1)} \\ \frac{(\ln b)^\beta p_3(\omega)}{\Gamma(\beta+1)} & \frac{(\ln b)^\beta p_4(\omega)}{\Gamma(\beta+1)} \end{pmatrix}$$

converges to 0, then problem (14.2) has a unique random solution.

Proof. Consider the operator $N : C([1, b], \mathbb{R}^m) \times C([1, b], \mathbb{R}^m) \times \Omega \to C([1, b], \mathbb{R}^m) \times C([1, b], \mathbb{R}^m)$ defined by

$$(x(\cdot, \omega), y(\cdot, \omega), \omega) \mapsto (N_1(x(t, \omega), y(t, \omega), \omega), N_2(x(t, \omega), y(t, \omega), \omega)),$$

where

$$N_1(x(t, \omega), y(t, \omega), \omega) = x_0(\omega) + \frac{1}{\Gamma(\alpha)} \int_1^t \left(\ln \frac{t}{s}\right)^{\alpha-1} f(s, x(s, \omega), y(s, \omega), \omega)\frac{ds}{s}$$

and

$$N_2(x(t, \omega), y(t, \omega), \omega) = y_0(\omega) + \frac{1}{\Gamma(\beta)} \int_1^t \left(\ln \frac{t}{s}\right)^{\beta-1} g(s, x(s, \omega), y(s, \omega), \omega)\frac{ds}{s}.$$

First we show that N is a random operator on $C([1, b], \mathbb{R}^m) \times C([1, b], \mathbb{R}^m)$. Since f and g are Carathéodory functions, $\omega \to f(t, x, y, \omega)$ and $\omega \to g(t, x, y, \omega)$ are measurable maps by Proposition 14.9, and we conclude that the maps

$$\omega \to N_1(x(t, \omega), y(t, \omega), \omega), \quad \omega \to N_2(x(t, \omega), y(t, \omega), \omega)$$

are measurable. As a result, N is a random operator on $C([1, b], \mathbb{R}^m) \times C([1, b], \mathbb{R}^m) \times \Omega$ into $C([1, b], \mathbb{R}^m) \times C([1, b], \mathbb{R}^m)$.

We next show that N satisfies all the conditions of Theorem 9.22 on $C([1, b], \mathbb{R}^m) \times C([1, b], \mathbb{R}^m)$. Let $(x(\cdot, \omega), y(\cdot, \omega)), (\widetilde{x}(\cdot, \omega), \widetilde{y}(\cdot, \omega)) \in C([1, b], \mathbb{R}^m) \times C([1, b], \mathbb{R}^m)$; then

$$\|N_1(t, x(t), y(t), \omega) - N_1(t, \widetilde{x}(t), \widetilde{y}(t), \omega)\|$$

$$= \left\|\frac{1}{\Gamma(\alpha)} \int_1^t \left(\ln \frac{t}{s}\right)^{\alpha-1} (f(s, x(s, \omega), y(s, \omega), \omega) - f(s, \widetilde{x}(s, \omega), \widetilde{y}(s, \omega), \omega))\frac{ds}{s}\right\|$$

$$\leq \frac{1}{\Gamma(\alpha)} \int_1^t \left(\ln \frac{t}{s}\right)^{\alpha-1} \|f(s, x(s, \omega), y(s, \omega), \omega) - f(s, \widetilde{x}(s, \omega), \widetilde{y}(s, \omega), \omega)\|\frac{ds}{s}$$

$$\leq \frac{1}{\Gamma(\alpha)} \int_1^t \left(\ln \frac{t}{s}\right)^{\alpha-1} p_1(\omega)\|x(s, \omega) - \widetilde{x}(s, \omega)\|\frac{ds}{s}$$

$$+ \frac{1}{\Gamma(\alpha)} \int_1^t \left(\ln \frac{t}{s}\right)^{\alpha-1} p_2(\omega)\|y(s, \omega) - \widetilde{y}(s, \omega)\|\frac{ds}{s}$$

$$\leq \frac{p_1(\omega)}{\Gamma(\alpha)} \int_1^t \left(\ln \frac{t}{s}\right)^{\alpha-1} \frac{ds}{s} \|x(\cdot,\omega) - \widetilde{x}(\cdot,\omega)\|_\infty$$

$$+ \frac{p_2(\omega)}{\Gamma(\alpha)} \int_1^t \left(\ln \frac{t}{s}\right)^{\alpha-1} \frac{ds}{s} \|y(\cdot,\omega) - \widetilde{y}(\cdot,\omega)\|_\infty$$

$$= \frac{p_1(\omega)}{\Gamma(\alpha)} \int_0^{\ln t} (\ln t - s)^{\alpha-1} ds \|x(\cdot,\omega) - \widetilde{x}(\cdot,\omega)\|_\infty$$

$$+ \frac{p_2(\omega)}{\Gamma(\alpha)} \int_0^{\ln t} (\ln t - s)^{\alpha-1} ds \|y(.,\omega) - \widetilde{y}(.,\omega)\|_\infty$$

$$\leq \frac{p_1(\omega)(\ln t)^\alpha}{\Gamma(\alpha+1)} \|x(\cdot,\omega) - \widetilde{x}(\cdot,\omega)\|_\infty + \frac{p_2(\omega)(\ln t)^\alpha}{\Gamma(\alpha+1)} \|y(\cdot,\omega) - \widetilde{y}(\cdot,\omega)\|_\infty.$$

Consequently,

$$\|N_1(x(\cdot,\omega), y(\cdot,\omega), \omega) - N_1(\widetilde{x}(\cdot,\omega), \widetilde{y}(\cdot,\omega), \omega))\|_\infty$$
$$\leq \|x(\cdot,\omega) - \widetilde{x}(\cdot,\omega)\|_\infty p_1(\omega) \tfrac{(\ln b)^\alpha}{\Gamma(\alpha+1)} + \|y(\cdot,\omega) - \widetilde{y}(\cdot,\omega)\|_\infty p_2(\omega) \tfrac{(\ln b)^\alpha}{\Gamma(\alpha+1)}.$$

Similarly, we obtain

$$\|N_2(x(\cdot,\omega), y(\cdot,\omega), \omega) - N_2(\widetilde{x}(\cdot,\omega), \widetilde{y}(\cdot,\omega), \omega)\|_\infty$$
$$\leq \|x(\cdot,\omega) - \widetilde{x}(\cdot,\omega)\|_\infty p_3(\omega) \tfrac{(\ln b)^\beta}{\Gamma(\beta+1)} + \|y(\cdot,\omega) - \widetilde{y}(\cdot,\omega)\|_\infty p_4(\omega) \tfrac{(\ln b)^\beta}{\Gamma(\beta+1)}.$$

Hence,

$$d(N(x(\cdot,\omega), y(\cdot,\omega), \omega), N(\widetilde{x}(\cdot,\omega), \widetilde{y}(\cdot,\omega), \omega))$$
$$\leq \widetilde{M}(\omega) d((x(\cdot,\omega), y(\cdot,\omega)), (\widetilde{x}(\cdot,\omega), \widetilde{y}(\cdot,\omega))),$$

where

$$d(x(\cdot,\omega), y(\cdot,\omega)) = \begin{pmatrix} \|x(\cdot,\omega) - y(\cdot,\omega)\|_\infty \\ \|x(\cdot,\omega) - y(\cdot,\omega)\|_\infty \end{pmatrix}.$$

Since for every $\omega \in \Omega$, $\widetilde{M}(\omega) \in \mathcal{M}_{2\times 2}(\mathbb{R}_+)$ converges to zero, then from Theorem 9.22, there exists a unique random fixed point of N which in turn is a solution of problem (14.2). $\square$

We recall Gronwall's lemma for singular kernels, whose proof can be found in [204, Lemma 7.1.1].

Lemma 14.15. *Let $v, a, \bar{a} : [1, b] \to [0, \infty)$ be continuous functions. If, for any $t \in [1, b]$,*

$$v(t) \leq a(t) + \int_1^t \left(\log \frac{t}{s}\right)^{\alpha-1} \frac{v(s)}{s} ds,$$

then there exists a constant $K = K(\beta)$ such that

$$v(t) \leq a(t) + \bar{a}(t) \int_1^t \left[\sum_{k=1}^\infty \frac{(\bar{a}(t)\Gamma(\alpha))^k}{\Gamma(k\alpha)} \left(\log \frac{t}{s}\right)^{\alpha-1} a(s)\right] \frac{ds}{s},$$

for every $t \in [1, b]$.

Next, we present an existence result that does not assume Lipschitz conditions. We need the following conditions:

(JH_2) For every $\omega \in \Omega$, the functions $f(\cdot, \cdot, \cdot, \omega)$ and $g(\cdot, \cdot, \cdot, \omega)$ are continuous, and $\omega \to f(\cdot, \cdot, \cdot, \omega)$ and $\omega \to g(\cdot, \cdot, \cdot, \omega)$ are measurable.

(JH_3) There exist measurable and bounded functions $\gamma_1, \gamma_2 : \Omega \to \mathbb{R}_+$ such that

$$\|f(t, x, y, \omega)\| \leq \gamma_1(\omega)(\|x\| + \|y\|), \quad \|g(t, x, y, \omega)\| \leq \gamma_2(\omega)(\|x\| + \|y\|),$$

for all $t \in [1, b]$, $\omega \in \Omega$ and $x, y \in \mathbb{R}^m$.

We prove an existence result for problem (14.2) by using a Leray-Schauder type random fixed point theorem in generalized Banach spaces.

Theorem 14.16. *Assume that (JH_1) and (JH_2) hold. Then the problem (14.2) has a random solution defined on $[1, b]$. Moreover, the solution set*

$$\begin{aligned} S(x_0, y_0) \;=\; & \{(x, y) : \Omega \to C([1, b], \mathbb{R}^m) \times C([1, b], \mathbb{R}^m) : (x(\cdot, \omega), y(\cdot, \omega)), \\ & \omega \in \Omega \text{ is a solution of (14.2)}\} \end{aligned}$$

is compact (i.e. for every $(x_n, y_n)_{n \in \mathbb{N}} \subset S(x_0, y_0)$ there exists a subsequence of $(x_n, y_n)_{n \in \mathbb{N}}$ converging to some element $(x, y) \in S(x_0, y_0)$).

Proof. Let $N : C([1, b], \mathbb{R}^m) \times C([1, b], \mathbb{R}^m) \times \Omega \to C([1, b], \mathbb{R}^m) \times C([1, b], \mathbb{R}^m)$ be the random operator defined in the proof of Theorem 13.5. In order to apply Theorem 9.27, we first show that N is completely continuous. The proof will be given in several steps.

Step 1 $N(\cdot, \cdot, \omega) = (N_1(\cdot, \cdot, \omega), N_2(\cdot, \cdot, \omega))$ *is continuous.* Let (x_n, y_n) be a sequence such that $(x_n, y_n) \to (x, y) \in C([1, b], \mathbb{R}^m) \times C([1, b], \mathbb{R}^m)$ as $n \to \infty$. Then

$$\begin{aligned} & \|N_1(x_n(\cdot, \omega), y_n(\cdot, \omega), \omega) - N_1(x(\cdot, \omega), y(\cdot, \omega), \omega)\|_\infty \\ & \leq \tfrac{(\ln b)^\alpha}{\Gamma(\alpha+1)} \|f(\cdot, x_n(\cdot, \omega), y_n(\cdot, \omega), \omega) - f(\cdot, x(\cdot, \omega), y(\cdot, \omega), \omega)\|_\infty. \end{aligned}$$

Since f is continuous,

$$\|N_1(x_n(\cdot, \omega), y_n(\cdot, \omega), \omega) - N_1(x(\cdot, \omega), y(\cdot, \omega), \omega)\|_\infty \to 0 \text{ as } n \to \infty.$$

Similarly,

$$\begin{aligned} & \|N_2(x_n(\cdot, \omega), y_n(\cdot, \omega), \omega) - N_2(x(\cdot, \omega), y(\cdot, \omega), \omega)\|_\infty \\ & \leq \tfrac{(\ln b)^\beta}{\Gamma(\beta+1)} \|g(\cdot, x_n(\cdot, \omega), y_n(\cdot, \omega), \omega) - g(\cdot, x(\cdot, \omega), y(\cdot, \omega), \omega)\|_\infty. \end{aligned}$$

Therefore,

$$\|N_2(x_n(\cdot, \omega), y_n(\cdot, \omega), \omega) - N_2(x(\cdot, \omega), y(\cdot, \omega), \omega)\|_\infty \to 0 \text{ as } n \to \infty,$$

so N is continuous.

Step 2. *N maps bounded sets into bounded sets in $C([1, b], \mathbb{R}^m) \times C([1, b], \mathbb{R}^m)$.* It will suffice to show that for any $q > 0$ there exists a positive constant l such that for each $(x, y) \in B_q = \{(x, y) \in C([1, b], \mathbb{R}) \times C([1, b], \mathbb{R}) : \|x\|_\infty \leq q, \ \|y\|_\infty \leq q\}$, we have

$$\|N(x, y, \omega)\|_\infty \leq l = (l_1, l_2).$$

For each $t \in [1, b]$, we have

$$\begin{aligned} & \|N_1(x(t, \omega), y(t, \omega), \omega)\| \\ & = \left\| x_0(\omega) + \tfrac{1}{\Gamma(\alpha)} \int_1^t (\ln \tfrac{t}{s})^{\alpha-1} f(s, x(s, \omega), y(s, \omega), \omega) \tfrac{ds}{s} \right\| \end{aligned}$$

$$\leq \|x_0(\omega)\| + \frac{\gamma_1(\omega)}{\Gamma(\alpha)} \int_1^b \|f(s, x(s, \omega), y(s, \omega), \omega)\| \frac{ds}{s}.$$

From (JH_2), we see that

$$\|N_1(x(\cdot, \omega), y(\cdot, \omega), \omega)\|_\infty \quad \leq \quad \|x_0(\omega)\| + \frac{2(\ln b)^\alpha q}{\Gamma(\alpha+1)} \gamma_1(\omega) := l_1.$$

Similarly, we have

$$\|N_2(x(\cdot, \omega), y(\cdot, \omega), \omega)\|_\infty \leq \|y_0(\omega)\| + \frac{2(\ln b)^\beta q}{\Gamma(\beta+1)} \gamma_2(\omega) := l_2.$$

Step 3. *N maps bounded sets into equicontinuous sets of $C([1, b], \mathbb{R}) \times C([1, b], \mathbb{R}^m)$.*
Let B_q be a bounded set in $C([1, b], \mathbb{R}^m) \times C([1, b], \mathbb{R}^m)$ as in Step 2. Let $r_1, r_2 \in J$ with $r_1 < r_2$ and $u \in B_q$. Then we have

$$\|N_1(x(r_2, \omega), y(r_2, \omega), \omega) - N_1(x(r_1, \omega), y(r_1, \omega), \omega)\|$$
$$\leq \frac{2q\gamma_1(\omega)}{\Gamma(\alpha)} \left[\int_{r_1}^{r_2} \left(\ln \frac{r_2}{s}\right)^{\alpha-1} \frac{ds}{s} + \int_1^{r_1} \left(\ln \frac{r_2}{s}\right)^{\alpha-1} - \left(\ln \frac{r_2}{s}\right)^{\alpha-1} \frac{ds}{s} \right].$$

Hence,

$$\|N_1(x(r_2, \omega), y(r_2, \omega), \omega) - N_1(x(r_1, \omega), y(r_1, \omega), \omega)\|$$
$$\leq \frac{2q\gamma_1(\omega)}{\Gamma(\alpha+1)} (\ln r_2 - \ln r_1)^\alpha + \frac{2q\gamma_1(\omega)}{\Gamma(\alpha+1)} \left[(\ln r_2)^\alpha - (\ln r_1)^\alpha \right],$$

and

$$\|N_2(x(r_2, \omega), y(r_2, \omega), \omega) - N_2(x(r_1, \omega), y(r_1, \omega), \omega)\|$$
$$\leq \frac{2q\gamma_1(\omega)}{\Gamma(\beta+1)} (\ln r_2 - \ln r_1)^\beta + \frac{2q\gamma_1(\omega)}{\Gamma(\beta+1)} \left[(\ln r_2)^\beta - (\ln r_1)^\beta \right].$$

The right-hand terms tends to zero as $|r_2 - r_1| \to 0$. As a consequence of Steps 1 to 3, together with the Arzelá-Ascoli theorem, we conclude that N maps B_q into a precompact set in $C([1, b], \mathbb{R}) \times C([1, b], \mathbb{R})$.
Step 4. It remains to show that

$$\mathcal{A}(\omega) = \{(x(\cdot, \omega), y(\cdot, \omega)) \in C([1, b], \mathbb{R}^m) \times C([1, b], \mathbb{R}^m) :$$
$$(x(\cdot, \omega), y(\cdot, \omega)) = \lambda(\omega) N(x(\cdot, \omega), y(\cdot, \omega), \omega), \lambda(\omega) \in (0, 1)\}$$

is bounded.
Let $(x, y) \in \mathcal{A}$. Then $x = \lambda(\omega) N_1(x, y)$ and $y = \lambda(\omega) N_2(x, y)$ for some $0 < \lambda < 1$. Thus, for $t \in [1, b]$, we have

$$\|x(t, \omega)\| \leq \|x_0(\omega)\| + \frac{1}{\Gamma(\alpha)} \int_1^t \left(\ln \frac{t}{s}\right)^{\alpha-1} \|f(s, x(s, \omega), y(s, \omega), \omega)\| \frac{ds}{s}$$
$$\leq \|x_0(\omega)\| + \frac{1}{\Gamma(\alpha)} \int_0^t \gamma_1(\omega) \left(\ln \frac{t}{s}\right)^{\alpha-1} (\|x(s, \omega)\| + \|y(s, \omega)\|) \frac{ds}{s}.$$

Hence,

$$\|x(t, \omega)\| \leq \|x_0(\omega)\| + \frac{\gamma_1(\omega)}{\Gamma(\alpha)} \int_1^t \left(\ln \frac{t}{s}\right)^{\alpha-1} (\|x(s, \omega)\| + \|y(s, \omega)\|) \frac{ds}{s}$$

and

$$\|y(t,\omega)\| \le \|y_0(\omega)\| + \frac{\gamma_2(\omega)}{\Gamma(\beta)} \int_0^t \left(\ln \frac{t}{s}\right)^{\beta-1} (\|x(s,\omega)\| + \|y(s,\omega)\|) \frac{ds}{s}.$$

Therefore,

$$\|x(t,\omega)\| + \|y(t,\omega)\| \le c + c \int_1^t \left(\ln \frac{t}{s}\right)^{\gamma-1} (\|x(s,\omega)\| + \|y(s,\omega)\|) \frac{ds}{s},$$

where

$$c = \|x_0(\omega)\| + \|y_0(\omega)\| + \frac{\gamma_1(\omega)}{\Gamma(\alpha)} + \frac{\gamma_2(\omega)}{\Gamma(\beta)}, \quad \gamma = \min(\alpha, \beta).$$

By Lemma 14.15, we have

$$\begin{aligned}
\|x(t,\omega)\| + \|y(t,\omega)\| &\le c + c \int_1^t \left[\sum_{k=1}^\infty \frac{(c\Gamma(\gamma))^k}{\Gamma(k\gamma)} \left(\ln \frac{t}{s}\right)^{k\gamma-1}\right] \frac{ds}{s} \\
&\le c + c \sum_{k=1}^\infty \frac{(c\Gamma(\gamma))^k}{\Gamma(k\gamma+1)} (\ln t)^{k\gamma} \\
&\le c \left[1 + \sum_{k=1}^\infty \frac{(c\Gamma(\gamma)(\ln t)^\gamma)^k}{\Gamma(k\gamma+1)}\right] \\
&\le c \left[1 + \sum_{k=1}^\infty \frac{(c\Gamma(\gamma)(\ln b)^\gamma)^k}{\Gamma(k\gamma+1)}\right] \\
&= c E_\gamma (c\Gamma(\gamma)(\ln b)^\gamma).
\end{aligned}$$

Hence,

$$\|x(\cdot,\omega)\|_\infty + \|y(\cdot,\omega)\|_\infty \le c E_\gamma (c\Gamma(\gamma)(\ln b)^\gamma) := K_*,$$

and so

$$\|x\|_\infty \le K_* \quad \text{and} \quad \|y\|_\infty \le K_*.$$

This shows that $\mathcal{A}$ is bounded.

As a consequence of Theorem 9.27 we conclude that N has a random fixed point $\omega \to (x(\cdot,\omega), y(\cdot,\omega))$ that is a solution to (14.2).

Step 5: *Compactness of the solution set.* Let $\{(x_n, y_n)\}_{n\in\mathbb{N}} \subset S(x_0, y_0)$ be a sequence. For every $n \in \mathbb{N}$ and for fixed $\omega \in \Omega$, we have

$$x_n(t,\omega) = x_0(\omega) + \frac{1}{\Gamma(\alpha)} \int_1^t \left(\ln \frac{t}{s}\right)^{\alpha-1} f(s, x_n(s,\omega), y_n(s,\omega), \omega) \frac{ds}{s}$$

and

$$y_n(t,\omega) = y_0(\omega) + \frac{1}{\Gamma(\beta)} \int_1^t (\ln \frac{t}{s})^{\beta-1} g(s, x_n(s,\omega), y_n(s,\omega), \omega) \frac{ds}{s}.$$

As in Steps 3 and 4, we can prove that the subsequence $\{(x_{n_k}, y_{n_k})\}_{k\in\mathbb{N}}$ of $\{(x_n, y_n)\}_{n\in\mathbb{N}}$ converges to some $(x(\cdot,\omega), y(\cdot,\omega)) \in C([1,b], \mathbb{R}^m) \times C([1,b], \mathbb{R}^m)$ such that

$$\omega \to x(t,\omega) \quad \text{and} \quad \omega \to y(t,\omega)$$

are measurable functions. Since $f(\cdot, \cdot, \cdot, \omega)$ and $g(\cdot, \cdot, \cdot, \omega)$ are continuous, for $t \in [1, b]$ we have

$$x(t, \omega) = x_0(\omega) + \frac{1}{\Gamma(\alpha)} \int_1^t \left(\ln \frac{t}{s} \right)^{\alpha-1} f(s, x(s, \omega), y(s, \omega), \omega) \frac{ds}{s},$$

and

$$y(t, \omega) = y_0(\omega) + \frac{1}{\Gamma(\beta)} \int_1^t (\ln \frac{t}{s})^{\beta-1} g(s, x(s, \omega), y(s, \omega), \omega) \frac{ds}{s}.$$

Thus, $S(x_0, y_0)$ is compact.   □

14.4 M^2-solutions

Our objective in this section is to apply the new concept of M^2- solutions to problem (14.2). To do this, we need some preliminary results that will be used throughout this section.

Let $(\Omega, \mathcal{F}, \mathbb{P})$ be a complete probability space with a filtration $(\mathcal{F} = \mathcal{F}_t)_{t \geq 0}$ satisfying the usual conditions (i.e., right continuous and $\mathcal{F}_0$ containing all $\mathbb{P}$-null sets). For a stochastic process $x : [1, b] \times \Omega \to \mathbb{R}^m$ we will write $x(t)$ (or simply x when no confusion is possible) instead of $x_t(\omega) = x(t, \omega)$. We say $x(\cdot, \cdot)$ is jointly measurable if the map $(t, \omega) \to x(t, \omega)$ is measurable as a map $B([1, b] \bigotimes \mathcal{F} \to \mathcal{B}(\mathbb{R}^m)$. For $\omega \in \Omega$, the path $t \to x(t, \omega)$ is called left-continuous if for each $t \in [1, b]$

$$x_s(\omega) \to x_t(\omega) \quad s \uparrow t.$$

A process $x(t, \omega)$ is stochastically continuous at a point $s \in [a, b]$ if for each $\epsilon > 0$

$$\lim_{t \to s} \mathbb{P} \{ \omega \in \Omega : \|x(t, \omega) - x(s, \omega)\| > \epsilon \} = 0.$$

Theorem 14.17. *([235]) If x is a stochastic process with state space $\mathbb{R}^m$ and all the paths of x are left-continuous (or right-continuous), then x is jointly measurable.*

If x and y are stochastic processes, we say that x is a modification of y if for each $t \in [1, b]$,

$$\mathbb{P}(\{\omega \in \Omega : x_t(\omega) = y_t(\omega)\}) = 1.$$

Theorem 14.18. *(Kolmogorov continuity theorem) [235] Suppose that $(\Omega, \mathcal{F}, \mathbb{P}, (x_t)_{t \geq 0})$, is a stochastic process with state space $\mathbb{R}^m$. If there are $\bar{\alpha}$, $\bar{\beta}$, $\sigma > 0$ such that*

$$\mathbb{E}\|x_t - x_s\|^{\bar{\alpha}} \leq \sigma|t - s|^{1+\bar{\beta}}, \quad t, s \in \mathbb{R}_+,$$

then the stochastic process has a continuous modification.

We introduce the notations:

Denote by $L^p(\Omega, \mathcal{F}, \mathbb{P}, \mathbb{R}^m)$, $p > 0$, the linear space of random variables (equivalence classes) $x : \Omega \to \mathbb{R}^m$ such that

$$\mathbb{E}\|x\|^p < \infty.$$

Let $M^p(1,b)$ be the space of (equivalence classes of) progressively measurable processes $x : [1,b] \times \Omega \to \mathbb{R}^m$ such that

$$\int_1^b \|x_t\|^2 dt < \infty, \quad \mathbb{P}, \text{ p.s } \omega \in \Omega, \text{ if } p = 0$$

and

$$\mathbb{E}\left(\int_1^b \|x_t\|^2 dt\right)^{\frac{p}{2}} < \infty, \quad \text{if } p > 0.$$

Note that the property of progressive measurability is independent of the choice of an element in an equivalence class x, and that for every $p \geq 0$,

$$M^p(1,b) \subset L^p(\Omega, \mathcal{F}, \mathbb{P}, L^p(1,b,\mathbb{R}^m)),$$

as a closed linear subspace. Hence, for each $p \in [1,\infty)$, the space $M^2(1,b)$ is a Banach space with respect to the norm

$$\|x\|_{M^p} = \left(\mathbb{E}\left(\int_1^b \|x_t\|^2 dt\right)^{\frac{p}{2}}\right)^{\frac{1}{p}}.$$

Moreover the space $M^2(1,b)$ is a Hilbert space.

Definition 14.19. *A pair x, $y \in M^2$ is called an M^2−solution of problem (14.2) if for $\mathbb{P}$, p.s $\omega \in \Omega$, and $t \in [1,b]$,*

$$x(t,\omega) = x_0(\omega) + \frac{1}{\Gamma(\alpha)} \int_1^t \left(\ln \frac{t}{s}\right)^{\alpha-1} f(s, x(s,\omega), y(s,\omega), \omega) \frac{ds}{s},$$

and

$$y(t,\omega) = y_0(\omega) + \frac{1}{\Gamma(\beta)} \int_1^t \left(\ln \frac{t}{s}\right)^{\beta-1} g(s, x(s,\omega), y(s,\omega), \omega) \frac{ds}{s}.$$

14.4.1 Existence and uniqueness of M^2−solutions

We wish to investigate the existence, uniqueness, modification continuity and stochastic continuity of M^2−solutions. Let us now introduce the following conditions that will be basic tools in the treatment of M^2−solutions.

$(\mathcal{JH}_4)$ Let f, $g : [1,b] \times \mathbb{R}^m \times \mathbb{R}^m \times \Omega \to \mathbb{R}^m$ be two functions such that $\omega \to f(\cdot,\cdot,\cdot,\omega)$, $g(\cdot,\cdot,\cdot,\omega)$ are measurable, $t \to f(t,\cdot,\cdot,\cdot)$, $g(t,\cdot,\cdot,\cdot)$ are continuous,

$$\|f(\cdot,0,0,\cdot)\|_{M^2} < \infty, \quad \text{and} \quad \|g(\cdot,0,0,\cdot)\|_{M^2} < \infty.$$

$(\mathcal{JH}_5)$ There exist positive real numbers c_1, c_2, c_3, c_4 such that for all x, $y,\widetilde{x}$, $\widetilde{y} \in \mathbb{R}^m$ and $\omega \in \Omega$,

$$\|f(t,x,y,\omega) - f(t,\widetilde{x},\widetilde{y},\omega)\| \leq c_1\|x - \widetilde{x}\| + c_2\|y - \widetilde{y}\|$$

and

$$\|g(t,x,y,\omega) - g(t,\widetilde{x},\widetilde{y},\omega)\| \leq c_3\|x - \widetilde{x}\| + c_4\|y - \widetilde{y}\|.$$

In this section we assume that $\alpha, \beta \in (\frac{1}{2}, 1)$, $\mathbb{E}\|x_0\|^2 < \infty$, and $\mathbb{E}\|y_0\|^2 < \infty$. We are now in a position to present the first result in this section.

Theorem 14.20. *Assume that the conditions $(\mathcal{JH}_4)$ and $(\mathcal{JH}_5)$ hold and the matrix $\widetilde{M}_* \in \mathcal{M}_{2\times 2}(\mathbb{R}_+)$ defined by*

$$\widetilde{M}_* = \begin{pmatrix} \dfrac{(\ln b)^\alpha \sqrt{2b}c_1}{\sqrt{2\alpha-1}\Gamma(\alpha)} & \dfrac{(\ln b)^\alpha \sqrt{2b}c_2}{\sqrt{2\alpha-1}\Gamma(\alpha)} \\ \dfrac{(\ln b)^\beta \sqrt{2b}c_3}{\sqrt{2\beta-1}\Gamma(\beta)} & \dfrac{(\ln b)^\beta \sqrt{2b}c_4}{\sqrt{2\beta-1}\Gamma(\beta)} \end{pmatrix}$$

converges to zero. Then problem (14.2) has unique solution in $M^2(1,b)$.

Proof. Consider the operator $N : M^2(1,b) \times M^2(1,b) \to M^2(1,b) \times M^2(1,b)$, defined by

$$(x(\cdot,\omega), y(\cdot,\omega)) \mapsto (N_1(x(t,\omega), y(t,\omega)), N_2(t, x(t,\omega), y(t,\omega)))$$

where

$$N_1(x(s,\omega), y(s,\omega)) \;=\; x_0(\omega) + \frac{1}{\Gamma(\alpha)} \int_1^t \left(\ln \frac{t}{s}\right)^{\alpha-1} f(s, x(s,\omega), y(s,\omega)) \frac{ds}{s}$$

and

$$N_2(x(t,\omega), y(t,\omega)) \;=\; y_0(\omega) + \frac{1}{\Gamma(\beta)} \int_1^t \left(\ln \frac{t}{s}\right)^{\beta-1} g(s, x(s,\omega), y(s,\omega)) \frac{ds}{s}.$$

First we show that N is a random operator on $M^2(1,b) \times M^2(1,b)$. Since f and g are Carathéodory functions, by Proposition 14.9, $\omega \to f(t, x, y, \omega)$ and $\omega \to g(t, x, y, \omega)$ are measurable maps, and so the maps

$$\omega \to N_1(x(t,\omega), y(t,\omega)), \quad \omega \to N_2(x(t,\omega), y(t,\omega))$$

are measurable. Also,

$$\int_1^b \|N_1(x(s,\omega), y(s,\omega))\|^2 ds$$
$$= \int_1^b \left[\left\|x_0(\omega) + \frac{1}{\Gamma(\alpha)} \int_0^t \left(\ln \frac{t}{s}\right)^{\alpha-1} f(s, x(s,\omega), y(s,\omega), \omega) \frac{ds}{s}\right\|^2\right] dt$$
$$\leq \frac{2}{\Gamma^2(\alpha)} \int_1^b \left[\left\|\int_0^t \left(\ln \frac{t}{s}\right)^{\alpha-1} f(s, x(s,\omega), y(s,\omega), \omega) \frac{ds}{s}\right\|^2\right] dt + 2 \int_1^b \|x_0(\omega)\|^2 dt$$
$$\leq 2(b-1)\|x_0(\omega)\|^2 + \frac{b(\ln b)^{2\alpha}}{(2\alpha-1)\Gamma^2(\alpha)} \int_1^b \|f(s, x(s,\omega), y(s,\omega), \omega)\|^2 ds$$
$$\leq 2(b-1)\|x_0(\omega)\|^2 + \frac{2b(\ln b)^{2\alpha}}{(2\alpha-1)\Gamma^2(\alpha)} \int_1^b \|f(s, 0, 0; \omega)\|^2 ds$$
$$+ \frac{2b(\ln b)^{2\alpha}}{(2\alpha-1)\Gamma^2(\alpha)} \int_1^b [c_1^2\|x(s,\omega)\|^2 + c_2^2\|y(s,\omega)\|^2] ds.$$

Therefore,

$$\int_1^b \|N_1(x(s,\omega), y(s,\omega))\|^2 ds$$
$$\leq 2(b-1)\|x_0(\omega)\|^2 + \frac{2b(\ln b)^{2\alpha}}{(2\alpha-1)\Gamma^2(\alpha)} \int_1^b \|f(s, 0, 0; \omega)\|^2 ds$$
$$+ \frac{2b(\ln b)^{2\alpha}}{(2\alpha-1)\Gamma^2(\alpha)} \int_1^b [c_1^2\|x(s,\omega)\|^2 + c_2^2\|y(s,\omega)\|^2] ds < \infty.$$

Consequently,

$$\mathbb{E} \int_1^b \|N_1(x(s,\omega), y(s,\omega))\|^2 ds$$

$$\leq 2(b-1)\mathbb{E}\|x_0(\omega)\|^2 + \frac{2b(\ln b)^{2\alpha}}{(2\alpha-1)\Gamma^2(\alpha)}\mathbb{E}\int_1^b \|f(s,0,0;\omega)\|^2 ds$$

$$+ \frac{2b(\ln b)^{2\alpha}}{(2\alpha-1)\Gamma^2(\alpha)}\mathbb{E}\int_1^b [c_1^2\|x(s,\omega)\|^2 + c_2^2\|y(s,\omega)\|^2]ds < \infty.$$

Similarly, we obtain

$$\int_1^b \|N_2(x(s,\omega),y(s,\omega))\|^2 ds$$

$$\leq 2(b-1)\|y_0(\omega)\|^2 + \frac{2b(\ln b)^{2\beta}}{(2\beta-1)\Gamma^2(\beta)}\int_1^b \|g(s,0,0,\omega)\|^2 ds$$

$$+ \frac{2b(\ln b)^{2\beta}}{(2\beta-1)\Gamma^2(\beta)}\int_1^b [c_3^2\|x(s,\omega)\|^2 + c_4^2\|y(s,\omega)\|^2]ds < \infty,$$

and

$$\mathbb{E}\int_1^b \|N_2(x(s,\omega),y(s,\omega))\|^2 ds$$

$$\leq 2(b-1)\mathbb{E}\|y_0(\omega)\|^2 + \frac{2b(\ln b)^{2\beta}}{(2\beta-1)\Gamma^2(\beta)}\mathbb{E}\int_1^b \|g(s,0,0,\omega)\|^2 ds$$

$$+ \frac{2b(\ln b)^{2\beta}}{(2\beta-1)\Gamma^2(\beta)}\mathbb{E}\int_1^b [c_3^2\|x(s,\omega)\|^2 + c_4^2\|y(s,\omega)\|^2]ds < \infty.$$

Therefore, N is a random operator on $M^2(1,b) \times M^2(1,b)$ into $M^2(1,b) \times M^2(1,b)$.

To show that N satisfies all the conditions of Theorem 8.1 on $M^2(1,b) \times M^2(1,b)$, let $(x(\cdot,\omega),y(\cdot,\omega)), (\widetilde{x}(\cdot,\omega),\widetilde{y}(\cdot,\omega)) \in M^2(1,b) \times M^2(1,b)$; then

$$\|N_1(t,x(t),y(t,\omega)) - N_1(t,\widetilde{x}(t),\widetilde{y}(t,\omega))\|^2$$

$$= \left\|\frac{1}{\Gamma(\alpha)}\int_1^t \left(\ln\frac{t}{s}\right)^{\alpha-1}\left[f(s,x(s,\omega),y(s,\omega)) - f(s,\widetilde{x}(s,\omega),\widetilde{y}(s,\omega))\right]\frac{ds}{s}\right\|^2$$

$$\leq \frac{2c_1^2}{\Gamma^2(\alpha)}\int_1^t \left(\ln\frac{t}{s}\right)^{2\alpha-2}\frac{ds}{s^2}ds \int_1^t \|x(s,\omega) - \widetilde{x}(s,\omega)\|^2 ds$$

$$+ \frac{2c_2^2}{\Gamma^2(\alpha)}\int_1^t \left(\ln\frac{t}{s}\right)^{2\alpha-2}\frac{ds}{s^2}\int_1^t \|y(s,\omega) - \widetilde{y}(s,\omega)\|^2 ds$$

$$= \frac{2c_1^2}{\Gamma^2(\alpha)}\int_0^{\ln t} (\ln t - s)^{2\alpha-2}\frac{ds}{e^s}\int_1^t \|x(s,\omega) - \widetilde{x}(s,\omega)\|^2 ds$$

$$+ \frac{2c_2^2}{\Gamma^2(\alpha)}\int_0^{\ln t} (\ln t - s)^{2\alpha-2}\frac{ds}{e^s}\int_1^t \|y(s,\omega) - \widetilde{y}(s,\omega)\|^2 ds$$

$$\leq \frac{2c_1^2(\ln t)^{2\alpha}}{(2\alpha-1)\Gamma^2(\alpha)}\int_1^t \|x(s,\omega) - \widetilde{x}(s,\omega)\|^2 ds$$

$$+ \frac{2c_2^2(\ln t)^{2\alpha}}{(2\alpha-1)\Gamma^2(\alpha)}\int_1^t \|y(s,\omega) - \widetilde{y}(s,\omega)\|^2 ds.$$

Hence,

$$\int_1^b \|N_1(x(s,\omega),y(s,\omega)) - N_1(\widetilde{x}(s,\omega),\widetilde{y}(s,\omega))\|^2 ds$$

$$\leq \frac{2bc_1^2(\ln b)^{2\alpha}}{(2\alpha-1)\Gamma^2(\alpha)}\int_1^b \|x(s,\omega) - \widetilde{x}(s,\omega)\|^2 ds$$

$$+ \frac{2bc_2^2(\ln b)^{2\alpha}}{(2\alpha-1)\Gamma^2(\alpha)}\int_1^b \|y(s,\omega) - \widetilde{y}(s,\omega)\|^2 ds,$$

and so

$$\mathbb{E}\left(\int_1^b \|N_1(x(s,\omega),y(s,\omega)) - N_1(\widetilde{x}(s,\omega),\widetilde{y}(s,\omega))\|^2 ds\right)$$

$$\leq \frac{2bc_1^2(\ln b)^{2\alpha}}{(2\alpha-1)\Gamma^2(\alpha)}\mathbb{E}\left(\int_1^b \|x(s,\omega)-\widetilde{x}(s,\omega)\|^2 ds\right)$$
$$+\frac{2bc_2^2(\ln b)^{2\alpha}}{(2\alpha-1)\Gamma^2(\alpha)}\mathbb{E}\left(\int_1^b \|y(s,\omega)-\widetilde{y}(s,\omega)\|^2 ds\right).$$

Similarly, we obtain

$$\mathbb{E}\left(\int_1^b \|N_1(x(s,\omega),y(s,\omega))-N_1(\widetilde{x}(s,\omega),\widetilde{y}(s,\omega))\|^2 ds\right)$$
$$\leq \frac{2bc_3^2(\ln b)^{2\beta}}{(2\beta-1)\Gamma^2(\alpha)}\mathbb{E}\left(\int_1^b \|x(s,\omega)-\widetilde{x}(s,\omega)\|^2 ds\right)$$
$$+\frac{2bc_4^2(\ln b)^{2\beta}}{(2\beta-1)\Gamma^2(\beta)}\mathbb{E}\left(\int_1^b \|y(s,\omega)-\widetilde{y}(s,\omega)\|^2 ds\right),$$

and so

$$\left(\mathbb{E}\left(\int_1^b \|N_1(x(s,\omega),y(s,\omega))-N_1(\widetilde{x}(s,\omega),\widetilde{y}(s,\omega))\|^2 ds\right)\right)^{\frac{1}{2}}$$
$$\leq \frac{\sqrt{2}bc_1(\ln b)^\alpha}{\sqrt{(2\alpha-1)}\Gamma(\alpha)}\left(\mathbb{E}\left(\int_1^b \|x(s,\omega)-\widetilde{x}(s,\omega)\|^2 ds\right)\right)^{\frac{1}{2}}$$
$$+\frac{\sqrt{2}bc_2(\ln b)^\alpha}{\sqrt{(2\alpha-1)}\Gamma(\alpha)}\left(\mathbb{E}\left(\int_1^b \|y(s,\omega)-\widetilde{y}(s,\omega)\|^2 ds\right)\right)^{\frac{1}{2}},$$

and

$$\left(\mathbb{E}\left(\int_1^b \|N_1(x(s,\omega),y(s,\omega))-N_1(\widetilde{x}(s,\omega),\widetilde{y}(s,\omega))\|^2 ds\right)\right)^{\frac{1}{2}}$$
$$\leq \frac{\sqrt{2}bc_3(\ln b)^\beta}{\sqrt{(2\beta-1)}\Gamma(\alpha)}\left(\mathbb{E}\left(\int_1^b \|x(s,\omega)-\widetilde{x}(s,\omega)\|^2 ds\right)\right)^{\frac{1}{2}}$$
$$+\frac{\sqrt{2}bc_4(\ln b)^\beta}{\sqrt{(2\beta-1)}\Gamma(\beta)}\left(\mathbb{E}\left(\int_1^b \|y(s,\omega)-\widetilde{y}(s,\omega)\|^2 ds\right)\right)^{\frac{1}{2}}.$$

Therefore,

$$d(N(x,y),N(\widetilde{x},\widetilde{y})\leq \widetilde{M}_* d((x,y),(\widetilde{x},\widetilde{y}),$$

where

$$d(x,y)=\left(\begin{array}{c}\|x-y\|_{M^2}\\ \|x-y\|_{M^2}\end{array}\right).$$

Since $\widetilde{M}_* \in \mathcal{M}_{n\times n}(\mathbb{R}_+)$ converges to zero, by Theorem 8.1 there exists a unique L^2–solution of problem (14.2). □

Theorem 14.21. *Assume that* $(\mathcal{JH}_4)$ *and* $(\mathcal{JH}_5)$ *hold. Then for each* x, $y \in M^2(1,b)$, *the processes*

$$z_t(\omega) = x_0(\omega)+\frac{1}{\Gamma(\alpha)}\int_1^t \left(\ln\frac{t}{s}\right)^{\alpha-1} f(s,x(s,\omega),y(s,\omega))\frac{ds}{s}, \quad \frac{1}{2}<\alpha<1,$$

and

$$\bar{z}_t(\omega) = y_0(\omega)+\frac{1}{\Gamma(\beta)}\int_1^t \left(\ln\frac{t}{s}\right)^{\beta-1} g(s,x(s,\omega),y(s,\omega))\frac{ds}{s}, \quad \frac{1}{2}<\beta<1,$$

have continuous modifications. Moreover, $t \to z_t, \bar{z}_t$ *are stochastically continuous.*

Proof. Let $x, y \in M^2(1, b)$ and $t, r \in [1, b]$ with $r < t$. By the Jensen and Hölder inequalities, we have

$$\sqrt{\|z_t(\omega) - z_r(\omega)\|}$$

$$\leq \sqrt{\frac{1}{\Gamma(\alpha)} \int_1^r \left[\left(\ln \frac{t}{s} \right)^{\alpha-1} - \left(\ln \frac{r}{s} \right)^{\alpha-1} \right] \|f(s, x(s, \omega), y(s, \omega), \omega)\| \frac{ds}{s}}$$

$$+ \sqrt{\frac{1}{\Gamma(\alpha)} \int_r^t \left(\ln \frac{t}{s} \right)^{\alpha-1} \|f(s, x(s, \omega), y(s, \omega), \omega)\| \frac{ds}{s}}$$

$$\leq \frac{1}{\Gamma(\alpha)} \int_1^r \sqrt{\frac{\left[\left(\ln \frac{t}{s} \right)^{\alpha-1} - \left(\ln \frac{r}{s} \right)^{\alpha-1} \right]}{s}} \|f(s, x(s, \omega), y(s, \omega), \omega)\| ds$$

$$+ \frac{1}{\Gamma(\alpha)} \int_r^t \sqrt{\left(\ln \frac{t}{s} \right)^{\alpha-1} \frac{\|f(s, x(s, \omega), y(s, \omega), \omega)\|}{s}} ds$$

$$\leq \frac{1}{\Gamma(\alpha)} \int_1^r \left[\left(\ln \frac{t}{s} \right)^{\alpha-1} - \left(\ln \frac{r}{s} \right)^{\alpha-1} \right] \frac{ds}{s} \int_1^r \|f(s, x(s, \omega), y(s, \omega))\| ds$$

$$+ \frac{1}{\Gamma(\alpha)} \int_r^t \left(\ln \frac{t}{s} \right)^{\alpha-1} \frac{ds}{s} \int_r^t \|f(s, x(s, \omega), y(s, \omega), \omega)\| ds$$

$$\leq \frac{2}{\Gamma(\alpha+1)} |(\ln t - \ln r)^\alpha| \int_1^b \|f(s, x(s, \omega), y(s, \omega))\| ds$$

$$+ \frac{1}{\Gamma(\alpha+1)} |(\ln t)^\alpha - (\ln r)^\alpha| \int_1^b \|f(s, x(s, \omega), y(s, \omega), \omega)\| ds$$

$$\leq \frac{2}{\Gamma(\alpha+1)} |(\ln t - \ln r)^\alpha| \int_1^b \left[c_1 \|x(s, \omega)\| + c_2 \|y(s, \omega)\| \right] ds$$

$$+ \frac{2}{\Gamma(\alpha+1)} |(\ln t - \ln r)^\alpha| \int_1^b \|f(s, 0, 0, \omega)\| ds$$

$$+ \frac{1}{\Gamma(\alpha+1)} |(\ln t)^\alpha - (\ln r)^\alpha| \int_1^b \left[c_1 \|x(s, \omega)\| + c_2 \|y(s, \omega)\| \right] ds$$

$$+ \frac{1}{\Gamma(\alpha+1)} |(\ln t)^\alpha - (\ln r)^\alpha| \int_1^b \|f(s, 0, 0, \omega)\| ds.$$

By Hölder's inequality, we obtain that

$$\sqrt{\|z_t(\omega) - z_r(\omega)\|}$$

$$\leq \frac{2b}{\Gamma(\alpha+1)} |(\ln t - \ln r)^\alpha| \int_1^b \left[c_1^2 \|x(s, \omega)\|^2 + c_2^2 \|y(s, \omega)\|^2 \right] ds$$

$$+ \frac{2b}{\Gamma(\alpha+1)} |(\ln t - \ln r)^\alpha| \int_1^b \|f(s, 0, 0, \omega)\|^2 ds$$

$$+ \frac{b}{\Gamma(\alpha+1)} |(\ln t)^\alpha - (\ln r)^\alpha| \int_1^b \left[c_1^2 \|x(s, \omega)\|^2 + c_2^2 \|y(s, \omega)\|^2 \right] ds$$

$$+ \frac{b}{\Gamma(\alpha+1)} |(\ln t)^\alpha - (\ln r)^\alpha| \int_1^b \|f(s, 0, 0, \omega)\|^2 ds$$

and

$$\sqrt{\|\bar{z}_t(\omega) - \bar{z}_r(\omega)\|}$$

$$\leq \frac{2b}{\Gamma(\beta+1)} \left| (\ln t - \ln r)^\beta \right| \int_1^b \left[c_3^2 \|x(s,\omega)\|^2 + c_4^2 \|y(s,\omega)\|^2 \right] ds$$

$$+ \frac{2b}{\Gamma(\beta+1)} \left| (\ln t - \ln r)^\beta \right| \int_1^b \|g(s,0,0,\omega)\|^2 ds$$

$$+ \frac{b}{\Gamma(\beta+1)} \left| (\ln t)^\beta - (\ln r)^\beta \right| \int_1^b \left[c_3^2 \|x(s,\omega)\|^2 + c_4^2 \|y(s,\omega)\|^2 \right] ds$$

$$+ \frac{b}{\Gamma(\beta+1)} \left| (\ln t)^\beta - (\ln r)^\beta \right| \int_1^b \|g(s,0,0,\omega)\|^2 ds.$$

Consequently

$$\mathbb{E}\sqrt{\|z_t(\omega) - z_r(\omega)\|} \leq \frac{2b}{\Gamma(\alpha+1)} |(\ln t - \ln r)^\alpha| \left[c_1^2 \|x\|_{M^2}^2 + c_2^2 \|y\|_{M^2}^2 \right]$$

$$+ \frac{2b}{\Gamma(\alpha+1)} |(\ln t - \ln r)^\alpha| \|f(\cdot,0,0,\cdot)\|_{M^2}^2$$

$$+ \frac{b}{\Gamma(\alpha+1)} |(\ln t)^\alpha - (\ln r)^\alpha| \left[c_1^2 \|x\|_{M^2}^2 + c_2^2 \|y\|_{M^2}^2 \right]$$

$$+ \frac{b}{\Gamma(\alpha+1)} |(\ln t)^\alpha - (\ln r)^\alpha| \|f(\cdot,0,0,\cdot)\|_{M^2}^2$$

and

$$\mathbb{E}\sqrt{\|\bar{z}_t(\omega) - \bar{z}_r(\omega)\|} \leq \frac{2b}{\Gamma(\beta+1)} \left| (\ln t - \ln r)^\beta \right| \left[c_3^2 \|x\|_{M^2}^2 + c_4^2 \|y\|_{M^2}^2 \right]$$

$$+ \frac{2b}{\Gamma(\beta+1)} \left| (\ln t - \ln r)^\beta \right| \|g(\cdot,0,0,\cdot)\|_{M^2}^2$$

$$+ \frac{b}{\Gamma(\beta+1)} \left| (\ln t)^\beta - (\ln r)^\beta \right| \left[c_3^2 \|x\|_{M^2}^2 + c_4^2 \|y\|_{M^2}^2 \right]$$

$$+ \frac{b}{\Gamma(\beta+1)} \left| (\ln t)^\beta - (\ln r)^\beta \right| \|g(\cdot,0,0,\cdot)\|_{M^2}^2.$$

Since $r < t$, this implies that

$$|(\ln t)^\alpha - (\ln r)^\alpha| = (\ln t)^\alpha - (\ln r)^\alpha$$

$$= \left(\frac{\ln t - \ln r}{2} + \frac{\ln t + \ln r}{2} \right)^\alpha - (\ln r)^\alpha$$

Using the fact that ξ^α is a convex function on $(0, \ln b]$, we obtain

$$(\ln t)^\alpha - (\ln r)^\alpha \leq \frac{1}{2}(\ln t - \ln r)^\alpha + \frac{1}{2}(\ln t + \ln r)^\alpha - (\ln r)^\alpha$$

$$= \frac{1}{2}(\ln t - \ln r)^\alpha + \frac{2^\alpha}{2}\left(\frac{\ln t}{2} + \frac{\ln r}{2} \right)^\alpha - (\ln r)^\alpha$$

$$\leq \frac{(\ln t - \ln r)^\alpha}{2} + \frac{(\ln t)^\alpha}{2^{2-\alpha}} + \frac{(\ln r)^\alpha}{2^{2-\alpha}} - (\ln r)^\alpha$$

$$\leq \frac{(\ln t - \ln r)^\alpha}{2} + \frac{(\ln t)^\alpha}{2} - \frac{(\ln r)^\alpha}{2}.$$

Hence

$$(\ln t)^\alpha - (\ln r)^\alpha \leq (\ln t - \ln r)^\alpha.$$

Applying, the Finite Increment Theorem to $\ln t - \ln r$, r, $t \in [1, b]$, we obtain

$$\mathbb{E}\sqrt{\|z_t(\omega) - z_r(\omega)\|} \leq \frac{2b}{\Gamma(\alpha+1)}|t-r|^\alpha\left[c_1^2\|x\|_{M^2}^2 + c_2^2\|y\|_{M^2}^2\right]$$
$$+\frac{2b}{\Gamma(\alpha+1)}|t-r|^\alpha\|f(\cdot,0,0,\cdot)\|_{M^2}^2$$
$$+\frac{b}{\Gamma(\alpha+1)}|t-r|^\alpha\left[c_1^2\|x\|_{M^2}^2 + c_2^2\|y\|_{M^2}^2\right]$$
$$+\frac{b}{\Gamma(\alpha+1)}|t-r|^\alpha\|f(\cdot,0,0,\cdot)\|_{M^2}^2$$

and

$$\mathbb{E}\sqrt{\|\bar{z}_t(\omega) - \bar{z}_r(\omega)\|} \leq \frac{2b}{\Gamma(\beta+1)}|t-r|^\beta\left[c_3^2\|x\|_{M^2}^2 + c_4^2\|y\|_{M^2}^2\right]$$
$$+\frac{2b}{\Gamma(\beta+1)}|t-r|^\beta\|g(\cdot,0,0,\cdot)\|_{M^2}^2$$
$$+\frac{b}{\Gamma(\beta+1)}|t-r|^\beta\left[c_3^2\|x\|_{M^2}^2 + c_4^2\|y\|_{M^2}^2\right]$$
$$+\frac{b}{\Gamma(\beta+1)}|t-r|^\beta\|g(\cdot,0,0,\cdot)\|_{M^2}^2.$$

Hence, there exist C, $\bar{C} > 0$ such that

$$\mathbb{E}\sqrt{\|z_t(\omega) - z_r(\omega)\|} \leq C|t-r|^{1+\alpha} \tag{14.4}$$

and

$$\mathbb{E}\sqrt{\|\bar{z}_t(\omega) - \bar{z}_r(\omega)\|} \leq C|t-r|^{1+\beta}. \tag{14.5}$$

By Kolmogrov's Theorem (Theorem 14.18), z_t and $\bar{z}_t$ have continuous modifications.

Now we show that z and $\bar{z}$ are stochastically continuous. Let t, $s \in [1, b]$ and $\epsilon > 0$ be given; then

$$\mathbb{P}(\{\omega \in \Omega : \|z_t(\omega) - z_r(\omega)\| > \epsilon\}) = \mathbb{P}(\{\omega \in \Omega : \sqrt{|z_t(\omega) - z_r(\omega)\|} > \sqrt{\epsilon}\}).$$

Using Markov's inequality, we obtain that

$$\mathbb{P}(\{\omega \in \Omega : \|z_t(\omega) - z_r(\omega)\| > \epsilon\}) \leq \frac{1}{\sqrt{\epsilon}}\mathbb{E}(\sqrt{|z_t(\omega) - z_r(\omega)\|})$$

and

$$\mathbb{P}(\{\omega \in \Omega : \|\bar{z}_t(\omega) - \bar{z}_r(\omega)\| > \epsilon\}) \leq \frac{1}{\sqrt{\epsilon}}\mathbb{E}(\sqrt{\|\bar{z}_t(\omega) - \bar{z}_r(\omega)\|}).$$

Hence, (14.4) and (14.5) imply that

$$\mathbb{P}(\{\omega \in \Omega : \|z_t(\omega) - z_r(\omega)\| > \epsilon\}) \leq \frac{C}{\sqrt{\epsilon}}|t-r|^{1+\alpha} \to 0 \text{ as } t \to r$$

and

$$\mathbb{P}(\{\omega \in \Omega : \|\bar{z}_t(\omega) - \bar{z}_r(\omega)\| > \epsilon\}) \leq \frac{1}{\sqrt{\epsilon}}|t-r|^{1+\beta} \to 0 \text{ as } t \to r.$$

$\square$

As a consequence of above theorem we can easily prove the following result.

Corollary 14.22. *Under the conditions of Theorem 14.20, every L^2-solution of problem (14.2) has a continuous modification and is stochastically continuous..*

For the existence of a modification of an M^2-solution of the problem (14.3), we assume, in addition to the Lipschitz condition $(\mathcal{JH}_5)$, that:

$(\mathcal{JH}_6)$ There exist positive constants $\bar{c}_i > 0$ $i = 1, \ldots, 6$ such that

$$\|f(t, x, y, \omega)\|^2 \leq \bar{c}_1 \|x\|^2 + \bar{c}_2 \|y\|^2 + \bar{c}_3$$

and

$$\|g(t, x, y, \omega)\|^2 \leq \bar{c}_4 \|x\|^2 + \bar{c}_5 \|y\|^2 + \bar{c}_6,$$

for each x, $y \in \mathbb{R}^m$, $t \in [1, b]$, $\mathbb{P}$, and a.e. $\omega \in \Omega$.

By some simple modifications of the proof of Theorem 14.20, we can prove the following result.

Theorem 14.23. *Suppose that $(\mathcal{JH}_5)$ and $(\mathcal{JH}_6)$ hold. Then every M^2-solution of problem (14.2) has a continuous modification and is stochastically continuous.*

Chapter 15

Existence Theory for Systems of Discrete Equations

15.1 Introduction and motivations

The idea of computing by recursion is as old as counting itself. It occurred in primitive form in the efforts of the Babylonians as early as 2000 B.C. to extract roots and in more explicit form around 450 B.C. in the Pythagoreans' study of figurative numbers, since in modern notation the triangular numbers satisfy the difference equation

$$t(n) = t(n-1) + n,$$

the square numbers the equation

$$s(n) = s(n-1) + n^2,$$

and so forth. The Pythagoreans also used a system of difference equations

$$x(n) = x(n-1) + 2y(n-1), \quad y(n) = x(n-1) + y(n-1)$$

to generate large solutions of Pell's equation,

$$x^2 - 2y^2 = 1,$$

and thereby approximations of $\sqrt{2}$. In his attempts to compute the circumference of a circle, Archimedes (about 250 B.C.) employed equations of the form

$$P(2n) = \frac{2p(n)P(n)}{p(n) + P(n)}, \quad p(2n) = \sqrt{p(n)P(2n)}$$

to compute the perimeters $P(n)$ and $p(n)$ of the circumscribed polygon of n sides and the inscribed polygon of n sides, respectively. Other familiar ancient discoveries about recurrence include the Euclidean algorithm and Zeno's paradox. Euclid also studied geometric series, although the general form of the sum was not obtained until around 1593 by Vieta. About 1202, Fibonacci formulated his famous rabbit problem that led to the Fibonacci sequence

$$1, 1, 2, 5, 8, 13, \ldots$$

However, it appears that the corresponding difference equation

$$F(n) = F(n-2) + F(n-1)$$

was first written down by Albert Girard around 1634 and solved by de Moivre in 1730. Bombelli studied the equation

$$y(n) = 2 + \frac{1}{y(n-1)}$$

in 1572, which is similar to the equation

$$z(n) = 1 + \frac{1}{z(n-1)}$$

satisfied by ratios of Fibonacci numbers. In order to approximate $\sqrt{2}$ Fibonacci also gave a rough definition for the concept of continued fractions that is intimately associated with difference equations. A more precise definition was formulated by Cataldi around 1613. (See Brezinski [2] for a lively discussion of the history of continued fractions.) The earliest known example of a difference equation in two indices, namely the equation

$$b_r(n+1) = b_r(n) + b_r(n-1)$$

for the binomial coefficients, can be traced back to Chia Hsien (1050) and Omar Khayyam (1100). The method of recursion was significantly advanced with the invention of mathematical induction by Francesco Maurolico in the sixteenth century and with its development by Fermat and Pascal in the seventeenth century. Sir Thomas Harriet (1560-1621) invented the calculus of finite differences, and Henry Briggs (1556-1630) applied it to the calculation of logarithms. It was rediscovered by Leibniz around 1672. Newton, Euler, LaGrange, Gauss, and many others used this calculus to study interpolation theory. The theory of finite differences was developed largely by Stirling in the early eighteenth century. Goldstine [8] gives a detailed historical description of the early work in this area. Meanwhile, an important class of nonlinear difference equations, which we now call Newton's method (known in primitive form by Vieta), was used by Newton around 1669 to study solutions of

$$y^3 - 2y - 5 = 0$$

and later in computations for Kepler's equation. In 1690, Raphson worked out a more systematic treatment of the method. Another important family of nonlinear difference equations consists of pairs of equations

Difference equations appear as natural descriptions of observed evolution phenomena because most measurements of time involving variables are discrete. They also appear in the applications of discretization methods for differential, integral and integro-differential equations. The application of the theory of difference equations is rapidly increasing to various fields, such as numerical analysis, control theory, finite mathematics, and computer sciences. In the foregoing example, the difference equation resulted immediately from modelling the real problem, without any differential equations being involved. At present, however, differential equations are the main "source" of difference equations. We mean the so-called difference procedures which are widely employed when finding approximate solutions of differential equations and which are represented by systems of difference equations, sometimes of a sufficiently high order.

Many statements concerning the theory of linear differential equations are also valid for the corresponding difference equations. A well-known example is the famous Poincare theorem on the asymptotic behavior of the solutions to difference equations which was published in 1885 (see Gelfand (1967) and van Strien (1978)). Another example, less well-known although it dates back to the beginning of our century, is the analytical theory of difference equations by G.D.Birkhoff. The Poincare theorem initiated the studies of Birkhoff and his students who formulated the general (in a certain sense) analytical theory of linear ordinary differential, difference, and q-difference equations (Birkhoff (1911, 1930, 1932); see also Maistrenko (1980)).

15.1.1 Cagan's model with backward-looking market participants

This subsection deals with Cagan's (1956) model of the money market with backward-looking market participants. Cagan assumed that the demand for real balances is decreasing in expected inflation. This is represented by M^d in

$$\frac{M^d(k)}{P(k)} = \left(\frac{P^e(k+1)}{P(k)}\right)^\sigma \tag{15.1}$$

where

$M(k)$ is time-k money demand,

$P(k)$ is the time-k price level,

$P^e(k+1)$ is market participants time-k expectation of time-$(k+1)$ inflation

- σ is a strictly negative constant.

As Cagan was writing before the development of the theory of rational expectations, he assumed that market participants have adaptive expectations: they believe that next period's inflation depends solely on this period's inflation. This can be represented as

$$\frac{P^e(k+1)}{P(k)} = \left(\frac{P(k)}{P(k-1)}\right)^\mu, \tag{15.2}$$

where μ is a constant of unknown sign. Given that time-k money demand is equal to the time-k money supply, $M(k)$, the equilibrium condition specifies that the supply of real balances is equal to a function of current inflation:

$$\frac{M(k)}{P(k)} = \left(\frac{P(k)}{P(k-1)}\right)^\alpha, \quad k \in \mathbb{Z} \tag{15.3}$$

where $\alpha = \sigma\mu$ is a constant that can be positive or negative. Taking the logarithm of both sides of equation (15.3) gives the linear specification

$$m(k) - p(k) = \alpha(p(k) - p(k-1)), \quad k \in \mathbb{Z}, \tag{15.4}$$

where lower case letters are the logarithm of upper case letters. If it is assumed that the money supply is constant at m then equation (15.4) can be written in the following form:

$$p(k) = ap(k) + d, \quad k \in \mathbb{Z},$$

where

$$a := \frac{\alpha}{1+\alpha} \quad \text{and} \quad d := \frac{m}{1+\alpha}.$$

15.1.2   Electronic model

Consider the electric circuit shown in the following figure

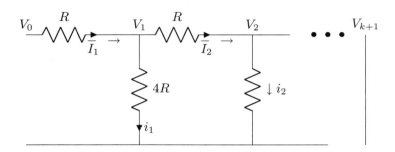

Assume that $V_0 = A$ is a given voltage and $V(K + 1) = 0$. Each resistance in the horizontal branch is equal to R and in the vertical branches equal to $4R$. We want to find the voltage $V(k)$ for $1 \leq k \leq K$. For this, according to Kirchhoff's current law, the sum of the currents flowing into a junction point is equal to the sum of the currents flowing away from the junction point. Applying this law at the junction point corresponding to the voltage $V(k + 1)$, we have

$$I(k + l) = I(k + 2) + i(k + 1).$$

Using Ohm's law, $I = \frac{V}{R}$, the above equation can be replaced by

$$\frac{V(k) - V(k + 1)}{R} = \frac{V(k + 1) - V(k + 2)}{R} - \frac{V(k + 1) - 0}{4R}$$

which, upon identifying $V(k)$ as $u(k)$, leads to the second order difference equation

$$4u(k + 2) - 9u(k + 1) + 4u(k) = 0, \quad k \in \mathbb{N}(0, K - 1),$$

and the boundary conditions are

$$u(0) = A, \quad u(K + 1) = 0.$$

15.2   Gronwall inequalities

In the study of ordinary differential equations and integral equations one often deals with certain integral inequalities. The Gronwall-Bellman inequality and its various linear and nonlinear generalizations are crucial in the discussion of existence, uniqueness, continuation, boundedness, stability and other qualitative properties of solutions to differential and integral equations and continuous dependence on initial data. To handle difference equations, some discrete Gronwall-Bellman type inequalities are needed. During the past few years, some investigators have established some useful and interesting discrete Gronwall-Bellman type inequalities, see Cheung (2004, 2006), Cheung and Ren (2006), Salem and Ralan (2004). Cheung (2004) has proved the following.

Theorem 15.1. *Let $p, q, f, u : \mathbb{N}(a) \to \mathbb{R}_+$ be nonnegative functions such that*

$$u(k) \le p(k) + q(k) \sum_{l=a}^{l=k-1} f(l)u(l), \quad \text{for all } k \in \mathbb{N}(a) := \{a, a+1, \ldots\}.$$

Then

$$u(k) \le p(k) + q(k) \sum_{l=a}^{l=k-1} p(l)f(l) \prod_{\tau=l+1}^{\tau=k-1} (1 + q(\tau)f(\tau)).$$

Proof. Define a function $v(k)$ on $\mathbb{N}(a)$ as follows

$$v(k) = \sum_{l=a}^{k-1} f(l)u(l).$$

Then, we have

$$\Delta v(k) = f(k)v(k), \quad v(a) = 0. \tag{15.5}$$

Since $u(k) \le p(k) + q(k)v(k)$ and $f(k) \ge 0$, from (15.5), we get

$$v(k+1) - (1 + q(k)f(k))v(k) \le p(k)f(k). \tag{15.6}$$

We multiply (15.6) by $\prod_{l=a}^{k}(1 + q(l)f(l))^{-1}$, to obtain

$$\Delta \prod_{l=a}^{k-1} (1 + q(l)f(l))^{-1} v(k) \le p(k)f(k) \prod_{l=a}^{k} (1 + q(l)f(l))^{-1}.$$

Summing the above inequality from a to $k-1$, and using $v(a) = 0$, we get

$$\prod_{l=a}^{k-1} (1 + q(l)f(l))^{-1} v(k) \le \sum_{l=a}^{k} p(l)f(l) \prod_{\tau=a}^{l} (1 + q(\tau)f(\tau))^{-1}.$$

Therefore

$$v(k) \le \sum_{l=a}^{k} p(l)f(l) \prod_{\tau=l+1}^{k-1} (1 + q(\tau)f(\tau)).$$

This implies that

$$p(k) + v(k) \le p(k) + \sum_{l=a}^{k} p(l)f(l) \prod_{\tau=l+1}^{k-1} (1 + q(\tau)f(\tau)).$$

Hence

$$u(k) \le p(k) + \sum_{l=a}^{k} p(l)f(l) \prod_{\tau=l+1}^{k-1} (1 + q(\tau)f(\tau)).$$

$\square$

Corollary 15.2. *In Theorem 15.1, assume that*

$$p(k) = p \in \mathbb{R}_+, \quad q(k) = q \in \mathbb{R}_+ \quad \text{for all} \quad k \in \mathbb{N}(a).$$

Then

$$u(k) \le p \prod_{l=a}^{k-1} (1 + qf(l)).$$

Corollary 15.3. *In Theorem 15.1, assume that $p(k)$ is nondecreasing and that $q(k) \geq 1$ for all $k \in \mathbb{N}(a)$. Then*

$$u(k) \leq p(k)q(k) \prod_{l=a}^{k-1}(1 + q(l)f(l)).$$

Theorem 15.4. *Let $p, q, f, u : \mathbb{N}(a) \to \mathbb{R}_+$ be nonnegative functions such that*

$$u(k) \leq p(k) + q(k) \sum_{l=a}^{l=k-1} E_i(k, u), \quad \text{for all } k \in \mathbb{N}(a).$$

where

$$E_i(k, u) = \sum_{l_1=a}^{k-1} f_{i1}(l_1) \sum_{l_2=a}^{l_1-1} f_{i1}(l_1) \ldots \sum_{l_i=a}^{l_{i-1}-1} f_{ii}(l_i)u(l_i).$$

Then

$$u(k) \leq p(k) + q(k) \sum_{l=a}^{l=k-1} \left[\sum_{i=1}^{r} \Delta E_i(k, p) \right] \prod_{\tau=l+1}^{k-1} \left[1 + \sum_{i=1}^{r} \Delta E_i(k, q) \right].$$

15.3 Cauchy discrete problem

In this section we are interested in investigating the nonlinear discrete system with initial condition:

$$\begin{cases} \Delta x(k) &= f(k, x(k), y(k)), \ k \in \mathbb{N}(a, b) := \{a, a+1, \ldots, b\}, \\ \Delta y(k) &= g(k, x(k), y(k)), \ k \in \mathbb{N}(a, b), \\ \\ x(a) &= x_0, \\ y(a) &= y_0, \end{cases} \quad (15.7)$$

where $f, g : \mathbb{N}(a, b) \times \mathbb{R}^m \to \mathbb{R}^m$ are given functions.

After the publication of the landmark paper of Hartman [156] in the year 1978, difference equations have become a major field of research. In the case where $g \equiv 0$ various mathematical results for difference equation (existence, asymptotic behavior, ...) have been obtained (see [3, 13, 15, 111, 181, 199, 214, 215] and the references therein).

In recent years, by using fixed point theory, topological degree theory (including continuation methods and coincidence degree theory), comparison methods and monotone iterative methods, the existence of solutions to difference equations have been extensively studied.

15.4 Existence and uniqueness

Let us introduce the following hypothesis:

(H_1) There exist nonnegative numbers a_i and b_i for each $i \in \{1, 2\}$

$$\begin{cases} |f(k, x, y) - f(k, \overline{x}, \overline{y})| \leq a_1|x - \overline{x}| + b_1|y - \overline{y}| \\ |g(k, x, y) - g(k, \overline{x}, \overline{y})| \leq a_2|x - \overline{x}| + b_2|y - \overline{y}| \end{cases}$$

for all $x, y, \overline{x}, \overline{y} \in \mathbb{R}^m$.

For our main consideration of Problem (15.7), a Perov fixed point is used to investigate the existence and uniqueness of solutions for a system of impulsive stochastic difference equations.

Theorem 15.5. *Assume that (H_1) is satisfied and the matrix*

$$M = (b-1) \begin{pmatrix} a_1 & b_1 \\ a_2 & b_2 \end{pmatrix} \in M_{2\times 2}(\mathbb{R}_+).$$

If M converges to zero, then the problem (15.7) has a unique solution.

Proof. Consider the operator $N: C(\mathbb{N}(a,b-1),\mathbb{R}^m) \times C(\mathbb{N}(a,b-1),\mathbb{R}^m) \to C(\mathbb{N}(a,b-1),\mathbb{R}^m) \times C(\mathbb{N}(a,b-1),\mathbb{R}^m)$ defined for $(x,y) \in C(\mathbb{N}(a,b-1),\mathbb{R}^m) \times C(\mathbb{N}(a,b-1),\mathbb{R}^m)$ by

$$N(x,y) = (N_1(x,y), N_2(x,y)), \tag{15.8}$$

where

$$N_1(x(k),y(k)) = x_0 + \sum_{l=a}^{k-1} f(l,x(l),y(l)), \quad k \in \mathbb{N}(a,b-1)$$

and

$$N_2(x(k),y(k)) = y_0 + \sum_{l=a}^{k-1} g(l,x(l),y(l)), \quad k \in \mathbb{N}(a,b-1).$$

We shall use Theorem 8.1 to prove that N has a fixed point. Indeed, let $(x,y),(\overline{x},\overline{y}) \in C(\mathbb{N}(a,b-1),\mathbb{R}^m) \times C(\mathbb{N}(a,b-1),\mathbb{R}^m)$. Then we have for each $k \in \mathbb{N}(a,b-1)$

$$|N_1(x(k),y(k)) - N_1(\overline{x}(l),\overline{y}(l))| = \left| \sum_{l=a}^{l=k-1} [f(l,x(l),y(l)) - f(l,\overline{x}(l),\overline{y}(l))] \right|.$$

Then

$$\|N_1(x,y) - N_1(\overline{x},\overline{y})\|_\infty \leq (b-1)a_1\|x - \overline{x}\|_\infty + (b-1)b_1\|y - \overline{y}\|_\infty.$$

Similarly we have

$$\|N_2(x,y) - N_2(\overline{x},\overline{y})\|_\infty \leq (b-1)a_2\|x - \overline{x}\|_\infty + (b-1)b_2\|y - \overline{y}\|_\infty.$$

Hence

$$\|N(x,y) - N(\overline{x},\overline{y})\|_\infty = \begin{pmatrix} \|N_1((x,y) - N_1(\overline{x},\overline{y})\|_\infty \\ \|N_2(x,y) - N_2(\overline{x},\overline{y})\|_\infty \end{pmatrix}$$

$$\leq (b-1) \begin{pmatrix} a_1 & b_1 \\ a_2 & b_2 \end{pmatrix} \begin{pmatrix} \|x - \overline{x}\|_\infty \\ \|y - \overline{y}\|_\infty \end{pmatrix}.$$

Therefore, for all $(x,y),(\overline{x},\overline{y}) \in C(\mathbb{N}(a,b-1),\mathbb{R}^m) \times C(\mathbb{N}(a,b-1),\mathbb{R}^m)$,

$$\|N(x,y) - N(\overline{x},\overline{y})\|_\infty \leq M \begin{pmatrix} \|x - \overline{x}\|_\infty \\ \|y - \overline{y}\|_\infty \end{pmatrix}.$$

From the Perov fixed point theorem, the mapping N has a unique fixed point $(x,y) \in C(\mathbb{N}(a,b-1),\mathbb{R}^m) \times C(\mathbb{N}(a,b-1),\mathbb{R}^m)$ which is the unique solution of problem (15.7). $\square$

Theorem 15.6. *Assume the following conditions are satisfied.*

(H_2) *There exist nonnegative valued functions* $\lambda_i, \gamma_i : \mathbb{N}(a) \to \mathbb{R}_+$ *for each* $i = 1, 2$, *such that*

$$\begin{cases} |f(k, x, y) - f(k, \overline{x}, \overline{y})| \leq \lambda_1(k)|x - \overline{x}| + \lambda_2(k)|y - \overline{y}| \\ |g(k, x, y) - g(k, \overline{x}, \overline{y})| \leq \gamma_1(k)|x - \overline{x}| + \gamma_2(k)|y - \overline{y}| \end{cases}$$

for all $x, y, \overline{x}, \overline{y} \in \mathbb{R}^m$.

(H_3) $h_1, h_2 : \mathbb{N}(a) \times \mathbb{R}^m \times \mathbb{R}^m \to \mathbb{R}^m$ *are functions such that*

$$|h_i(k, x, y)| \leq \mu_1(k), \quad i = 1, 2,$$

where μ_i *are nonnegative functions defined on* $\mathbb{N}(a)$.

Then, for the solutions $(x(k, x_0), y(k, y_0))$ *and* $(u(k, u_0), v(k, v_0))$ *on* $\mathbb{N}(a)$ *of the initial value problem (15.7) and*

$$\begin{cases} \Delta u(k) = h_1(k, u(k), v(k)) + f(k, u(k, u_0), v(k, v_0)), & k \in \mathbb{N}(a), \\ \Delta v(k) = h_2(k, u(k), v(k)) + g(k, u(k, u_0), v(k, v_0)), & k \in \mathbb{N}(a), \\ u(a) = u_0, \\ v(a) = v_0, \end{cases} \tag{15.9}$$

the following inequalities hold:

$$|x(k, x_0) - u(k, u_0)| \leq \left(|x_0 - u_0| + |y_0 - v_0| + \sum_{l=a}^{l=k-1} \mu(l) \right) \prod_{l=a}^{l=k-1} (1 + \lambda(k)),$$

and

$$|y(k, x_0) - v(k, v_0)| \leq \left(|y_0 - v_0| + |x_0 - u_0| + \sum_{l=a}^{l=k-1} \mu(l) \right) \prod_{l=a}^{l=k-1} (1 + \lambda(k)),$$

where

$$\lambda(k) = \lambda_1(k) + \lambda_2(k) + \gamma_1(k) + \gamma_2(k), \quad \mu(k) = \mu_1(k) + \mu_2(k), \quad k \in \mathbb{N}(a).$$

Proof. The problems (15.7) and (15.9) are equivalent to

$$\begin{cases} x(k, x_0) = x_0 + \sum_{l=a}^{k-1} f(l, x(l, x_0), y(l, y_0)), & k \in \mathbb{N}(a) \\ y(k, y_0) = y_0 + \sum_{l=a}^{k-1} g(l, x(l, x_0), y(l, y_0)), & k \in \mathbb{N}(a), \end{cases}$$

and

$$\begin{cases} u(k, x_0) = u_0 + \sum_{l=a}^{k-1} (h_1(l, u(l, u_0), v(l, v_0)) + f(l, u(l, u_0), v(l, v_0))), \\ \qquad k \in \mathbb{N}(a), \\ v(k, y_0) = v_0 + \sum_{l=a}^{k-1} (h_1(l, u(l, u_0), v(l, v_0)) + g(l, u(l, x_0), v(l, v_0))), \\ \qquad k \in \mathbb{N}(a). \end{cases}$$

We find that

$$
\begin{cases}
\begin{aligned}
x(k, x_0) - u(k, u_0) &= x_0 - u_0 + \sum_{l=a}^{k-1}(f(l, x(l, x_0), y(l, y_0)) \\
&\quad - f(l, u(l, u_0), v(l, v_0))) \\
&\quad - \sum_{l=a}^{k-1} h_1(k, u(k, u_0), v(k, v_0)) \\
y(k, y_0) - v(k, v_0) &= y_0 - v_0 + \sum_{l=a}^{k-1}(g(l, x(l, x_0), y(l, y_0)) \\
&\quad - g(l, u(l, u_0), v(l, v_0))) \\
&\quad - \sum_{l=a}^{k-1} h_2(l, u(l, u_0), v(l, v_0)).
\end{aligned}
\end{cases}
$$

Then

$$
w(k) \le |x_0 - u_0| + |y_0 - v_0| + \sum_{l=a}^{k-1} \lambda(k) w(k) + \sum_{l=a}^{k-1} \mu(l)
$$

where

$$
w(k) = |x(k, x_0) - u(k, u_0)| + |y(k, y_0) - v(k, u_0)|, \quad k \in \mathbb{N}(a)
$$

and

$$
\lambda(k) = \lambda_1(k) + \lambda_2(k) + \gamma_1(k) + \gamma_2(k), \quad \mu(k) = \mu_1(k) + \mu_2(k), \quad k \in \mathbb{N}(a).
$$

From Corollary 15.3, we get

$$
w(k) \le \left(|x_0 - u_0| + |y_0 - v_0| + \sum_{l=a}^{k-1} \mu(l) \right) \prod_{l=a}^{k-1}(1 + \lambda(l)).
$$

Hence

$$
|x(k, x_0) - u(k, u_0)| \le \left(|x_0 - u_0| + |y_0 - v_0| + \sum_{l=a}^{l=k-1} \mu(l) \right) \prod_{l=a}^{l=k-1}(1 + \lambda(k)),
$$

and

$$
|y(k, x_0) - v(k, v_0)| \le \left(|y_0 - v_0| + |x_0 - u_0| + \sum_{l=a}^{l=k-1} \mu(l) \right) \prod_{l=a}^{l=k-1}(1 + \lambda(k)).
$$

$\square$

Now we consider the following Cauchy problem with parameter

$$
\begin{cases}
\Delta x(k) &= f(k, x(k), y(k), \alpha), \ k \in \mathbb{N}(a), \\
\Delta y(k) &= g(k, x(k), y(k), \alpha), \ k \in \mathbb{N}(a), \\
x(a) &= x_0, \\
y(a) &= y_0,
\end{cases}
\tag{15.10}
$$

where $\alpha \in \mathbb{R}^m$ is a parameter such that $|\alpha - \alpha_0| \le \delta$ and α_0 is a fixed vector in $\mathbb{R}^m$ and $f, g : \mathbb{N}(a) \times \mathbb{R}^n \times \mathbb{R}^n \times \mathbb{R}^m \to \mathbb{R}^n$ are given functions.

Theorem 15.7. *For fixed $\alpha_0 \in \mathbb{R}^m$ and $\delta > 0$ such that $|\alpha - \alpha_0| \leq \delta$ the functions f and g satisfy the following conditions:*

(H_4) *There exist nonnegative valued functions $\lambda_i, \gamma_i, \mu_i : \mathbb{N}(a) \to \mathbb{R}_+$ for each $i = 1, 2$, such that*

$$\begin{cases} |f(k, x, y, \alpha) - f(k, \overline{x}, \overline{y}, \alpha)| \leq \lambda_1(k)|x - \overline{x}| + \lambda_2(k)|y - \overline{y}| \\ |g(k, x, y, \alpha) - g(k, \overline{x}, \overline{y}, \alpha)| \leq \gamma_1(k)|x - \overline{x}| + \gamma_2(k)|y - \overline{y}| \end{cases}$$

and

$$\begin{cases} |f(k, x, y, \alpha) - f(k, x, y, \alpha_1)| \leq \mu_1(k)|\alpha_1 - \alpha_2| \\ |g(k, x, y, \alpha_1) - g(k, x, y, \alpha_2)| \leq \mu_2(k)|\alpha_1 - \alpha_2|. \end{cases}$$

Then, for the solutions $(x(k, x_1, \alpha_1), y(k, y_1, \alpha_1)$ and $(u(k, u_2, \alpha_2), v(k, v_2, \alpha_2)$ of (15.10), the following inequalities hold

$$|x(k, x_1, \alpha_1) - u(k, x_2, \alpha_2)|$$

$$\leq \left(|x_1 - u_2| + |y_1 - v_2| + |\alpha_1 - \alpha_2| + \sum_{l=a}^{l=k-1} \mu(l) \right) \times \prod_{l=a}^{l=k-1} (1 + \lambda(k))$$

and

$$|y(k, y_1, \alpha_1) - v(k, u_2, \alpha_2)|$$

$$\leq \left(|x_1 - u_2| + |y_1 - v_2| + |\alpha_1 - \alpha_2| + \sum_{l=a}^{l=k-1} \mu(l) \right) \times \prod_{l=a}^{l=k-1} (1 + \lambda(k)),$$

where

$$\lambda(k) = \lambda_1(k) + \lambda_2(k) + \gamma_1(k) + \gamma_2(k), \quad \mu(k) = \mu_1(k) + \mu_2(k), \; k \in \mathbb{N}(a).$$

15.5   Existence and compactness of solution sets

In ordinary differential equations the Arzela-Ascoli theorem plays an important role. In this section we give the discrete version of the Arzela-Ascoli theorem. The topology on $\mathbb{N}(0, b + 1)$ will be the discrete topology. Let $(E, | \cdot |)$ be a Banach space. We denote the space of continuous functions on $\mathbb{N}(0, b + 1)$ by

$$C(\mathbb{N}(0, b + 1), E) = \{ y : \mathbb{N}(0, b + 1) \to E \mid y \text{ is continuous} \}$$

which is a Banach space when equipped with the norm

$$\|y\|_\infty = \sup_{k \in \mathbb{N}(0, b+1)} |y(k)|.$$

Now we state and prove the discrete Arzela-Ascoli Theorem.

Theorem 15.8. *Let A be a closed subset of $C(\mathbb{N}(0, b + 1), E)$. If Ω is uniformly bounded and the set*

$$\{ y(k) : y \in \Omega \}$$

is relatively compact for each $k \in \mathbb{N}(0, b + 1)$, then Ω is compact.

Proof. We need only show that every sequence in Ω has a Cauchy subsequence. Let $\Omega_1 = \{f_{1,1}, f_{1,2}, \ldots\}$ be any sequence in Ω. Notice the sequence $\{f_{1,i}(0)\}$, $i = 1, 2, \ldots$ has a convergent subsequence and let $\Omega_2 = \{f_{2,1}, f_{2,2}, \ldots\}$ denote this subsequence. For $\{f_{2,i}(1)\}$, $i = 1, 2, \ldots$ let $\Omega_3 = \{f_{3,1}, f_{3,2}, \ldots\}$ be the subsequence of Ω_2 such that $\{f_{3,i}(1)\}$, $i = 1, 2, \ldots$ converges. Since Ω_3 is a subsequence of Ω_2 then $\{f_{3,i}(0)\}$, $i = 1, 2, \ldots$ also converges. Continue this process to get a list of sequences

$$\Omega_1, \Omega_2, \ldots \Omega_{b+2}, \Omega_{b+3}.$$

in which each sequence is a subsequence of the one directly on the left of it and for each k, the sequence $\Omega_k = \{f_{k,1}, f_{k,2}, \ldots\}$ has the property that $\{f_{k,i}(k-2)\}, i = 1, 2, \ldots$ is a convergent sequence. Thus for each $k \in \mathbb{N}(0, b+1)$, the sequence $\{f_{b+3,i}(k)\}$ is convergent. Then since $\{f_{T+3,i}(k)\}$ is Cauchy for each $k \in \mathbb{N}(0, b+1)$, and since $\mathbb{N}(0, b+1)$ is finite, we have that there exists $n_0 \in \mathbb{N}$ independent of k such that

$$m, n \geq n_0 \Rightarrow |f_{b+3,m}(k) - f_{b+3,n}(k)| < \epsilon, \quad k \in \mathbb{N}(0, b+1).$$

Thus Ω_{b+3} is Cauchy. $\qquad\square$

We also need the following characterization for relatively compact sets in $BC(\mathbb{N}, E)$, which is the discrete version of the Przeradzki theorem [253].

Theorem 15.9. *A set $\Omega \subset BC(\mathbb{N}, E)$ is relatively compact if the following conditions hold:*

(i) for every $k \in \mathbb{N}$ the set $\{y(k) : x \in \Omega\}$ is relatively compact in E,

(ii) for every $\epsilon > 0$ there exists $N' \in \mathbb{N} \setminus \{0\}$ and $\delta > 0$ such that if $x, y \in \Omega$ with $|x(N') - y(N')| \leq \delta$ then $|x(k) - y(k)| \leq \epsilon$ for all $k \in \{N', N' + l, \ldots\}$.

We shall also need the following existence principles.

Theorem 15.10. *Let $f, g : \mathbb{N}(a, b-1) \times \mathbb{R}^n \times \mathbb{R}^n \longrightarrow \mathbb{R}^n$ be continuous functions. Assume that condition*

(H_5) There exist $p_1, p_2 \in C(\mathbb{N}(a, b-1), \mathbb{R}_+)$ such that

$$|f(k, x, y)| \leq p_1(k)(|x| + |y|), \quad k \in \mathbb{N}(a, b-1), \quad (x, y) \in \mathbb{R}^n \times \mathbb{R}^n,$$

and

$$|g(k, x, y)| \leq p_2(k)(|x| + |y|), \quad k \in \mathbb{N}(a, b-1), \quad (x, y) \in \mathbb{R}^n \times \mathbb{R}^n,$$

holds. Then the problem (15.7) has at least one solution. Moreover, the solution set $S(x_0, y_0)$ is compact and the multivalued map $S : (x_0, y_0) \multimap S(x_0, y_0)$ is u.s.c.

Proof. Clearly, the fixed points of N are solutions to (15.7), where N is defined in (15.8). In order to apply Theorem 8.14, we first show that N is completely continuous. The proof will be given in several steps.

- **Step 1.** $N = (N_1, N_2)$ is continuous.
  Let (x_m, y_m) be a sequence such that $(x_m, y_m) \to (x, y) \in C(\mathbb{N}(a, b-1), \mathbb{R}^n) \times C(\mathbb{N}(a, b-1), \mathbb{R}^n)$ as $m \to \infty$. Then

$$|N_1(x_m(k), y_m(k)) - N_1(x(k), y(k))| = \left| \sum_{l=a}^{k-1} f(l, x_n(l), y_n(l)) \right.$$

$$-\sum_{l=1}^{k-1} f(l, x(l), y(l))\bigg|$$

$$\leq \bigg|\sum_{l=a}^{b} f(l, x_n(l), y_n(l))$$

$$-\sum_{l=1}^{b} f(l, x(l), y(l))\bigg|.$$

Similarly

$$|N_2(x_n(k), y_n(k)) - N_2(x(k), y(k))| \leq \bigg|\sum_{l=a}^{b} g(l, x_m(l), y_m(l))$$

$$-\sum_{l=1}^{b} g(l, x(l), y(l))\bigg|.$$

Since f and g are continuous functions, we get, as $m \to \infty$,

$$\|N_1(x_m, y_m) - N_1(x, y)\|_\infty \leq \sum_{l=a}^{b} |f(l, x_m(l), y_m(l)) - f(l, x(l), y(l))|$$

$$\to 0,$$

and

$$\|N_2(x_m, y_m) - N_2(x, y)\|_\infty \leq \sum_{l=a}^{b} |g(l, x_m(l), y_m(l)) - g(l, x(l), y(l))|$$

$$\to 0.$$

- **Step 2.** N maps bounded sets into bounded sets in $C(\mathbb{N}(a, b-1), \mathbb{R}^n) \times C(\mathbb{N}(a, b-1), \mathbb{R}^n)$.

Indeed, it is enough to show that for any $q > 0$ there exists a positive constant l such that for each $(x, y) \in B_q = \{(x, y) \in C(\mathbb{N}(a, b-1), \mathbb{R}^n) \times C(\mathbb{N}(a, b-1), \mathbb{R}^n) : \|x\|_\infty \leq q, \|y\|_\infty \leq q\}$, we have

$$\|N(x, y)\|_\infty \leq l = (l_1, l_2).$$

Then for each $k \in \mathbb{N}(a, b-1)$, we get

$$|N_1(x(k), y(k))| = |x_0 + \sum_{l=a}^{k-1} f(l, x_n(l), y_n(l))|$$

$$\leq |x_0| + \sum_{l=1}^{b} f(l, x(l), y(l))|.$$

Therefore

$$\|N_1(x, y)\|_\infty \leq |x_0| + 2q \sum_{k=1}^{b} p_1(k) := l_1.$$

Similarly, we have

$$\|N_2(x, y)\|_\infty \leq |y_0| + 2q \sum_{k=1}^{b} p_2(k) := l_2.$$

Moreover, for each $k \in \mathbb{N}(a, b-1)$, we have

$$\{N_1(x(k), y(k)) : (x, y) \in B_q\}, \quad \{N_2(x(k), y(k)) : (x, y) \in B_q\}$$

are relatively compact in $\mathbb{R}^n$. Then, as a consequence of Theorem 15.8, we conclude that $\overline{N(B_q)}$ is compact.

As a consequence of Steps 1 and 2, $N : C(\mathbb{N}(a, b-1), \mathbb{R}^n) \to C(\mathbb{N}(a, b-1), \mathbb{R}^n)$ is completely continuous.

Step 3. It remains to show that

$$\begin{aligned}
\mathcal{A} = \ & \{(x, y) \in C(\mathbb{N}(a, b-1), \mathbb{R}^n) \times C(\mathbb{N}(a, b-1), \mathbb{R}^n) : \\
& (x, y) = \lambda N(x, y), \lambda \in (0, 1)\}
\end{aligned}$$

is bounded.

Let $(x, y) \in \mathcal{A}$. Then $x = \lambda N_1(x, y)$ and $y = \lambda N_2(x, y)$ for some $0 < \lambda < 1$. Thus, for $k \in \mathbb{N}(a, b-1)$, we have

$$\begin{aligned}
|x(k)| & \leq |x_0| + \sum_{l=1}^{k-1} |f(l, x(l), y(l))| \\
& \leq |x_0| + \sum_{l=a}^{k-1} p_1(l)(|x(l)| + |y(l)|) ds,
\end{aligned}$$

and

$$|y(k)| \leq |x_0| + \sum_{l=a}^{k-1} p_2(l)(|x(l)| + |y(l)|) ds.$$

Therefore

$$|x(k)| + |y(k)| \leq |x_0| + |y_0| + \sum_{l=a}^{k-1} p(l)(|x(l)| + |y(l)|),$$

where

$$p(k) = p_1(k) + p_2(k), \quad k \in \mathbb{N}(a, b-1).$$

By Theorem 15.1, we have

$$|x(k)| + |y(k)| \leq (|x_0| + |y_0|) \left(1 + \sum_{l=a}^{k-1} p(k) \prod_{l+1}^{k-1} (1 + p(\tau))\right).$$

Hence

$$\|x\|_\infty + \|y\|_\infty \leq (|x_0| + |y_0|) \left(1 + \sum_{l=a}^{b} p(k) \prod_{l+1}^{b} (1 + p(\tau))\right).$$

This shows that $\mathcal{A}$ is bounded.

As a consequence of Theorem 8.14 we deduce that N has a fixed point (x, y) which is a solution to the problem (15.7).

Step 4: Compactness of the solution set. For each $(x_0, y_0) \in \mathbb{R}^n \times \mathbb{R}^n$, let

$$S(x_0, y_0) = \{(x, y) \in C(\mathbb{N}(a, b-1), \mathbb{R}^n) \times C(\mathbb{N}(a, b-1), \mathbb{R}^n) :$$

(x, y) is a solution of problem (15.7)}.

From Step 3, there exists $\widetilde{M}$ such that for every $(x, y) \in S((x_0, y_0))$, $\|x\|_\infty \leq \widetilde{M}$; $\|y\|_\infty \leq \widetilde{M}$. Since N is completely continuous, $N(S(x_0, y_0))$ is relatively compact in $(x, y) \in C(\mathbb{N}(a, b - 1), \mathbb{R}^n) \times C(\mathbb{N}(a, b - 1), \mathbb{R}^n)$. Let $(x, y) \in S(x_0, y_0)$); then $(x, y) = N(x, y)$ hence $S(x_0, y_0) \subset \overline{N(S(x_0, y_0))}$. It remains to prove that $S(x_0, y_0)$ is a closed subset in $(x, y) \in C(\mathbb{N}(a, b - 1), \mathbb{R}^n) \times C(\mathbb{N}(a, b - 1), \mathbb{R}^n)$. Let $\{(x_m, y_m) : m \in \mathbb{N}\} \subset S(x_0, y_0)$ be such that $(x_m, y_m)_{m \in \mathbb{N}}$ converges to (x, y). For every $m \in \mathbb{N}$, and $k \in \mathbb{N}(a, b - 1)$

$$x_m(k) = x_0 + \sum_{l=a}^{k-1} f(l, x_m(l), y_m(l)). \tag{15.11}$$

and

$$y_m(k) = y_0 + \sum_{l=a}^{k-1} g(l, x_m(l), y_m(l)). \tag{15.12}$$

Set

$$z_1(k) = x_0 + \sum_{l=a}^{k-1} f(l, x(l), y(l)) \tag{15.13}$$

and

$$z_2(k) = y_0 + \sum_{l=a}^{k-1} g(l, x(l), y(l)). \tag{15.14}$$

Since f and g are continuous functions, we can prove that

$$x(k) = x_0 + \sum_{l=a}^{k-1} f(l, x(l), y(l)), \quad k \in \mathbb{N}(a, b - 1),$$

and

$$y(k) = y_0 + \sum_{l=a}^{k-1} g(l, x(l), y(l)), \quad k \in \mathbb{N}(a, b - 1).$$

Therefore $(x, y) \in S(x_0, y_0)$ which yields that $S(x_0, y_0)$ is closed, hence a compact subset in $C(\mathbb{N}(a, b - 1), \mathbb{R}^n) \times C(\mathbb{N}(a, b - 1), \mathbb{R}^n)$.

Finally, we prove that $S(\cdot)$ is u.s.c. by proving that the graph of S

$$\Gamma_S := \{(\bar{x}, \bar{y}, x, y) : y \in S(\bar{x}, \bar{y})\}$$

is closed. Let $(\bar{x}_m x, \bar{y}_m, x_m, y_m) \in \Gamma_S$ be such that $(\bar{x}_m x, \bar{y}_m, x_m, y_m) \to (\bar{x}, \bar{y}, x, y)$ as $m \to \infty$. Since $(x_m, y_m) \in S(\bar{x}_m, \bar{y}_m)$, then

$$x_m(k) = \bar{x}_m + \sum_{l=a}^{k-1} f(l, x_m(l), y_m(l)), \quad k \in \mathbb{N}(a, b - 1),$$

and

$$y_m(k) = \bar{y}_m + \sum_{l=a}^{k-1} g(l, x_m(l), y_m(l)), \quad k \in \mathbb{N}(a, b - 1).$$

Arguing as in Step 2, we can prove that

$$x(k) = \bar{x} + \sum_{l=a}^{k-1} f(l, x(l), y(l)), \quad k \in \mathbb{N}(a, b - 1),$$

and

$$y(k) = \bar{y} + \sum_{l=a}^{k-1} g(l, x(l), y(l)), \quad k \in \mathbb{N}(a, b-1).$$

Thus, $(x, y) \in S(\bar{x}, \bar{y})$. Now, we show that S maps bounded sets into relatively compact sets of $C(\mathbb{N}(a, b-1), \mathbb{R}^n) \times C(\mathbb{N}(a, b-1), \mathbb{R}^n)$. Let B be a bounded set in $\mathbb{R}^n \times \mathbb{R}^n$ and let $\{(x_m, y_m)\} \subset S(B)$. Then there exists $\{(\bar{x}_m, \bar{y}_m)\} \subset B$ such that

$$x_m(k) = \bar{x}_m + \sum_{l=a}^{k-1} f(l, x_m(l), y_m(l)), \quad k \in \mathbb{N}(a, b-1),$$

and

$$y_m(k) = \bar{y}_m + \sum_{l=a}^{k-1} g(l, x_m(l), y_m(l)), \quad k \in \mathbb{N}(a, b-1).$$

Since $\{(\bar{x}_m, \bar{y}_m)\}$ is a bounded sequence, there exists a subsequence of $\{(\bar{x}_m, \bar{y}_m)\}$ converging to $(\bar{x}, \bar{y})$. As in Step 2, we can show that $\{(x_m, y_m) : m \in \mathbb{N}\}$ is uniformly bounded. As a consequence of Theorem 8.25, we conclude that there exists a subsequence of $\{(x_m, y_m)\}$ converging to (x, y) in $C(\mathbb{N}(a, b-1), \mathbb{R}^n) \times C(\mathbb{N}(a, b-1), \mathbb{R}^n)$. By the continuity of f and g, we can prove that

$$x(k) = \bar{x} + \sum_{l=a}^{k-1} f(l, x(l), y(l)), \quad k \in \mathbb{N}(a, b-1),$$

and

$$y(k) = \bar{y} + \sum_{l=a}^{k-1} g(l, x(l), y(l)), \quad k \in \mathbb{N}(a, b-1).$$

Thus, $(x, y) \in \overline{S(B)}$. This implies that $S(\cdot)$ is *u.s.c.* $\qquad\square$

Now we shall consider the discrete system (15.7) where $f, g : \mathbb{N}(a) \times \mathbb{R}^n \times \mathbb{R}^n \to \mathbb{R}^n$ are continuous functions. For the second result we shall also use the following discrete version of Avramescu criteria of compactness in $BC(\mathbb{N}, \mathbb{R}^n)$.

Theorem 15.11. *A set $\Omega \subset BC(\mathbb{N}, E)$ is relatively compact if the following conditions hold:*

(i) *for every $k \in \mathbb{N}$ the set $\{y(k) : x \in \Omega\}$ is relatively compact in E,*

(ii) *the functions from Ω are equiconvergent at infinity, i.e. for every $\epsilon > 0$ there exists $k(\epsilon) \in \mathbb{N}$ such that if $|y(k) - y(\infty)| \le \epsilon$ for all $k > k_\epsilon$ and $y \in \Omega$.*

Theorem 15.12. *Let Let $f, g : \mathbb{N}(a, b-1) \times \mathbb{R}^n \times \mathbb{R}^n \longrightarrow \mathbb{R}^n$ be continuous functions satisfying*

(H_6) *There exist $p_1, p_2 \in C(\mathbb{N}(a, b-1), \mathbb{R}_+)$ such that*

$$|f(k, x, y)| \le p_1(k)(|x| + |y|), \quad k \in \mathbb{N}(a, b-1), \quad (x, y) \in \mathbb{R}^n \times \mathbb{R}^n,$$

and

$$|g(k, x, y)| \le p_2(k)(|x| + |y|), \quad k \in \mathbb{N}(a, b-1), \quad (x, y) \in \mathbb{R}^n \times \mathbb{R}^n,$$

with

$$\sum_{l=a}^{\infty} (p_1(k) + p_2(k)) < \infty.$$

Then problem (15.10) has at least one solution. Moreover, the solution set $S(x_0, y_0)$ is compact and the multivalued map $S : (x_0, y_0) \multimap S(x_0, y_0)$ is u.s.c.

Proof. Consider the operator $N : BC(\mathbb{N}(a), \mathbb{R}^n) \times C(\mathbb{N}(a), \mathbb{R}^n) \to BC(\mathbb{N}(a), \mathbb{R}^n)$ defined for $(x, y) \in C(\mathbb{N}(a), \mathbb{R}) \times C(\mathbb{N}(a), \mathbb{R}^n)$ by

$$N(x, y) = (N_1(x, y), N_2(x, y)), \tag{15.15}$$

where

$$N_1(x(k), y(k)) = x_0 + \sum_{l=a}^{k-1} f(l, x(l), y(l)), \quad k \in \mathbb{N}(a)$$

and

$$N_2(x(k), y(k)) = y_0 + \sum_{l=a}^{k-1} g(l, x(l), y(l)), \quad k \in \mathbb{N}(a).$$

In order to apply Lemma 8.15, we first show that N is completely continuous. The proof will be given in several steps.

Step 1. $N = (N_1, N_2)$ is continuous.
Let (x_m, y_m) be a sequence such that $(x_m, y_m) \to (x, y) \in BC(\mathbb{N}(a), \mathbb{R}^n) \times BC(\mathbb{N}(a), \mathbb{R}^n)$ as $m \to \infty$. Then

$$|N_1(x_n(k), y_n(k)) - N_1(x(k), y(k))| \leq \sum_{l=a}^{k-1} |(f(l, x_n(l), y_n(l)) - f(l, x(l), y(l)))|.$$

Similarly

$$|N_2(x_m(k), y_m(k)) - N_2(x(k), y(k))| \leq \sum_{l=a}^{k-1} |(g(l, x_n(l), y_n(l)) - g(l, x(l), y(l)))|.$$

Using the condition (H_6), for every $\epsilon > 0$, there exists $b(\epsilon) \in \mathbb{N}$ such that

$$\sum_{l=b(\epsilon)}^{\infty} 2q(p_1(k) + p_2(k)) < \frac{\epsilon}{2},$$

where

$$\|x_m\| \leq q, \quad \|y_m\| \leq q, \quad \text{for each } m \in \mathbb{N}.$$

Hence

$$\|N_1(x_m, y_m) - N_1(x, y)\|_\infty \leq \sum_{l=a}^{b(\epsilon)-1} |f(l, x_m(l), y_m(l)) - f(l, x(l), y(l))|$$

$$+ \sum_{l=b(\epsilon)}^{\infty} |f(l, x_m(l), y_m(l)) - f(l, x(l), y(l))|$$

$$\leq \sum_{l=a}^{b(\epsilon)-1} |f(l, x_m(l), y_m(l)) - f(l, x(l), y(l))|$$

$$+2q \sum_{l=b(\epsilon)}^{\infty} (p_1(k) + p_2(k))$$

$$\leq \sum_{l=a}^{b(\epsilon)-1} |f(l, x_m(l), y_m(l))$$

$$-f(l, x(l), y(l))| + \frac{\epsilon}{2}$$

and

$$\|N_2(x_m, y_m) - N_2(x, y)\|_\infty \leq \sum_{l=a}^{b(\epsilon)-1} |g(l, x_m(l), y_m(l)) - g(l, x(l), y(l))|$$

$$+ \sum_{l=b(\epsilon)}^{\infty} |g(l, x_m(l), y_m(l)) - g(l, x(l), y(l))|$$

$$\leq \sum_{l=a}^{b(\epsilon)-1} |g(l, x_m(l), y_m(l)) - g(l, x(l), y(l))|$$

$$+2q \sum_{l=b(\epsilon)}^{\infty} (p_1(k) + p_2(k))$$

$$\leq \sum_{l=a}^{b(\epsilon)-1} |g(l, x_m(l), y_m(l))$$

$$-g(l, x(l), y(l))| + \frac{\epsilon}{2}.$$

Since f, g are continuous functions, we get

$$\|N_1(x_m, y_m) - N_1(x, y)\|_\infty \leq \sum_{l=a}^{b(\epsilon)-1} |f(l, x_m(l), y_m(l)) - f(l, x(l), y(l))| + \frac{\epsilon}{2}$$

$$\to 0 \text{ as } m \to \infty$$

and

$$\|N_2(x_m, y_m) - N_2(x, y)\|_\infty \leq \sum_{l=a}^{b(\epsilon)-1} |g(l, x_m(l), y_m(l)) - g(l, x(l), y(l))| + \frac{\epsilon}{2}$$

$$\to 0 \text{ as } m \to \infty.$$

Step 2. We now show that $N(B_q)$ is equiconvergent at ∞, i.e., for every $\epsilon > 0$ there exists $k(\epsilon) \in \mathbb{N}$ such that

$$|N_i(x(k), y(k)) - N_i(x(\infty), y(\infty))| \leq \epsilon \text{ for all } k > k_\epsilon \text{ and } (x, y) \in B_q, i = 1, 2,$$

where

$$B_q = \{(x, y) \in BC(\mathbb{N}(a), \mathbb{R}^n) \times BC(\mathbb{N}(a), \mathbb{R}^n) : \|x\|_\infty \leq q, \|y\|_\infty \leq q\}.$$

Letting $(x, y) \in B_q$, then

$$|N_1(x(k), y(k)) - N_1(x(\infty), y(\infty))| \leq \sum_{l=k}^{\infty} |f(l, x_(l), y(l))|$$

$$\leq \ 2q \sum_{l=k}^{\infty} (p_1(k) + p_2(k))$$

and

$$|N_2(x(k), y(k)) - N_2(x(\infty), y(\infty))| \ \leq \ \sum_{l=k}^{\infty} |g(l, x_(l), y(l))|$$

$$\leq \ 2q \sum_{l=k}^{\infty} (p_1(k) + p_2(k)).$$

Since $2q \sum_{l=a}^{\infty} (p_1(k) + p_2(k)) < \infty$, there exists $k(\epsilon) \in \mathbb{N}$ such that

$$2q \sum_{l=k}^{\infty} (p_1(k) + p_2(k)) \leq \epsilon \qquad \text{for all } k > k(\epsilon).$$

Hence, for all $k > k(\epsilon)$ and $(x, y) \in B_q$, and $i = 1, 2$,

$$|N_i(x(k), y(k)) - N_i(x(\infty), y(\infty))| \leq \epsilon.$$

Then $N(B_q)$ is equiconvergent. As in theorem we can easily prove that $N(B_q)$ is uniformly bounded and for each $b \in \mathbb{N}$ and $k \in N(a, b - 1)$, the set

$$\{N(x(k), y(k)) : \ (x, y) \in B_q\}$$

is relatively compact in $\mathbb{R}^m \times \mathbb{R}^n$. With Theorem 15.11, we conclude that N is completely continuous.

Step 3. *A priori bounds on solutions.* Let $(x, y) \in BC(\mathbb{N}(a), \mathbb{R}^n)$ be such that $(x, y) = N(x, y)$. Thus, for $k \in \mathbb{N}(a)$, we have

$$|x(k)| \ \leq \ |x_0| + \sum_{l=1}^{k-1} |f(l, x(l), y(l))|$$

$$\leq \ |x_0| + \sum_{l=a}^{k-1} p_1(l)(|x(l)| + |y(l)|)ds,$$

and

$$|y(k)| \ \leq \ |x_0| + \sum_{l=a}^{k-1} p_2(l)(|x(l)| + |y(l)|)ds.$$

Therefore

$$|x(k)| + |y(k)| \ \leq \ |x_0| + |y_0| + \sum_{l=a}^{k-1} p(l)(|x(l)| + |y(l)|),$$

where

$$p(k) = p_1(k) + p_2(k), \quad k \in \mathbb{N}(a).$$

By Theorem 15.1, we have

$$|x(k)| + |y(k)| \leq (|x_0| + |y_0|) \left(1 + \sum_{l=a}^{k-1} p(k) \prod_{l+1}^{k-1} (1 + p(\tau)) \right).$$

Hence

$$\|x\|_\infty + \|y\|_\infty \le (|x_0| + |y_0|)\left(1 + \sum_{l=a}^{\infty} p(k) \prod_{l+1}^{b}(1 + p(\tau))\right) = R.$$

Finally, let

$$U := \{y \in BC(\mathbb{N}(a), \mathbb{R}^n) : (\|x\|_\infty, \|y\|_\infty) < (R+1, R+1)\}$$

and consider the operator $N : \overline{U} \to BC(\mathbb{N}(a), \mathbb{R}^n)$. From the choice of U, there is no $y \in \partial U$ such that $y \in \lambda N(y)$ for some $\lambda \in (0, 1)$. As a consequence of the version of the nonlinear alternative of Leray-Schauder in generalized Banach space (Lemma 8.15), N has a fixed point (x, y) in U which is a solution of problem (15.10).

Step 4. Arguing as in the proof of Theorem 15.10, Step 4, we can prove that the solutions set $S(x_0, y_0)$ of problem (15.10) is compact, where the multivalued operator $S(x_0, y_0) : \mathbb{R}^n \times \mathbb{R}^n \to \mathcal{P}(BC(\mathbb{N}(a), \mathbb{R}^n))$ is defined by

$$S(x_0, y_0) = \{(x, y) \in BC(\mathbb{N}(a), \mathbb{R}^m) :$$
$$(x, y) \text{ is solution of the problem (15.10)}\}.$$

$\square$

15.6 Systems of difference equations with infinite delay

Here we are concerned with the existence and uniqueness of bounded solutions in some state space of sequences for a system of semilinear functional difference equations with infinite delay. Several aspects of the theory of functional difference equations can be understood as a proper generalization of the theory of ordinary difference equations. However, the fact that the state space for functional difference equations is infinite dimensional requires the development of methods and techniques coming from functional analysis (e.g., theory of semigroups of operators on Banach spaces, spectral theory, fixed point theory, etc.). Some important contributions to the study of the mathematical aspects of such equations have been undertaken in [3, 4, 84] and the references therein.

Abstract retarded functional difference equations in phase space have great importance in applications. Consequently, the theory of difference equations with infinite delay has drawn the attention of several authors. Qualitative analysis, discrete maximal regularity, exponential dichotomy, and periodicity have received much attention; see [5, 7, 11, 40, 54, 55, 76, 88–91, 94, 112, 149, 151, 211, 264–266, 271]. For more information on functional difference equations, we suggest also [72, 87, 99, 210, 221, 278].

We consider the following system of linear functional difference equations

$$\begin{cases} x(n+1) &= A_1(n, x_n, y_n), \ n \ge 0, \\ y(n+1) &= A_2(n, x_n, y_n), \ n \ge 0, \end{cases} \tag{15.16}$$

and its perturbation

$$\begin{cases} x(n+1) &= A_1(n, x_n, y_n) + f_1(n, x_n, y_n), \ n \ge 0, \\ y(n+1) &= A_2(n, x_n, y_n) + f_2(n, x_n, y_n), \ n \ge 0, \\ x(0) &= \varphi \in \mathcal{B}, \\ y(0) &= \psi \in \mathcal{B}, \end{cases} \tag{15.17}$$

where $A_1, A_2 : \mathbb{Z}^+ \times \mathcal{B} \times \mathcal{B} \to \mathbb{C}^r$ are bounded linear maps with respect to the variables x_n and y_n; f_1, f_2 are $\mathbb{C}^r-$valued functions defined on the product space $\mathbb{Z}^+ \times X \times X$ under suitable conditions; $\mathcal{B}$ denotes an abstract phase space that we will explain briefly below, X is an appropriate Banach space. The notation, $x_\cdot$, denotes the $\mathcal{B}-$valued function defined by $n \to x_n$, where x_n is the history function, which is defined by $x_n(m) = x(n + m)$ for all $m \in \mathbb{Z}^-$.

Next, we are concerned with the following homogeneous retarded linear functional equations,

$$\begin{cases} x(n+1) & = L_1(x_n, y_n), \ n \geq 0, \\ y(n+1) & = L_2(x_n, y_n), \ n \geq 0, \end{cases} \tag{15.18}$$

and their perturbations, along with initial conditions, defined by the semilinear difference equation with infinite delay

$$\begin{cases} x(n+1) & = L_1(x_n, y_n) + g_1(n, x_n, y_n), \ n \geq 0, \\ y(n+1) & = L_2(x_n, y_n) + g_2(n, x_n, y_n), \ n \geq 0, \\ x(0) & = \varphi \in \mathcal{B}, \\ y(0) & = \psi \in \mathcal{B}, \end{cases} \tag{15.19}$$

where $L_1, L_2 : \mathcal{B} \times \mathcal{B} \to \mathbb{C}^r$ are bounded operators and $g_1, g_2 : \mathbb{Z}^+ \times \mathcal{B} \times \mathcal{B} \to \mathbb{C}^r$ are given functions.

15.6.1 Definitions and fundamental results

Here we present notations and provide some auxiliary results that we will need in all sections of this chapter. The phase space $\mathcal{B} = \mathcal{B}(\mathbb{Z}^-, \mathbb{C}^r)$ is a Banach space with a norm denoted by $\| \cdot \|_\mathcal{B}$ which is a subfamily of functions from $\mathbb{Z}^-$ into $\mathbb{C}^r$ and it is assumed to satisfy the following axioms.

Axiom (A): There are a positive constant J and nonnegative functions $N(\cdot)$ and $M(\cdot)$ on $\mathbb{Z}^+$ with the property that if $x : \mathbb{Z}^+ \to \mathbb{C}^r$ is a function such that if $x_0 \in \mathcal{B}$, then for all $n \in \mathbb{Z}^+$

(i) $x_n \in \mathcal{B}$;

(ii) $J|x_n| \leq \|x_n\|_\mathcal{B} \leq N(n) \sup_{0 \leq s \leq n} |x(s)| + M(n)\|x_0\|_\mathcal{B}$.

Denote by $B(\mathbb{Z}^-, \mathbb{C}^r)$ the set of bounded functions from $\mathbb{Z}^-$ to $\mathbb{C}^r$.

Axiom (B): The inclusion map $i : (B(Z^-, \mathbb{C}^r), \| \cdot \|_\infty) \to (\mathcal{B}, \| \cdot \|_\mathcal{B})$ is continuous, i.e., there is a constant $d > 0$ such that $\|\varphi\|_\mathcal{B} \leq d\|\varphi\|_\infty$ for all $\varphi \in \mathcal{B}(\mathbb{Z}^-, \mathbb{C}^r)$.

Hereafter, $\mathcal{B}$ will denote a phase space satisfying Axioms (A) and (B).

For any $n \geq \tau$ we define the bounded linear operator $U(n, \tau) : \mathcal{B} \to \mathcal{B}$ by $U(n, \tau)\varphi = x_n(\tau, \varphi, 0)$ for $\varphi \in \mathcal{B}$, where $x(\cdot, \tau, \varphi, 0)$ denotes the solution of the homogeneous linear system (15.16). The operator $U(n, \tau)$ is called the solution operator of the homogeneous linear system (15.16).

Definition 15.13. *[76] We say that equation (15.16) (or its solution operator $U(n, \tau), n, \ \tau \in \mathbb{Z}^+$) has an exponential dichotomy on $\mathcal{B}$ with data $(\alpha, K, P(\cdot))$, if the solution operator $U(n, \tau)$ satisfies the following property: there are positive constants α, K, and a projection operator $P(n), n \in \mathbb{Z}^+$, in $\mathcal{B}$, such that if $Q(n) = I - P(n)$, where I is the identity operator, then:*

(i) $U(n, \tau)P(\tau) = P(n)U(n, \tau), n \geq \tau.$

(ii) *The restriction* $U(n,\tau)|Range(Q(\tau)), n \geq \tau$, *is an isomorphism from* $Range(Q(\tau))$ *onto* $Range(Q(n))$, *and then we define* $(U(\tau,n))$ *as its inverse mapping.*

(iii) $\|U(n,\tau)\varphi\|_{\mathcal{B}} \leq Ke^{-\alpha(n-\tau)}\|\varphi\|_{\mathcal{B}}, n \geq \tau, \varphi \in P(\tau)\mathcal{B}$.

(iv) $\|U(n,\tau)\varphi\|_{\mathcal{B}} \leq Ke^{\alpha(n-\tau)}\|\varphi\|_{\mathcal{B}}, \tau > n, \varphi \in Q(\tau)\mathcal{B}$,

We denote by $\Gamma(n,s)$ the Green function associated with (15.16), that is,

$$\Gamma(t,s) = \begin{cases} U(n,s+1)P(s+1) & n-1 \geq s, \\ -U(n,s+1)Q(s+1) & s > n-1. \end{cases} \tag{15.20}$$

Denote by X the Banach space of all bounded functions $\eta : \mathbb{Z}^+ \to \mathcal{B}$ endowed with the norm

$$\|\eta\| = \sup_{n \geq 0} \|\eta\|_{\mathcal{B}}. \tag{15.21}$$

For any number $1 \leq p < \infty$, we consider the following spaces:

$$l^p(\mathbb{Z}^+, \mathcal{B}) = \{\xi : \mathbb{Z}^+ \to \mathcal{B}/\|\xi(n)\|_p^p = \sum_{n=0}^{\infty} \|\xi(n)\|_{\mathcal{B}}^p < \infty\},$$

$$l^\infty(\mathbb{Z}^+, \mathcal{B}) = \{\xi : \mathbb{Z}^+ \to \mathcal{B}/\|\xi\|_\infty = \sup_{n \in \mathbb{Z}^+} \|\xi(n)\|_{\mathcal{B}} < \infty\},$$

$$l_\beta^\infty(\mathbb{Z}^+, \mathcal{B}) = \{\xi : \mathbb{Z}^+ \to \mathcal{B}/\|\xi\|_\beta = \sup_{n \in \mathbb{Z}^+} \|\xi(n)\|_{\mathcal{B}}e^{-\beta n} < \infty\},$$

$$l_\beta^1(\mathbb{Z}^+, \mathbb{C}^r) = \{\varphi : \mathbb{Z}^- \to \mathbb{C}^r/\|\varphi\|_{1,\beta} = \sum_{n=0}^{\infty} |\varphi(n)| \, e^{-\beta n} < \infty\},$$

$$l^p(\mathbb{Z}^+, \mathbb{C}^r) = \{\varphi : \mathbb{Z}^+ \to \mathbb{C}^r/\|\varphi\|_p^p = \sum_{n=0}^{\infty} |\varphi(n)|^p < \infty\}.$$

Lemma 15.14. *Let* $\tau \in \mathbb{Z}^+$. *Assume that the function* $z : [\tau, \infty) \to \mathcal{B}$ *satisfies the relation*

$$z(n) = U(n,\tau)z(\tau) + \sum_{s=\tau}^{n-1} U(n,s+1)E^0(P(s)), n \geq \tau, \tag{15.22}$$

where

$$E^0(t) = \begin{cases} Id & \text{the unite matrix }, t = 0 \\ 0 & \text{the zero matrix}, \ t < 0 \end{cases}$$

and define a function $y : \mathbb{Z} \to \mathbb{C}^r$ *by*

$$y(n) = \begin{cases} (z(n))(0), & n \geq \tau, \\ (z(\tau))(n-\tau), & n < \tau. \end{cases} \tag{15.23}$$

Then $y(n)$ *satisfies the equation*

$$y(n+1) = L(n,y_n) + P(n), \ n \geq \tau, \tag{15.24}$$

together with relation $y_n = z(n), \ n \geq \tau$.

Definition 15.15. *[91] We say that system (15.16) has discrete maximal regularity if, for each $h_1, h_2 \in l^p(\mathbb{Z}^+, \mathbb{C}^r)$ $(1 \le p < \infty)$ and each $\varphi, \psi \in P(0)\mathcal{B}$, the solution z of the boundary value problem*

$$\begin{cases} z(n+1) & = A_1(n, z_n, \tilde{z}_n) + h_1(n), \ n \ge 0, \\ \tilde{z}(n+1) & = A_2(n, z_n, \tilde{z}_n) + h_2(n), \ n \ge 0, \\ P(0)z(0) & = \varphi \in \mathcal{B}, \\ P(0)\tilde{z}(0) & = \psi \in \mathcal{B}, \end{cases} \qquad (15.25)$$

satisfies $z., \tilde{z}. \in l^p(\mathbb{Z}^+, \mathcal{B})$.

Theorem 15.16. *[76] Assume that system (15.16) has an exponential dichotomy on $\mathcal{B}$ with data $(\alpha, K, P(\cdot))$. Then system (15.17) has discrete maximal regularity.*

Theorem 15.17. *Assume that system (15.16) has an exponential dichotomy with data $(\alpha, K, P(\cdot))$. Then, for any $h_1, h_2 \in l^p(\mathbb{Z}^+, \mathbb{C}^r)$ $(1 \le p < \infty)$ and each $\varphi, \psi \in P(0)\mathcal{B}$, the boundary value problem (15.25) has a unique solution $(z, \tilde{z})$ so that $z., \tilde{z}. \in l^p(\mathbb{Z}^+, \mathcal{B})$.*

Theorem 15.18. *(Exponential boundedness of the solution operator) [76]. Assume that*

(H_1) $\{A_1(n, \cdot, \cdot)\}$ and $\{A_2(n, \cdot, \cdot)\}$ are uniformly bounded sequences of bounded linear operators mapping $\mathcal{B} \times \mathcal{B}$ into $\mathbb{C}^r$. There are constants $M_i, N_i > 1, i = 1, 2$ such that

$$|A_1(n, \varphi, \psi)| \le M_1\|\varphi\|_{\mathcal{B}} + N_1\|\psi\|_{\mathcal{B}}, \text{ for all } n \in \mathbb{Z}^+ \text{ and } \varphi, \psi \in \mathcal{B},$$

and

$$|A_2(n, \varphi, \psi)| \le M_2\|\varphi\|_{\mathcal{B}} + N_2\|\psi\|_{\mathcal{B}}, \text{ for all } n \in \mathbb{Z}^+ \text{ and } \varphi, \psi \in \mathcal{B},$$

(H_2) The functions $N(\cdot)$ and $M(\cdot)$ given in Axiom A are bounded.

Then, there are positive constants λ and δ such that

$$\|U(n, m)\|_{\mathcal{B}} \le \lambda e^{\delta(n-m)}, \ n \ge m \ge 0. \qquad (15.26)$$

Proposition 15.19. *[76] Under the conditions $(H_1) - (H_2)$, if system (15.16) has an exponential dichotomy with data $(\alpha, K, P(.))$, then*

(i) $\sup_{n \in \mathbb{Z}^+} \|P(n)\|_{\mathcal{B}} < \infty$.

(ii) $Range(P(n)) = \{\varphi \in \mathcal{B} : e^{-\eta(n-m)}U(n, m)\varphi$ is bounded for $n \ge m\}$ for any $0 < \eta < \alpha$.

(iii) Let $\widehat{P}(0)$ be a projection such that $Range(\widehat{P}(0)) = Range(P(0))$. Then (15.17) has an exponential dichotomy on $\mathbb{Z}^+$ with data $(\alpha, \widehat{K}, \widehat{P}(\cdot))$, where

$$\widehat{P}(n) = P(n) + U(n, 0)\widehat{P}(0)U(0, n)Q(n),$$

$$\widehat{K} = (K + K^2\|\widehat{P}(0)\|_{\mathcal{B}}) \sup_{m \ge 0}(1 + \|P(n)\|_{\mathcal{B}}).$$

In addition, we have

$$\sup_{m \ge 0} \|\widehat{P}(m)\|_{\mathcal{B}} \le (1 + K^2\|\widehat{P}(0) - P(0)\|_{\mathcal{B}}) \sup_{m \ge 0}(1 + \|P(m)\|_{\mathcal{B}}). \qquad (15.27)$$

Also one has

$$\widehat{P}(n) = P(n) + o(1), \text{ as } n \to \infty. \qquad (15.28)$$

Definition 15.20. *[167]. A sequence $\xi \in l^\infty(\mathbb{Z}^+, \mathcal{B})$ is called (discrete) S-asymptotically ω-periodic if there is $\omega \in \mathbb{Z}^+\backslash\{0\}$ such that $\lim_{n\to\infty}(\xi(n + \omega) - \xi(n)) = 0$. In this case we say that ω is an asymptotic period of ξ.*

In this work the notation $SAP_\omega(\mathcal{B})$ stands for the subspace of $l^\infty(\mathbb{Z}^+, \mathcal{B})$ consisting of all the (discrete) $S-$asymptotically ω-periodic sequences. From [167], $SAP_\omega(\mathcal{B})$ is a Banach space.

Definition 15.21. *[167] A strongly continuous function $F : \mathbb{Z}^+ \to \mathcal{L}(\mathcal{B})$ is said to be strongly S-asymptotically periodic if, for each $\varphi \in \mathcal{B}$, there is $\omega_\varphi \in \mathbb{Z}^+\backslash\{0\}$ such that $F(.)\varphi$ is S-asymptotically ω_φ-periodic.*

Definition 15.22. *[167]. A continuous function $g : \mathbb{Z}^+ \times \mathcal{B} \to \mathbb{C}^r$ is said to be uniformly S-asymptotically ω-periodic on bounded sets if, for every bounded subset B of $\mathcal{B}$, the set $\{g(n, \varphi) : n \in \mathbb{Z}^+, \varphi \in B\}$ is bounded and $\lim_{n\to\infty}(g(n, \varphi) - g(n + \omega, \varphi)) = 0$ uniformly on $\varphi \in B$.*

Definition 15.23. *[167]. A function $g : \mathbb{Z}^+ \times \mathcal{B} \to \mathbb{C}^r$ is called uniformly asymptotically continuous on bounded sets if, for every $\epsilon > 0$ and every bounded subset B of $\mathcal{B}$, there are $K_{\epsilon,B} \geq 0$ and $\delta_{\epsilon,B} \geq 0$ such that $|g(n, \varphi) - g(n, \psi)| < \epsilon$, for all $n \geq K_{\epsilon,B}$ and all $\varphi, \psi \in B$ with $\|\varphi - \psi\| < \delta_{\epsilon,B}$.*

Lemma 15.24. *[167]. Let $g : \mathbb{Z}^+ \times \mathcal{B} \to \mathbb{C}^r$ be uniformly S-asymptotically ω-periodic on bounded sets and asymptotically uniformly continuous on bounded sets. Let $\xi : \mathbb{Z}^+ \to \mathcal{B}$ be a discrete S-asymptotically ω-periodic function. Then the function $g(\cdot, \xi(\cdot))$ is discrete $S-$asymptotically ω-periodic.*

15.6.2 Boundedness of solutions

In this subsection, we are concerned with the study of the existence of bounded solutions for the semilinear difference equation with infinite delay via discrete maximal regularity.

Theorem 15.25. *[225] Assume that system (15.16) has exponential dichotomy on $\mathcal{B}$, and in addition to conditions (H_1) and (H_2), suppose that the following conditions hold:*

(H_3) *The functions $f_i(n, \cdot, \cdot) : l^p(\mathbb{Z}^+, \mathcal{B}) \times l^p(\mathbb{Z}^+, \mathcal{B}) \to \mathbb{C}^r$, $i = 1, 2$ satisfy, for all $x, y, \overline{x}, \overline{y} \in l^p(\mathbb{Z}^+, \mathcal{B})$ and $n \in \mathbb{Z}^+$,*

$$|f_1(n, x, y) - f_1(n, \overline{x}, \overline{y})| \leq a_1(n)\|x - \overline{x}\|_p + b_1(n)\|y - \overline{y}\|_p$$

and

$$|f_2(n, x, y) - f_2(n, \overline{x}, \overline{y})| \leq a_2(n)\|x - \overline{x}\|_p + b_2(n)\|y - \overline{y}\|_p$$

where $a_i, b_i \in l^p(\mathbb{Z}^+)$, $i = 1, 2$.

(H_4) *$f_1(\cdot, 0, 0), f_2(\cdot, 0, 0) \in l^p(\mathbb{Z}^+, \mathbb{C}^r)$.*

(H_5) *The matrix $M \in \mathcal{M}_{2\times 2}(\mathbb{R}_+)$ such that*

$$M = 2dK(1 - e^{-\alpha})^{-1} \sup_{n\in\mathbb{Z}^+}(1 + \|P(n)\|_\mathcal{B}) \begin{pmatrix} \|a_1\|_p & \|b_1\|_p \\ \|a_2\|_p & \|b_2\|_p \end{pmatrix}$$

converges to zero.

Then, for each $\varphi, \psi \in P(0)\mathcal{B}$ there is a unique bounded solution (x, y) of system (15.17) with $P(0)x_0 = \varphi$, $P(0)y_0 = \psi$, such that $(x., y.) \in l^p(\mathbb{Z}^+, \mathcal{B}) \times l^p(\mathbb{Z}^+, \mathcal{B})$.

Proof. Let ξ, η be sequences in $l^p(\mathbb{Z}^+, \mathcal{B})$. Using conditions (H_3) and (H_4) we obtain that the function $F(\cdot) = f_1(\cdot, \xi, \eta)$ is in $l^p(\mathbb{Z}^+, \mathbb{C}^r)$, and we have

$$
\begin{aligned}
\|F\|_p^p &= \sum_{n=0}^{\infty} |f_1(n, \xi, \eta)|^p \\
&\leq \sum_{n=0}^{\infty} (|f_1(n, \xi, \eta) - f_1(n, 0, 0)| + |f_1(n, 0, 0)|)^p \\
&\leq 2^p \sum_{n=0}^{\infty} |f_1(n, \xi, \eta) - f_1(n, 0, 0)|^p + 2^p \sum_{n=0}^{\infty} |f_1(n, 0, 0)|^p \\
&\leq 2^p \sum_{n=0}^{\infty} (a_1^p(n)\|\xi\|_{\mathcal{B}}^p + b_1^p(n)\|\eta\|_{\mathcal{B}}^p) + 2^p \|f_1(n, 0, 0)\|_p^p.
\end{aligned}
$$

Hence
$$
\|F\|_p \leq 2(\|a_1\|_p\|\xi\|_p + \|b_1\|_p\|\eta\|_p + \|f_1(\cdot, 0, 0)\|_p).
$$

Similarly we obtain that the function $G(\cdot) = f_2(\cdot, \xi, \eta)$ is in $l^p(\mathbb{Z}^+, \mathbb{C}^r)$, and

$$
\begin{aligned}
\|G\|_p^p &= \sum_{n=0}^{\infty} |f_2(n, \xi, \eta)|^p \\
&\leq \sum_{n=0}^{\infty} (|f_2(n, \xi, \eta) - f_2(n, 0, 0)| + |f_2(n, 0, 0)|)^p \\
&\leq 2^p \sum_{n=0}^{\infty} |f_2(n, \xi, \eta) - f_2(n, 0, 0)|^p + 2^p \sum_{n=0}^{\infty} |f_2(n, 0, 0)|^p \\
&\leq 2^p \sum_{n=0}^{\infty} (a_2^p(n)\|\xi\|_{\mathcal{B}}^p + b_2^p(n)\|\eta\|_{\mathcal{B}}^p) + 2^p \|f_2(n, 0, 0)\|_p^p.
\end{aligned}
$$

Hence
$$
\|G\|_p \leq 2(\|a_2\|_p\|\xi\|_p + \|b_2\|_p\|\eta\|_p + \|f_2(\cdot, 0, 0)\|_p).
$$

If $\varphi, \psi \in P(0)\mathcal{B}$, then by Theorem 15.16, system (15.17) has discrete maximal regularity, and so the Cauchy system

$$
\begin{cases}
z(n+1) &= A_1(n, z_n, \widetilde{z}_n) + F(n), \ n \in \mathbb{Z}^+, \\
\widetilde{z}(n+1) &= A_2(n, z_n, \widetilde{z}_n) + G(n), \ n \in \mathbb{Z}^+, \\
P(0)z_0 &= \varphi, \\
P(0)\widetilde{z}_0 &= \psi,
\end{cases}
\tag{15.29}
$$

has a unique solution $(z, \widetilde{z}) \in l^p(\mathbb{Z}^+, \mathcal{B}) \times l^p(\mathbb{Z}^+, \mathcal{B})$, which is given by

$$
z_n = H_1(\xi(n), \eta(n)) = U(n, 0)P(0)\varphi + \sum_{s=0}^{\infty} \Gamma(n, s) f_1(s, \xi, \eta)
\tag{15.30}
$$

and

$$
\widetilde{z}_n = H_2(\xi(n), \eta(n)) = U(n, 0)P(0)\psi + \sum_{s=0}^{\infty} \Gamma(n, s) f_2(s, \xi, \eta).
\tag{15.31}
$$

We now show the operator $H : l^p(\mathbb{Z}^+, \mathcal{B}) \times l^p(\mathbb{Z}^+, \mathcal{B}) \to l^p(\mathbb{Z}^+, \mathcal{B}) \times l^p(\mathbb{Z}^+, \mathcal{B})$ has a unique fixed point, defined as follows:

$$H(\xi(n), \eta(n)) = (H_1(\xi(n), \eta(n)), H_2(\xi(n), \eta(n))), \ (\xi, \eta) \in l^p(\mathbb{Z}^+, \mathcal{B}) \times l^p(\mathbb{Z}^+, \mathcal{B}).$$

Let $\xi_1, \eta_1, \xi_2, \eta_2 \in l^p(\mathbb{Z}^+, \mathcal{B})$. We have that

$$\|H_1(\xi_1, \eta_1) - H_1(\xi_2, \eta_2)\|_p$$

$$= \left[\sum_{n=0}^{\infty} \left\| \sum_{s=0}^{\infty} \Gamma(n, s) \left[f_1(s, \xi_1, \eta_1) - f_1(s, \xi_2, \eta_2) \right] \right\|_{\mathcal{B}}^p \right]^{\frac{1}{p}}$$

$$\leq dK \sup_{n \in \mathbb{Z}^+} (1 + \|P(n)\|_{\mathcal{B}})$$

$$\times \left[\sum_{n=0}^{\infty} \left(\sum_{s=0}^{\infty} e^{-\alpha|n-(s+1)|} \left[a_1(s) \|\xi_1 - \xi_2\|_{\mathcal{B}} + b_1(s) \|\eta_1 - \eta_2\|_{\mathcal{B}} \right] \right)^p \right]^{\frac{1}{p}}$$

$$\leq 2dK \sup_{n \in \mathbb{Z}^+} (1 + \|P(n)\|_{\mathcal{B}})$$

$$\times \sum_{n=0}^{\infty} e^{-\alpha n} \sum_{s=0}^{\infty} e^{\alpha(s+1)} (\|a_1\|_p \|\xi_1 - \xi_2\|_p + \|b_1\|_p \|\eta_1 - \eta_2\|_p)$$

$$\leq 2dK(1 - e^{-\alpha})^{-1} \sup_{n \in Z^+} (1 + \|P(n)\|_{\mathcal{B}})$$

$$\times (\|a_1\|_p \|\xi_1 - \xi_2\|_p + \|b_1\|_p \|\eta_1 - \eta_2\|_p).$$

Hence,

$$\|H_1(\xi_1, \eta_1) - H_1(\xi_2, \eta_2)\|_p \ \leq \ 2dK(1 - e^{-\alpha})^{-1} \sup_{n \in \mathbb{Z}^+} (1 + \|P(n)\|_{\mathcal{B}})$$

$$\times \ (\|a_1\|_p \|\xi_1 - \xi_2\|_p + \|b_1\|_p \|\eta_1 - \eta_2\|_p).$$

Similarly,

$$\|H_2(\xi_1, \eta_1) - H_2(\xi_2, \eta_2)\|_p \ \leq \ 2dK(1 - e^{-\alpha})^{-1} \sup_{n \in \mathbb{Z}^+} (1 + \|P(n)\|_{\mathcal{B}})$$

$$\times \ (\|a_2\|_p \|\xi_1 - \xi_2\|_p + \|b_2\|_p \|\eta_1 - \eta_2\|_p).$$

Then,

$$\|H(\xi_1, \eta_1) - H(\xi_2, \eta_2)\|_p$$

$$\leq 2dK(1 - e^{-\alpha})^{-1} \sup_{n \in \mathbb{Z}^+} (1 + \|P(n)\|_{\mathcal{B}}) \begin{pmatrix} \|a_1\|_p & \|b_1\|_p \\ \|a_2\|_p & \|b_2\|_p \end{pmatrix} \begin{pmatrix} \|\xi_1 - \xi_2\|_p \\ \|\eta_1 - \eta_2\|_p \end{pmatrix}.$$

By (H_5) and Theorem 8.1, it follows that H has a unique fixed point $(\xi, \eta) \in l^p(\mathbb{Z}^+, \mathcal{B}) \times l^p(\mathbb{Z}^+, \mathcal{B})$.

Let (ξ, η) be the unique fixed point of H. Then we have

$$\|\xi\|_p \ = \ \left[\sum_{n=0}^{\infty} \left\| U(n, 0)P(0)\varphi + \sum_{s=0}^{\infty} U(n, s+1) f_1(s, \xi, \eta) \right\|_{\mathcal{B}}^p \right]^{\frac{1}{p}}$$

$$\leq \ \left[\sum_{n=0}^{\infty} \|U(n, 0)P(0)\varphi\|_{\mathcal{B}}^p \right]^{\frac{1}{p}} + \left[\sum_{n=0}^{\infty} \left\| \sum_{s=0}^{\infty} U(n, s+1) f_1(s, \xi, \eta) \right\|_{\mathcal{B}}^p \right]^{\frac{1}{p}}$$

$$\leq \quad K \left[\sum_{j=0}^{\infty} e^{-\alpha p j} \right]^{1/p} \|\varphi\|_{\mathcal{B}}$$

$$+2dK \sup_{m\geq 0}(1 + \|P(m)\|_{\mathcal{B}})(1 - e^{-\alpha})^{-1}$$

$$\times \left[\sum_{s=0}^{\infty} |f_1(s, \xi, \eta)|^p \right]^{1/p}$$

$$\leq \quad K(1 - e^{-\alpha})^{-1}\|\varphi\|_{\mathcal{B}}$$

$$+2dK \sup_{m\geq 0}(1 + \|P(m)\|_{\mathcal{B}})(1 - e^{-\alpha})^{-1}$$

$$\times (\|a_1\|_p\|\xi\|_p + \|b_1\|_p\|\eta\|_p + \|f_1(., 0)\|_p),$$

and

$$\|\eta\|_p \quad \leq \quad K(1 - e^{-\alpha})^{-1}\|\psi\|_{\mathcal{B}}$$

$$+2dK \sup_{m\geq 0}(1 + \|P(m)\|_{\mathcal{B}})(1 - e^{-\alpha})^{-1}$$

$$\times (\|a_2\|_p\|\xi\|_p + \|b_2\|_p\|\eta\|_p + \|f_2(., 0)\|_p),$$

so

$$\begin{pmatrix} \|\xi\|_p \\ \|\eta\|_p \end{pmatrix}$$

$$\leq 2Kd \sup_{m\geq 0}(1 + \|P(m)\|_{\mathcal{B}}(1 - e^{-\alpha})^{-1} \begin{pmatrix} \|a_1\|_p & \|b_1\|_p \\ \|a_2\|_p & \|b_2\|_p \end{pmatrix} \begin{pmatrix} \|\xi\|_p \\ \|\eta\|_p \end{pmatrix}$$

$$+ K(1 - e^{-\alpha})^{-1} \begin{pmatrix} \|\varphi\|_p \\ \|\psi\|_p \end{pmatrix} + 2dK \sup_{m\geq 0}(1 + \|P(m)\|_{\mathcal{B}})$$

$$\times (1 - e^{-\alpha})^{-1} \begin{pmatrix} \|f_1(\cdot, 0)\|_p \\ \|f_2(\cdot, 0)\|_p \end{pmatrix}.$$

Then

$$\begin{pmatrix} \|\xi\|_p \\ \|\eta\|_p \end{pmatrix} \quad \leq \quad (Id - M)^{-1} \left[K(1 - e^{-\alpha})^{-1} \begin{pmatrix} \|\varphi\|_p \\ \|\psi\|_p \end{pmatrix} \right.$$

$$\left. + 2dK \sup_{m\geq 0}(1 + \|P(m)\|_{\mathcal{B}})(1 - e^{-\alpha})^{-1} \begin{pmatrix} \|f_1(\cdot, 0)\|_p \\ \|f_2(\cdot, 0)\|_p \end{pmatrix} \right].$$

□

15.6.3   Weighted boundedness and asymptotic behavior

We have the following result about weighted bounded solutions.

Theorem 15.26. *[225] Assume that conditions $(H_1) - (H_2)$ hold. Let λ and δ be the constants of Theorem 15.18. In addition, suppose that the following conditions hold:*

(C_1) *The functions $f_i(n, \cdot, \cdot) : \mathcal{B} \times \mathcal{B} \to \mathbb{C}^r$, $i = 1, 2$ satisfy, for all $x, y, \overline{x}, \overline{y} \in \mathcal{B}$ and $n \in \mathbb{Z}^+$,*

$$|f_1(n, x, y) - f_1(n, \overline{x}, \overline{y})| \quad \leq \quad a_1(n)\|x - \overline{x}\|_{\mathcal{B}} + b_1(n)\|y - \overline{y}\|_{\mathcal{B}},$$

and

$$|f_2(n,x,y) - f_2(n,\overline{x},\overline{y})| \leq a_2(n)\|x - \overline{x}\|_{\mathcal{B}} + b_2(n)\|y - \overline{y}\|_{\mathcal{B}}$$

where $a_i, b_i \in l^1(\mathbb{Z}^+)$, $i = 1, 2$;

(C_2) $f_1(\cdot, 0, 0), f_2(\cdot, 0, 0) \in l^1_\delta(\mathbb{Z}^+, \mathbb{C}^r)$;

(C_3) The matrix $\widehat{M} \in \mathcal{M}_{2\times 2}(\mathbb{R}_+)$ defined by

$$\widehat{M} = \lambda d e^{-\delta} \left(\begin{array}{cc} \|a_1\|_1 & \|b_1\|_1 \\ \|a_2\|_1 & \|b_2\|_1 \end{array} \right)$$

is convergent to zero.

Then, there is an unique weighted bounded solution (x, y) of system (15.17) with $x_0 = 0, y_0 = 0$.

Proof. We define the operator $\Omega = (\Omega_1, \Omega_2)$ on $l^\infty_\delta(\mathbb{Z}^+, \mathcal{B})$ by

$$\Omega_1(\xi(n), \eta(n)) = \sum_{s=0}^{n-1} U(n, s+1) f_1(s, \xi(s), \eta(s)), \ \xi, \eta \in l^\infty_\delta$$

and

$$\Omega_2(\xi(n), \eta(n)) = \sum_{s=0}^{n-1} U(n, s+1) f_2(s, \xi(s), \eta(s)), \ \xi, \eta \in l^\infty_\delta.$$

We now show that the operator $\Omega : l^\infty_\delta(\mathbb{Z}^+, \mathcal{B}) \times l^\infty_\delta(\mathbb{Z}^+, \mathcal{B}) \to l^\infty_\delta(\mathbb{Z}^+, \mathcal{B}) \times l^\infty_\delta(\mathbb{Z}^+, \mathcal{B})$ has a unique fixed point. We observe that Ω is well defined. In fact, we obtain

$$\|\Omega_1(\xi, \eta)\|_{\mathcal{B}} e^{-\delta n}$$

$$\leq \lambda d e^{-\delta} \sum_{s=0}^{n-1} |f_1(s, \xi(s), \eta(s))| e^{-\delta s}$$

$$\leq \lambda d e^{-\delta} \left[\sum_{s=0}^{n-1} (a_1(s)\|\xi(s)\|_{\mathcal{B}} e^{\delta s} + b_1(s) e^{\delta s} \|\eta(s)\|_{\mathcal{B}} + \sum_{s=0}^{n-1} |f_1(\cdot, 0, 0)| e^{\delta s}) \right]$$

and

$$\|\Omega_2(\xi, \eta)\|_{\mathcal{B}} e^{-\delta n}$$

$$\leq \lambda d e^{-\delta} \sum_{s=0}^{n-1} |f_2(s, \xi(s), \eta(s))| e^{-\delta s}$$

$$\leq \lambda d e^{-\delta} \left[\sum_{s=0}^{n-1} (a_2(s)\|\xi(s)\|_{\mathcal{B}} e^{\delta s} + b_2(s) e^{\delta s} \|\eta(s)\|_{\mathcal{B}} + \sum_{s=0}^{n-1} |f_2(\cdot, 0, 0)| e^{\delta s}) \right],$$

and so

$$\|\Omega_1(\xi, \eta)\|_\delta \leq \lambda d e^{-\delta}[\|a_1\|_1 \|\xi(s)\|_\delta + \|b_1\|_1 \|\eta(s)\|_\delta + \|f_1(\cdot, 0, 0)\|_{1,\delta}]$$

and

$$\|\Omega_2(\xi, \eta)\|_\delta \leq \lambda d e^{-\delta}[\|a_2\|_1 \|\xi(s)\|_\delta + \|b_2\|_1 \|\eta(s)\|_\delta + \|f_2(\cdot, 0, 0)\|_{1,\delta}].$$

Hence, the space l_β^∞ is invariant under Ω. Next let (ξ, η) and $(\overline{\xi}, \overline{\eta})$ be in $l_\delta^\infty \times l_\delta^\infty$. Then

$$\|\Omega_1(\xi(s), \eta(s)) - \Omega_1(\overline{\xi}(s), \overline{\eta}(s))\|_\delta \leq \lambda d e^{-\delta} \left[\|a_1\|_1 \|\xi - \overline{\xi}\|_\delta + \|b_1\|_1 \|\eta - \overline{\eta}\|_\delta \right]$$

and

$$\|\Omega_2(\xi(s), \eta(s)) - \Omega_2(\overline{\xi}(s), \overline{\eta}(s))\|_\delta \leq \lambda d e^{-\delta} \left[\|a_2\|_1 \|\xi - \overline{\xi}\|_\delta + \|b_2\|_1 \|\eta - \overline{\eta}\|_\delta \right].$$

Then

$$\|\Omega(\xi, \eta) - \Omega(\overline{\xi}, \overline{\eta})\|_\delta \leq \lambda d e^{-\delta} \begin{pmatrix} \|a_1\|_1 & \|b_1\|_1 \\ \|a_2\|_1 & \|b_2\|_1 \end{pmatrix} \times \begin{pmatrix} \|\xi - \overline{\xi}\|_\delta \\ \|\eta - \overline{\eta}\|_\delta \end{pmatrix}.$$

It follows that Ω has a unique fixed point $(\xi, \eta) \in l_\delta^\infty(\mathbb{Z}^+, \mathcal{B}) \times l_\delta^\infty(\mathbb{Z}^+, \mathcal{B})$. The uniqueness of the solution is reduced to the uniqueness of the fixed point of the map Ω. Let (ξ, η) be the unique fixed point of Ω. Then we have

$$\|\xi\|_\delta \leq \lambda d e^{-\delta} [\|a_1\|_1 \|\xi\|_\delta + \|b_1\|_1 \|\eta\|_\delta + \|f_1(\cdot, 0, 0)\|_{1,\delta}].$$

and

$$\|\eta\|_\delta \leq \lambda d e^{-\delta} [\|a_2\|_1 \|\xi\|_\delta + \|b_2\|_1 \|\eta\|_\delta + \|f_2(\cdot, 0, 0)\|_{1,\delta}].$$

Then

$$\begin{pmatrix} \|\xi\|_\delta \\ \|\eta\|_\delta \end{pmatrix} \leq \lambda d e^{-\delta} \begin{pmatrix} \|a_1\|_1 & \|b_1\|_1 \\ \|a_2\|_1 & \|b_2\|_1 \end{pmatrix} \begin{pmatrix} \|\xi\|_\delta \\ \|\eta\|_\delta \end{pmatrix} + \lambda d e^{-\delta} \begin{pmatrix} \|f_1(\cdot, 0, 0)\|_{1,\delta} \\ \|f_2(\cdot, 0, 0)\|_{1,\delta} \end{pmatrix}.$$

So,

$$\begin{pmatrix} \|\xi\|_\delta \\ \|\eta\|_\delta \end{pmatrix} \leq \lambda d e^{-\delta} (Id - \widehat{M})^{-1} \begin{pmatrix} \|f_1(\cdot, 0, 0)\|_{1,\delta} \\ \|f_2(\cdot, 0, 0)\|_{1,\delta} \end{pmatrix}.$$

$\square$

15.6.4   Asymptotic periodicity

The next result ensures the existence and uniqueness of a discrete S-asymptotically ω-periodic solution for the problem (15.19).

Theorem 15.27. *[225] Assume that the solution operator of (15.18) is strongly S-asymptotically ω-periodic semigroup. Let $g_1, g_2 : \mathbb{Z}^+ \times \mathcal{B} \times \mathcal{B} \to \mathbb{C}^r$ be functions such that $g_1(\cdot, 0, 0)$ and $g_2(\cdot, 0, 0)$ are summable in $\mathbb{Z}^+$ and there exist summable functions $a_i, b_i \in l^1(\mathbb{Z}^+)$, $i = 1, 2$, such that*

$$|g_1(n, x, y) - g_1(n, \overline{x}, \overline{y})| \leq a_1(n) \|x - \overline{x}\|_\mathcal{B} + b_1(n) \|y - \overline{y}\|_\mathcal{B},$$

and

$$|g_2(n, x, y) - g_2(n, \overline{x}, \overline{y})| \leq a_2(n) \|x - \overline{x}\|_\mathcal{B} + b_2(n) \|y - \overline{y}\|_\mathcal{B},$$

for all $x, y, \overline{x}, \overline{y} \in \mathcal{B}$ and $n \in \mathbb{Z}^+$. Then there is a unique discrete S-asymptotically ω-periodic solution of the problem (15.19) for every $\varphi, \psi \in \mathcal{B}$.

Proof. We define the operator T on the space $SAP_\omega(\mathcal{B})$ by

$$T(\xi(n), \eta(n)) = (T_1(\xi(n), \eta(n)), T_2(\xi(n), \eta(n))),$$

where

$$T_1(\xi(n), \eta(n)) = U(n)\varphi + \sum_{s=0}^{n-1} U(n-1-s)f_1(s, \xi(s), \eta(s))$$

and

$$T_2(\xi(n), \eta(n)) = U(n)\psi + \sum_{s=0}^{n-1} U(n-1-s)f_2(s, \xi(s), \eta(s))$$

for all $(\xi, \eta) \in SAP_\omega(\mathcal{B})$. Then we can write

$$\nu_1(n) = \sum_{s=0}^{n-1} U(n-1-s)f_1(s, \xi(s), \eta(s)), \text{ for all } (\xi, \eta) \in SAP_\omega(\mathcal{B})$$

and

$$\nu_2(n) = \sum_{s=0}^{n-1} U(n-1-s)f_2(s, \xi(s), \eta(s)), \text{ for all } (\xi, \eta) \in SAP_\omega(\mathcal{B}).$$

We shall prove that T is well defined. We note that the functions

$$T(\cdot)\varphi, T(\cdot)\psi \in SAP_\omega(\mathcal{B}).$$

Moreover, the semigroup $U(n)$ is uniformly bounded in $\mathbb{Z}^+$. We get

$$\|\nu_1\|_\infty \leq Md[\|a_1\|_1\|\xi\|_\infty + \|b_1\|_1\|\eta\|_\infty + \|f_1(., 0, 0)\|_1]$$

and

$$\|\nu_2\|_\infty \leq Md[\|a_2\|_1\|\xi\|_\infty + \|b_2\|_1\|\eta\|_\infty + \|f_2(., 0, 0)\|_1].$$

On the other hand, we have

$$\left\| \sum_{s=n_1}^{m} U(m-s)f_1(s, \xi(s), \eta(s)) \right\| \leq M \left[\left(\sum_{s=n_1}^{\infty} a_1(s) \right) \|\xi\|_\infty \right.$$
$$\left. + \left(\sum_{s=n_1}^{\infty} b_1(s) \right) \|\eta\|_\infty + \left(\sum_{s=n_1}^{\infty} |f_1(s, 0, 0)| \right) \right]$$

and

$$\left\| \sum_{s=n_1}^{m} U(m-s)f_2(s, \xi(s), \eta(s)) \right\| \leq M \left[\left(\sum_{s=n_1}^{\infty} a_2(s) \right) \|\xi\|_\infty \right.$$
$$\left. + \left(\sum_{s=n_1}^{\infty} b_2(s) \right) \|\eta\|_\infty + \left(\sum_{s=n_1}^{\infty} |f_2(s, 0, 0)| \right) \right].$$

Hence we obtain that

$$\lim_{m \to \infty} \sum_{s=n_1}^{m} U(m-s)f_1(s, \xi(s), \eta(s)) = 0$$

and

$$\lim_{m\to\infty} \sum_{s=n_1}^{m} U(m-s)f_2(s,\xi(s),\eta(s)) = 0.$$

Taking into account that $T(.)$ is S-asymptotically ω-periodic and

$$\begin{aligned}
\nu_1(n+\omega) - \nu_1(n) &= \sum_{s=0}^{n_1-1} [U(n-1-s+\omega) - U(n-1-s)]f_1(s,\xi(s),\eta(s)) \\
&+ \sum_{s=n_1}^{n_1-1+\omega} U(n-1-s+\omega)f_1(s,\xi(s),\eta(s)) \\
&- \sum_{s=n_1}^{n_1-1} U(n-1-s)f_1(s,\xi(s),\eta(s)),
\end{aligned}$$

we obtain that

$$\lim_{n\to\infty} \nu_1(n+\omega) - \nu_1(n) = 0.$$

Similarly, we have

$$\begin{aligned}
\nu_2(n+\omega) - \nu_2(n) &= \sum_{s=0}^{n_1-1} [U(n-1-s+\omega) - U(n-1-s)]f_2(s,\xi(s),\eta(s)) \\
&+ \sum_{s=n_1}^{n_1-1+\omega} U(n-1-s+\omega)f_2(s,\xi(s),\eta(s)) \\
&- \sum_{s=n_1}^{n_1-1} U(n-1-s)f_2(s,\xi(s),\eta(s)),
\end{aligned}$$

and hence

$$\lim_{n\to\infty} \nu_2(n+\omega) - \nu_2(n) = 0.$$

$\square$

15.6.5   Volterra difference system with infinite delay

We apply our previous result to Volterra difference systems with infinite delay. Volterra difference equations can be considered as natural generalization of difference equations. During the last few years Volterra difference equations have emerged vigorously in several applied fields, and currently there is wide interest in developing the qualitative theory for such equations.

Let γ be a positive real number and let $A(n)$ and $K(n)$ be $r \times r$ matrices defined for $n \in \mathbb{Z}^+$, $s \in \mathbb{Z}^+$ such that

$$\sum_{n=0}^{\infty} |K(n)|e^{\gamma n} < +\infty$$

and

$$\|A\|_\infty = \sup_{n\geq 0} |A(n)| < \infty.$$

We consider the following Volterra difference system with infinite delay:

$$x(n+1) = \sum_{s=-\infty}^{n} A(n)K(n-s)x(s), \quad n \geq 0. \tag{15.32}$$

This equation is viewed as a functional difference equation on the phase space $\mathcal{B}_\gamma$, where $\mathcal{B}_\gamma$ is defined as follows:

$$\mathcal{B}_\gamma = \mathcal{B}_\gamma(\mathbb{Z}^-, \mathbb{C}^r) = \{\varphi : \mathbb{Z}^- \to \mathbb{C}^r : \sup_{n \in \mathbb{Z}^+} \frac{\varphi(-n)}{e^{\gamma n}} < +\infty\} \tag{15.33}$$

with the norm:

$$\|\varphi\|_{\mathcal{B}_\gamma} = \sup_{n \in \mathbb{Z}^+} \frac{\varphi(-n)}{e^{\gamma n}}, \ \varphi \in \mathcal{B}_\gamma. \tag{15.34}$$

Next, we consider the following Volterra difference system with infinite delay:

$$\begin{cases} x(n+1) &= \sum_{s=-\infty}^{n} A(n)K(n-s)x(s) + a_1(n)x(n) + a_2(n)y(n), \ n \geq 0, \\ y(n+1) &= \sum_{s=-\infty}^{n} A(n)K(n-s)y(s) + b_1(n)x(n) + b_2(n)y(n), \ n \geq 0, \\ P(0)x_0 &= \varphi, \\ P(0)y_0 &= \psi, \end{cases} \tag{15.35}$$

We recall that the Volterra system (15.35) is viewed as retarded functional difference equations on the phase space $\mathcal{B}_\gamma$.

As consequence of Theorem 15.25 we have the following result.

Theorem 15.28. *[225] Assume that System (15.32) has an exponential dichotomy, and $a_i, b_i \in l^p(\mathbb{Z}^+)$, $i = 1, 2$. Then for each $\varphi, \psi \in P(0)\mathcal{B}_\gamma$ there is a unique bounded solution (x, y) of the system (15.35) such that $(x_\cdot, y_\cdot) \in l^p(\mathbb{Z}^+, \mathcal{B}_\gamma) \times l^p(\mathbb{Z}^+, \mathcal{B}_\gamma)$; in particular, $(x, y) \in l^p(\mathbb{Z}^+, \mathbb{C}^r) \times l^p(\mathbb{Z}^+, \mathbb{C}^r)$.*

Here

$$f_1(n, x_n, y_n) = a_1(n)x(n) + a_2(n)y(n) \ ; f_2(n, x_n, y_n) = b_1(n)x(n) + b_2(n)y(n),$$

$$A_1(n, x_n, y_n) = \sum_{s=-\infty}^{n} A(n)K(n-s)x(s)$$

$$A_2(n, x_n, y_n) = \sum_{s=-\infty}^{n} A(n)K(n-s)y(s).$$

Proof. Clearly, $\{A_1(n, \cdot, \cdot)\}$ and $\{A_2(n, \cdot, \cdot)\}$ are uniformly bounded sequences of bounded linear operators mapping $\mathcal{B} \times \mathcal{B}$ into $\mathbb{C}^r$. Here

$$\|x_n\|_{\mathcal{B}_\gamma} = \sup_{s \in \mathbb{Z}^+} \frac{x_n(-s)}{e^{\gamma s}} \ \leq \ \sup_{s \in \mathbb{Z}^+} \|x(n-s)\|$$

$$\leq \ \sup_{0 \leq s \leq n} \|x(s)\|,$$

and the functions f_1 and f_2 satisfy, for all $x, y, \overline{x}, \overline{y} \in l^p(\mathbb{Z}^+, \mathcal{B})$ and $n \in \mathbb{Z}^+$, we have

$$\begin{aligned} |f_1(n, x, y) - f_1(n, \overline{x}, \overline{y})| &\leq a_1(n)\|x - \overline{x}\|_p + a_2(n)\|y - \overline{y}\|_p, \\ |f_2(n, x, y) - f_2(n, \overline{x}, \overline{y})| &\leq b_1(n)\|x - \overline{x}\|_p + b_2(n)\|y - \overline{y}\|_p, \end{aligned}$$

where $a_i, b_i \in l^p(\mathbb{Z}^+)$, $i = 1, 2$.
Then

$$M = 2dK(1 - e^{-\alpha})^{-1} \sup_{n \in \mathbb{Z}^+}(1 + \|P(n)\|_\mathcal{B}) \begin{pmatrix} \|a_1\|_p & \|a_2\|_p \\ \|b_1\|_p & \|b_2\|_p \end{pmatrix}.$$

Therefore, all the conditions of Theorem 15.25 are satisfied. If M converges to zero, then problem (15.35) has a unique bounded solution (x, y). $\qquad\square$

15.7   Boundary value problems

Continuous boundary value problems associated with second-order nonlinear differential equations have a long history and many different techniques have been developed to establish various qualitative features of the solutions. For the remainder of this chapter, we shall be interested in questions of existence and uniqueness of solutions for certain problems associated with singular equations with deviating arguments. Such problems arise in the study of variational problems in control theory and other areas of applied mathematics.

15.8   Second order boundary value problems

We are now concerned with solutions for the system,

$$
\begin{cases}
\Delta^2 x(k) - f(k, x(k), y(k)) = 0, & k \in \mathbb{N}(0, b), \\
\Delta^2 y(k) - g(k, x(k), y(k)) = 0, & k \in \mathbb{N}(0, b), \\
\alpha_0 x(0) - \beta_0 \Delta x(0) = 0, \\
\gamma_0 x(b+1) + \delta_0 \Delta x(b+1) = 0, \\
\bar{\alpha}_0 y(0) - \bar{\beta}_0 \Delta y(0) = 0, \\
\bar{\gamma}_0 x(b+1) + \bar{\delta}_0 \Delta x(b+1) = 0,
\end{cases}
\tag{15.36}
$$

where $\beta_0, \delta_0, \bar{\beta}_0, \bar{\delta}_0 \in \mathbb{R}\backslash\{0\}$, $\alpha_0, \gamma_0, \bar{\alpha}_0, \bar{\gamma}_0 \in \mathbb{R}$, $f, g : \mathbb{N}(a, b) \times \mathbb{R}^m \times \mathbb{R}^m \to \mathbb{R}^m$ are continuous functions.

Lemma 15.29. *A function $x \in C(\mathbb{N}(0, b), \mathbb{R}^m)$ is a solution of problem*

$$
\begin{cases}
\Delta^2 x(k) = -h(k), & k \in \mathbb{N}(0, b), \\
\alpha_0 x(0) - \beta_0 \Delta x(0) = 0, \\
\gamma_0 x(b+1) + \delta_0 \Delta x(b+1) = 0.
\end{cases}
\tag{15.37}
$$

where $f \in C(\mathbb{N}(0, b), \mathbb{R}^m)$, and $\alpha_0 \gamma_0 (b+1) + \alpha_0 \delta_0 + \beta_0 \gamma_0 \neq 0$ if and only if

$$
x(k) = \sum_{l=0}^{b} G(k, i) h(i), \ k \in \mathbb{N}(0, b),
$$

where

$$
G(k, i) = \begin{cases}
\frac{(\beta_0 + \alpha_0 (i+1))(\delta_0 + \gamma_0 (b+1-k))}{\alpha_0 \gamma_0 (b+1) + \alpha_0 \delta_0 + \beta_0 \gamma_0}, & i \in \{0, \ldots, k-1\}, \\[2mm]
\frac{(\beta_0 + \alpha_0 k)(\delta_0 + \gamma_0 (b-i))}{\alpha_0 \gamma_0 (b+1) + \alpha_0 \delta_0 + \beta_0 \gamma_0}, & i \in \{k, \ldots, b\}.
\end{cases}
$$

Proof. Let $x \in C(\mathbb{N}(0, b), \mathbb{R}^m)$ be a solution of problem (15.37), then

$$
\Delta^2 x(k) = h(k) \Rightarrow \Delta x(k+1) - \Delta x(k) = -h(k).
$$

In particular,

$$
\begin{array}{lll}
i = 0, & \Delta x(1) - \Delta x(0) & = -h(0) \\
i = 1, & \Delta x(2) - \Delta x(1) & = -h(1) \\
\cdots \\
i = k - 1, & \Delta x(k) - \Delta x(k-1) & = -h(k-1).
\end{array}
$$

By summing the above equations, we get

$$\Delta x(k) = \Delta x(0) - \sum_{i=0}^{k-1} h(i).$$ (15.38)

Thus

$$
\begin{aligned}
i = 0, \quad & x(1) - x(0) & = \quad & \Delta x(0) - 0 \\
i = 1, \quad & x(2) - x(1) & = \quad & \Delta x(0) - h(0) \\
& \cdots \\
i = k - 1, \quad & x(k) - x(k-1) & = \quad & \Delta x(0) - h(0) - h(1) \ldots - h(k-2).
\end{aligned}
$$

Hence

$$x(k) = x(0) + k\Delta x(0) - \sum_{i=0}^{k-1}(k - i - 1)h(i).$$ (15.39)

From (15.38) and (15.39), we have

$$x(b+1) = x(0) + (b+1)\Delta x(0) - \sum_{i=0}^{b}(k - i - 1)h(i).$$ (15.40)

and

$$\Delta x(b+1) = \Delta x(0) - \sum_{i=0}^{b} h(i).$$ (15.41)

Since

$$\alpha_0 x(0) - \beta_0 \Delta x(0) = 0, \quad \gamma_0 x(b+1) + \delta_0 \Delta x(b+1) = 0,$$

then

$$\Delta x(0) = \frac{\alpha_0}{\beta_0}, \quad \Delta x(b+1) = \frac{\gamma_0}{\delta_0}.$$

On the other hand, from (15.40) and (15.41) we have

$$
\begin{aligned}
x(0) \quad = \quad & \frac{\delta_0 \beta_0}{\alpha_0 \gamma_0 + \gamma_0 \beta_0 + \gamma_0 \alpha_0 (b+1)} \sum_{i=0}^{b} h(i) \\
& + \frac{\gamma_0 \beta_0}{\alpha_0 \gamma_0 + \gamma_0 \beta_0 + \gamma_0 \alpha_0 (b+1)} \sum_{i=0}^{b} (b - i)h(i),
\end{aligned}
$$

and

$$x(k) = x(0)\left(\frac{\beta_0 + k\alpha_0}{\beta_0}\right) - \sum_{i=0}^{k-1}(k - i - 1)h(i).$$

By the above relations, we get

$$
\begin{aligned}
x(k) &= \frac{\delta_0(\beta_0 + k\alpha_0)}{\alpha_0\gamma_0 + \gamma_0\beta_0 + \gamma_0\alpha_0(b+1)} \sum_{i=0}^{b} h(i) \\
&+ \frac{\gamma_0(\beta_0 + k\alpha_0)}{\alpha_0\gamma_0 + \gamma_0\beta_0 + \gamma_0\alpha_0(b+1)} \sum_{i=0}^{b} (b-i)h(i) \\
&- \sum_{i=0}^{k-1} (k-i-1)h(i) \\
&= \frac{\beta_0 + k\alpha_0}{\alpha_0\gamma_0 + \gamma_0\beta_0 + \gamma_0\alpha_0(b+1)} \sum_{i=k}^{b} (\delta_0 + \gamma_0(b-i))h(i) \\
&+ \frac{\beta_0 + k\alpha_0}{\alpha_0\gamma_0 + \gamma_0\beta_0 + \gamma_0\alpha_0(b+1)} \sum_{i=0}^{k-1} (\delta_0 + \gamma_0(b-i))h(i) \\
&- \sum_{i=0}^{k-1} (k-i-1)h(i).
\end{aligned}
$$

This implies that

$$
x(k) = \sum_{l=0}^{b} G(k,i)h(i), \quad k \in \mathbb{N}(0,b).
$$

$\square$

We now establish existence and uniqueness of solutions for problems of type (15.36).

Theorem 15.30. *Assume that (H_1) is satisfied and the matrix*

$$
\bar{M} = G_* \begin{pmatrix} a_1 & b_1 \\ a_2 & b_2 \end{pmatrix} \in \mathcal{M}_{2\times 2}(\mathbb{R}_+),
$$

where

$$
\sup\{|G(i,j)| : (i,j) \in \mathbb{N}(0,b) \times \mathbb{N}(0,b)\}.
$$

If $\bar{M}$ converges to zero. Then the problem (15.36) has a unique solution.

Proof. Consider the operator $\bar{N} : C(\mathbb{N}(0,b),\mathbb{R}^m) \times C(\mathbb{N}(0,b),\mathbb{R}^m) \to C(\mathbb{N}(0,b),\mathbb{R}^m)$ defined for $(x,y) \in C(\mathbb{N}(0,b),\mathbb{R}^m) \times C(\mathbb{N}(0,b),\mathbb{R}^m)$ by

$$
\bar{N}(x,y) = (\bar{N}_1(x,y), \bar{N}_2(x,y)), \tag{15.42}
$$

where

$$
\bar{N}_1(x(k),y(k)) = \sum_{l=0}^{b} G(k,l)f(l,x(l),y(l)), \quad k \in \mathbb{N}(0,b)
$$

and

$$
\bar{N}_2(x(k),y(k)) = \sum_{l=0}^{b} G(k,l)g(l,x(l),y(l)), \quad k \in \mathbb{N}(0,b).
$$

Using the same reasoning used in Theorem 15.5, we obtain that the operator $\bar{N}$ has a unique fixed point which is a solution of problem (15.36). $\square$

We now state an analogue of Theorem 15.10 for boundary value problems.

Theorem 15.31. *Assume that*

$(\bar{H}_*)$ *There exist* $p_1, p_2 \in C(\mathbb{N}(0,b), \mathbb{R}_+)$ *and* $\alpha, \beta \in (0,1)$ *such that*

$$|f(k,x,y)| \le p_1(k)(|x|+|y|)^\alpha, \quad k \in \mathbb{N}(0,b), \quad (x,y) \in \mathbb{R}^m \times \mathbb{R}^m,$$

and

$$|g(k,x,y)| \le p_2(k)(|x|+|y|)^\beta, \quad k \in \mathbb{N}(0,b), \quad (x,y) \in \mathbb{R}^m \times \mathbb{R}^m.$$

If

$$\alpha_0 \gamma_0 (b+1) + \alpha_0 \delta_0 + \beta_0 \gamma_0 \ne 0$$

and

$$\bar{\alpha}_0 \bar{\gamma}_0 (b+1) + \bar{\alpha}_0 \bar{\delta}_0 + \bar{\beta}_0 \bar{\gamma}_0 \ne 0,$$

then problem (15.36) has at least one solution. Moreover, the solution set

$$\bar{S} = \{(x,y) \in C(\mathbb{N}(0,b), \mathbb{R}^m) \times (\mathbb{N}(0,b), \mathbb{R}^m) : (x,y) \text{ is a solution of } (15.36)\}$$

is compact.

15.9 Multiplicity of solutions for nth order boundary value problems

In this section we consider the following higher order boundary value problem:

$$\begin{cases} -\Delta^n x(k) &= f(k, x, \Delta x, \ldots, \Delta^{n-1}x, y, \Delta y, \ldots, \Delta^{n-1}y)), \\ & \qquad k \in \mathbb{N}(n-1, b-1), \\ -\Delta^2 y(k) &= g(k, x, \Delta x, \ldots, \Delta^{n-1}x, y, \Delta y, \ldots, \Delta^{n-1}y)), \\ & \qquad k \in \mathbb{N}(n-1, b-1), \\[2mm] \Delta^i x(0) &= 0, \ 0 \le i \le n-2 \\ \Delta^p x(b+n-p) &= 0, \\ \Delta^i y(0) &= 0, \ 0 \le i \le n-2 \\ \Delta^p y(b+n-p) &= 0, \end{cases} \tag{15.43}$$

where $f, g : N(0, b+n-1) \times \mathbb{R}^{n-1} \times \mathbb{R}^{n-1}$ are continuous functions.

Lemma 15.32. *A function* $x \in C(\mathbb{N}(n-1, b-1), \mathbb{R})$ *is a solution of problem*

$$\begin{cases} \Delta^n x(k) = -h(k), & k \in \mathbb{N}(n-1, b-1), \\[2mm] \Delta^i x(0) = 0, & 0 \le i \le n-2, \\ \Delta^p x(b+n-p) = 0, \end{cases} \tag{15.44}$$

where $h \in C(\mathbb{N}(0, b+n-1), \mathbb{R})$, *if and only if*

$$x(k) = \sum_{l=0}^{b} G(k,i)h(i), \ k \in \mathbb{N}(0,b),$$

where

$$G(k,i) = \frac{1}{(n-1)!} \begin{cases} \frac{k^{n-1}(b+n-p-l-1)^{n-p-1}}{(b+n+p)^{n-p-1}} - (k-l-1)^{n-1}, \\ i \in \{0, \ldots, k-n\}, \\ \\ \frac{k^{n-1}(b+n-p-l-1)^{n-p-1}}{(b+n+p)^{n-p-1}}, \\ i \in \{k-n+1, \ldots, b\}. \end{cases}$$

We now give a few lemmas concerning the properties of G.

Lemma 15.33. *For $(k,i) \in \mathbb{N}(0, b+n) \times \mathbb{N}(0, b)$, we have*

$$G(k,i) \leq L(b+n-p-1-i)^{n-1},$$

where

$$L = \frac{(b+n)^{n-1}}{(n-1)!(b+n-p)^{n-p-1}}.$$

Lemma 15.34. *For $(k,i) \in \mathbb{N}(n-1, b+n-p) \times \mathbb{N}(0, b)$, we have*

$$G(k,i) \geq MG(b+n-p, i),$$

where

$$M = \min_{i \in \mathbb{N}(0,b)} \frac{G(b-n, i)}{G(b+n-p, i)}.$$

Throughout, it is assumed that there exist $h, \bar{h} : [0, \infty) \times [0, \infty) \to (0, \infty)$ and $h_1, h_2, \bar{h}_1, \bar{h}_2 : \mathbb{N}(0, b) \to \mathbb{R}$ such that for every $u, \bar{u} \in [0, \infty)$ we have:

(H_7)

$$h_1(k) \leq \frac{f(k, u, u_1, \ldots, u_{n-1}, u, u_1, \ldots, u_{n-1})}{h(u, \bar{u})} \leq h_2(k),$$

and

$$\bar{h}_1(k) \leq \frac{g(k, u, u_1, \ldots, u_{n-1}, u, u_1, \ldots, u_{n-1})}{\bar{h}(u, \bar{u})} \leq \bar{h}_2(k).$$

(H_8) $h_1, \bar{h}_1$ are nonnegative on $\mathbb{N}(0, b)$ and are not identically zero on $\mathbb{N}(0, b)$.

Now we present the result of this section which establishes the existence of three positive solutions to the boundary value problem (15.43).

Theorem 15.35. *Suppose $0 < \alpha < \beta < M^{-1}\beta \leq \gamma$ are given such that $h, \bar{h}$ satisfies the following condition:*

(H_8) $h(u, \bar{u}) < \frac{a}{q}, \bar{h}(u, \bar{u}) < \frac{\alpha}{q},$ *for $0 \leq u; \bar{u} \leq \alpha$*

(H_9) $h(u, \bar{u}) \geq \frac{\beta}{r}, \bar{h}(u, \bar{u}) \geq \frac{\beta}{r},$ *for $\beta \leq u, \bar{u} \leq M^{-1}\beta$*

(H_{10}) $h(u, \bar{u}) \leq \frac{\gamma}{q}, \bar{h}(u, \bar{u}) \leq \frac{\gamma}{q},$ *for $0 \leq u; \bar{u} \leq \gamma$.*

Then, the boundary value problem (15.43) has three positive solutions.

Proof. Let the cone $\mathcal{C} \subset C_d(\mathbb{N}(n-1, b-1), \mathbb{R}) \times C_d(\mathbb{N}(n-1, b-1), \mathbb{R})$ be defined by

$$\mathcal{C} = \{(x, y) \in C_d(\mathbb{N}(0, b+n-1), \mathbb{R}) \times C_d(\mathbb{N}(0, b+n-1), \mathbb{R}) : \\ x(k) \geq 0, \ y(k) \geq 0, \ k \in \mathbb{N}(n-1, b-1)\},$$

where

$$C_d(\mathbb{N}(n-1, b-1), \mathbb{R}) = \{y \in C(\mathbb{N}(n-1, b-1), \mathbb{R}) :$$
$$|dy(k)| < \infty, \ y(k) \geq 0, \ k \in \mathbb{N}(n-1, b-1)\}$$

is the Banach space with the norm

$$\|y\|_\infty = \sup\{d|y(k)| : k \in \mathbb{N}(0, b+n-1)\}.$$

From Lemma 15.33 all the solutions of problem (15.43) are fixed points of the operator $N : \mathcal{C} \to C_d(\mathbb{N}(0, b+n-1), \mathbb{R}) \times C_d(\mathbb{N}(0, b+n-1), \mathbb{R})$ defined for $(x, y) \in C_d(\mathbb{N}(0, b+n-1), \mathbb{R}) \times C_d(\mathbb{N}(0, b+n-1), \mathbb{R})$ by

$$N(x, y) = (N_1(x, y), N_2(x, y)), \tag{15.45}$$

where

$$N_1(x(k), y(k)) = \sum_{l=0}^{b} G(k, l) f(k, x, \Delta x, \ldots, \Delta^{n-1}x, y, \Delta y, \ldots, \Delta^{n-1}y)),$$
$$k \in \mathbb{N}(0, b+n-1),$$

and

$$N_2(x(k), y(k)) = \sum_{l=0}^{b} G(k, l) g(k, x, \Delta x, \ldots, \Delta^{n-1}x, y, \Delta y, \ldots, \Delta^{n-1}y)),$$
$$k \in \mathbb{N}(0, b+n-1).$$

By the standard argument used in Theorem 15.10 we can prove that N is completely continuous. Now we show that $N(\mathcal{C}) \subset \mathcal{C}$. Indeed, if $(x, y) \in \mathcal{C}$, then from (H_7), we have

$$\sum_{l=0}^{b} G(k, l) h_1(k) \leq \sum_{l=0}^{b} G(k, l) f(k, x(k), \Delta x(k), \ldots, \Delta^{n-1}x(k),$$
$$y(k), \Delta y(k), \ldots, \Delta^{n-1}y(k))$$
$$\leq \sum_{l=0}^{b} G(k, l) h_2(k),$$

and

$$\sum_{l=0}^{b} G(k, l) \bar{h}_1(k) \leq \sum_{l=0}^{b} G(k, l) g(k, x(k), \Delta x(k), \ldots, \Delta^{n-1}x(k),$$
$$y(k), \Delta y(k), \ldots, \Delta^{n-1}y(k))$$
$$\leq \sum_{l=0}^{b} G(k, l) \bar{h}_2(k).$$

It follows from (H_8) that, for $(x, y) \in \mathcal{C}$,

$$N(x(k), y(k)) \geq 0, \quad k \in \mathbb{N}(0, b+n-1) \Rightarrow N(x, y) \in \mathcal{C}.$$

Define $\sigma : \mathcal{C} \to [0, \infty) \times [0, \infty)$ by

$$\sigma(x, y) = (\min_{k \in \mathbb{N}(n-1, b+n-1)} x(k), \min_{k \in \mathbb{N}(n-1, b+n-1)} y(k))$$

It is clear that σ is a nonnegative continuous concave functional on $\sigma : \mathcal{C}$ and $\sigma(x,y) \leq \binom{\|x\|_\infty}{\|y\|_\infty}$.

Now we are in the position to show that the conditions of the Leggett-Williams Theorem are met. To see that condition $(C1)$ of the Leggett-Williams Theorem is met, we let

$$(x(k), y(k)) = \left(\frac{\beta + M^{-1}\beta}{2}, \frac{\beta + M^{-1}\beta}{2} \right),$$

then

$$\sigma(x,y) = \left(\frac{\beta + M^{-1}\beta}{2}, \frac{\beta + M^{-1}\beta}{2} \right) > \binom{\beta}{\beta} = \beta_* \quad \text{and} \quad \|(x,y)\|_\infty \leq \beta_*.$$

Thus

$$(x,y) \in \mathcal{C}(\sigma, \beta_*, \bar{\beta}_*)$$

and

$$(x,y) \in \{(x,y) \in \mathcal{C}(\sigma, \beta_*, \bar{\beta}_*) : \sigma(x,y) > \beta_*\},$$

where

$$\bar{\beta}_* = \binom{M^{-1}\beta}{M^{-1}\beta}.$$

Next, let $(x,y) \in \mathcal{C}(\sigma, \beta_*, \bar{\beta}_*)$. Then

$$\sigma(N(x,y)) = \left(\min_{k \in \mathbb{N}(n-1, b+n-1)} N_1(x(k), y(k)), \min_{k \in \mathbb{N}(n-1, b+n-1)} N_2(x(l), y(l)) \right).$$

Thus

$$\min_{k \in \mathbb{N}(n-1, b+n-1)} N_1(x(k), y(k)) \geq \min_{k \in \mathbb{N}(n-1, b+n-1)} \sum_{l=0}^{b} G(k,l) h_1(x(l), y(l))$$

$$\geq M \sum_{l=0}^{b} G(b+n-p, l) h_1(x(l), y(l))$$

$$> \beta,$$

and

$$\min_{k \in \mathbb{N}(n-1, b+n-1)} N_2(x(k), y(k)) > \beta.$$

Hence

$$\sigma(N(x,y)) > \bar{\beta}_*.$$

Thus Condition $(C1)$ holds.

Let $(x,y) \in \mathcal{C}$ such that $\|(x,y)\| \leq \alpha_* = \binom{\alpha}{\alpha}$. By Lemma 15.33, for each $k \in \mathbb{N}(0, b+n-1)$ we have

$$N_1(x(k), y(k)) \leq \sum_{l=0}^{b} G(k,l) h_1(k) h(x(l), y(l))$$

$$\leq L \sum_{l=0}^{b} (b+n-p-1)^{(n-p-1)} h_1(k) h(x(l), y(l))$$

$$< \alpha$$

and

$$
\begin{aligned}
N_2(x(k), y(k)) &\leq \sum_{l=0}^{b} G(k,l)\bar{h}_1(k)\bar{h}(x(l), y(l)) \\
&\leq L\sum_{l=0}^{b}(b+n-p-1)^{(n-p-1)}\bar{h}_1(k)\bar{h}(x(l), y(l)) \\
&< \alpha.
\end{aligned}
$$

Therefore, Condition (C2) of the Leggett-Williams Theorem holds.

Finally we show that Condition (C3) is also satisfied. Indeed, let $(x, y) \in \mathcal{C}(\sigma, \bar{\beta}_*, \gamma_*)$, $\gamma_* = \begin{pmatrix} \gamma \\ \gamma \end{pmatrix}$, with

$$
\|N(x,y)\|_\infty > \bar{\beta}_*.
$$

From Lemma 15.34, we get

$$
\begin{aligned}
\min_{k\in\mathbb{N}(n-1,b+n-1)} N_1(x(k), y(k)) &\geq M\sum_{l=0}^{b} G(b+n-p,l)\bar{f}(l) \\
&= M\|N_1(x,y)\|_\infty \\
&> \beta,
\end{aligned}
$$

where $\bar{f}(k) = (k, x(k), \Delta x(k), \ldots, \Delta^{n-1}x(k), y(k), \Delta y(k), \ldots, \Delta^{n-1}y(k))$ and

$$
\begin{aligned}
\min_{k\in\mathbb{N}(n-1,b+n-1)} N_2(x(k), y(k)) &\geq M\sum_{l=0}^{b} G(b+n-p,l)\bar{g}(l) \\
&= M\|N_2(x,y)\|_\infty \\
&> \beta,
\end{aligned}
$$

where

$$
\bar{g}(k) = g(k, x(k), \Delta x(k), \ldots, \Delta^{n-1}x(k), y(k), \Delta y(k), \ldots, \Delta^{n-1}y(k)).
$$

Hence the hypotheses of the Leggett-William's theorem type are satisfied, and then Problem (15.43) has at least three positive solutions.

$\square$

Chapter 16

Discrete Inclusions

16.1 Cauchy problem for discrete inclusions

In this section, our main objective is to establish sufficient conditions for the local and global existence of solutions to the following system of first order discrete inclusions,

$$\begin{cases} \Delta x(k) & \in F(k, x(k), y(k)), \ k \in \mathbb{N}(0, b), \\ \Delta y(k) & \in G(k, x(k), y(k)), \ k \in \mathbb{N}(0, b), \\ x(0) & = x_0, \\ y(0) & = y_0, \end{cases} \qquad (16.1)$$

where $F, G : \mathbb{N}(0, b) \times \mathbb{R}^m \times \mathbb{R}^m \longrightarrow \mathcal{P}(\mathbb{R}^m)$ are given multifunctions.

16.2 Existence and compactness result

In this subsection, we present a global existence result and prove the compactness of the solution set for the problem (16.1) by using a nonlinear alternative for multivalued maps combined with a compactness argument. The nonlinearity is *u.s.c.* with respect to the spatial variable and satisfies a linear growth condition.

Theorem 16.1. *Suppose* $F, G : \mathbb{N}(0, b) \times \mathbb{R}^m \to \mathcal{P}_{cp,cv}(\mathbb{R}^m)$ *such that* $(x, y) \to F(k, x, y)$ *and* $(x, y) \to G(k, x, y)$ *are u.s.c. In addition, assume*

$(\mathcal{M}_1)$ *there exist a continuous functions* $\psi_1, \psi_2 : \mathbb{N}(0, b) \longrightarrow \mathbb{R}_+$

$$\|F(k, x, y)\|_{\mathcal{P}} \le \psi_1(k)(\|x\| + \|y\|), \ \text{for each} \ k \in \mathbb{N}(0, b) \ \text{and} \ x \in \mathbb{R}^m,$$

and

$$\|G(k, x, y)\|_{\mathcal{P}} \le \psi_2(k)(\|x\| + \|y\|), \ \text{for each} \ k \in \mathbb{N}(0, b) \ \text{and} \ x \in \mathbb{R}^m.$$

Then problem (16.1) has at least one solution. Moreover, the solution set $S_{F,G}(x_0, y_0)$ *is compact and the multivalued map* $S_{F,G} : (a, \bar{b}) \multimap S_{F,G}(a, b)$ *is u.s.c.*

Proof of Theorem 16.1. Set $E = C(\mathbb{N}(0, b), \mathbb{R}^m)$.

Step 1. Existence of solutions. Consider the operator $N : E \to \mathcal{P}(E)$ defined for $y \in E$ by

$$N(y) = \left\{ (h_1, h_2) \in E \times E : (h_1(k), h_2(k)) = \begin{cases} x_0 + \sum_{l=0}^{k} v_1(l), & k \in \mathbb{N}(0, b), \\ y_0 + \sum_{l=0}^{k} v_2(l), & k \in \mathbb{N}(0, b), \end{cases} \right.$$

where

$$v_1 \in S_{F,x,y} = \{v \in C(\mathbb{N}(0,b), \mathbb{R}^m) : v(k) \in F(k, x(k), y(k)), k \in \mathbb{N}(0,b)\},$$

and

$$v_2 \in S_{G,x,y} = \{v \in C(\mathbb{N}(0,b), \mathbb{R}^m) : v(k) \in G(k, x(k), y(k)), k \in \mathbb{N}(0,b)\}.$$

It is clear that the fixed points of the operator N are solutions of Problem (16.1). We shall show that N satisfies the assumptions of Lemma 6.7. Also, notice that since $S_{F,x,y}$ and $S_{G,x,y}$ are convex (because F has convex values), then N takes convex values.

Claim 1. *N sends bounded sets in E into bounded sets in $E \times E$.* Let $r_1, r_2 > 0$, $B_{r_1,r_2} := \{(x,y) \in E \times E : \|x\|_\infty \le r_1, \|y\|_\infty \le r_2\}$ be a bounded set in $E \times E$. Let $(x,y) \in B_{r_1,r_2}$. Then there exist $v_1 \in S_{F,x,y}$ and $v_2 \in S_{G,x,y}$ such that, for each, $k \in \mathbb{N}(0,b)$, we have

$$h_1(k) = x_0 + \sum_{l=0}^{k} v_1(l)$$

and

$$h_2(k) = y_0 + \sum_{l=0}^{k} v_2(l).$$

From $(\mathcal{M}_1)$, we obtain

$$\|h_1\|_\infty \le \|x_0\| + (r_1 + r_2) \sum_{l=0}^{b} \psi_1(k)$$

and

$$\|h_2\|_\infty \le \|y_0\| + (r_1 + r_2) \sum_{l=0}^{b} \psi_2(k).$$

This shows that N sends bounded sets into bounded sets in $E \times E$.

Claim 2. We now show that the set

$$N(B_{r_1,r_2})(k) = \{(h_1(k), h_2(k)) \in \mathbb{R}^m \times \mathbb{R}^m : (x,y) \in B_{r_1,r_2}, (h_1, h_2) \in N(x,y)\}$$

is relatively compact in $\mathbb{R}^m \times \mathbb{R}^m$. From Claim 1, we have that $N(B_{r_1,r_2})(k)$ is bounded in $\mathbb{R}^m \times \mathbb{R}^m$ and since using $(x,y) \to F(k,x,y)$ and $(x,y) \to G(k,x,y)$ are u.s.c. for each $\mathbb{N}(0,b)$, then $N(B_{r_1,r_2})(k)$ is closed. Hence $N(B_{r_1,r_2})(k)$ is compact in $\mathbb{R}^m \times \mathbb{R}^m$.

With Theorem 15.8 and Claims 1-2, we conclude that N is completely continuous.

Claim 3. *N is u.s.c.* To this end, we show that N has a closed graph. Let $(h_1^n, h_2^n) \in N(x_n, y_n)$ such that $(h_1^n, h_2^n) \longrightarrow (h_1, h_2)$ and $(x_n, y_n) \longrightarrow (x,y)$, as $n \to \infty$. Then there exists $M > 0$ such that $\|x_n\| \le M$ and $\|y_n\| \le M$. We shall prove that $(h_1, h_2) \in N(x,y)$. $(h_1^n, h_2^n) \in N(x_n, y_n)$ means that there exist $v_1^n \in S_{F,x_n,y_n}$ and $v_2^n \in S_{G,x_n,y_n}$ such that for each $k \in \mathbb{N}(0,b)$

$$h_1^n(k) = x_0 + \sum_{l=0}^{k} v_1^n(k)$$

and

$$h_2^n(k) = x_0 + \sum_{l=0}^{k} v_2^n(k).$$

$(\mathcal{M}_1)$ implies that

$$(v_1^n, v_1^n) \in B \quad = \quad \{(x,y) \in E \times E :$$

$$\|x\|_\infty \le 2M \sum_{l=0}^{b} \psi_1(k), \|y\|_\infty \le 2M \sum_{l=0}^{b} \psi_2(k)\}.$$

Then there exists a compact set $\Omega \subset E \times E$ such that

$$(v_1^n, v_2^n)_{n \in \mathbb{N}} \subset \Omega, \quad n \in \mathbb{N}.$$

Thus there exists a subsequence of $(v_1^n, v_2^n)_{n \in \mathbb{N}}$ such that

$$(v_1^n, v_2^n) \to (v_1, v_2) \quad \text{as } n \to \infty$$

and

$$(v_1^n(k), v_2^n(k)) \in (F(k, x_n(k), y_n(k)), G(k, x_n(k), y_n(k))), \text{ for } k \in \mathbb{N}(0, b).$$

Since the maps $F(k, \cdot, \cdot)$ and $G(k, \cdot, \cdot)$ are u.s.c. for each $k \in \mathbb{N}(0, b)$, this implies that

$$v_1(k) \in F(k, x(k), y(k)), \quad v_2(k) \in G(k, x(k), y(k)).$$

We can do this for each $k \in \mathbb{N}(0, b)$, so $(h_1, h_2) \in N(x, y)$.

Claim 4. *A priori bounds on solutions.* Let $(x, y) \in E \times E$ be such that $(x, y) \in N(x, y)$. Then there exists $(v_1, v_2) \in S_{F,x,y} \times S_{G,x,y}$ such that

$$x(k) = x_0 + \sum_{l=0}^{k} v_1(k), \ k \in \mathbb{N}(0, b)$$

and

$$y(k) = y_0 + \sum_{l=0}^{k} v_2(k), \ k \in \mathbb{N}(0, b).$$

Therefore

$$\begin{aligned}
\|x(k)\| &\le \|x_0\| + \sum_{l=0}^{k} \|v_1(k)\| \\
&\le \|x_0\| + \sum_{l=0}^{k} \|F(k, x(k), y(k))\|_{\mathcal{P}} \\
&\le \|y_0\| + \sum_{l=0}^{k} \psi_1(k)(\|x(k)\| + \|y(k)\|)
\end{aligned}$$

and

$$\|y(k)\| \le \|y_0\| + \sum_{l=0}^{k} \psi_2(k)(\|x(k)\| + \|y(k)\|).$$

Then

$$\|x(k)\| + \|y(k)\| \le \|x_0\| + \|y_0\| + (\|\psi_1\|_\infty + \|\psi_2\|_\infty) \sum_{l=0}^{k} (\|x(k)\| + \|y(k)\|).$$

By Lemma 15.1, we get

$$\|x\|_\infty + \|y\|_\infty \le (\|x_0\| + \|y_0\|)(1 + (b+1)(1 + (\|\psi_1\|_\infty + \|\psi_2\|_\infty))^{b+1}) := \widetilde{M}.$$

Finally, let

$$U := \{(x, y) \in E \times E : \|x\|_\infty < \widetilde{M} + 1, \quad \|y\|_\infty < \widetilde{M} + 1\}$$

and consider the operator $N : \overline{U} \to \mathcal{P}_{cv,cp}(E \times E)$. From the choice of U, there is no $(x, y) \in \partial U$ such that $(x, y) \in \lambda N(x, y)$ for some $\lambda \in (0, 1)$. As a consequence of the multivalued version of the nonlinear alternative of Leray-Schauder (Lemma 6.7), N has a fixed point $(x, y) \in U$ which is a solution of problem (16.1).

Step 2: Compactness of the solution set. For each $(a, \bar{b}) \in \mathbb{R}^m \times \mathbb{R}^m$, let

$$S_{F,G}(a, \bar{b}) = \{(x, y) \in E \times E : (x, y) \text{ is a solution of problem } (16.1)\}.$$

From Step 1, there exists $\widetilde{M}$ such that for every $y \in S_{F,G}(a, \bar{b})$, $\|x\|_\infty \le \widetilde{M}$, $\|y\|_\infty \le \widetilde{M}$. Since N is completely continuous, $N(S_{F,G}(a, b))$ is relatively compact in $E \times E$. Let $(x, y) \in S_{F,G}(a, \bar{b})$; then $(x, y) \in N(x, y)$ hence $S_{F,G}(a, \bar{b}) \subset \overline{N(S_{F,G}(a, \bar{b}))}$. It remains to prove that $S_{F,G}(a, \bar{b})$ is a closed subset in $E \times E$. Let $\{(x_n, y_n) : n \in \mathbb{N}\} \subset S_{F,G}(a, \bar{b})$ be such that $((x_n, y_n))_{n \in \mathbb{N}}$ converges to (x, y). For every $n \in \mathbb{N}$, there exist $v_n, \bar{v}_n$ such that $v_n(k) \in F(k, x_n(k), y_n(k))$, $\bar{v}_n(k) \in G(k, x_n(k), y_n(k))$, $k \in \mathbb{N}(0, b)$ and

$$x_n(k) = a + \sum_{l=0}^{k} v_n(k), \tag{16.2}$$

$$y_n(k) = \bar{b} + \sum_{l=0}^{k} v_n(k). \tag{16.3}$$

Arguing as in Claim 3, we can prove that there exist v and $\bar{v}$ such that $v_n(k) \in F(k, x_n(k), y_n(k))$, $\bar{v}_n(k) \in G(k, x_n(k), y_n(k))$, $k \in \mathbb{N}(0, b)$ and

$$x(k) = a + \sum_{l=0}^{k} v(k)., \quad k \in \mathbb{N}(0, b),$$

and

$$y(k) = \bar{b} + \sum_{l=0}^{k} \bar{v}(k)., \quad k \in \mathbb{N}(0, b).$$

Therefore $(x, y) \in S_{F,G}(a, \bar{b})$ which yields that $S_{F,G}(a, \bar{b})$ is closed, and hence a compact subset in $E \times E$. Finally, we prove that $S_F(\cdot, \cdot)$ is u.s.c. by proving that the graph of $S_{F,G}$

$$\Gamma_{S_{F,G}} := \{(a, \bar{b}, x, y) : y \in S_{F,G}(a, \bar{b})\}$$

is closed. Let $(a_n, \bar{b}_n, x_n, y_n) \in \Gamma_{S_{F,G}}$ be such that $(a_n, \bar{b}_n, x_n, y_n) \to (a, \bar{b}, x, y)$ as $n \to \infty$. Since $(x_n, y_n) \in S_{F,G}(a_n, \bar{b}_n)$, there exist $v_n \in S_{F,x_n,y_n}$ and $\bar{v}_n \in S_{G,x_n,y_n}$ such that

$$x_n(k) = a_n + \sum_{l=0}^{k} v_n(l), \quad k \in \mathbb{N}(0, b)$$

and

$$y_n(k) = \bar{b}_n + \sum_{l=0}^{k} \bar{v}_n(l), \quad k \in \mathbb{N}(0, b).$$

Arguing as in Claim 4, we can prove that there exist $v \in S_{F,x,y}$ and $\bar{v} \in S_{G,x,y}$ such that

$$x(k) = a + \sum_{l=0}^{k} v(l), \quad k \in \mathbb{N}(0, b)$$

and

$$y(k) = \bar{b} + \sum_{l=0}^{k} \bar{v}(l), \quad k \in \mathbb{N}(0, b).$$

Thus, $(x, y) \in S_{F,G}(a, \bar{b})$. Now, we show that $S_{F,G}$ maps bounded sets into relatively compact sets of $E \times E$. Let B be a bounded set in $\mathbb{R}^m \times \mathbb{R}^m$ and let $\{(x_n, y_n)\} \subset S_{F,G}(B)$. Then there exists $\{(a_n, \bar{b}_n)\} \subset B$ such that

$$x_n(k) = a_n + \sum_{l=0}^{k} v_n(k), \quad k \in \mathbb{N}(0, b),$$

and

$$y_n(k) = \bar{b}_n + \sum_{l=0}^{k} \bar{v}_n(k), \quad k \in \mathbb{N}(0, b),$$

where $v_n \in S_{F,x_n,y_n}$, $\bar{v}_n \in S_{G,x_n,y_n}$, $n \in \mathbb{N}$. Since $\{(a_n, \bar{b}_n)\}$ is a bounded sequence, there exists a subsequence of $\{(a_n, \bar{b}_n)\}$ converging to (a, b). As in Claims 2-3, we can show that $\{(x_n, y_n) : n \in \mathbb{N}\}$ is relatively compact in $E \times E$, then there exists a subsequence of $\{(x_n, y_n)\}$ converging to (x, y) in $E \times E$. By a similar argument of Claim 4, we can prove that

$$x(k) = a + \sum_{l=0}^{k} v(l), \quad k \in \mathbb{N}(0, b)$$

and

$$y(k) = \bar{b} + \sum_{l=0}^{k} \bar{v}(l), \quad k \in \mathbb{N}(0, b),$$

where $v \in S_{F,G,y}$, $\bar{v} \in S_{G,x,y}$. Thus, $(x, y) \in \overline{S_{F,G}(B)}$. This implies that $S_{F,G}(\cdot)$ is u.s.c., ending the proof of Theorem 16.1. $\qquad\square$

Chapter 17

Semilinear System of Discrete Equations

In this chapter, we establish several results on the existence and uniqueness of solutions for a class of semilinear systems of difference equations with initial and boundary conditions. The approach is based on fixed point theory in vector-valued Banach spaces. Several aspects of the theory of semilinear difference equations can be understood as a proper generalization of the theory of ordinary difference equations. However, the fact that the state space for functional difference equations is infinite dimensional requires the development of methods and techniques coming from functional analysis (e.g., theory of semigroups of operators on Banach spaces, spectral theory, fixed point theory etc.). Some important contributions to the study of the mathematical aspects of such equations have been undertaken in [3, 13, 55, 84, 166].

We consider the semilinear discrete system of the form

$$\begin{cases} x(t) & = A(t)x(t) + f_1(t, x(t), y(t)), \ k \in \mathbb{N}(a, b), \\ y(t) & = A(t)y(t) + f_2(t, x(t), y(t)), \ k \in \mathbb{N}(a, b), \\ x(a) & = x_0, \\ y(a) & = y_0, \end{cases} \tag{17.1}$$

where $\mathbb{N}(a, b) = \{a, a+1, \ldots, b+1\}$, $f_1, f_2 : \mathbb{N}(a, b) \times X \to X$ are given functions and with a variable linear operator $A(t)$ in a Banach space X.

Later, we will study the boundary-value problem

$$\begin{cases} x(t) & = A(t)x(t) + f_1(t, x(t), y(t)), \ k \in \mathbb{N}(0, b), \\ y(t) & = A(t)y(t) + f_2(t, x(t), y(t)), \ k \in \mathbb{N}(0, b), \\ L_1(x(0)) & = l_1 \in X, \\ L_2(y(0)) & = l_2 \in X, \end{cases} \tag{17.2}$$

where $L_1, L_2 : C(\mathbb{N}(0, b), X) \to X$ are bounded linear operators.

17.1   Existence and uniqueness results

Consider the equation

$$x(k + 1) = A(k)x(k) \ (k = 0, 1, \ldots), \tag{17.3}$$

with a variable linear operator $A(k)$ on a Banach space X. Then the linear operator

$$U(k, s) : X \to X \ (k, s = 0, 1, \ldots),$$

defined by

$$U(k, j) = A(k-1) \cdots A(j) \ (k = j+1, j+2, \ldots) \text{ and } U(j, j) = I \ (j = 0, 1, \ldots), \tag{17.4}$$

will be called be the evolution operator of equation (17.3). Recall that I is the identity operator on X.

It is simple to check that the evolution operator has the following properties:

$$U(k, j) = U(k, t)U(t, j) \ (k \geq t \geq j; j = 0, 1, \ldots)$$

and

$$U(k + 1, j) = A(k)U(k, j) \ (k \geq j; j = 0, 1, \ldots).$$

Lemma 17.1. *Let $x(k)$ be a solution of equation (17.3). Then*

$$x(k) = U(k, s)x(s) \ (k \geq s). \tag{17.5}$$

The operator $U(k) = U(k, 0)$ will be called the Cauchy operator of equation (17.3). If $A(k) = A$ is a constant operator, then

$$U(k) = A^k \text{ and } U(k, s) = U(k - s) = A^{k-s} \ (k \geq s).$$

A solution of the nonhomogeneous equation

$$x(k + 1) = A(k)x(k) + f(k, x(k)) \ (k = 0, 1, \ldots),$$

with a given sequence $\{f(k, x(k)) \in X\}_{k=0}^{\infty}$, can be represented in the form

$$x(k) = U(k, 0)x_0 + \sum_{t=0}^{k-1} U(k, t + 1)f(k, x(k)), \ k = 1, \ldots. \tag{17.6}$$

Let us introduce the following hypotheses for f_1, $f_2 : \mathbb{N}(a, b) \times X \times X \to X$:

$(\mathcal{J}_1)$ There exist nonnegative numbers a_i and b_i for each $i = 1, 2$ such that

$$|f_1(k, x, y) - f_1(k, \overline{x}, \overline{y})| \leq a_1|x - \overline{x}| + b_1|y - \overline{y}|$$

and

$$|f_2(k, x, y) - f_2(k, \overline{x}, \overline{y})| \leq a_2|x - \overline{x}| + b_2|y - \overline{y}|,$$

for all x, y, $\overline{x}$, $\overline{y} \in X$.

$(\mathcal{J}_2)$ There exists a positive number Q such that

$$\|U(k, s)\| \leq Q, \text{ for all } k, s \in \mathbb{N}(a, b).$$

$(\mathcal{J}_3)$ The evolution semigroup $\{U(k, s)\}_{k-s>0}$ is compact in X.

$(\mathcal{J}_4)$ There exist p_1, p_2, $\bar{p}_1$, $\bar{p}_2 \in C(\mathbb{N}(a, b), \mathbb{R}_+)$ such that

$$|f_1(k, x, y)| \leq p_1(k)(|x| + |y|) + \bar{p}_1(k), \ k \in \mathbb{N}(a, b), \ (x, y) \in X \times X,$$

and

$$|f_2(k, x, y)| \leq p_2(k)(|x| + |y|) + \bar{p}_2(k), \ k \in \mathbb{N}(a, b), \ (x, y) \in X \times X.$$

We can now prove our first uniqueness result in this section.

Theorem 17.2. *Assume that $(\mathcal{J}_1) - (\mathcal{J}_2)$ are satisfied and that the matrix*

$$M = Qb \begin{pmatrix} a_1 & b_1 \\ a_2 & b_2 \end{pmatrix} \in M_{2 \times 2}(\mathbb{R}_+)$$

converges to zero. Then the problem (17.1) has a unique solution.

Proof. Let $N : C(\mathbb{N}(a,b), X) \times C(\mathbb{N}(a,b), X) \to C(\mathbb{N}(a,b), X) \times C(\mathbb{N}(a,b), X)$ be the operator defined by

$$N(x,y) = (N_1(x,y), N_2(x,y)), \quad (x,y) \in C(\mathbb{N}(a,b), X) \times C(\mathbb{N}(a,b), X)$$

where

$$N_1(x(k), y(k)) = U(k,a)x_0 + \sum_{l=a}^{k-1} U(k, l+1) f_1(l, x(l), y(l)), \quad k \in \mathbb{N}(a,b)$$

and

$$N_2(x(k), y(k)) = U(k,a)y_0 + \sum_{l=a}^{k-1} U(k, l+1) f_2(l, x(l), y(l)), \quad k \in \mathbb{N}(a,b).$$

We shall use Theorem 8.1 to prove that N has a unique fixed point.
Let $(x,y), (\overline{x}, \overline{y}) \in C(\mathbb{N}(a,b), X) \times C(\mathbb{N}(a,b), X)$. Then we have for each $k \in \mathbb{N}(a,b)$

$$|N_1(x(k), y(k)) - N_1(\overline{x}(k), \overline{y}(k))|$$

$$= \left| \sum_{l=s}^{k-1} U(k, l+1)[f_1(l, x(l), y(l)) - f_1(l, \overline{x}(l), \overline{y}(l))] \right|,$$

so

$$\|N_1(x,y) - N_1(\overline{x}, \overline{y})\|_\infty \leq Qba_1\|x - \overline{x}\|_\infty + Qbb_1\|y - \overline{y}\|_\infty.$$

Similarly, we have

$$\|N_2(x,y) - N_2(\overline{x}, \overline{y})\|_\infty \leq Qba_2\|x - \overline{x}\|_\infty + Qbb_2\|y - \overline{y}\|_\infty.$$

Hence,

$$\|N(x,y) - N(\overline{x}, \overline{y})\|_\infty = \begin{pmatrix} \|N_1(x,y) - N_1(\overline{x}, \overline{y})\|_\infty \\ \|N_2(x,y) - N_2(\overline{x}, \overline{y})\|_\infty \end{pmatrix}$$

$$\leq Qb \begin{pmatrix} a_1 & b_1 \\ a_2 & b_2 \end{pmatrix} \begin{pmatrix} \|x - \overline{x}\|_\infty \\ \|y - \overline{y}\|_\infty \end{pmatrix}.$$

Therefore,

$$\|N(x,y) - N(\overline{x}, \overline{y})\|_\infty \leq M \begin{pmatrix} \|x - \overline{x}\|_\infty \\ \|y - \overline{y}\|_\infty \end{pmatrix},$$

for all $(x,y), (\overline{x}, \overline{y}) \in C(\mathbb{N}(a,b), X) \times C(\mathbb{N}(a,b), X)$. From the fixed point theorem 8.1, the mapping N has a unique fixed $(x,y) \in C(\mathbb{N}(a,b), X) \times C(\mathbb{N}(a,b), X)$ which in turn is a unique solution of problem (17.1). □

Now we give an existence result based on a Schaefer type fixed point theorem in a vector-valued Banach space.

Theorem 17.3. *Let f_1, $f_2 : \mathbb{N}(a,b) \times X \times X \to X$ be continuous functions for which $(\mathcal{J}_2) - (\mathcal{J}_4)$ hold. Then the problem (17.1) has at least one solution.*

Proof. Let N be the operator defined in the proof of Theorem 17.2.

Step 1: $N = (N_1, N_2)$ *is continuous.* Let (x_m, y_m) be a sequence such that $(x_m, y_m) \to (x, y) \in C(\mathbb{N}(a,b), X) \times C(\mathbb{N}(a,b), X)$ as $m \to \infty$. Then

$$|N_1(x_m(k), y_m(k)) - N_1(x(k), y(k))|$$

$$= \left| \sum_{l=a}^{k-1} [f_1(l, x_m(l), y_m(l)) - f_1(l, x(l), y(l))] \right|$$

$$\leq \sum_{l=a}^{b} |f_1(l, x_m(l), y_m(l)) - f_1(l, x(l), y(l))|$$

and

$$|N_2(x_m(k), y_m(k)) - N_2(x(k), y(k))|$$

$$\leq \sum_{l=a}^{b} |f_2(l, x_m(l), y_m(l)) - f_2(l, x(l), y(l))|.$$

Since f_1 and f_2 are continuous functions,

$$\|N_1(x_m, y_m) - N_1(x, y)\|_\infty \to 0, \text{ as } m \to \infty$$

and

$$\|N_2(x_m, y_m) - N_2(x, y)\|_\infty \to 0, \text{ as } m \to \infty.$$

Thus, N is continuous.

Step 2: N *maps bounded sets into bounded sets in* $C(\mathbb{N}(a,b), X) \times C(\mathbb{N}(a,b), X)$. It suffices to show that for any $q > 0$ there exists a positive constant l such that for each $(x, y) \in B_q$, where

$$B_q = \{(x, y) \in C(\mathbb{N}(a,b), X) \times C(\mathbb{N}(a,b), X) : \|x\|_\infty \leq q, \|y\|_\infty \leq q\},$$

we have

$$\|N(x, y)\|_\infty \leq l := (l_1, l_2).$$

For each $k \in \mathbb{N}(a,b)$, we obtain

$$\|N_1(x, y)\|_\infty \leq |x_0| + 2q \sum_{k=a}^{b} p_1(k) := l_1.$$

Similarly, we have

$$\|N_2(x, y)\|_\infty \leq |y_0| + 2q \sum_{k=a}^{b} p_2(k) := l_2.$$

Observe that from $(\mathcal{J}_4)$, there exist $Q_1, Q_2 > 0$ such that

$$|f_1(k, x, y)| \leq 2q \|p_1\|_\infty + \|\bar{p}_1\|_\infty, \text{for all } k \in \mathbb{N}(a,b), \ x, y \in C(\mathbb{N}(a,b), X),$$

and

$$|f_2(k, x, y)| \leq 2q \|p_2\|_\infty + \|\bar{p}_2\|_\infty, \text{ for all } k \in \mathbb{N}(a,b), \ x, y \in C(\mathbb{N}(a,b), X).$$

Since $\{U(k, s)\}_{k-s>0}$ is compact, for each $k, s \in \mathbb{N}(a,b)$, we have that

$$\{N_1(x(k), y(k)) : (x, y) \in B_q\} \quad \text{and} \quad \{N_2(x(k), y(k)) : (x, y) \in B_q\}$$

are relatively compact in X. In the case where $k = s$, we have

$$\{N_1(x(k), y(k)) : (x, y) \in B_q\} = \{x_0\}$$

and

$$\{N_2(x(k), y(k)) : (x, y) \in B_q\} = \{y_0\}.$$

By the Theorem 15.8, we see that $\overline{N(B_q \times B_q)}$ is compact. As consequence of Steps 1 to 2, N is completely continuous.

Step 3: It remains to show that

$$\mathcal{A} = \{(x, y) \in C(\mathbb{N}(a, b), X) \times C(\mathbb{N}(a, b), X) : (x, y) = \lambda N(x, y), \lambda \in (0, 1)\}$$

is bounded. Let $(x, y) \in \mathcal{A}$. Then $x = \lambda N_1(x, y)$ and $y = \lambda N_2(x, y)$ for some $0 < \lambda < 1$. Thus, for $k \in \mathbb{N}(a, b)$, we have

$$|x(k)| \leq |x_0| + \sum_{l=a}^{l=k-1} |f_1(l, x(l), y(l))|$$

$$\leq |x_0| + \sum_{l=a}^{b} \bar{p}_1(l) + \sum_{l=a}^{k-1} p_1(l)(|x(l)| + |y(l)|).$$

and

$$|y(k)| \leq |y_0| + \sum_{l=a}^{b} \bar{p}_2(l) + \sum_{l=a}^{k-1} p_2(l)(|x(l)| + |y(l)|).$$

Therefore

$$|x(k)| + |y(k)| \leq |x_0| + |y_0| + \sum_{l=a}^{b} \bar{p}_1(l) + \sum_{l=a}^{b} \bar{p}_2(l) + \sum_{l=a}^{k-1} p(l)(|x(l)| + |y(l)|),$$

where

$$p(k) = p_1(k) + p_2(k), \ k \in \mathbb{N}(a, b).$$

By Theorem 15.1, we have

$$|x(k)| + |y(k)| \leq \left(|x_0| + |y_0| + \sum_{l=a}^{b} \bar{p}_1(l) + \sum_{l=a}^{b} \bar{p}_2(l) \right)$$

$$\times \left(1 + \sum_{l=a}^{k-1} p(l) \prod_{l+1}^{k-1} (1 + p(\tau)) \right),$$

so

$$\|x\|_\infty + \|y\|_\infty \leq \left(|x_0| + |y_0| + \sum_{l=a}^{b} \bar{p}_1(l) + \sum_{l=a}^{b} \bar{p}_2(l) \right)$$

$$\times \left(1 + \sum_{l=a}^{b} p(l) \prod_{l+1}^{b} (1 + p(\tau)) \right).$$

This shows that $\mathcal{A}$ is bounded. As a consequence of Theorem 8.23, we deduce that N has a fixed point (x, y) that is a solution to problem (17.1). $\square$

17.2   Boundary value problems

The following condition will be needed in the sequel.

$(\mathcal{L})$ The operators $\widetilde{L}_1$, $\widetilde{L}_2 : X \to X$ defined by

$$\widetilde{L}_1(x) = L_1(U(\cdot,0)x), \quad \widetilde{L}_2(x) = L_2(U(\cdot,0)x), \quad x \in X,$$

have bounded inverses $\widetilde{L}_1^{-1}, \widetilde{L}_2^{-1} : X \to X$.

Lemma 17.4. *Under condition* $(\mathcal{L})$, *the mild solution* $(x,y) \in C(\mathbb{N}(0,b), X) \times C(\mathbb{N}(0,b), X)$ *of problem (17.2) can be written as*

$$x(k) = \widetilde{L}_1^{-1}\left(l_1 - L_1 \sum_{i=0}^{k-1} U(k,i+1)f_1(i, x(i), y(i))\right)$$

$$+ \sum_{i=0}^{k-1} U(k,i+1)f_1(i, x(i), y(i))$$

and

$$y(k) = \widetilde{L}_2^{-1}\left(l_2 - L_2 \sum_{i=0}^{k-1} U(k,i+1)f_2(i, x(i), y(i))\right)$$

$$+ \sum_{i=0}^{k-1} U(k,i+1)f_2(i, x(i), y(i)).$$

Proof. The function $(x,y) \in C(\mathbb{N}(0,b), X) \times C(\mathbb{N}(0,b), X)$ is a mild solution of (17.2) if and only if

$$x(k) = U(k,0)x(0) + \sum_{i=0}^{k-1} U(k,i+1)f_1(i, x(i), y(i)),$$

$$L_1(U(k,0)x(0)) + L\left(\sum_{t=0}^{k-1} U(k,t+1)f_1(t, x(t), y(t))\right) = l_1,$$

and

$$y(k) = U(k,0)y(0) + \sum_{i=0}^{k-1} U(k,i+1)f_2(i, x(i), y(i)),$$

$$L_2(U(k,0)y(0)) + L_2\left(\sum_{t=0}^{k-1} U(k,t+1)f_2(t, x(t), y(t))\right) = l_2.$$

Hence,

$$\widetilde{L}_1(U(k,0)x(0)) = l_1 - L_1\left(\sum_{t=0}^{k-1} U(k,t+1)f_2(t, x(t), y(t))\right),$$

and

$$\widetilde{L}_2(U(k,0)y(0)) = l_2 - L_2\left(\sum_{t=0}^{k-1} U(k,t+1)f_2(t, x(t), y(t))\right).$$

Since $\widetilde{L}_1$ and $\widetilde{L}_2$ are invertible operators, we have

$$x(0) = \widetilde{L}_1^{-1}l - U(k,0)\widetilde{L}_1^{-1}\left(L_1 \sum_{i=0}^{k-1} U(k,i+1)f_1(i,x(i)) \right)$$

and

$$y(0) = \widetilde{L}_2^{-1}l - U(k,0)\widetilde{L}_2^{-1}\left(L_2 \sum_{i=0}^{k-1} U(k,i+1)f_2(i,x(i)) \right).$$

Hence

$$
\begin{aligned}
x(k) \;=\;& U(k,0)\widetilde{L}_1^{-1}\left(l_1 - L_1 \sum_{i=0}^{k-1} U(k,i+1)f_1(i,x(i),y(i)) \right) \\
&+ \sum_{i=0}^{k-1} U(k,i+1)f_1(i,x(i),y(i))
\end{aligned}
$$

and

$$
\begin{aligned}
y(k) \;=\;& U(k,0)\widetilde{L}_2^{-1}\left(l_2 - L_2 \sum_{i=0}^{k-1} U(k,i+1)f_2(i,x(i),y(i)) \right) \\
&+ \sum_{i=0}^{k-1} U(k,i+1)f_2(i,x(i),y(i)).
\end{aligned}
$$

□

We next present our first existence result in this section.

Theorem 17.5. *Let f_1 $f_2 : \mathbb{N}(0,b)\times X \times X \to X$ be continuous functions and let conditions $(\mathcal{J}_2)$, $(\mathcal{J}_3)$ and $(\mathcal{L})$ be satisfied. Assume that*

$(\mathcal{J}_5)$ *There exist p_3, p_4, $\bar{p}_3$, $\bar{p}_4 \in C(\mathbb{N}(0,b),\mathbb{R}_+)$ and γ_1, $\gamma_2 \in [0,1)$ such that*

$$|f_1(k,x,y)| \le p_3(k)(|x|+|y|)^\alpha + \bar{p}_3(k), \ \ k \in \mathbb{N}(0,b), \ (x,y) \in X \times X,$$

and

$$|f_2(k,x,y)| \le p_4(k)(|x|+|y|)^\beta + \bar{p}_4(k), \ \ k \in \mathbb{N}(0,b), \ (x,y) \in X \times .$$

Then the problem (17.2) has at least one solution.

Proof. Define the operator $\bar{N} : C(\mathbb{N}(0,b),X) \times C(\mathbb{N}(0,b),X) \to C(\mathbb{N}(0,b),X)$ by

$$\bar{N}(x,y) = (\bar{N}_1(x,y), \bar{N}_2(x,y)), \ (x,y) \in C(\mathbb{N}(0,b),X) \times C(\mathbb{N}(0,b),X),$$

where

$$
\begin{aligned}
\bar{N}_1(x(k),y(k)) \;=\;& U(k,0)\widetilde{L}_1^{-1}\left(l_1 - L_1 \sum_{i=0}^{k-1} U(k,i+1)f_1(i,x(i),y(i)) \right) \\
&+ \sum_{i=0}^{k-1} U(k,i+1)f_1(i,x(i),y(i)),
\end{aligned}
$$

and

$$\bar{N}_2(x(k), y(k)) = U(k,0)\widetilde{L}_2^{-1}\left(l_2 - L_2\sum_{i=0}^{k-1}U(k,i+1)f_2(i,x(i),y(i))\right)$$

$$+ \sum_{i=0}^{k-1}U(k,i+1)f_2(i,x(i),y(i)).$$

Step 1: $N = (\bar{N}_1, \bar{N}_2)$ *is continuous.* Let (x_m, y_m) be a sequence such that $(x_m, y_m) \to (x,y) \in C(\mathbb{N}(0,b), X) \times C(\mathbb{N}(0,b), X)$ as $m \to \infty$. Then

$$|N_1(x_m(k), y_m(k)) - N_1(x(k), y(k))|$$

$$= |U(k,0)\widetilde{L}_1^{-1}\left(L_1\sum_{i=0}^{k-1}U(k,i+1) \times [f_1(i,x_m(i),y_m(i)) - f_1(i,x(i),y(i))]\right)$$

$$+ \sum_{i=0}^{k-1}U(k,i+1) \times [f_1(i,x_m(i),y_m(i)) - f_1(i,x(i),y(i))]|$$

and

$$|N_2(x_m(k), y_m(k)) - N_2(x(k), y(k))|$$

$$= |U(k,0)\widetilde{L}_2^{-1}\left(L_2\sum_{i=0}^{k-1}U(k,i+1) \times [f_2(i,x_m(i),y_m(i)) - f_2(i,x(i),y(i))]\right)$$

$$+ \sum_{i=0}^{k-1}U(k,i+1) \times [f_2(i,x_m(i),y_m(i)) - f_2(i,x(i),y(i))]|.$$

Hence,

$$\|N_1(x_m, y_m) - N_1(x,y)\|_\infty \leq Q^2\|\widetilde{L}_1^{-1}\|_{B(X)}\|L_1\|_{B(X)}\sum_{i=0}^{b}|f_1(i,x_m(i),y_m(i))$$

$$-f_1(i,x(i),y(i))|$$

$$+Q\sum_{i=0}^{b}|f_1(i,x_m(i),y_m(i)) - f_1(i,x(i),y(i))|$$

and

$$\|N_2(x_m, y_m) - N_2(x,y)\|_\infty \leq Q^2\|\widetilde{L}_2^{-1}\|_{B(X)}\|L_2\|_{B(X)}\sum_{i=0}^{b}|f_2(i,x_m(i),y_m(i))$$

$$-f_2(i,x(i),y(i))|$$

$$+Q\sum_{i=0}^{b}|f_2(i,x_m(i),y_m(i)) - f_2(i,x(i),y(i))|.$$

Since f_1 and f_2 are continuous,

$$\|N_1(x_m, y_m) - N_1(x,y)\|_\infty \to 0, \text{ as } m \to \infty$$

and

$$\|N_2(x_m, y_m) - N_2(x,y)\|_\infty \to 0, \text{ as } m \to \infty,$$

so N is continuous.

Step 2: *N maps bounded sets into bounded sets in* $C(\mathbb{N}(0,b), X) \times C(\mathbb{N}(0,b), X)$. It suffices to show that for any $q > 0$ there exists a positive constant M such that for each $x \in B_1, y \in B_2$, where

$$B_1 = \{x \in C(\mathbb{N}(0,b), X) : \|x\|_\infty \le q\}$$

and

$$B_2 = \{y \in C(\mathbb{N}(0,b), X) : \|y\|_\infty \le q\},$$

we have

$$\|N(x,y)\|_\infty \le M = (M_1, M_2).$$

Then for each $k \in \mathbb{N}(0,b)$,

$$\|N_1(x,y)\|_\infty \le Q\left[\|\widetilde{L}_1^{-1}\|_{B(X)}\|l_1\| + 2q^{\gamma_1}Q\|L_1\|_{B(X)}\sum_{k=1}^{b} p_3(k) + \sum_{k=1}^{b}\bar{p}_3(k)\right]$$
$$+2q^{\gamma_1}Q\sum_{i=0}^{b}\bar{p}_3(k).$$

Similarly,

$$\|N_2(x,y)\|_\infty \le Q\left[\|\widetilde{L}_2^{-1}\|_{B(X)}\|l_2\| + 2q^{\gamma_2}Q\|L_2\|_{B(X)}\sum_{k=1}^{b} p_4(k) + \sum_{k=1}^{b}\bar{p}_4(k)\right]$$
$$+2q^{\gamma_2}Q\sum_{i=0}^{b}\bar{p}_4(k).$$

By $(\mathcal{J}_3)$, we can easily prove that, for each $k \in \mathbb{N}(0,b)$, we have

$$\{N_1(x(k),y(k)) : (x,y) \in B\} \text{ and } \{N_2(x(k),y(k)) : (x,y) \in B\}$$

are relatively compact in X. Then as a consequence of Theorem 15.8, we conclude that $\overline{N(B)}$ is compact. As consequence of Steps 1 to 2, N is completely continuous.

Step 3: It remains to show that $\mathcal{A} = \{(x,y) \in C(\mathbb{N}(0,b), X) \times C(\mathbb{N}(0,b), X) : (x,y) = \lambda N(x,y), \lambda \in (0,1)\}$ is bounded. Let $(x,y) \in \mathcal{A}$. Then $x = \lambda N_1(x,y)$ and $y = \lambda N_2(x,y)$ for some $0 < \lambda < 1$. Thus, for $k \in \mathbb{N}(0,b)$, we have

$$\|x(k)\| \le Q\Big[\|\widetilde{L}_1^{-1}\|_{B(X)}\|l_1\| + (Q\|\widetilde{L}_1^{-1}\|_{B(X)}\|L_1\|_{B(X)} + 1) \times$$
$$\left(\sum_{i=0}^{b} p_3(i)(|x(i)| + |y(i)|)^{\gamma_1} + \sum_{i=0}^{b}\bar{p}_3(i)\right)\Big]$$

and

$$\|y(k)\| \le Q\Big[\|\widetilde{L}_2^{-1}\|_{B(X)}\|l_2\| + (Q\|\widetilde{L}_2^{-1}\|_{B(X)}\|L_2\|_{B(X)} + 1) \times$$
$$\left(\sum_{i=0}^{k-1} p_4(i)(|x(i)| + |y(i)|)^{\gamma_2} + \sum_{i=0}^{b}\bar{p}_3(i)\right)\Big].$$

Therefore

$$\|x\|_\infty + \|y\|_\infty \le Q_1 + Q_2\sum_{i=0}^{b} p_*(i)(\|x\|_\infty + \|y\|_\infty)^{\max(\gamma_1, \gamma_2)}$$

where

$$Q_1 = Q(\|\widetilde{L}_1^{-1}\|_{B(X)}\|l_1\| + \|\widetilde{L}_2^{-1}\|_{B(X)}\|l_2\| + Q_2 M),$$
$$Q_2 = Q(\|\widetilde{L}_1^{-1}\|_{B(X)}\|L_1\|_{B(X)} + \|\widetilde{L}_2^{-1}\|_{B(X)}\|L_2\|_{B(X)} + 2),$$

and

$$M = \sum_{i=0}^{b}(\bar{p}_3(i) + p_4(i)), \ \ \|p_*(k)\| = p_3(k) + p_4(k), \ \ k \in \mathbb{N}(0, b).$$

Hence,

$$\|x\|_\infty + \|y\|_\infty \ \leq \ \max(1, C),$$

where

$$C = \left[Q_1 + Q_2 \sum_{i=0}^{b} p_*(i) \right]^{\frac{1}{1 - \max(\gamma_1, \gamma_2)}}.$$

This shows that A is bounded. As a consequence of Theorem 8.23, N has a fixed point (x, y) that is a solution to problem (17.2). $\qquad\square$

Using measure of noncompactness arguments, we have a second existence result.

Theorem 17.6. *Let f_1, $f_2 : \mathbb{N}(0, b) \times X \times X \to X$ be continuous functions. In addition to $(\mathcal{J}_2)$ and $(\mathcal{J}_5)$ assume that*

$(\mathcal{J}_6)$ *There exist nonnegative numbers a_i and b_i for each $i \in \{1, 2\}$ such that for every bounded set $B_1 \times B_2 \subset X \times X$ we have*

$$\alpha(f_1(k, B_1, B_2)) \leq a_1\alpha(B_1) + a_2\alpha(B_2), \ \ k \in \mathbb{N}(0, b), \ (x, y) \in X \times X,$$

and

$$\alpha(f_2(k, B_1, B_2)) \leq b_1\alpha(B_1) + b_2\alpha(B_2), \ \ k \in \mathbb{N}(0, b), \ (x, y) \in X \times X,$$

where α is Kuratowski's measure of noncompactness.

If the matrix

$$M = Qb \begin{bmatrix} (Q\|\widetilde{L}_1^{-1}\|_{B(X)}\|L_1\|_{B(X)} + 1)a_1 & (Q\|\widetilde{L}_1^{-1}\|_{B(X)}\|L_1\|_{B(X)} + 1)a_2 \\ (Q\|\widetilde{L}_2^{-1}\|_{B(X)}\|L_2\|_{B(X)} + 1)b_1 & (Q\|\widetilde{L}_2^{-1}\|_{B(X)}\|L_2\|_{B(X)} + 1)b_2 \end{bmatrix}$$

converges to zero, then the problem (17.2) has a solution.

Proof. Let $\bar{N} : C(\mathbb{N}(0, b), X) \times C(\mathbb{N}(0, b), X) \to C(\mathbb{N}(0, b), X)$ be the operator defined in the proof of Theorem 17.5. As in that proof we can easily show that $\bar{N}$ is continuous, transforms bounded sets into bounded sets, and

$$\mathcal{A} = \{(x, y) \in C(\mathbb{N}(0, b), X) \times C(\mathbb{N}(0, b), X) : (x, y) = \lambda\bar{N}(x, y), \lambda \in (0, 1)\}$$

is bounded. In order to apply Theorem 8.23, we will show that $\bar{N}$ is β-condensing where

$$\beta(B_1 \times B_2) = (\beta_1(B_1), \beta_2(B_2)),$$

$$\beta_1(B) = \beta_2(B) = \sup\{\alpha(B(k)) : k \in \mathbb{N}(0, b)\}, \ B \in C(\mathbb{N}(0, b), X)$$

and

$$\beta_*(B_1 \times B_2) = (\alpha_1(B_1), \alpha_2(B_2)).$$

To show this, we see that for each $k \in \mathbb{N}(0, b)$,

$$\beta_*(N(B_1 \times B_2)(k))$$

$$= \left(\begin{array}{c} \alpha_1(\{N_1(x(k), y(k)); x \in B_1, y \in B_2\}) \\ \alpha_2(\{N_2(x(k), y(k)); x \in B_1, y \in B_2\}) \end{array} \right)$$

$$\leq \left(\begin{array}{c} Q^2\|G^{-1}\|\|L\| \sum_{i=s}^{k-1} \alpha_1(\{f_1(i, x(i), y(i)), x \in B_1, y \in B_2\}) \\ Q^2\|G^{-1}\|\|L\| \sum_{i=s}^{k-1} \alpha_2(\{f_2(i, x(i), y(i)), x \in B_1, y \in B_2\}) \end{array} \right)$$

$$+ \left(\begin{array}{c} Q \sum_{i=s}^{k-1} \alpha_1(\{f_1(i, x(i), y(i)), x \in B_1, y \in B_2\}) \\ Q \sum_{i=s}^{k-1} \alpha_2(\{f_2(i, x(i), y(i)), x \in B_1, y \in B_2\}) \end{array} \right)$$

$$\leq \left(\begin{array}{c} Q^2\|G^{-1}\|\|L\|b\alpha_1(f_1(k, B_1, B_2)) + Qb\alpha_1(f_1(k, B_1, B_2)) \\ Q^2\|G^{-1}\|\|L\|b\alpha_2(f_2(k, B_1, B_2)) + Qb\alpha_2(f_2(k, B_1, B_2)) \end{array} \right)$$

$$\leq \left(\begin{array}{c} Q^2\|G^{-1}\|\|L\|b(a_1\beta_1(B_1) + a_2\beta_2(B_2)) \\ +Qb(a_1\beta_1(B_1) + a_2\beta_2(B_2)) \\ \\ Q^2\|G^{-1}\|\|L\|b(b_1\beta_1(B_1) + b_2\beta_2(B_2)) \\ +Qb(b_1\beta_1(B_1) + b_2\beta_2(B_2)) \end{array} \right)$$

$$\leq Qb \left(\begin{array}{c} Q\|G^{-1}\|\|L\| + 1 \\ Q\|G^{-1}\|\|L\| + 1 \end{array} \right) \left(\begin{array}{cc} a_1 & a_2 \\ b_1 & b_2 \end{array} \right) \left(\begin{array}{c} \beta_1(B_1) \\ \beta_2(B_2) \end{array} \right).$$

Hence,

$$\beta(N(B_1 \times B_2)) \leq M\beta(B_1 \times B_2).$$

As a consequence of Theorem 8.23, $\bar{N}$ has a fixed point (x, y) that is a solution to (17.2).

$\square$

Chapter 18

Discrete Boundary Value Problems

In this chapter, we will be concerned with solutions of the nth order difference equation

$$w(t+n) = f(t, w(t), \dots, w(t+n-1)), \ t \in \mathbb{Z}, \ n \geq 2, \tag{18.1}$$

satisfying nonlocal boundary conditions

$$w(t_j + i) = w_{ij}, \ 0 \leq i \leq m_j - 1, \ 1 \leq j \leq k-1,$$

$$w(t_k + i) - ds \sum_{p=1}^{m} \alpha_{ip} w(\eta_{ip}) = w_{ik}, \ 0 \leq i \leq m_k - 1, \tag{18.2}$$

where $2 \leq k \leq n$, $m \in \mathbb{N}$, $m_1, \dots, m_k$ are positive integers such that $\sum_{i=1}^{k} m_i = n$, $t_1 < t_1 + m_1 - 1 < t_2 < t_2 + m_2 - 1 < \cdots < t_{k-1} < t_{k-1} + m_{k-1} - 1 < \eta_{01} < \eta_{01} + 1 < \cdots < \eta_{0p} < \eta_{0m} + 1 < \eta_{11} < \eta_{11} + 1 < \cdots < \eta_{m_k-1,0} < \eta_{m_k-1,0} + 1 < \cdots < \eta_{m_k-1,m} < \eta_{m_k-1,m} + 1 < t_k$ in $\mathbb{Z}$, and $\alpha_{01}, \dots, \alpha_{m_k-1,m}, w_{01}, \dots, w_{m_k-1,k} \in \mathbb{R}$.

We assume throughout this chapter that for (18.1):

(i) $f(t, d_1, d_2, \dots, d_n) : \mathbb{Z} \times \mathbb{R}^n \to \mathbb{R}$ is continuous,

(ii) $\frac{\partial f}{\partial d_i}(t, d_1, d_2, \dots, d_n) : \mathbb{Z} \times \mathbb{R}^n \to \mathbb{R}$ are continuous, $i = 1, 2, \dots, n$, and

(iii) The equation $d_{n+1} = f(t, d_1, d_2, \dots, d_n)$ can be solved for d_1 as a continuous function of $d_2, d_3, \dots, d_{n+1}$, for all $t \in \mathbb{Z}$.

Remark 18.1. *We observe condition (iii) implies that solutions of initial value problems for (18.1) exist and are unique on all of $\mathbb{Z}$.*

The heart of this work utilizes continuous dependence of solutions of (18.1) to find differences with respect boundary points and derivatives with respect to boundary values. After imposing a few disconjugacy-type conditions on (18.1), we will see that derivatives of a solution, $w(t)$, to (18.1) relate to the following linear nth order difference equation

$$z(t+n) = \sum_{i=1}^{n} \frac{\partial f}{\partial d_i}(t, w(t), \dots, w(t+n-1)) z(t+i-1) \tag{18.3}$$

called the *variational equation along $w(t)$*.

The motivation for the research conducted in this work on the relationship between a solution to a differential or difference equation and the associated variational equation can trace its origin to 1964 when Hartman proved a theorem he attributed to Peano about initial value problems for differential equations [155]. Since then, Henderson and several others have extended and redefined these results in various ways including boundary value problems for both differential and difference equations. For differential equations results, we point the reader to [110, 159, 161, 207]. For results on difference equations, we reference [45, 92, 93, 156, 239]. Also, interest in multipoint and nonlocal boundary value problems has grown significantly in recent years as can be seen in [12, 30, 113, 186] which is why we have incorporated nonlocality into our boundary conditions. Also, related to the results of this chapter is the recent paper by Hopkins *et al.* [172].

18.1   Initial value problems

This section is devoted to results for initial value problems for difference equations. First, we will present a continuous dependence result for difference equations. Next, we present a difference equation analogue of the theorem mentioned above that Hartman attributed to Peano. The theorem will play a crucial role in proving the main results of this paper. The proofs of both theorems are omitted as they follow along the standard path for initial value problems for differential equations. Lastly, we provide a result involving differences of solutions to an initial value problem. The proof is very similar to those given in [92] and [93] and is omitted.

We will denote the unique solution of the initial value problem (18.1) satisfying the initial conditions

$$u(t_0 + i - 1) = c_i, \ 1 \leq i \leq n, \tag{18.4}$$

where $t_0 \in \mathbb{Z}$ and $c_1, c_2, \ldots, c_n \in \mathbb{R}$, by

$$u(t) = u(t, t_0, c_1, c_2, \ldots, c_n). \tag{18.5}$$

Theorem 18.2. *[Continuous Dependence with Respect to Initial Values] Assume conditions* (i) *and* (iii) *hold. Let* $t_0 \in \mathbb{Z}$ *and* $c_1, c_2, \ldots, c_n \in \mathbb{R}$ *be given. Then for all* $\epsilon > 0$ *and for all* $k \in \mathbb{N}$, *there exists* $\delta(\epsilon, k, t_0, c_1, \ldots, c_n) > 0$ *such that* $|c_i - e_i| < \delta$, $1 \leq i \leq n$, *implies* $|u(t, t_0, c_1, \ldots, c_n) - u(t, t_0, e_1, \ldots, e_n)| < \epsilon$ *for* $m \in [t_0 - k, t_0 + k]$ *and* $e_1, e_2, \ldots, e_n \in \mathbb{R}$.

Theorem 18.3. *[Differentiation with Respect to Initial Values] Assume* (i), (ii), *and* (iii) *hold. Let* $t_0 \in \mathbb{Z}$ *and* $c_1, c_2, \ldots, c_n \in \mathbb{R}$ *be given. Then, for* $j = 1, 2, \ldots, n$, $\beta_j := \frac{\partial u}{\partial c_j}(t, t_0, c_1, \ldots, c_n)$ *exists and is the solution of the variational equation* (18.3) *along* $u(t, t_0, c_1, \ldots, c_n)$; *i.e.,*

$$\beta_j(t + n) = \sum_{i=1}^{n} \frac{\partial f}{\partial d_i}(t, u(t), u(t+1), \ldots, u(t+n-1))\beta_j(t + i - 1)$$

satisfying the initial conditions

$$\beta_j(t_0 + i - 1) = \delta_{ij}, \ 1 \leq i \leq n.$$

Theorem 18.4. *[Differences with Respect to Initial Points] Assume* (i), (ii), *and* (iii) *hold. Let* $t_0 \in \mathbb{Z}$ *and* $c_1, c_2, \ldots, c_n \in \mathbb{R}$ *be given. Then*

$$\gamma(t) := \Delta_{t_0} u(t, t_0, c_1, \ldots, c_n) = u(t, t_0 + 1, c_1, \ldots, c_n) - u(t, t_0, c_1, \ldots, c_n)$$

is the solution of the nth order linear difference equation

$$\gamma(t + n) = \sum_{r=1}^{n} A_r(t)\gamma(t + r - 1),$$

satisfying the initial conditions

$$\gamma(t_0 + i) = -\Delta_t u(t, t_0 + 1, c_1, \ldots, c_n)|_{t=t_0+i}, \ 0 \leq i \leq n - 1,$$

where

$$A_r(t) = \int_0^1 \frac{\partial f}{\partial d_r}(t, w(t, t_0 + 1, c_1, \ldots, c_n), w(t + 1, t_0 + 1, c_1, \ldots, c_n), \ldots,$$
$$sw(t + r - 1, t_0 + 1, c_1, \ldots, c_n) + (1 - s)w(t + r - 1, t_0, c_1, \ldots,$$
$$c_n), \ldots, w(t + n - 1, t_0, c_1, \ldots, c_n))ds.$$

18.2 Nonlocal boundary value problems

In order to form a correlation between the work in the last section and nonlocal boundary value problems, we must first establish that solutions of (18.1) are unique. To accomplish this, we use Hartman's definition of a generalized zero in [156].

Definition 18.5. *Let $v : \mathbb{Z} \to \mathbb{R}$. We say v has a generalized zero at $n_0 \in \mathbb{Z}$ provided either $v(n_0) = 0$ or there exists $k \in \mathbb{N}$ such that $(-1)^k v(n_0 - k)v(n_0) > 0$ and if $k > 1$, $v(n_0 - k + 1) = \cdots = v(n_0 - 1) = 0$.*

Remark 18.6. *For the remainder of the work, we denote a solution of (18.1), (18.2) by $w(t, t_1, \ldots, t_k, \eta_{01}, \ldots, \eta_{m_k-1,m}, \alpha_{01}, \ldots, \alpha_{m_k-1,m}, w_{01}, \ldots, w_{m_k-1,m})$ and to ease the burdensome notation, we will write solutions as $w(t, \cdot)$ where "$\cdot$" represents the emphasized variable.*

Next, we define uniqueness properties properties which are disconjugacy-type properties for difference equations.

Definition 18.7. *Let $n \geq 2$, $2 \leq k \leq n$, $m \in \mathbb{N}$, and $m_1, \ldots, m_k$ be positive integers such that $\sum_{i=1}^{k} m_i = n$. The nonlinear difference equation (18.1) is said to satisfy Property (U) on $\mathbb{Z}$ if, whenever $w_1(t)$ and $w_2(t)$ are solutions of (18.1) such that $w_1(t) - w_2(t)$ has a generalized zero at $t_j + i$ for each $0 \leq i \leq m_j - 1$, $1 \leq j \leq k - 1$, and $[w_1(t) - w_2(t)] - \sum_{p=1}^{m} \alpha_{ip}[w_1(\eta_{ip}) - w_2(\eta_{ip})]$ has a generalized zero at $t_k + i$ for each $0 \leq i \leq m_k - 1$, where $t_1 < t_1 + m_1 - 1 < t_2 < t_2 + m_2 - 1 < \cdots < t_{k-1} < t_{k-1} + m_{k-1} - 1 < \eta_{01} < \eta_{01} + 1 < \cdots < \eta_{0p} < \eta_{0m} + 1 < \eta_{11} < \eta_{11} + 1 < \cdots < \eta_{m_k-1,0} < \eta_{m_k-1,0} + 1 < \cdots < \eta_{m_k-1,m} < \eta_{m_k-1,m} + 1 < t_k$ in $\mathbb{Z}$ and $\alpha_{01}, \ldots, \alpha_{m_k-1,m} \in \mathbb{R}$, then $w_1(t) \equiv w_2(t)$ on $\mathbb{Z}$.*

Definition 18.8. *Let $n \geq 2$, $2 \leq k \leq n$, $m \in \mathbb{N}$, and $m_1, \ldots, m_k$ be positive integers such that $\sum_{i=1}^{k} m_i = n$. The linear difference equation*

$$s(t + n) = \sum_{i=1}^{n} N_i(t)s(t + i - 1) \tag{18.6}$$

is said to satisfy Property (U) on $\mathbb{Z}$, provided there is no nontrivial solution $s(t)$ of (18.1) such that $s(t)$ has a generalized zero at $t_j + i$ for each $0 \leq i \leq m_j - 1$, $1 \leq j \leq k - 1$, and $s(t) - \sum_{p=1}^{m} \alpha_{ip} s(\eta_{ip})$ has a generalized zero at $t_k + i$ for each $0 \leq i \leq m_k - 1$ where $t_1 < t_1 + m_1 - 1 < t_2 < t_2 + m_2 - 1 < \cdots < t_{k-1} < t_{k-1} + m_{k-1} - 1 < \eta_{01} < \eta_{01} + 1 < \cdots < \eta_{0p} < \eta_{0m} + 1 < \eta_{11} < \eta_{11} + 1 < \cdots < \eta_{m_k-1,0} < \eta_{m_k-1,0} + 1 < \cdots < \eta_{m_k-1,m} < \eta_{m_k-1,m} + 1 < t_k$ in $\mathbb{Z}$, and $\alpha_{01}, \ldots, \alpha_{m_k-1,m} \in \mathbb{R}$.

We now state a result establishing, under Property (U), the continuous dependence of solutions with respect to boundary values.

Theorem 18.9. *[Continuous Dependence with Respect to Boundary Values] Assume conditions (i) and (iii) hold and that (18.1) satisfies Property (U) on $\mathbb{Z}$. Let $y(t)$ be a solution of (18.1) and let $n \geq 2$, $2 \leq k \leq n$, $m \in \mathbb{N}$, and $m_1, \ldots, m_k$ such that $\sum_{i=1}^{k} m_i = n$. Also, let $t_1 < t_1 + m_1 - 1 < t_2 < t_2 + m_2 - 1 < \cdots < t_{k-1} < t_{k-1} + m_{k-1} - 1 < \eta_{01} < \eta_{01} + 1 < \cdots < \eta_{0p} < \eta_{0m} + 1 < \eta_{11} < \eta_{11} + 1 < \cdots < \eta_{m_k-1,0} < \eta_{m_k-1,0} + 1 < \cdots < \eta_{m_k-1,m} < \eta_{m_k-1,m} + 1 < t_k$ in $\mathbb{Z}$ and $\alpha_{01}, \ldots, \alpha_{m_k-1,m} \in \mathbb{R}$ be given. Then, there exists $\epsilon > 0$ such that, if $\delta_{01}, \ldots, \delta_{m_k-1,k}, \beta_{01}, \ldots, \beta_{m_k-1,m} \in \mathbb{R}$ with $|\delta_{ij}| < \epsilon$, $0 \leq i \leq m_j - 1$, $1 \leq j \leq k$*

and $|\alpha_{ip} - \beta_{ip}| < \epsilon$, $0 \le i \le m_k - 1$ *and* $1 \le p \le m$, *the nonlocal boundary value problem* (18.1) *satisfying*

$$w(t_j + i) = y(t_j + i) + \delta_{ij}, \ 0 \le i \le m_j - 1, \ 1 \le j \le k - 1,$$

$$w(t_k + i) - \sum_{p=1}^{m} \alpha_{ip} w(\eta_{ip}) = y(t_k + i) - \sum_{p=1}^{m} \beta_{ip} y(\eta_{ip}) + \delta_{ik}, \ 0 \le i \le m_k - 1,$$

has a unique solution:

$$w\Big(t, t_1, \ldots, t_k, \eta_{01}, \ldots, \eta_{m_k-1,m}, \beta_{01}, \ldots, \beta_{m_k-1,m}, y(t_1) + \delta_{01}, \ldots,$$

$$y(t_1 + m_1 - 1) + \delta_{m_1-1,1}, \ldots, y(t_k) - \sum_{p=1}^{m} \beta_{0p} y(\eta_{0p}) + \delta_{0k}, \ldots,$$

$$y(t_k + m_k - 1) - \sum_{p=1}^{m} \beta_{m_k-1,p} y(\eta_{m_k-1,p}) + \delta_{m_k-1,k}\Big).$$

Moreover, as $\epsilon \to 0$, *this solution converges to* $y(t)$ *on* $\mathbb{Z}$.

The proof relies on the Brouwer Invariance of Domain Theorem, which we state here.

Theorem 18.10. *If* U *is an open subset of* $\mathbb{R}^n$, *and* $\phi : U \to \mathbb{R}^n$ *is one to one and continuous on* U, *then* ϕ *is a homeomorphism and* $\phi(U)$ *is an open subset of* $\mathbb{R}^n$.

The main idea behind the proof of Theorem 18.9 is to redefine the boundary value problem as an initial value problem and then utilize the well-known results in Theorem 18.2. The way to do this is by defining a function that maps an n-tuple of initial values to an n-tuple of boundary values. One can then show the function satisfies the conditions of Theorem 18.10 using Property (U) and Theorem 18.2 which allows for perturbation of boundary values in the range since the range is an open set. Afterward, it is a matter of utilizing the conclusion of Theorem 18.10 and Theorem 18.2 to show that the initial values associated with the perturbed solution to the initial value problem converge uniformly to the initial values of the solution of the boundary value problem as $\epsilon \to 0$. For more details, we refer the reader to a typical proof which can be found in [93] and [162].

18.3   Differentiation of solutions with respect to boundary conditions

With the ideas and strategies of the previous sections, we are now able to provide the major results of this chapter. The first result pertains to derivatives with respect to boundary values and the other pertains to differences with respect to boundary points.

Theorem 18.11. *Let* $n \ge 2$, $2 \le k \le n$, *and* $m \in \mathbb{N}$ *be given, and let* $m_1, \ldots, m_k$ *be positive integers such that* $\sum_{i=1}^{k} m_i = n$. *Assume conditions* (i), (ii), *and* (iii) *are satisfied, that* (18.1) *satisfies Property (U) on* $\mathbb{Z}$, *and that the variational equation* (18.3) *satisfies Property (U) along solutions of* (18.1). *Suppose*

$$w(t) = w(t, t_1, \ldots, t_k, \eta_{01}, \ldots, \eta_{m_k-1,m}, \alpha_{01}, \ldots, \alpha_{m_k-1,m}, w_{01}, \ldots, w_{m_k-1,k})$$

is the solution of (18.1) on $\mathbb{Z}$ where

$$w(t_j + i) = w_{ij}, \quad 0 \leq i \leq m_j - 1, \ 1 \leq j \leq k - 1,$$

$$w(t_k + i) - \sum_{p=1}^{m} \alpha_{ip} w(\eta_{ip}) = w_{ik}, \quad 0 \leq i \leq m_k - 1,$$

$t_1 < t_1 + m_1 - 1 < t_2 < t_2 + m_2 - 1 < \cdots < t_{k-1} + m_{k-1} - 1 < \eta_{01} < \eta_{01} + 1 < \cdots < \eta_{0p} <$
$\eta_{0m} + 1 < \eta_{11} < \eta_{11} + 1 < \cdots < \eta_{m_k-1,0} < \eta_{m_k-1,0} + 1 < \cdots < \eta_{m_k-1,m} < \eta_{m_k-1,m} + 1 < t_k$
in $\mathbb{Z}$, and $\alpha_{01}, \ldots, \alpha_{m_k-1,m}, w_{01}, \ldots, w_{m_k-1,k} \in \mathbb{R}$. Then,

(a) for $0 \leq q \leq m_k - 1$ and $1 \leq l \leq m$, $p_{ql}(t) := ds\frac{\partial w}{\partial \alpha_{ql}}(t)$ exists on $\mathbb{Z}$ and is the solution of (18.3) along $w(t)$ satisfying

$$\begin{aligned}
p_{ql}(t_j + i) &= 0, & 0 \leq i \leq m_j - 1, \ 1 \leq j \leq k - 1, \\
p_{ql}(t_k + i) &- \sum_{p=1}^{m} \alpha_{ip} p_{ql}(\eta_{ip}) = 0, & 0 \leq i \leq m_k - 1, \ i \neq q, \\
p_{ql}(t_k + q) &- \sum_{p=1}^{m} \alpha_{qp} p_{ql}(\eta_{qp}) = w(\eta_{ql}).
\end{aligned}$$

(b) for $1 \leq l \leq k - 1$ and $0 \leq q \leq m_l - 1$, $z_{ql}(t) := ds\frac{\partial w}{\partial w_{ql}}(t)$ exists on $\mathbb{Z}$ and is the solution of (18.3) along $w(t)$ satisfying

$$\begin{aligned}
z_{ql}(t_j + i) &= 0, & 0 \leq i \leq m_j - 1, \ 1 \leq j \leq k - 1, \ j \neq l, \\
z_{ql}(t_l + i) &= 0, & 0 \leq i \leq m_l - 1, \ i \neq q, \\
z_{ql}(t_l + q) &= 1, & \\
z_{ql}(t_k + i) &- \sum_{p=1}^{m} \alpha_{ip} z_{ql}(\eta_{ip}) = 0, & 0 \leq i \leq m_k - 1,
\end{aligned}$$

and for $0 \leq q \leq m_k - 1$, $z_{qk}(t) := ds\frac{\partial w}{\partial w_{qk}}(t)$ exists on $\mathbb{Z}$ and is the solution of (18.3) along $w(t)$ satisfying

$$\begin{aligned}
z_{qk}(t_j + i) &= 0, & 0 \leq i \leq m_j - 1, \ 1 \leq j \leq k - 1, \\
z_{qk}(t_k + i) &- \sum_{p=1}^{m} \alpha_{ip} z_{qk}(\eta_{ip}) = 0, & 0 \leq m_k - 1, \ i \neq q, \\
z_{qk}(t_k + q) &- \sum_{p=1}^{m} \alpha_{qp} z_{qk}(\eta_{qp}) = 1.
\end{aligned}$$

Proof. We will only present the proof of part (b) as part (a) follows along the same lines. First, let $\epsilon > 0$ be as in Theorem 18.9, and let $0 < |h| < \epsilon$ be given. Fix $1 \leq l \leq k - 1$, $0 \leq q \leq m_l - 1$, and consider the quotient,

$$\begin{aligned}
z_{qlh}(t) &= \frac{1}{h}[w(t, w_{ql} + h) \\
&\quad - w(t, w_{ql})].
\end{aligned}$$

Notice that for $0 \leq i \leq m_j - 1$, $1 \leq j \leq k - 1$, $j \neq l$, and $h \neq 0$,

$$z_{qlh}(t_j + i) = \frac{1}{h}[w(t_j + i, w_{ql} + h) - w(t_j + i, w_{ql})]$$

$$= [w_{ij} - w_{ij}]$$
$$= 0,$$

for $0 \le i \le m_l - 1$, $i \ne q$, and $h \ne 0$,

$$z_{qlh}(t_l + i) = \frac{1}{h}[w(t_l + i, w_{ql} + h) - w(t_l + i, w_{ql})]$$
$$= [w_{il} - w_{il}] = 0,$$

and for $0 \le i \le m_k - 1$ and $h \ne 0$,

$$z_{qlh}(t_k + i) \quad - \quad \sum_{p=1}^{m} \alpha_{ip} z_{qlh}(\eta_{ip})$$

$$= \frac{1}{h}\left[(w(t_k + i, w_{ql} + h) - \sum_{p=1}^{m} \alpha_{ip} w(\eta_{ip}, w_{ql} + h)) \right.$$

$$\left. - w(t_k + i, w_{ql}) + \sum_{p=1}^{m} \alpha_{ip} w(\eta_{ip}, w_{ql}) \right]$$

$$= [w_{ik} - w_{ik}]$$
$$= 0.$$

Also, we have, for $h \ne 0$,

$$z_{qlh}(t_l + q) = \frac{1}{h}[w(t_l + q, w_{ql} + h,) - w(t_l + q, w_{ql})]$$
$$= \frac{1}{h}[w_{ql} + h - w_{ql}]$$
$$= 1.$$

Thus, for $z_{qlh}(t)$ the nonlocal boundary conditions are satisfied, but it remains to show that $z_{qlh}(t)$ solves the variational equation. To this end, we view $w(t)$ in terms of the solution of an initial value problem of (18.1).
For $m_l \le i \le n - 1$, let

$$\sigma_i = w(t_l + i, w_{ql}),$$

and

$$\epsilon_i = \epsilon_i(h) = w(t_l + i, w_{ql} + h) - \sigma_i.$$

Then, by Theorem 18.9, $\epsilon_i \to 0$ as $h \to 0$ for $m_l \le i \le n - 1$. Thus, we have $w(t) = u(t, t_l, w_{0l}, \ldots, w_{ql}, \ldots, w_{m_l-1,l}, \sigma_{m_l}, \sigma_{m_l+1}, \ldots, \sigma_{n-1})$, and

$$z_{qlh}(t) = \frac{1}{h}[u(t, t_l, w_{0l}, \ldots, w_{ql} + h, \ldots, w_{m_l-1,l}, \sigma_{m_l} + \epsilon_{m_l},$$
$$\sigma_{m_l+1} + \epsilon_{m_l+1}, \ldots, \sigma_{n-1} + \epsilon_{n-1})$$
$$- u(t, t_l, w_{0l}, \ldots, w_{ql}, \ldots, w_{m_l-1,l}, \sigma_{m_l}, \sigma_{m_l+1}, \ldots, \sigma_{n-1})].$$

Now we implement a telescoping sum to yield

$$z_{qlh}(t) = \frac{1}{h}[u(t, t_l, w_{0l}, \ldots, w_{ql} + h, \ldots, w_{m_l-1,l}, \sigma_{m_l} + \epsilon_{m_l},$$

$$\sigma_{m_l+1} + \epsilon_{m_l+1}, \ldots, \sigma_{n-1} + \epsilon_{n-1})$$
$$-u(t, t_1, w_{0l}, \ldots, w_{ql}, \ldots, w_{m_l-1,l}, \sigma_{m_l} + \epsilon_{m_l},$$
$$\sigma_{m_l+1} + \epsilon_{m_l+1}, \ldots, \sigma_{n-1} + \epsilon_{n-1})]$$
$$+[u(t, t_1, w_{0l}, \ldots, w_{ql}, \ldots, w_{m_l-1,l}, \sigma_{m_l} + \epsilon_{m_l},$$
$$\sigma_{m_l+1} + \epsilon_{m_l+1}, \ldots, \sigma_{n-1} + \epsilon_{n-1})$$
$$-u(t, t_1, w_{0l}, \ldots, w_{ql}, \ldots, w_{m_l-1,l}, \sigma_{m_l},$$
$$\sigma_{m_l+1} + \epsilon_{m_l+1}, \ldots, \sigma_{n-1} + \epsilon_{n-1})]$$
$$+ - \cdots$$
$$+[u(t, t_1, w_{0l}, \ldots, w_{ql}, \ldots, w_{m_l-1,l}, \sigma_{m_l}, \sigma_{m_l+1}, \ldots, \sigma_{n-1} + \epsilon_{n-1})$$
$$-u(t, t_1, w_{0l}, \ldots, w_{ql}, \ldots, w_{m_l-1,l}, \sigma_{m_l}, \sigma_{m_l+1}, \ldots, \sigma_{n-1})].$$

Thus, by using Theorem 18.3 and the Mean Value Theorem, we have

$$z_{qlh}(t) = \beta_q(t, u(t, t_l, w_{0l}, \ldots, w_{ql} + \bar{h}, \ldots, w_{m_l-1,l}, \sigma_{m_l} + \epsilon_{m_l},$$
$$\sigma_{m_l+1} + \epsilon_{m_l+1}, \ldots, \sigma_{n-1} + \epsilon_{n-1})$$
$$+ \frac{\epsilon_{m_l}}{h} \beta_{m_l}(t, u(t, t_l, w_{0l}, \ldots, w_{ql}, \ldots, w_{m_l-1,l}, \sigma_{m_l} + \bar{\epsilon}_{m_l},$$
$$\sigma_{m_l+1} + \epsilon_{m_l+1}, \ldots, \sigma_{n-1} + \epsilon_{n-1})$$
$$+ \frac{\epsilon_{m_l+1}}{h} \beta_{m_l+1}(t, u(t, t_l, w_{0l}, \ldots, w_{ql}, \ldots, w_{m_l-1,l}, \sigma_{m_l},$$
$$\sigma_{m_l+1} + \bar{\epsilon}_{m_l+1}, \ldots, \sigma_{n-1} + \epsilon_{n-1})$$
$$+ \cdots$$
$$+ \frac{\epsilon_{n-1}}{h} \beta_{n-1}(t, u(t, t_l, w_{0l}, \ldots, w_{ql}, \ldots, w_{m_l-1,l}, \sigma_{m_l},$$
$$\sigma_{m_l+1}, \ldots, \sigma_{n-1} + \bar{\epsilon}_{n-1}),$$

where $w_{ql} + \bar{h}$ is between w_{ql} and $w_{ql} + h$, $\sigma_i + \bar{\epsilon}_i$ is between σ_i and $\sigma_i + \epsilon_i$ for $m_l \le i \le n-1$, and $\beta_j(t_l + i) = \delta_{ij}$, for $0 \le i, j \le n-1$, and solves (18.3). Hence, for $\lim_{h\to 0} z_{qlh}(t)$ to exist, we need $\lim_{h\to 0} \frac{\epsilon_i}{h}$ to exist for each $m_l \le i \le n-1$. Now, from the construction of $z_{qlh}(t)$,

$$z_{qlh}(t_j + i) = 0, \qquad 0 \le i \le m_j - 1, \ 1 \le j \le k-1, \ j \ne l,$$

and

$$z_{qlh}(t_k + i) - \sum_{p=1}^{m} \alpha_{ip} z_{qlh}(\eta_{ip}) = 0, \qquad 0 \le i \le m_k - 1.$$

Hence, we have a system of $n - m_l$ linear equations with $n - m_l$ unknowns:

$$-\beta_q(t_j + i, u(t, t_l, w_{0l}, \ldots, w_{ql} + \bar{h}, \ldots, w_{m_l-1,l}, \sigma_{m_l} + \epsilon_{m_l},$$
$$\sigma_{m_l+1} + \epsilon_{m_l+1}, \ldots, \sigma_{n-1} + \epsilon_{n-1})$$
$$= \frac{\epsilon_{m_l}}{h} \beta_{m_l}(t_j + i, u(t, t_l, w_{0l}, \ldots, w_{ql}, \ldots, w_{m_l-1,l}, \sigma_{m_l} + \bar{\epsilon}_{m_l},$$
$$\sigma_{m_l+1} + \epsilon_{m_l+1}, \ldots, \sigma_{n-1} + \epsilon_{n-1})$$
$$+ \frac{\epsilon_{m_l+1}}{h} \beta_{m_l+1}(t_j + i, u(t, t_l, w_{0l}, \ldots, w_{ql}, \ldots, w_{m_l-1,l}, \sigma_{m_l},$$
$$\sigma_{m_l+1} + \bar{\epsilon}_{m_l+1}, \ldots, \sigma_{n-1} + \epsilon_{n-1})$$
$$+ \cdots$$

$$+ \frac{\epsilon_{n-1}}{h} \beta_{n-1}(t_j + i, u(t, t_l, w_{0l}, \ldots, w_{ql}, \ldots, w_{m_l-1,l}, \sigma_{m_l},$$

$$\sigma_{m_l+1}, \ldots, \sigma_{n-1} + \bar{\epsilon}_{n-1})),$$

$$0 \le i \le m_j - 1, \ 1 \le j \le k - 1, \ j \ne l,$$

and

$$-\beta_q(t_k + i, u(t, t_l, w_{0l}, \ldots, w_{ql} + \bar{h}, \ldots, w_{m_l-1,l}, \sigma_{m_l} + \epsilon_{m_l},$$

$$\sigma_{m_l+1} + \epsilon_{m_l+1}, \ldots, \sigma_{n-1} + \epsilon_{n-1})$$

$$+ \sum_{p=1}^{m} \alpha_{ip} \beta_q(\eta_{ip}, u(t, t_l, w_{0l}, \ldots, w_{ql} + \bar{h}, \ldots, w_{m_l-1,l}, \sigma_{m_l} + \epsilon_{m_l},$$

$$\sigma_{m_l+1} + \epsilon_{m_l+1}, \ldots, \sigma_{n-1} + \epsilon_{n-1})$$

$$= ds \frac{\epsilon_{m_l}}{h} \Bigg[\beta_{m_l}(t_k + i, u(t, t_l, w_{0l}, \ldots, w_{ql}, \ldots, w_{m_l-1,l}, \sigma_{m_l} + \bar{\epsilon}_{m_l},$$

$$\sigma_{m_l+1} + \epsilon_{m_l+1}, \ldots, \sigma_{n-1} + \epsilon_{n-1}))$$

$$- \sum_{p=1}^{m} \alpha_{ip} \beta_{m_l}(\eta_{ip}, u(t, t_l, w_{0l}, \ldots, w_{ql}, \ldots, w_{m_l-1,l}, \sigma_{m_l} + \bar{\epsilon}_{m_l},$$

$$\sigma_{m_l+1} + \epsilon_{m_l+1}, \ldots, \sigma_{n-1} + \epsilon_{n-1})) \Bigg] + \cdots$$

$$+ ds \frac{\epsilon_{n-1}}{h} \Bigg[\beta_{n-1}(t_k + i, u(t, t_l, w_{0l}, \ldots, w_{ql}, \ldots, w_{m_l-1,l}, \sigma_{m_l},$$

$$\sigma_{m_l+1}, \ldots, \sigma_{n-1} + \bar{\epsilon}_{n-1}))$$

$$- \sum_{p=1}^{m} \alpha_{ip} \beta_{n-1}(\eta_{ip}, u(t, t_l, w_{0l}, \ldots, w_{ql}, \ldots, w_{m_l-1,l}, \sigma_{m_l},$$

$$\sigma_{m_l+1}, \ldots, \sigma_{n-1} + \bar{\epsilon}_{n-1})) \Bigg], \quad 0 \le i \le m_k - 1.$$

From now on, we will, at times, suppress the arguments of β, the subscripts of η and α, and the limits of summation to make the notation easier to read. Note that $u(\cdot)$ is not necessarily the same within the system of equations. Therefore, we consider the following matrix along the solution $u(t)$.

$$M :=$$

$$\begin{pmatrix}
\beta_{m_l}(t_1, u(t)) & \beta_{m_l+1}(t_1, u(t)) & \cdots & \beta_{n-1}(t_1, u(t)) \\
\beta_{m_l}(t_1 + 1, u(t)) & \beta_{m_l+1}(t_1 + 1, u(t)) & \cdots & \beta_{n-1}(t_1 + 1, u(t)) \\
\vdots & \vdots & \ddots & \vdots \\
\beta_{m_l}(t_1 + m_1 - 1, u(t)) & \beta_{m_l+1}(t_1 + m_1 - 1, u(t)) & \cdots & \beta_{n-1}(t_1 + m_1 - 1, u(t)) \\
\vdots & \vdots & \ddots & \vdots \\
\beta_{m_l}(t_{l-1} + m_{l-1} - 1, u(t)) & \beta_{m_l+1}(t_{l-1} + m_{l-1} - 1, u(t)) & \cdots & \beta_{n-1}(t_{l-1} + m_{l-1} - 1, u(t)) \\
\beta_{m_l}(t_{l+1}, u(t)) & \beta_{m_l+1}(t_{l+1}, u(t)) & \cdots & \beta_{n-1}(t_{l+1}, u(t)) \\
\vdots & \vdots & \ddots & \vdots \\
\beta_{m_l}(t_k, u(t)) - & \beta_{m_l+1}(t_k, u(t)) - & \cdots & \beta_{n-1}(t_k, u(t)) - \\
\sum \alpha \beta_{m_l}(\eta, u(t)) & \sum \alpha \beta_{m_l+1}(\eta, u(t)) & & \sum \alpha \beta_{n-1}(\eta, u(t)) \\
\vdots & \vdots & \ddots & \vdots \\
\beta_{m_l}(t_k + m_k - 1, u(t)) - & \beta_{m_l+1}(t_k + m_k - 1, u(t)) - & \cdots & \beta_{n-1}(t_k + m_k - 1, u(t)) - \\
\sum \alpha \beta_{m_l}(\eta, u(t)) & \sum \alpha \beta_{m_l+1}(\eta, u(t)) & \cdots & \sum \alpha \beta_{n-1}(\eta, u(t))
\end{pmatrix}$$
.

We claim $\det(M) \neq 0$. Suppose to the contrary that $\det(M) = 0$. Then there exist $p_i \in \mathbb{R}$, $m_l \leq i \leq n-1$, not all zero such that

$$
p_{m_l}
\begin{pmatrix}
\beta_{m_l}(t_1, u(t)) \\
\beta_{m_l}(t_1 + 1, u(t)) \\
\vdots \\
\beta_{m_l}(t_{l-1} + m_{l-1} - 1, u(t)) \\
\beta_{m_l}(t_{l+1}, u(t)) \\
\vdots \\
\beta_{m_l}(t_k, u(t)) - \\
\sum \alpha \beta_{m_l}(\eta, u(t)) \\
\vdots \\
\beta_{m_l}(t_k + m_k - 1, u(t)) - \\
\sum \alpha \beta_{m_l}(\eta, u(t))
\end{pmatrix}
+ \cdots + p_{n-1}
\begin{pmatrix}
\beta_{n-1}(t_1, u(t)) \\
\beta_{n-1}(t_1 + m_1 - 1, u(t)) \\
\vdots \\
\beta_{n-1}(t_{l-1} + m_{l-1} - 1, u(t)) \\
\beta_{n-1}(t_{l+1}, u(t)) \\
\vdots \\
\beta_{n-1}(t_k, u(t)) - \\
\sum \alpha \beta_{n-1}(\eta, u(t)) \\
\vdots \\
\beta_{n-1}(t_k + m_k - 1, u(t)) - \\
\sum \alpha \beta_{m_l}(\eta, u(t))
\end{pmatrix}
=
\begin{pmatrix}
0 \\
0 \\
\vdots \\
0 \\
0 \\
\vdots \\
0 \\
\\
\vdots \\
0
\end{pmatrix}.
$$

Set

$$\omega(t, u(t)) := p_{m_l}\beta_{m_l}(t, u(t)) + \cdots + p_{n-1}\beta_{n-1}(t, u(t)).$$

Then, $\omega(t, u(t))$ is a nontrivial solution of (18.3), but

$$\omega(t_j + i, u(t)) = 0, \; 0 \leq i \leq m_j - 1, \; 1 \leq j \leq k - 1,$$

and

$$\omega(t_k + i, u(t)) - \sum_{p=1}^{m} \alpha_{ip}\omega(\eta_{ip}, u(t)) = 0, \; 0 \leq i \leq m_k - 1,$$

which when coupled with Property (U) implies $\omega(t, u(t)) \equiv 0$ a contradiction. Hence, $\det(M) \neq 0$. Thus, as a result of continuous dependence, for $h \neq 0$ and sufficiently small, $\det(M(h)) \neq 0$ implying $M(h)$ has an inverse where $M(h)$ is the appropriately defined matrix from the system of equations. Therefore, for each $m_l \leq i \leq n-1$, we can solve for $\epsilon_i(h)/h$, by using Cramer's rule:

$$\frac{\epsilon_i(h)}{h} = \frac{1}{|M(h)|} \times$$

$$
\begin{vmatrix}
\beta_{m_l}(t_1) & \cdots & \beta_{i-2}(t_1) & -\beta_q(t_1) & \beta_i(t_1) & \cdots & \beta_{n-1}(t_1) \\
\vdots & \ddots & \vdots & \vdots & \vdots & \ddots & \vdots \\
\beta_{m_l}(t_1 + m_1 - 1) & \cdots & \beta_{i-2}(t_1 + m_1 - 1) & -\beta_q(t_1 + m_1 - 1) & \beta_i(t_1 + m_1 - 1) & \cdots & \beta_{n-1}(t_1 + m_1 - 1) \\
\vdots & \ddots & \vdots & \vdots & \vdots & \ddots & \vdots \\
\beta_{m_l}(t_k) & & \beta_{i-2}(t_k) & -\beta_q(t_k) & \beta_i(t_k) & & \beta_{n-1}(t_k) \\
-\sum \alpha \beta_{m_l}(\eta) & \cdots & -\sum \alpha \beta_{i-2}(\eta) & +\sum \alpha \beta_q(\eta) & -\sum \alpha \beta_i(\eta) & \cdots & -\sum \alpha \beta_{n-l}(\eta) \\
\vdots & \ddots & \vdots & \vdots & \vdots & \ddots & \vdots \\
\beta_{m_l}(t_k + m_k - 1) & & \beta_{i-2}(t_k + m_k - 1) & -\beta_q(t_k + m_k - 1) & \beta_i(t_k + m_k - 1) & & \beta_{n-1}(t_k + m_k - 1) \\
-\sum \alpha \beta_{m_l}(\eta) & \cdots & -\sum \alpha \beta_{i-2}(\eta) & +\sum \alpha \beta_q(\eta) & -\sum \alpha \beta_i(\eta) & \cdots & -\sum \alpha \beta_{n-l}(\eta)
\end{vmatrix}
$$

Note that as $h \to 0$, $\det(M(h)) \to \det(M)$, and so for $m_l \leq i \leq n - 1$, $\epsilon_i(h)/h \to \det(M_i)/\det(M) := B_i$ as $h \to 0$, where M_i is the $n - m_l \times n - m_l$ matrix found by replacing the appropriate column of the matrix defining M by

$$\text{col}\Big[-\beta_q(t_1, u(t)), \ldots, -\beta_q(t_1 + m_1 - 1, u(t)), \ldots,$$

$$-\beta_q(t_{l-1}, u(t)), \ldots, -\beta_q(t_{l-1} + m_{l-1} - 1, u(t)), -\beta_q(t_{l+1}, u(t)), \ldots,$$

$$-\beta_q(t_1 + m_{l+1} - 1, u(t)), \ldots,$$

$$-\beta_q(t_k, u(t)) + \sum_{p=1}^{m} \alpha_{0p}\beta_q(\eta_{0p}, u(t)), \ldots,$$

$$-\beta_q(t_k + m_k - 1, u(t)) + \sum_{p=1}^{m} \alpha_{m_k-1,p}\beta_q(\eta_{m_k-1,p}, u(t))\Big].$$

Now let $z_{ql}(t) = ds \lim_{h \to 0} z_{qlh}(t)$, and note by construction of $z_{qlh}(t)$,

$$z_{ql}(t) = \frac{\partial w}{\partial w_{ql}}(t).$$

Furthermore,

$$z_{ql}(t) = \lim_{h \to 0} z_{qlh}(t) = \sum_{i=m_l}^{n-1} B_i\beta_i(t, u(t)),$$

which is a solution of the variational equation (18.3) along $w(t)$. In addition,

$$z_{ql}(t_j + i) = \lim_{h \to 0} z_{qlh}(t_j + i) = 0,\ 0 \le i \le m_j - 1,\ 1 \le j \le k - 1,\ j \ne l,$$

$$z_{ql}(t_l + i) = \lim_{h \to 0} z_{qlh}(t_l + i) = 0,\ 0 \le i \le m_l - 1,\ i \ne q,$$

$$z_{ql}(t_l + q) = \lim_{h \to 0} z_{qlh}(t_l + q) = 1,$$

and

$$z_{ql}(t_k + i) - ds \sum_{p=1}^{m} \alpha_{ip} z_{ql}(\eta_{ip}) = ds \lim_{h \to} \left[z_{qlh}(t_k + i) - ds \sum_{p=1}^{m} \alpha_{ip} z_{qlh}(\eta_{ip}) \right]$$

$$= 0,\ 0 \le i \le m_k - 1.$$

This completes the argument for $\frac{\partial w}{\partial w_{ql}}$. The proof for $\frac{\partial w}{\partial w_{qk}}$ is completed similarly. $\square$

Theorem 18.12. *Let $n \ge 2$, $2 \le k \le n$, and $m \in \mathbb{N}$ be given and let $m_1, \ldots, m_k$ be positive integers such that $\sum_{i=1}^{k} m_i = n$. Assume conditions* (i), (ii), *and* (iii) *hold and that* (18.1) *satisfies Property (U) on $\mathbb{Z}$. Suppose*

$$w(t) = w(t, t_1, \ldots, t_k, \eta_{01}, \ldots, \eta_{m_k-1,m}, \alpha_{01}, \ldots, \alpha_{m_k-1,m}, w_{01}, \ldots, w_{m_k-1,k})$$

is the solution of (18.1) *on $\mathbb{Z}$ where*

$$w(t_j + i) = w_{ij},\ 0 \le i \le m_j - 1,\ 1 \le j \le k - 1,$$

$$w(t_k + i) - \sum_{p=1}^{m} \alpha_{ip} w(\eta_{ip}) = w_{ik},\ 0 \le i \le m_k - 1,$$

$t_1 < t_1 + m_1 - 1 < t_2 < t_2 + m_2 - 1 < \cdots < t_{k-1} + m_{k-1} - 1 < \eta_{01} < \eta_{01} + 1 < \cdots < \eta_{0p} <$
$\eta_{0m} + 1 < \eta_{11} < \eta_{11} + 1 < \cdots < \eta_{m_k-1,0} < \eta_{m_k-1,0} + 1 < \cdots < \eta_{m_k-1,m} < \eta_{m_k-1,m} + 1 < t_k$
in $\mathbb{Z}$, and $\alpha_{01}, \ldots, \alpha_{m_k-1,m}, w_{01}, \ldots, w_{m_k-1,k} \in \mathbb{R}$. Then,

(a) for $1 \le l \le k - 1$,

$$\nu_l(t) := \Delta_{t_l} w(t, t_1, \ldots, t_k, \eta_{01}, \ldots, \eta_{m_k-1,m},$$

$$\alpha_{01}, \ldots, \alpha_{m_k-1,m}, w_{01}, \ldots, w_{m_k-1,k})$$

$$= \quad w(t, t_1, \ldots, t_l + 1, \ldots, t_k, \eta_{01}, \ldots, \eta_{m_k-1,m},$$
$$\alpha_{01}, \ldots, \alpha_{m_k-1,m}, w_{01}, \ldots, w_{m_k-1,k})$$
$$-w(t, t_1, \ldots, t_l, \ldots, t_k, \eta_{01}, \ldots, \eta_{m_k-1,m},$$
$$\alpha_{01}, \ldots, \alpha_{m_k-1,m}, w_{01}, \ldots, w_{m_k-1,k})$$

is a solution of the linear difference equation

$$\nu_l(t+n) = \sum_{r=1}^{n} A_{lr}(t)\nu_l(t+r-1),$$

where for $1 \leq r \leq n$,

$$A_{lr}(t) \quad = \quad \int_0^1 \frac{\partial f}{\partial d_r}(t, w(t, t_1, \ldots, t_l, \ldots, t_k, \eta_{01}, \ldots, \eta_{m_k-1,m},$$
$$\alpha_{01}, \ldots, \alpha_{m_k-1,m}, w_{01}, \ldots, w_{m_k-1,k}),$$
$$w(t+1, t_1, \ldots, t_l, \ldots, t_k, \eta_{01}, \ldots, \eta_{m_k-1,m},$$
$$\alpha_{01}, \ldots, \alpha_{m_k-1,m}, w_{01}, \ldots, w_{m_k-1,k}), \ldots,$$
$$sw(t+r-1, t_1, \ldots, t_l + 1, \ldots, t_k, \eta_{01}, \ldots, \eta_{m_k-1,m},$$
$$\alpha_{01}, \ldots, \alpha_{m_k-1,m}, w_{01}, \ldots, w_{m_k-1,k})$$
$$+(1-s)w(t+r-1, t_1, \ldots, t_l, \ldots, t_k, \eta_{01}, \ldots, \eta_{m_k-1,m},$$
$$\alpha_{01}, \ldots, \alpha_{m_k-1,m}, w_{01}, \ldots, w_{m_k-1,k}), \ldots,$$
$$w(t+n-1, t_1, \ldots, t_l + 1, \ldots, t_k, \eta_{01}, \ldots, \eta_{m_k-1,m},$$
$$\alpha_{01}, \ldots, \alpha_{m_k-1,m}, w_{01}, \ldots, w_{m_k-1,k}))ds,$$

with boundary conditions

$$\nu_l(t_j + i) \quad = \quad 0, \ 0 \leq i \leq m_j - 1, \ 1 \leq j \leq k - 1, \ j \neq l,$$
$$\nu_l(t_l + i) \quad = \quad -\Delta_t w(t, t_1, \ldots, t_l + 1, \ldots, t_k, \eta_{01}, \ldots, \eta_{m_k-1,m},$$
$$\alpha_{01}, \ldots, \alpha_{m_k-1,m}, w_{01}, \ldots, w_{m_k-1,k})|_{t=t_l+i-1}, \ 0 \leq i \leq m_l,$$
$$\nu_l(t_k + i) \quad - \quad \sum_{p=1}^{m} \alpha_{ip}\nu_l(\eta_{ip}) = 0, \ 0 \leq i \leq m_k - 1,$$

(b)

$$\nu_k(t) : \quad = \quad \Delta_{t_k} w(t, t_1, \ldots, t_k, \eta_{01}, \ldots, \eta_{m_k-1,m},$$
$$\alpha_{01}, \ldots, \alpha_{m_k-1,m}, w_{01}, \ldots, w_{m_k-1,k})$$
$$= \quad w(t, t_1, \ldots, t_{k-1}, t_k + 1, \eta_{01}, \ldots, \eta_{m_k-1,m},$$
$$\alpha_{01}, \ldots, \alpha_{m_k-1,m}, w_{01}, \ldots, w_{m_k-1,k})$$
$$-w(t, t_1, \ldots, t_{k-1}, t_k, \eta_{01}, \ldots, \eta_{m_k-1,m},$$
$$\alpha_{01}, \ldots, \alpha_{m_k-1,m}, w_{01}, \ldots, w_{m_k-1,k})$$

is a solution of the linear difference equation

$$\nu_k(t+n) = \sum_{r=1}^{n} A_{kr}(t)\nu_k(t+r-1),$$

where for $1 \leq r \leq n$,

$$
\begin{aligned}
A_{kr}(t) \;=\; & \int_0^1 \frac{\partial f}{\partial d_r}(t, w(t, t_1, \ldots, t_k, \eta_{01}, \ldots, \eta_{m_k-1,m}, \\
& \alpha_{01}, \ldots, \alpha_{m_k-1,m}, w_{01}, \ldots, w_{m_k-1,k}), \\
& w(t+1, t_1, \ldots, t_k, \eta_{01}, \ldots, \eta_{m_k-1,m}, \\
& \alpha_{01}, \ldots, \alpha_{m_k-1,m}, w_{01}, \ldots, w_{m_k-1,k}), \ldots, \\
& sw(t+r-1, t_1, \ldots, t_{k-1}, t_k+1, \eta_{01}, \ldots, \eta_{m_k-1,m}, \\
& \alpha_{01}, \ldots, \alpha_{m_k-1,m}, w_{01}, \ldots, w_{m_k-1,k}) \\
& +(1-s)w(t+r-1, t_1, \ldots, t_{k-1}, t_k, \eta_{01}, \ldots, \eta_{m_k-1,m}, \\
& \alpha_{01}, \ldots, \alpha_{m_k-1,m}, w_{01}, \ldots, w_{m_k-1,k}), \ldots, \\
& w(t+n-1, t_1, \ldots, t_k+1, \eta_{01}, \ldots, \eta_{m_k-1,m}, \\
& \alpha_{01}, \ldots, \alpha_{m_k-1,m}, w_{01}, \ldots, w_{m_k-1,k}))ds,
\end{aligned}
$$

with boundary conditions

$$
\nu_k(t_i+i) \;=\; 0, \; 0 \leq i \leq m_j-1, \; 1 \leq j \leq k-1,
$$

$$
\begin{aligned}
\nu_k(t_k+i) \;-\; & \sum_{p=1}^{m} \alpha_{ip}\nu_k(\eta_{ip}) \\
=\; & -\Delta_t w(t, t_1, \ldots, t_{k-1}, t_k+1, \eta_{01}, \ldots, \eta_{m_k-1,m}, \\
& \alpha_{01}, \ldots, \alpha_{m_k-1,m}, w_{01}, \ldots, w_{m_k-1,k}))|_{t=t_k+i}, \\
& 0 \leq i \leq m_k-1.
\end{aligned}
$$

Furthermore,

(c) for $0 \leq q \leq m_k-1$ and $1 \leq l \leq m$,

$$
\begin{aligned}
\xi_{ql}(t) \;:\;=\; & \Delta_{\eta_{ql}} w(t, t_1, \ldots, t_k, \eta_{01}, \ldots, \eta_{m_k-1,m}, \\
& \alpha_{01}, \ldots, \alpha_{m_k-1,m}, w_{01}, \ldots, w_{m_k-1,k}) \\
=\; & w(t, t_1, \ldots, t_k, \eta_{01}, \ldots, \eta_{ql}+1, \ldots, \eta_{m_k-1,m}, \\
& \alpha_{01}, \ldots, \alpha_{m_k-1,m}, w_{01}, \ldots, w_{m_k-1,k}) \\
& -w(t, t_1, \ldots, t_k, \eta_{01}, \ldots, \eta_{ql}, \ldots, \eta_{m_k-1,m}, \\
& \alpha_{01}, \ldots, \alpha_{m_k-1,m}, w_{01}, \ldots, w_{m_k-1,k})
\end{aligned}
$$

is a solution of the linear difference equation

$$
\xi_{ql}(t+n) = \sum_{r=1}^{n} A_{qlr}(t)\xi_{ql}(t+r-1),
$$

where

$$
\begin{aligned}
A_{qlr}(t) \;=\; & \int_0^1 \frac{\partial f}{\partial d_r}(t, w(t, t_1, \ldots, t_k, \eta_{01}, \ldots, \eta_{ql}, \ldots, \eta_{m_k-1,m}, \\
& \alpha_{01}, \ldots, \alpha_{m_k-1,m}, w_{01}, \ldots, w_{m_k-1,k}), \\
& w(t+1, t_1, \ldots, t_k, \eta_{01}, \ldots, \eta_{ql}, \ldots, \eta_{m_k-1,m}, \\
& \alpha_{01}, \ldots, \alpha_{m_k-1,m}, w_{01}, \ldots, w_{m_k-1,k}), \ldots, \\
& sw(t+r-1, t_1, \ldots, t_k, \eta_{01}, \ldots, \eta_{ql}+1, \ldots, \eta_{m_k-1,m},
\end{aligned}
$$

$$\alpha_{01}, \ldots, \alpha_{m_k-1,m}, w_{01}, \ldots, w_{m_k-1,k})$$
$$+(1-s)w(t+r-1, t_1, \ldots, t_k, \eta_{01}, \ldots, \eta_{ql}, \ldots, \eta_{m_k-1,m},$$
$$\alpha_{01}, \ldots, \alpha_{m_k-1,m}, w_{01}, \ldots, w_{m_k-1,k}), \ldots,$$
$$w(t+n-1, t_1, \ldots, t_k, \eta_{01}, \ldots, \eta_{ql}+1, \ldots, \eta_{m_k-1,m},$$
$$\alpha_{01}, \ldots, \alpha_{m_k-1,m}, w_{01}, \ldots, w_{m_k-1,k}))ds,$$

with boundary conditions

$$\xi_{ql}(t_j + i) = 0, \ 0 \le i \le m_j - 1, \ 1 \le j \le k-1,$$

$$\xi_{ql}(t_k + i) - \sum_{p=1}^{m} \alpha_{ip}\xi_{ql}(\eta_{ip}) = 0, \ 0 \le i \le m_k - 1, \ i \ne q,$$

$$\xi_{ql}(t_k + q) - \sum_{p=1}^{m} \alpha_{ip}\xi_{ql}(\eta_{ip})$$

$$= \alpha_{ql}\Delta_t(t, t_1, \ldots, t_k, \eta_{01}, \ldots, \eta_{ql}+1, \ldots, \eta_{m_k-1,m},$$
$$\alpha_{01}, \ldots, \alpha_{m_k-1,m}, w_{01}, \ldots, w_{m_k-1,k})|_{t=\eta_{ql}}.$$

Proof. The proofs of part (a), (b), and (c) are very similar. To that end, we will verify part (a) and leave the remaining parts to the reader. Let $1 \le l \le k$, and we use a telescoping sum, the Mean Value Theorem, and difference calculus to obtain,

$$
\begin{aligned}
\nu_l(t+n) &= w(t+n, t_l+1) - w(t+n, t_l) \\
&= f(t, w(t, t_l+1), w(t+1, t_l+1), \ldots, w(t+n-1, t_l+1)) \\
&\quad -f(t, w(t, t_l), w(t+1, t_l), \ldots, w(t+n-1, t_l)) \\
&= [f(t, w(t, t_l+1), w(t+1, t_l+1), \ldots, w(t+n-1, t_l+1)) \\
&\quad -f(t, w(t, t_l), w(t+1, t_l+1), \ldots, w(t+n-1, t_l+1))] \\
&\quad +[f(t, w(t, t_l), w(t+1, t_l+1), \ldots, w(t+n-1, t_l+1)) \\
&\quad -f(t, w(t, t_l), w(t+1, t_l), \ldots, w(t+n-1, t_l+1))] + -\cdots \\
&\quad +[f(t, w(t, t_l), w(t+1, t_l), \ldots, w(t+n-1, t_l+1)) \\
&\quad -f(t, w(t, t_l), w(t+1, t_l), \ldots, w(t+n-1, t_l))] \\
&= \int_0^1 \frac{\partial f}{\partial d_1}(t, sw(t, t_l+1) + (1-s)w(t, t_l), w(t+1, t_l+1), \\
&\quad \ldots, w(t+n-1, t_l+1))ds \times (w(t, t_l) - w(t, t_l+1)) \\
&\quad + \int_0^1 \frac{\partial f}{\partial d_2}(t, w(t, t_l), sw(t+1, t_l+1) + (1-s)w(t+1, t_l), \\
&\quad \ldots, w(t+n-1, t_l+1))ds \times (w(t+1, t_l+1) - w(t, t_l)) + \cdots \\
&\quad + \int_0^1 \frac{\partial f}{\partial d_n}(t, w(t, t_l), w(t+1, t_l), \ldots, sw(t+n-1, t_l+1) \\
&\quad +(1-s)w(t+n-1, t_l))ds \times (w(t+n-1, t_l+1) \\
&\quad -w(t, t_l+1)) \\
&= A_{l1}\nu_l(t) + A_{l2}\nu_l(t+1) + \cdots + A_{ln}\nu_l(t+n-1).
\end{aligned}
$$

All that remains is to verify the boundary conditions. For $0 \le i \le m_l - 1, 1 \le l \le k-1$,

$$
\begin{aligned}
\nu_l(t_l + i) &= w(t_l + i, t_l + 1) - w(t_l + i, t_l) \\
&= w(t_l + i, t_l + 1) - w(t_l + i + 1, t_l + 1)
\end{aligned}
$$

$$+w(t_l + i + 1, t_l + 1) - w(t_l + i, t_l)$$
$$= w_{i-1,l} - w_{i-1,l} - \Delta_t w(t, t_l + 1)|_{t=t_l+i}$$
$$= -\Delta_t w(t, t_l + 1)|_{t=t_l+i}.$$

Also, for $0 \leq i \leq m_j - 1$, $1 \leq j \leq k - 1$, $j \neq l$,

$$\nu_l(t_j + i) = w(t_j + i, t_l + 1) - w(t_j + i, t_l) = [w_{ij} - w_{ij}] = 0,$$

and for $0 \leq i \leq m_k - 1$,

$$\nu_l(t_k + i) - \sum_{p=1}^{m} \alpha_{ip} \nu_l(\eta_{ip})$$
$$= \left[w(t_k + i, t_l + 1) - \sum_{p=1}^{m} \alpha_{ip} w(\eta_{ip}, t_l + 1) \right]$$
$$- \left[w(t_k + i, t_l) - \sum_{p=1}^{m} \alpha_{ip} w(\eta_{ip}, t_l) \right]$$
$$= [w_{ik} - w_{ik}]$$
$$= 0.$$

$\square$

Bibliography

[1] M. J. Ablowitz and P. A. Clarkson, *Solitons, Nonlinear Evolution Equations and Inverse Scattering*, Cambridge University Press, 1991.

[2] E. Acerbi and G. Mingione, Regularity results for a class of functionals with nonstandard growth, *Arch. Ration. Mech. Anal.* **156** (2001), 121-140.

[3] R. P. Agarwal, *Difference Equations and Inequalities. Theory, Methods, and Applications*, Second edition. Monographs and Textbooks in Pure and Applied Mathematics, 228. Marcel Dekker, Inc., New York, 2000.

[4] R. P. Agarwal and P. J. Y. Andwong, *Advanced topics in difference equations, vol. 404 of Mathematics and its Applications*. Kluwer Academic Publishers Group, Dordrecht, 1997.

[5] R. P. Agarwal, C. Cuevas and M. Frasson, Semilinear functional difference equations with infinite delay, *Math. Comput. Model.* **55**, (2012), 1083-1105.

[6] R. P. Agarwal and D. O'Regan, A note on the existence of multiple fixed points for multivalued maps with applications, *J. Differential Equations* **160** (2000), 389–403.

[7] R. P. Agarwal, D. O'Regan and P. J. Y. Wong, Constant-sign periodic and almost periodic solutions of a system of difference equations, *Comput. Math. Appl.* **50**, (2005), 1725-1754.

[8] R. P. Agarwal, D. O'Regan, and P. J. Y. Wong, Constant-sign solutions of a system of integral equations with integrable singularities, *J. Integral Equations Appl.*, **19** (2007), 117-142.

[9] R. P. Agarwal, D. O'Regan, and P. J. Y. Wong, Constant-sign solutions for singular systems of Fredholm integral equations, *Math. Methods Appl. Sci.*, **33** (2010), 1783-1793.

[10] R. P. Agarwal, D. O'Regan, and P. J. Y. Wong, Constant-sign solutions for systems of singular integral equations of Hammerstein type, *Math. Comput. Modelling*, **50** (2009), 999-1025.

[11] R. P. Agarwal and J. Popenda, Periodic solutions of first order linear difference equations, *Math. Comput. Modelling*, **22**, 1 (1995), 11-19.

[12] R. P. Agarwal and M. Sambandham, Multipoint boundary value problems for general discrete systems, *Dynam. Systems Appl.* **6** (1997), 469-492.

[13] R. P. Agarwal and P. J. Y. Wong, *Advanced Topics in Difference Equations*, Kluwer, Dordrecht, 1997.

[14] Z. Agur, L. Cojocaru, G. Mazaur, R. M. Anderson and Y. L. Danon, Pulse mass measles vaccination across age cohorts, *Proc. Nat. Acad. Sci. USA.* **90** (1993), 11698–11702.

[15] C. D. Ahlbrandt and A. C. Peterson, *Discrete Hamiltonian Systems: Difference Equations, Continued Fractions, and Riccati Equations*, Kluwer, Boston, 1996.

[16] R. R. Akhmerov, M. I. Kamenskii, A. S. Potapov, A. E. Rodkina, and B. N. Sadovskii, *Measures of Noncompactness and Condensing Operators*, Bikhäuser, Boston, 1992.

[17] F. J. Almgren, Jr, Approximation of rectifiable currents by Lipschitz Q-valued functions, In: *Seminar on Minimal Submanifolds*, ed. by E. Bombier, Ann. Math. Stud. **103**, 243-259, Princeton University Press, Princeton, 1983.

[18] F. J. Almgren, Jr, Almgren's big regularity paper. Q-valued functions minimizing Dirichlet's integral and the regularity of area-minimizing rectifiable currents up to codimension 2, with a preface by J. E. Taylor and V. Scheffer, World Scientific Monograph Series in Mathematics, **1**, World Scientific Publishing, River Edge, NJ, 2000.

[19] L. Ambrosio, M. Gobbino, and D. Pallara, Approximation problems for curvature varifolds, *J. Geom. Anal.* **8** (1998), 1-19.

[20] L. Ambrosio and P. Tilli, *Selected Topics on Analysis in Metric Spaces*, Oxford University Press, Oxford, 2004.

[21] D. R. Anderson, R. I. Avery and J. Henderson, Functional expansion-compression fixed point theorem of Leggett-Williams type, *Electron. J. Differential Equations* **2010** (2010), 1-9.

[22] D. R. Anderson, R. I. Avery and J. Henderson, Some fixed point theorems of Leggett-Williams type, *Rocky Mountain J. Math.* **41** (2011), 371-386.

[23] D. R. Anderson, and J. Henderson and X. Liu, Omitted ray fixed point theorem, *J. Fixed Point Theory Appl.* **17** (2015), 313-330.

[24] J. Andres and L. Górniewicz, *Topological Fixed Point Principles for Boundary Value Problems*, Kluwer, Dordrecht, 2003.

[25] G. Anichini, G. Conti, and P. Zecca, Approximation of nonconvex set valued mappings, *Boll. Un. Mat. Ital. C6*, **4** (1985), 145-154.

[26] G. Anichini, G. Conti, and P. Zecca, Approximation and selection theorem for nonconvex multifunctions in infinite dimensional space, *Boll. Un. Mat. Ital. B* **(7)4** (1990), 411–422.

[27] Q. H. Ansari, *Metric Spaces. Including Fixed Point Theory and Set-valued Maps*, Narosa Publishing House, New Delhi, 2010.

[28] J. Appell, E. De Pascale, H. T. Nguyen, and P. P. Zabreiko, *Multi-Valued Superpositions, Dissertationaes Math.*, **345**, 1995.

[29] W. Arendt, Semi groups and evolution equations: functional calculus, regularity and kernel estimates. In *Evolutionary equations.* Vol. I, Handbook of Differential Equations, North-Holland, Amsterdam, 2004, pp. 1-85.

[30] A. Ashyralyev, I. Karatay and P. E. Sobolevskii, On well-posedness of the nonlocal boundary value problem for parabolic difference equations, *Discrete Dyn. Nat. Soc.* (2004), 273-286.

[31] J. P. Aubin, *Viability Theory*, Birkhäuser, Boston, 1991.

[32] J. P. Aubin, *Neural Networks and Qualitative Physics: A Viability Approch*, Cambridge University Press, 1996.

[33] J. P. Aubin and A. Cellina, *Differential Inclusions*, Springer-Verlag, Berlin, 1984.

[34] J. P. Aubin and H. Frankowska, *Set-Valued Analysis,* Birkhauser, Boston, 1990.

[35] R. I. Avery, A generalization of the Leggett-Williams fixed point theorem, *Math. Sci. Res. Hot-Line* **3** (1999), 9-14.

[36] C. Avramescu, A fixed point theorem for multivalued mapping, *Electron. J. Qual. Theory Differ. Equ.* **2004** (2004), No. 17, 1–10.

[37] C. Avramescu and C. Vladimirescu, Fixed point theorems of Krasnosel'skii type in a space of continuous functions, *Fixed Point Theory* **5** (2004), 181–195.

[38] A. V. Bäcklund, *Concerning surfaces with constant negative curvature,* (trans. by E. A. Coddington), New Ear, Lancaster, PA, 1905.

[39] Z. Bai and W. Ge, Existence of three positive solutions for some second-order boundary value problems, *Compt. Mathe. Appli.* **48** (2004), 699-707.

[40] C. T. H Baker and Y. Song, Periodic solutions of discrete Volterra equations, *Math. Comput. Simulation*, **64**, (2004), 521-542.

[41] C. S. Barroso, Krasnosel'skii's fixed point theorem for weakly continuous maps, *Nonlinear Anal.* **55** (2003), 25–31.

[42] C. S. Barroso and E. V. Teixeira, A topological and geometric approach to fixed points results for sum of operators and applications, *Nonlinear Anal.* **60** (2005), 625–650.

[43] H. H. Bauschke and P. L. Combettes, *Convex Analysis and Monotone Operator Theory in Hilbert Spaces*, Springer, New York, 2011.

[44] G. Beer, On a theorem of Cellina for set valued functions, *Rocky Mountain J. Math.* **18** (1988), 37-47.

[45] M. Benchohra, S. Hamani, J. Henderson, S. K. Ntouyas and A. Ouahab, Differentiation and differences for solutions of nonlocal boundary value problems for second order difference equations, *Internat. J. Difference Equ.* **2** (2007), 37-47.

[46] M. Benchohra, J. Henderson and S. K. Ntouyas, *Impulsive Differential Equations and Inclusions*, Contemporary Mathematics and Its Applications, 2, Hindawi, New York, 2006.

[47] C. Berge, *Topological Spaces*, Macmillan, New York 1963.

[48] C. Berge, *Théorie Générale des Jeux a n Personnes*, Gauthier-Villars, Paris (1957).

[49] H. Berrezoug, J. Henderson, and A. Ouahab, Existence and uniqueness of solutions for a system of impulsive differential equations on the half-line, *J. Nonlinear Funct. Anal.* **2017** (2017), Article ID 38, pp. 1-16.

[50] A. T. Bharucha-Reid, *Random Integral Equations*, New York, Academic Press, 1972.

[51] A. T. Bharucha-Reid, Fixed point theorems in probabilistic analysis, *Bull. Amer. Math. Soc.* **82** (1976), 641–657.

[52] L. Bianchi, *Lezioni di geometria differenziale*, Vol. 1, Enrico Spoerri, Pisa 1922.

[53] I. Bihari, A generalisation of a lemma of Bellman and its application to uniqueness problems of differential equations, *Acta Math. Acad. Sci. Hungar.*, **7** (1956), 81-94.

[54] S. Blunck, Analyticity and discrete maximal regularity on L^p-spaces, *J. Funct. Anal.* **183**, (2001), 211-230.

[55] S. Blunck, Maximal regularity of discrete and continuous time evolution equations, *Studia Math.* **146**, (2001), 157-176.

[56] O. Bolojan-Nica, G. Infante, and R. Precup, Existence results for systems with coupled nonlocal initial conditions, *Nonlinear Anal.* **94** (2014), 231–242.

[57] M. Boriceanu, Krasnosel'skii-type theorems for multivalued operators, *Fixed Point Theory* **9** (2008), 35–45.

[58] G. Borisovich, B. D. Gel'man, A. D. Myshkis, and V. V. Obukhovskiĭ, Topological Methods in the Fixed-Point Theory of Multi-Valued Maps, *Uspekhi Mat. Nauk* **35** (1980), no. 1, 59-126 (in Russian); *Russian Math. Surveys* **35**, (1980), no. 1, 65-143. (Engl. transl.)

[59] G. Borisovich, B. D. Gel'man, A. D. Myshkis, and V. V. Obukhovskiĭ, *Introduction to the Theory of Multivalued Mappings*, Voronezh, Voronezh. Gos. Univ. (in Russian) English translation from Itogi Nauki i Tekhnild, Seriya Matematicheskii Analiz, **19** (1982), 127–230.

[60] N. Bourbaki, *Topologie générale*, Diffusion C.C.L.S. Paris, 1958.

[61] A. Bressan, Directionally continuous selections and differential inclusions, *Funkcial. Ekvac.* **31** (1988), 459-470.

[62] A. Bressan and G. Colombo, Extensions and selections of maps with decomposable values, *Studia Math.* **90** (1988), 69-86.

[63] H. Brezis, *Sobolev Spaces and Partial Differential Equations*, Universitext, Springer, New York, 2011.

[64] R. F. Brown, M. Furi, L. Górniewicz, and B. Jiang, *Handbook of Topological Fixed Point Theory.* Springer, Dordrecht, 2005.

[65] S. Budişan, Generalizations of Karasonselskii's fixed point theorem in cones, *Stud. Univ.Babeş-Bolyai Math.* **56** (2011), 165-171.

[66] D. Bump, *Lie groups.* Second edition, Graduate Texts in Mathematics, **225**, Springer, New York, 2013.

[67] T. A. Burton, A fixed point theorem of Krasnosel'skii, *Appl. Math. Lett.* **11** (1998), 85–88.

[68] T. A. Burton and C. Kirk, A fixed point theorem of Krasnosel'skii-Schaefer type, *Math. Nachr.* **189** (1998), 23–31.

[69] P. L. Butzer, A. A. Kilbas, and J. J. Trujillo, Compositions of Hadamard-type fractional integration operators and the semigroup property, *J. Math. Anal. Appl.* **269** (2002), 387–400.

[70] P. L. Butzer, A. A. Kilbas, and J.J. Trujillo, Fractional calculus in the Mellin setting and Hadamard-type fractional integrals, *J. Math. Anal. Appl.* **269** (2002), 1–27.

[71] P. L. Butzer, A. A. Kilbas, and J. J. Trujillo, Mellin transform analysis and integration by parts for Hadamard-type fractional integrals, *J. Math. Anal. Appl.* **270** (2002), 1–15.

[72] A. Caicedo, C. Cuevas, G. M. Mophou, and G. M. N'Guérékata, Asymptotic behavior of solutions of some semilinear functional diferential equations with infinite delay in Banach spaces, *J. Franklin Inst.* **349**, (2012), 1-24.

[73] M. Caputo, *Elasticità e Dissipazione*. Zanichelli, Bologna (1969).

[74] M. Caputo, Linear models of dissipation whose Q is almost frequency independent, Part **II**, *Geophys. J. R. Astr. Soc.* **13** (1967), 529–539; Reprinted in: *Fract. Calc. Appl. Anal.* **11**, No 1 (2008), 3–14.

[75] M. Caputo and F. Mainardi, Linear models of dissipation in anelastic solids. *Riv. Nuovo Cimento (Ser. II)* **1** (1971), 161–198.

[76] F. Cardoso and C. Cuevas, Exponential dichotomy and boundedness for retarded functional difference equations, *J. Difference Equ. Appl.* **15**, (2009), 261-290.

[77] B. Cascales, V. Kadets, and J. Rodriguez, Measurability and selections of multifunctions in Banach spaces, *J. Convex Anal.* **17** (2010), 229-240.

[78] C. Castaing, Sur un nouvelle classe d'équation d'évolution dans les espaces de Hilbert, exposé no 10. *Sém. Anal. Convexe*, University of Montpellier, page 24 pages, 1983.

[79] C. Castaing and M. Valadier, Convex Analysis and Measurable Multifunctions, *Lecture Notes in Mathematics* **580** Springer-Verlag, Berlin, 1977.

[80] A. Cellina, A theorem on the approximation of compact set-valued mappings, *Atti Accad. Naz. Lincei Rend. Cl. Sci. Fis. Mat. Natur.* (8) **47** (1969), 149–163.

[81] Y. Chalco-Cano, J. J. Nieto, A. Ouahab and H. Román-Flores, Solution set for fractional differential equations with Riemann-Liouville derivative, *Fract. Calc. Appl. Anal*, **6** (2013), 682-694.

[82] Y. Q. Chen, Fixed points for convex continuous mappings in topological vector spaces, *Proc. Amer. Math. Soc.* **129** (2001), 2157–2162.

[83] F. H. Clarke, Yu. S. Ledyaev, R. J. Sten, and P. R. Wolenski, *Nonsmooth Analysis and Control Theory*, Springer, New York, 1998.

[84] C. Corduneanu, Almost periodic discrete processes, *Libertas Math.* **2** (1982), 159-169.

[85] C. Corduneanu, *Integral Equations and Stability of Feedback Systems*, Academic Press, New York, 1973.

[86] H. Covitz and S. B. Nadler Jr., Multivalued contraction mappings in generalized metric spaces, *Israel J. Math.* **8** (1970), 5-11.

[87] C. Cuevas, Weighted convergent and bounded solutions of Volterra difference systems with infinite delay, *J. Differ. Equations Appl.* **6** (2000), 461-480.

[88] C. Cuevas and L. Del Campo, An asymptotic theory for retarded functional difference equations, *Comput.Math. Appl.* **49**, (2005), 841-855.

[89] C. Cuevas and M. Pinto, Convergent solutions of linear functional difference equations in phase space, *J. Math. Anal. Appl.* **277**, (2003), 324-341.

[90] C. Cuevas, and C. Vidal, Discrete dichotomies and asymptotic behavior for abstract retarded functional difference equations in phase space, *J. Difference Equ. Appl.* **8**, (2002), 603-640.

[91] C. Cuevas and C. Vidal, A note on discrete maximal regularity for functional difference equations with infinite delay, *Adv. Difference Equ.* (2006), Art. 97614, 1-11.

[92] A. Datta, Differences with respect to boundary points for right focal boundary conditions, *J. Differ. Equations Appl.* **4** (1998), pp. 571-578.

[93] A. Datta and J. Henderson, Differentiation of solutions of difference equations with respect to right focal boundary values, *Panamer. Math. J.* **2** (1992), 1-16.

[94] B. De Andrade and C. Cuevas, S-asymptotically ω−periodic and asymptotically ω−periodic solutions to semi-linear Cauchy problems with non-dense domain, *Nonlinear Anal.* **72**, (2010), 3190–3208.

[95] F. S. De Blasi, Characterization of certain classes of semicontinuous multifunctions by continous approximations, *J. Math. Anal. Appl. Vol 106*, (1985), 1-18.

[96] F. S. De Blasi and G. Pianigiani, *Hausdorff measurable multifunctions, J. Math. Anal. Appl. Vol 228*, (1998), 1–15.

[97] K. Deimling, *Multivalued Differential Equations*, Walter De Gruyter, Berlin, 1992.

[98] K. Deimling, *Nonlinear Functional Analysis*, Springer, Berlin, 1992.

[99] L. Del Campo, M. Pinto, and C. Vidal, Almost and asymptotically almost periodic solutions of abstract retarded functional difference equations in phase space, *J. Difference Equ. Appl.* **17**, (2011) 915-934.

[100] C. De Lellis, C. R. Grisanti, and P. Tilli, Regular selections for multiple-valued functions, *Ann. Mat. Pura Appl., Ser. IV* **183** (2004), No. 1, 79-95.

[101] C. De Lellis and E. N. Spadaro, Q-valued functions revisited *Mem. Am. Math. Soc.* **991**, i-v, 79p, 2011.

[102] M. M. Deza and E. Deza, *Encyclopedia of Distances*, Springer, New York, 2009.

[103] K. Diethelm, *The Analysis of Fractional Differential Equations. An Application-Oriented Exposition Using Differential Operators of Caputo Type*, Lecture Notes in Mathematics **2004**, Springer, Berlin, 2010.

[104] K. Diethelm and A. D. Freed, On the solution of nonlinear fractional order differential equations used in the modeling of viscoplasticity, in *Scientifice Computing in Chemical Engineering II-Computational Fluid Dynamics, Reaction Engineering and Molecular Properties* (F. Keil, W. Mackens, H. Voss, and J. Werther, Eds.), pp. 217-224, Springer, Heidelberg, 1999.

[105] S. Djebali, L. Gorniewicz, and A. Ouahab,*Solutions Sets for Differential Equations and Inclusions*, De Gruyter Series in Nonlinear Analysis and Applications **18**, de Gruyter, Berlin, 2013.

[106] S. Djebali and Z. Sahnoun, Nonlinear alternatives of Schauder and Krasnosel'skii types with applications to Hammerstein integral equations in L^1 spaces, *J. Differential Equations* **249** (2010), 2061–2075.

[107] J. Dugundji, *Topoloy*, 8th ed. Allyn and Bacon, Boston, 1973.

[108] J. Dugundji and A. Granas, *Fixed Point Theory*, Springer, New York, 2003.

[109] D. E. Edmunds and J. Rakosnik, Sobolev embedding with variable exponent, *Studia Math.* **143** (2000), 267-293.

[110] J. Ehrke, J. Henderson, C. Kunkel, and Q. Sheng, Boundary data smoothness for solutions of nonlocal boundary value problems for second order differential equations, *J. Math. Anal. Appl.* **333** (2007), 191-203.

[111] S. Elaydi, *An Introduction to Difference Equations*, Springer, New York, 1996.

[112] S. Elaydi, S. Murakami and E. Kamiyama, Asymptotic equivalence for difference equations with infinite delay, *J. Differ. Equations Appl.* **5**, (1999), 1-23.

[113] P. W. Eloe, Maximum principles for a family of nonlocal boundary value problems, *Adv. Difference Equ.* (2004), 201-210.

[114] U. Enz, Discrete mass, elementary length, and a topological invariant as a consequence of a relativistic invariant variational principle, *Phys. Rev.* **131** (1963), 1392-1394.

[115] X. L. Fan and Q. H. Zhang, Existence of solutions for $p(x)$-Laplacian Dirichlet problem, *Nonlinear Anal.* **52** (2003), 1843-1852.

[116] X. L. Fan and D. Zhao, On the spaces $L^{p(x)}$ and $W^{m,p(x)}$, *J. Math. Anal. Appl.* **263** (2001), 424-446.

[117] A. F. Filippov, *Differential Equations with Discontinuous Right-hand Sides*, Kluwer Academic Publishers, Dordrecht, 1988.

[118] J. Fiš, *Iterated Function and Multifunction Systems, Attractors and their Basins of Attraction*, Ph.D. Thesis, Palacký University, Olomouc, 2002 (in Czech).

[119] H. Frankowska, Local controllability and infinitesimal generators of semi-groups of set-valued maps, *SIAM J. Control Optim.* **25** (1987), 412-431.

[120] H. Frankowska, A priori estimates for operational differential inclusions, *J. Differential Equations* **84** (1990), 100-128.

[121] M. Fréchet, La notion d'écart et le calcul fonctionnel, *C. R. Acad. Sci. Paris* **140** (1905), 772-774.

[122] M. Fréchet, *Sur quelques points du calcul fonctionnel* (Thèse), Rendiconti Circolo Mat. Palermo **22** (1906), 1-74.

[123] J. Frenkel and T. Kontorova, Propagation of dislocation in crystals, *J. Phys. USSR* **1** (1939), 137.

[124] M. Frigon, Théorèmes d'existence de solutions d'inclusions différentielles. [Existence theorems for solutions of differential inclusions] Topological methods in differential equations and inclusions (Montreal, PQ, 1994), 51–87, *NATO Adv. Sci. Inst. Ser. C Math. Phys. Sci.*, **472**, Kluwer, Dordrecht, 1995.

[125] Y. Y. Gambo, F. Jarad, D. Baleanu, and T. Abdeljawad, On Caputo modification of the Hadamard fractional derivatives *Adv. Difference Equ.* **2014**, (2014), 12 pp.

[126] J. Garcia-Falset, Existence of fixed points for the sum of two operators, *Math. Nachr.* **12** (2010), 1726–1757.

[127] J. Garcia-Falset, K. Latrach, E. Moreno-Gálvez, and M. A Taoudi, Schaefer-Krasnosel'skii fixed points theorems using a usual measure of weak noncompactness, *J. Differential Equations* **352** (2012), 3436–3452.

[128] J. Garcia-Falset and O. Muñiz-Pérez, Fixed point theory for 1-set weakly contractive and pseudocontractive mappings. *Appl. Math. Comput.* **219** (2013), 6843–6855.

[129] L. Gaul, P. Klein, and S. Kemple, Damping description involving fractional operators, *Mech. Systems Signal Processing* **5** (1991), 81–88.

[130] W. G. Glockle and T. F. Nonnenmacher, A fractional calculus approach of self-similar protein dynamics. *Biophys. J.* **68** (1995), 46–53.

[131] J. Goblet, A Peano type theorem for a class of nonconvex-valued differential inclusions, *Set-Valued Anal* **16** (2008), 913-921.

[132] J. Goblet, A selection theory for multiple-valued functions in the sense of Almgren, *Ann. Acad. Sci. Fenn., Math.* **31**, (2006), 297-314.

[133] J. Goblet, Lipschitz extension of multiple Banach-valued functions in the sense of Almgren, *Houston J. Math.* **35** (2009), No. 1, 223-231.

[134] J. Goblet and W. Zhu, Regularity of Dirichlet nearly minimizing multiple-valued functions *J. Geom. Anal.* **18** (2008), No. 3, 765-794.

[135] J. A. Goldstein, *Semigroups of Linear Operators and Applications*, Oxford Univ. Press, New York, 1985.

[136] L. Górniewicz, Homological methods in fixed point theory of multivalued maps, *Dissertations Math.* **129** (1976), 1-71.

[137] L. Górniewicz, *Topological Approach to Differential Inclusions*, Topological Methods in Differential and Inclusions (A. Granas and M. Frigon,eds.) NATO ASI Series C, Vol. 472, Kluwer, Dordrecht, 1995.

[138] L. Górniewicz, *Topological Fixed Point Theory of Multivalued Mappings*, Mathematics and its Applications, **495**, Kluwer, Dordrecht, 1999.

[139] L. Górniewicz, Differential inclusions theory initialed by Cracow Mathematical School, *Annal. Math. Sil.* **25** (2011), 7-25.

[140] L. Górniewicz, A. Granas, and W. Kryszewski, Sur la méthode de l'homotopie dans la théorie des points fixes pour les applications multivoques. Partie II: L'indice dans les ANR-s compacts, *C. R. Acad. Sci. Paris Sér I Math.*, **308**, (1989), no. 14, 449-452.

[141] L. Górniewicz, A. Granas, and W. Kryszewski, On the homotopy method in the fixed point index theory of multivalued mappings of compact absolute neighborhood retracts, *J. Math. Anal. Appl.* **161** (1991), 457–473.

[142] L. Górniewicz and M. Lassonde, Approximation and fixed points for compositions of R_δ, *Topology Appl.* **55** (1994), 239–250.

[143] J. R. Graef, J. Henderson, and A. Ouahab, Multivalued versions of a Krasnosel'skii type fixed point theorem, *J. Fixed Point Theory Appl* **19** (2017), 1059–1082.

[144] J. R. Graef, J. Henderson, and A. Ouahab, Some Krasnosel'skii type random fixed point theorems, *J. Nonlinear Funct. Anal.* **2017** (2017), Article ID 46, pp. 1–34.

[145] J. R. Graef, J. Henderson, and A. Ouahab, *Impulsive Differential Inclusions, A Fixed Point Approach*, De Gruyter Series in Nonlinear Analysis and Applications Vol. **20**, De Gruyter, Berlin, 2013.

[146] J. R. Graef, J. Henderson, and A. Ouahab, Fractional differential inclusions in the Almgren sense, *Fract. Calc. Appl. Anal.* **18** (2015), 673-686.

[147] D. J. Guo and V. Lakshmikantham, *Nonlinear Problems in Abstract Cones*, Notes and Reports in Mathematics in Science and Engineering, **5**, Academic Press, Boston, 1988.

[148] J. Hadamard, Essai sur l'étude des fonctions donnees par leur development de Taylor, *J. Mat. Pure Appl. Ser.* **8** (1892), 101–186.

[149] A. Halanay, Solutions périodiques et presque-périodiques des systèmes d'équations aux differences finies, *Arch. Rational Mech. Anal.* **12** (1963), 134-149.

[150] A. Halanay and D. Wexler, *Teoria Calitativa a systeme cu Impulduri, Editura Republicii Socialiste Romania*, Bucharest, 1968.

[151] Y. Hamaya, Existence of an almost periodic solution in a difference equation with infinite delay, *J. Difference Equ. Appl.* **9**, (2003), 227–237.

[152] O. Hanš, Random fixed point theorems. 1957 Transactions of the first Prague conference on information theory, statistical decision functions, random processes, Liblice, November 1956, pp. 105–125, Czechoslovak Academy of Sciences, Prague.

[153] O. Hanš, Random operator equations, *Proceedings of Fourth Berkeley Symposium on Mathematics, Statistics and Probability*, University of California Press, Berkeley, 1961, **II** 185–202.

[154] O. Hanš and A. Špacek, Random fixed point approximation by differentiable trajectories. 1960 Transactions of the second Prague conference on information theory, pp. 203–213, Academic Press, New York, 1960.

[155] P. Hartman, *Ordinary Differential Equations*, Wiley, New York, 1964.

[156] P. Hartman, Difference equations: Disconjugacy, principal solutions, Green's functions, complete monotonicity, *Trans. Amer. Math. Soc.* **246** (1978), 1-30.

[157] F. Hausdorff, *Grundzüge der Mengenlehre*, Verlag Von Veit & Company, Leipzig (1914), Reprinted by Chelsea Publishing Company, New York, 1949.

[158] S. Heikkila and V. Lakshmikantham, *Monotone Iterative Techniques for Discontinuous Nonlinear Differential Equations*, Marcel Dekker, New York, 1994.

[159] J. Henderson, Disconjugacy, disfocality, and differentiation with respect to boundary conditions, *J. Differ. Equations Appl.* **121** (1987), pp. 1-9.

[160] J. Henderson, *Boundary Value Problems for Functional Differential Equations*, World Scientific, Singapore, 1995.

[161] J. Henderson, B. Hopkins, E. Kim, and J. W. Lyons, Boundary data smoothness for solutions of nonlocal boundary value problems for nth order differential equations, *Involve* **1** (2008), 167-181.

[162] J. Henderson and L. Lee, Continuous dependence and differentiation of solutions of finite difference equations, *Internat. J. Math. Math. Sci.* **14** (1991), 747-756.

[163] J. Henderson, S. K. Ntouyas, and I. K. Purnaras, Positive solutions for systems of second order four-point nonlinear boundary value problems, *Commun. Appl. Anal.*, **12** (2008), 29-40.

[164] J. Henderson and A. Ouahab, Differential inclusions in the Almgren sense on unbounded domains, *Ann. Pol. Math.* **110**, (2014), 91-99.

[165] J. Henderson, A. Ouahab, and M. Seghier, Random solutions to a system of fractional differential equations via the Hadamard fractional derivative, *Eur. Phys. J. Special Topics* **226** (2017), 3525-3549 (2017).

[166] J. Henderson, A. Ouahab, and M. Slimani, Existence results of system of semilinear discrete equations, *Int. J. Difference Equ.* **12**, (2017), 235-253.

[167] H. R. Henriquez, M. Pierri and P. Taboas, On S-asymptotically ω-periodic functions on Banach spaces and applications, *J. Math. Anal. Appl.* **343**, (2008), 1119-1130.

[168] D. Henry, *Geometric Theory of Semilinear Parabolic Partial Differential Equations*, Springer, Berlin, 1989.

[169] F. Hiai and H. Umegaki, Integrals, conditional expectations, and martingales of multivalued functions, *J. Multivariate Anal.* **7** (1977), 149-182.

[170] C. J. Himmelberg, Measurable relations, *Fund. Math.*, **87** (1975), 59-71.

[171] C. Himmelberg and F. Van Vleck, One the topological triviality of solution sets, *Rocky Mountain J. Math.* **10** (1980), 621-252.

[172] B. Hopkins, E. Kim, J. W. Lyons, and K. Speer, Boundary data smoothness for solutions of nonlocal boundary value problems for second order difference equations, *Comm. Appl. Nonlinear Anal.* **16** (2009), 1-12.

[173] Sh. Hu and N. Papageorgiou, *Handbook of Multivalued Analysis, Volume I: Theory*, Kluwer, Dordrecht, 1997.

[174] S. Itoh, Random fixed point theorems with an application to random differential equations in Banach spaces, *J. Math. Anal. Appl.* **67** (1979), 261-273.

[175] K. Jachmann and J. Wirth, Diagonalisation schemes and applications, *Ann. Mat. Pura Appl.* **189** (2010), 571-590.

[176] M. Q. Jacobs, Measurable multivalued mappings and Lusin's theorem, *Trans. Amer. Math. Soc.*, **134**, (1968), 471-481.

[177] F. Jarad, T. Abdeljawad, and D. Baleanu, Caputo-type modification of the Hadamard fractional derivatives, *Adv. Difference Equ.* **2012** (2012), 8 pp.

[178] J. Jarnik and J. Kurzweil, On conditions on right hand sides of differential relations, *Časopis. Pěst. Mat.*, **102**, (1977), 334-349.

[179] B. S. Jensen, *The Dynamic Systems of Basic Economic Growth Models*, Kluwer, Dordrecht, 1994.

[180] M. Kamenskii, V. Obukhovskii, and P. Zecca, *Condensing Multivalued Maps and Semilinear Differential Inclusions in Banach Spaces*, Walter de Gruyter & Co., Berlin, 2001.

[181] W. G. Kelley and A. C. Peterson, *Difference Equations: An Introduction with Applications*, Academic Press, New York, 1991.

[182] A. A. Kilbas, Hadamard-type fractional calculus, *J. Korean Math. Soc.* **38** (2001), 1191–1204.

[183] A. A. Kilbas and J. J. Trujillo, Hadamard-type integrals as G-transforms, *Integral Transforms and Special Functions* **14** (2003), 413–427.

[184] A. A. Kilbas, H. M. Srivastava and J. J. Trujillo, *Theory and Applications of Fractional Differential Equations*, North-Holland Mathematics Studies, **204**, Elsevier, Amsterdam, 2006.

[185] M. Kisielewicz, *Differential Inclusions and Optimal Control*, Kluwer, Dordrecht, 1991.

[186] L. Kong and Q. Kong, Positive solutions of nonlinear $m-$point boundary value problems on a measure chain, *J. Difference Equ. Appl.* **9** (2003), 121-133.

[187] O. Kovacik and J. Rakosnik, On spaces $L^{p(x)}$ and $W^{1,p(x)}$, *Czechoslovak Math. J.* **41** (1991), 592-618.

[188] M. A. Krasnosel'skii, Some problems of nonlinear analysis, *Amer. Math. Soc. Transl. Ser.* (2) **10** (1958), 345–409.

[189] M. A. Krasnosel'skii, *Topological Methods in the Theory of Nonlinear Integral Equations*, MacMillan, New York, 1964.

[190] M. A. Krasnosel'skii, *Positive Solutions of Operator Equations*, Noordhoff, Groningen, 1964.

[191] E. Kruger-Thiemr, Formal theory of drug dosage regiments, *J. Theoret. Biol.* **13** (1966), 212–235.

[192] W. Kryszewski, Homotopy invariants for set-valued maps; homotopy-approximation approach, *Fixed point theory and applications*, Pitman Res. Notes Math. Ser., **252**, (1991), 269-284.

[193] C. Kuratowski, *Topology*, Vol. I, Academic Press, New York, 1966.

[194] C. Kuratowski, *Topology*, Vol. II, Academic Press, New York, 1968.

[195] K. Kuratowski, Operations on semicontinuous set-valued mappings, in *Sere. 1962-1963 Analisi, Algebra, Geometriea et Topol.*, Vol. 2, Roma (1965), pp. 449-461.

[196] D. R. Kurepa, Tableaux ramifiés d'ensembles. Espaces pseudo-distanciés, *C. R. Acad. Sci. Paris* **198** (1934), 1563-1565.

[197] G. S. Ladde and V. Lakshmikantham, *Random Differential Inequalities*, Academic Press, New York, 1980.

[198] V. Lakshmikantham, D. D. Bainov, and P. S. Simeonov, *Theory of Impulsive Differential Equations*, World Scientific, Singapore, 1989.

[199] V. Lakshmikantham and D. Trigiante, *Difference Equations: Numerical Methods and Applications*, Academic Press, New York, 1986.

[200] A. Lasota and Z. Opial, An application of the Kakutani-Ky Fan theorem in the theory of ordinary differential equations, *Bull. Acad. Pol. Sci. Ser. Sci. Math. Astronom. Phys.* **13** (1965), 781-786.

[201] A. Lasota and J. A. Yorke, The generic property of existence of solutions of differential equations in Banach spaces, *J. Diff. Equs.* **13**, (1973), 1–12.

[202] J. M. Lasry and R. Robert, *Analyse Non Linéaire Multivoque, Publ. No. 7611, Centre de Recherche de Mathématique de la Décision,* Université de Dauphine, Paris, 1–190.

[203] R. W. Leggett and L. R. Williams, Multiple positive fixed points of nonlinear operators on ordered Banach spaces, *Indiana Univ. Math. J.* **28** (1979), 673-688.

[204] S. Y. Lin, Generalized Gronwall inequalities and their applications to fractional differential equations, *J. Inequal. Appl.* **2013** (2013), No. 549, 9 pp.

[205] V. Lupulescu and S. K. Ntouyas, Random fractional differential equations, *Int. Electron. J. Pure Appl. Math.* **4** (2012), 119–136.

[206] V. Lupulescu, D. O'Regan, and G. Rahman, Existence results for random fractional differential equations, *Opuscula Math.* **34** (2014), 813–825.

[207] W. Lyons, Differentiation of solutions of nonlocal boundary value problems with respect to boundary data, *Electron. J. Qual. Theory Differ. Equ.*, **2011** (2011), No. 51, 11 pp.

[208] F. Mainardi, Fractional calculus: Some basic problems in continuum and statistical mechanics, in *Fractals and Fractional Calculus in Continuum Mechanics* (A. Carpinteri and F. Mainardi, Eds.), pp. 291-348, Springer, Heidelberg, 1997.

[209] H. Marchaud, Sur les champs de demi-cônes et les équations difféentielles du premier ordre, *Bull. Sci. Math.* **62** (1938), l-38.

[210] H. Matsunaga and S. Murakami, Some invariant manifolds for functional difference equations with infinite delay, *J. Difference Equ. Appl.* **10**, (2004), 661-689.

[211] H. Matsunaga and S. Murakami, Asymptotic behavior of solutions of functional difference equations, *J. Math. Anal. Appl.* **305**, (2005), 391-410.

[212] L. McLinden, An application of Ekeland's theorem to minimax problems, *Nonlinear Anal.* **6** (1982), 189-196.

[213] F. Metzler, W. Schick, H. G. Kilian, and T. F. Nonnenmacher, Relaxation in filled polymers: A fractional calculus approach, *J. Chem. Phys.* **103** (1995), 7180-7186.

[214] R. E. Mickens, *Difference Equations: Theory and Applications*, Van Nostrand Reinhold, New York, 2nd edition, 1990.

[215] R. E. Mickens, *Nonstandard Finite Difference Models of Differential Equations*, World Scientific, Singapore, 1994.

[216] K. S. Miller and B. Ross, *An Introduction to the Fractional Calculus and Differential Equations*, John Wiley, New York, 1993.

[217] V. D. Milman and A. A. Myshkis, On the stability of motion in the presence of impulses, *Sib. Math. J.* (in Russian) **1** (1960), 233–237.

[218] M. Moshinsky, Sobre los problemas de condiciones a la frontiera en una dimension de caracteristicas discontinuas, *Bol. Soc. Mat. Mexicana* **7** (1950), 1-25.

[219] A. Mukherjea, Transformations aléatoires separables. Théorème du point fixe aléatoire, *C. R. Acad. Sei. Paris Sér.* A-B **263** (1966), 393-395.

[220] A. Mukherjea, *Random Transformations of Banach Spaces*, Ph. D. Dissertation, Wayne State University, Detroit, 1968.

[221] S. Murakami, Representation of solutions of linear functional difference equations in phase space. In *Proceedings of the Second World Congress of Nonlinear Analysts, Part 2 (Athens, 1996) (1997)*, Vol. 30, pp. 1153-1164.

[222] J. D. Murray, *Mathematical Biology*, Springer, Berlin, 1989.

[223] J. Musielak, *Orlicz spaces and modular spaces*, Lecture Notes in Mathematics, Vol. **1034**, Springer, Berlin, 1983.

[224] H. Nakano, *Modulared Semi-ordered Linear Spaces*, Maruzen Co., Ltd., Tokyo, 1950.

[225] J. J. Nieto, A. Ouahab, and M.A. Slimani, Existence and boundedness of solutions for systems of difference equations with infinite delay, Glas. Mat. **53** (2018), 123–141.

[226] K. Nikodem, *K*-convex and *K*-concave set-valued functions, *Zeszyty Nauk. Politech. Lódz. Mat.* **559** (1989), 1–75.

[227] D. O'Regan, Fixed point theory for weakly contractive maps with applications to operator inclusions in Banach space relative to weak topology, *Z. Anal. Anwend.* **17** (1998), 282–296.

[228] D. O'Regan, Fixed point theorems for weakly sequentially closed maps, *Arch. Math. Brno* **36** (2000), 61–70.

[229] D. O'Regan, Y. J. Cho and Y. Q. Chen, *Topological Degree Theory and Applications*, Series in Mathematical Analysis and Applications, Vol 10, Chapman & Hall, London, 2006.

[230] A. Oustaloup, *La dérivation non entière. Théorie, synthèse applications*, Hermes, Paris, 1995.

[231] W. Padgett and C. Tsokos, *Random Integral Equations with Applications to Life Science and Engineering*, Academic Press, New York, 1976.

[232] P. K. Palamides, Positive and monotone solutions of an *m*-point boundary value problem, *Electron. J. Differential Equations* **18** (2002), 1-16.

[233] N. S. Papageorgiou, Random fixed point theorems for measurable multifunctions in Banach spaces, *Proc. Amer. Math. Soc.* **97** (1986), 507–514.

[234] N. S. Papageorgiou, On the structure of the solution set of evolution inclusions with time-dependent subdifferentials, *Rend. Sem. Mat. Univol. Padova*, **97**, (1997), 163–186.

[235] E. Pardoux and A. Rascanu, *Stochastic Differential Equations, Backward SDEs, Partial Differential Equations*, Stochastic Modelling and Applied Probability, 69. Springer, Heidelberg, 2014.

[236] A. Pazy, *Semigroups of Linear Operators and Applications to Partial Differential Equations,* Springer, New York, 1983.

[237] A. I. Perov, On the Cauchy problem for a system of ordinary differential equations, *Pviblizhen. Met. Reshen. Differ. Uvavn.* **2** (1964), 115-134. (in Russian).

[238] A. I. Perov, A. V. Kibenko, On a certain general method for investigation of boundary value problems, *Izv. Akad. Nauk SSSR, Ser. Mat.*, **30** (1966), 249-264. (in Russian).

[239] A. C. Peterson, Existence and uniqueness theorems for nonlinear difference equations, *J. Math. Anal. Appl.* **125** (1987), 185-191.

[240] I. R. Petre and A. Petrusel, Krasnoselskii's theorem in generalized Banach spaces and applications, *Electron. J. Qual. Theory Differ. Equ.* **2012** (2012), No. 85, 20 pp.

[241] A. Petruşel, Multivalued operators and fixed points, *Pure Math. Appl.* **9** (1998), 165–170.

[242] A. Petruşel, *Operatorial Inclusions*, House of the Book of Science, Cluj-Napoca, 2002.

[243] C. Pierson-Gorez, *Problèmes aux Limites Pour des Equations Différentielles avec Impulsions,* Ph.D. Thesis, University of Louvain-la-Neuve, 1993 (in French).

[244] I. Podlubny, *Fractional Differential Equations*, Academic Press, San Diego, 1999.

[245] S. I. Pohožaev, Eigenfunctions of the equation $\Delta u + \lambda f(u) = 0$, *Soviet Math. Dokl.* **5** (1965), 1408-1411.

[246] B. L. S. Prakasa Rao, Stochastic integral equations of mixed type II. *J. Math. Physical Sci.* **7** (1973), 245-260.

[247] R. Precup, *Methods in Nonlinear Integral Equations*, Kluwer, Dordrecht, 2000.

[248] R. Precup, A vector version of Krasnosel'skii's fixed point theorem in cones and positive periodic solutions of nonlinear systems, *J. Fixed Point Theory Appl.* **2** (2007), 141-151.

[249] R. Precup, Componentwise compression-expansion conditions for systems of nonlinear operator equations and applications, Mathematical Models in Engineering, Biology and Medicine, 284-293, American Institue of Physics Conference Proceedings 1124, American Institue of Physics, Melville, NY, 2009.

[250] R. Precup, Existence, localization and multiplicity results for positive radial solutions of semilinear elliptic systems, *J. Math. Anal. Appl.*, **352** (2009), 48-56.

[251] R. Precup, The role of matrices that are convergent to zero in the study of semilinear operator systems, *Math. Comp. Modelling* **49** (2009), 703-708.

[252] R. Precup and A. Viorel, Existence results for systems of nonlinear evolution equations, *Int. J. Pure Appl. Math. IJPAM*, **47** (2008), 199-206.

[253] B. Przeradzki, The existence of bounded solutions for differential equations in Hilbert spaces, *Ann. Polon. Math.* **56** (1992), 103–121.

[254] L. E. Rybinski, An application of the continuous selection theorem of the study of fixed point of multivalued mapping, *J. Math. Anal. Appl.* **153** (1990), 391–396.

[255] T. Rzeżuchowski, Scorza-Dragoni type theorem for upper semicontinous multivalued functions, *Bull. Acd. Polon. Sci. SéR. Sci. Math. Astr. Phys.*, **28** (1980), 61-66.

[256] S. G. Samko, A. A. Kilbas and O. I. Marichev, *Fractional Integrals and Derivatives, Theory and Applications*, Gordon and Breach, London, 1993.

[257] A. M. Samoilenko and N. A. Perestyuk, *Impulsive Differential Equations*, World Scientific, Singapore, 1995.

[258] I. Sharapudinov, On the topology of the space $L^{p(t)}([0,1])$, *Math. Notes* (1979), 796-806.

[259] M. L. Sinacer, J. J. Nieto, and A. Ouahab, Random fixed point theorem in generalized Banach space and applications, *Random Oper. Stoch. Equ.* **24** (2016), 93-112.

[260] A. Skorohod, *Random Linear Operators*, Reidel, Boston, 1985.

[261] T.H. Skyrme, A non-linear field theory, *Proc. Roy. Soc. A* **260** (1961), 127-138.

[262] G. V. Smirnov, *Introduction to the Theory of Differential Inclusions,* Graduate Studies in Mathematics 41, American Mathematewski, Sur la semi-continuité inférieure du "tendeur" d'un ensemble compact, variant d'une façon continue, *Bull. Acad Polon. Sci.*, **9**, (1961), 869–872.

[263] B. Solomon, A new proof of the closure theorem for integral currents, *Indiana Univ. Math. J.* **33** (1984), 393-418.

[264] Y. Song, Periodic and almost periodic solutions of functional difference equations with finite delay, *Adv. Difference Equ. (2007), Art. ID* 68023, 15 pp.

[265] Y. Song, Asymptotically almost periodic solutions of nonlinear Volterra difference equations with unbounded delay, *J. Difference Equ. Appl.* **14**, (2008), 971-986.

[266] Y. Song, Positive almost periodic solutions of nonlinear discrete systems with finite delay, *Comput.Math. Appl.* **58**, (2009), 128-134.

[267] T. T. Soong, *Random Differential Equations in Science and Engineering*, Academic Press, New York, 1973.

[268] A. Špaček, Zulfallige Gleichungen, *Czechoslovak, Math. J.* **5** (1995), 462-466.

[269] J. L Strand, *Random Ordinary Differential Equations*, Reidel, Boston, 1985.

[270] W. A. Strauss, Existence of solitary waves in higher dimensions, *Commun. Math. Phys.* **55** (1977), 149-172.

[271] S. Sugiyama, On periodic solutions of difference equations, *Bull. Sci. Engrg. Res. Lab.Waseda Univ.* **52** (1971), 89-94.

[272] F. Sullivan, A characterization of complete metric spaces, *Proc. Amer. Math. Soc.* **83** (1981), 345-346.

[273] F. Terkelsen, A short proof of Fan's fixed point theorem. *Proc. Amer. Math. Soc.* **42** (1974), 643–644.

[274] J. M. A. Toledano, T. D. Benavides, and G. L. Azedo, *Measures of Noncompactness in Metric Fixed Point Theory*, Birkhäuser, Basel, 1997.

[275] A. A. Tolstonogov, *Differential Inclusions in a Banach Space*, Kluwer, Dordrecht, 2000.

[276] I. Tsenov, Generalization of the problem of best approximation of a function in the space L^s, *Uch. Zap. Dagestan Gos. Univ.* **7** (1961), 25-37.

[277] R. S. Varga, *Matrix Iterative Analysis*, Springer Series in Computational Mathematics, **27**, Springer, Berlin, 2000.

[278] C. Vidal, Existence of periodic and almost periodic solutions of abstract retarded functional difference equations in phase spaces, *Adv. Difference Equ.* (2009), Art. ID 380568, 19 pp.

[279] A. Viorel, *Contributions to the Study of Nonlinear Evolution Equations*, Ph.D. thesis, Babeş-Bolyai University Cluj-Napoca, 2011.

[280] J. Von Neumann, *A model of general economic equilibrium*, Collected Works, **VI**, Pergamon Press, Oxford, (1963), 29-37.

[281] I. I. Vrabie, C_0-*semigroups and applications*, North-Holland Mathematics Studies, 191. North-Holland Publishing Co., Amsterdam, 2003.

[282] H. Vu, N. N. Phung, and N. Phuong, On fractional random differential equations with delay, *Opuscula Math.* **36** (2016), 541–556.

[283] M. Wadati, Electromagnetically induced transparency and soliton propagations, *J. Phys. Soc. Jpn.* **77**, 024003 (2008), 4 pages.

[284] T. Ważewski, Sur la semi-continuité inférieure du "tendeur" d'un ensemble compact, variant d'une façon continue, *Bull. Acad Polon. Sci.*, **9**, (1961), 869–872.

[285] T. Ważewski, Sur une condition équivalente à l'équation au contingent, *Bull. Acad. Polon. Sci. Ser. Sci. Math. Astronom. Phys* **9** (1961), 865–867.

[286] T. Ważewski, On an optimal control problem, in *Differential Equations and Applications, Conference Proceedings Prague*, 1962, 1963, 229–242.

[287] J. R. L. Webb and G. Infante, Positive solutions of nonlocal boundary value problems: a unified approach, *J. London Math. Soc.* **74** (2006), 673-693.

[288] A. Weil, *Intégration dans les Groupes Topologiques et ses Applications*, Actualités scientifiques et industrielles, **869**, Hermann & Cie., Paris, 1940.

[289] G. B. Witham, *Linear and Nonlinear Waves*, Wiley, New York, 1974.

[290] T. Xiang and R. Yuan, A class of expansive-type Krasnosel'skii fixed point theorems, *Nonlinear Anal.* **71** (2009), 3229–3239.

[291] P. P. Zabrejko, *K*-metric and *K*-normed linear spaces: survey, *Collect. Math.* **48** (1997), 825–859.

[292] S. C. Zaremba, Sur une extension de la notion d'équation différentielle, *C. R. Acad. Sci. Paris* **199**, (1934), 545–548.

[293] S. C. Zaremba, Sur les équations au paratingent, *Bull. Sci. Math.* **60** (1936), 139–160.

[294] Z. Zhao Multiple fixed points of a sum operator and applications, *J. Math. Anal. Appl.* **360** (2009) 1-6.

[295] V. V. Zhikov, Averaging of functionals of the calculus of variations and elasticity theory, *Math. USSR. Izv.* **29** (1987), 33–66.

[296] Q. J. Zhu, On the solution set of differential inclusions in Banach space, *J. Differential Equations* **93** (1991), 213–237.

Index

拓扑方法可以说是区分经典数学和近现代数学的标志之一,我们以三位大家的名言为证:

今天,拓扑的天使和抽象代数的精灵为每一个数学领域的灵魂而斗争.

——H. Weyl

不可压缩流体湍流的统计力学要求对可测函数变换的研究,这又与一定的代数有关,其中还能引入拓扑学.

——Kampe De Feriet

数学对具有代数和拓扑结构的抽象集合进行研究.数学越来越成为综合的和多价的.它的进展主要是靠对每一个理论的最初结果进行扩展,目的在于推广性质和统一那些具有公共结构的理论——如 Schwarz 所说的"合成与统一".

——L. Felix

本书就是一部原版引进的专门讲拓扑方法的数学专著,中文书名或可译为《微分方程与包含的拓扑方法》.本书一共有三位作者.第一位是约翰·R.格雷夫(John R. Graef),美国人,田纳西大学查塔努加分校的数学教授,此前曾在密西西比州立大学任教.第二位是约翰尼·亨德森(Johnny Henderson),美国人,贝勒大学杰出的数学教授,曾在奥本大学和密苏里科技大学担任教职,是美国数学学会的初始成员.第三位是阿卜杜勒加尼·奥哈比(Abdelghani Ouahab),阿尔及利亚人,阿尔及利亚吉拉利·利亚贝斯大学西迪贝尔数学实验室的数学教授.

本书作者在序言中写道:

数学家有各种各样的动机去研究具有速度的动力系统,该速度只由系统状态来决定.从广义上讲,本项研究的一部分动机涉及替换微分方程,例如

$$y'=f(y)$$

其中 f 是一个单值函数,利用微分包含

$$y' \in F(y) \tag{0.1}$$

其中 F 是一个多值函数.微分不等式

$$y'^i \leqslant f^i(x_1,\cdots,x_n), i=1,\cdots,n$$

的一个系统可以被看作一个微分包含.如果一个隐式微分方程

$$f(y,y')=0$$

已经给出了,那么我们可以将 $F(y)=\{v:f(y,v)=0\}$ 化简为微分包含.微分包含被用来研究右手边相对未知的常微分方程.

例 0.1 考虑右边的一个不连续的方程

$$y'=1-2\text{sgn}\,y$$

其中

$$\text{sgn}\,y=\begin{cases}+1, & y>0 \\ -1, & y<0\end{cases}$$

以上问题的经典解由

$$y(t)=\begin{cases}3t+c_1, & y<0 \\ -t+c_2, & y>0\end{cases}$$

给出.

随着 t 的增长,经典解趋向于直线 $y=0$,但它不能沿着这条直线继续,因为这样得到的映射 $y(t)=0$ 并不满足通常意义上的方程(对于 $y'(t)=0$,右侧的值为 $1-2\text{sgn}\,0=1$).因此,没有从 $y(0)=0$ 开始的初始值问题的经典解.

因此,需要对解的概念进行概括.为了绕过不连续的右手边非线性,这个问题可以被认为是一个微分包含

$$y'(t) \in F(y(t)), \quad \text{a. e. } t \in [0,\infty), y(0)=y_0 \tag{0.2}$$

其中,$F: \mathbf{R}^n \rightarrow \mathscr{P}(\mathbf{R}^n)$ 是 $\mathbf{R}^n$ 的所有子集的集合中的一个向量集值映射,可以通过多种方式定义.F 的最简单的凸定义是通过所谓的 Filippov 正则化[117]获得的

$$F(y)=\bigcap_{\varepsilon>0}\overline{\text{co}(f(\{y \in \mathbf{R}^n \mid \|y\| \leqslant \varepsilon\} \backslash M))}$$

其中,conv $F(y)$ 是 f 的凸包,M 是一个空集(比如 $\mu(M)$,其中 μ 表示 $\mathbf{R}^n$ 中的 Lebesgue 测度),且 ε 是以 y 为圆心的球的半径.

在 20 世纪 30 年代,微分包含是由波兰和法国的数学家 Zaremba 和 Marchaud[209,292,293] 提出的.他们最感兴趣的是存在性结果和一些定性性质.Zaremba 和 Marchaud 分别研究了所谓的仿切方程和拟切方程.在 20 世纪 60

年代,Ważewski[284-286]证明,人们可以更多地使用"经典的"解,而不是拟切解或仿切解,因为这些解在 Carathéodory 的意义上被理解为几乎在任何地方都满足式(0.1)的绝对连续函数. 微分包含的一个最重要的例子来自控制理论. Filippov 和 Ważewski 考虑了控制系统

$$y'(t) = f(y, u), u \in U$$

其中,u 是一个控制参数. 它出现在控制系统和微分包含

$$y' \in F(y, U) = \bigcup_{u \in U} f(y, u)$$

中,拥有相同的轨迹. 如果控制集合依赖于 y,也就是 $U = U(y)$,那么我们得到了微分包含

$$y' \in F(y, U(y))$$

一个控制系统和对应的微分包含之间的对等是用来证明最优控制理论中的存在定理的中心思想. 由于经济、社会和生物系统的动力学是多值的,因此微分包含在具有滞后性的宏观系统中充当自然模型. 微分包含是推广常微分方程的概念. 因此,微分方程所考虑的所有问题,例如解的存在性、解的连续性、对初始条件和参数的依赖性,都存在于微分包含理论中. 由于微分包含通常从给定点开始有许多解,因此出现了新问题,例如研究解的集合的拓扑性质、选择具有给定性质的解、评估可达集(生存理论)等. 为了解决以上问题,人们发展了特殊的数学方法. 近几十年来,微分包含一直是许多研究人员深入研究的主题. 例如参考文献 [24,33,34,138,173,180,275]. 如需进一步阅读并获得详细信息,请参阅 Andres 和 Górniewicz 的文献 [24]、Aubin 和 Celina 的文献 [33]、Aubin 和 Frankowska 的文献[34]、Deimling 的文献[97]、Górniewicz 的文献[138]、Hu 和 Papageorgiou 的文献 [173],Kamenskii 的文献 [180] 以及 Tolstonogov 的文献 [275].

本书内容安排如下. 第 1 章我们介绍了本书其余部分将使用的多值分析的基本背景. 除了介绍多值函数的上半连续性和下半连续性、弱连续性、开映射、闭映射和紧映射以及半连续映射的和与积等基本概念,还解释了书中要使用的大部分符号. 此处包含了多值映射的示例. 第 2 章从描述 Hausdorff-Pompeiu 度量拓扑开始,讨论了与 Vietoris 拓扑的关系,并给出了完全度量空间上映射的一些不动点定理.

多值映射的可测性是第 3 章的重点. 可测选择函数的存在在本章中起着重要作用. 在这方面,Lusin 性质和 Scorza-Dragoni 性质在这里也发挥着重要作用. 第 4 章专门研究连续选择定理,特别关注了 Michael 选择定理. 线性多值算子是第 5 章的重要主题.

第 6 章介绍了各种不动点定理,包括 Kakutani 定理、Fan 定理、Leray 和 Schauder 定理、Tychonoff 和 Krasnosel'skii 定理,以及 Krasnosel'skii 定理的许多变体,包括 Fréchet 空间中的相关内容. 在本章的讨论中引入了非紧性测度的测量. 还验证了两个多值算子之和以及拓扑向量空间中的算子的不动点定理. 第 7 章研究了广义度量和 Banach 空间. 该章还引入了收敛矩阵的概念.

第 8 章研究了向量度量空间中的不动点定理. 包括 Schaefer 不动点定理的各种版本, Leary-Schauder 型的非线性备选方案, Perov 定理以及 Covitz 和 Nadler 不动点定理.

第 9 章研究了半直线上的脉冲微分方程系统.

第 10 章致力于随机不动点定理, 特别是 Krasnosel'skii 和 Perov 型的定理.

第 11 章介绍了对半群的研究, 第 12 章研究了微分包含的线性和半线性系统, 具有脉冲影响的包含也被包括在内.

微分方程的随机系统在第 13 章中进行了研究. 我们特别调查了二阶耦合系统的 Cauchy 问题和边值问题. 在第 14 章中研究了涉及 Hadamard 分数导数的随机分数阶微分方程.

第 15 章作者把注意力转向了微分方程. 初始值问题的解的存在性被认为是解空间的紧致性问题.

第 16 章讨论了离散包含的解的存在性和解空间的紧致性. 第 17 章讨论了离散方程的半线性系统. 第 18 章讨论了二阶差分方程组的边值问题.

书中包含了超过 300 篇相关参考文献的参考书目.

近年微分包含在国内的研究者日众, 比如在控制方面. 动态系统一般用 $\dot{x} = f(x, u)$ 来描述, 其中的 $f(x, u)$ 是一个确定的函数. 如果系统存在莫名其妙的不确定性, 即对于一个确定的输入 $u = u(t)$ 和确定的初始状态 $x(t_0) = x_0$, 系统的状态 $x(t)$ 不再是一个时间 t 的确定函数, 也就是说 $f(x, u)$ 不再是一个函数, 而是一个集合值映射, 即

$$F(x, u) = \{f_1(x, u), f_2(x, u), \cdots, f_p(x, u)\}$$

(为了区别函数, 用 $F(x, u)$ 表示集合值映射). 相应地, 系统的模型就要改写成 $\dot{x} \in F(x, u)$. $\dot{x} \in F(x, u)$ 就是微分包含, 是一种新的模型. 人们不禁猜想, 研究的模型换成微分包含是不是意味着新一代控制理论的诞生? 回答: "不知道." 因为尽管目前还未见端倪, 但是要否定也为时过早.

应用微分包含来描述控制系统一方面是由于实际需要, 另一方面也是由于微分包含理论的发展. 国内没有看到考察微分包含的历史的研究和报道. 根据 J-P Aubin 和 A. Cellina 在 *Differential Inclusions—Set-Valued Maps and Viability Theory* 一书中的介绍, 微分包含的雏形最早出现在 20 世纪 30 年代. 在 20 世纪 50 年代以后获得了数学界的广泛垂青, 因而硕果累累. 上述由 J-P Aubin 和 A. Cellina 撰写的著作于 1984 年由 Springer-Verlag 在德国出版, 该书几乎囊括了 20 世纪 80 年代之前获得的所有重要结论, 被认为是至今为止微分包含理论的经典. 该书主要讲述了 Hilbert 空间上的微分包含理论. 秦松喜将其上半部分进行了编译, 并由厦门大学出版社出版. G. V. Smirnov 在 2002 年出版了 *Introduction to the Theory of Differential Inclusions*, 这是第一本讲述有限维空间上微分包含的专著, 是研究微分包含的控制理论不可不读的经典.

从 20 世纪 70 年代起, 控制理论就涉足微分包含, 大致算来也有 40 多年历史了. 控制理论采用微分包含模型的研究主要有三个阶段: 从 20 世纪 70 年代起, 主要工作是将极大值原理推广到微分包含, 尽管有所建树, 但是不能算有实质性的突破. 从 20 世纪后期起,

对微分包含系统的能控性有较多的研究,但是对于微分包含系统,能控性在系统设计中的作用还很不清楚,因此其结论没有受到足够的重视.进入 21 世纪,出现了对线性多胞体系统的控制和对 Luré 型微分包含系统的观测方面的研究.这些工作是真正可以算作控制系统设计的内容,就此而言,这些成果给微分包含控制理论带来了突破性的进展.

本书的目录为:

其实我们数学工作室最初的核心业务是出版高端的奥林匹克数学竞赛书籍,但许多竞赛试题早已"越界"到了高数领域.以我们的邻居越南为例,就出现了下列试题:

已知连续函数 $f: \mathbf{R} \to (0, +\infty)$ 满足

$$\lim_{x \to -\infty} f(x) = \lim_{x \to +\infty} f(x) = 0$$

证明:(1)$f(x)$ 在 $\mathbf{R}$ 上有最大值;

(2)存在两个实数列 $\{x_n\}$,$\{y_n\}$,满足对于所有的正整数 n,均有 $x_n < y_n$,$f(x_n) = f(y_n)$,且 $\{x_n\}$,$\{y_n\}$ 收敛于同一个极限.

(2019 年越南数学奥林匹克)

证明 (1)由于 $f(0) > 0$,且

$$\lim_{x \to -\infty} f(x) = \lim_{x \to +\infty} f(x) = 0$$

则存在一个足够小的负数 a 和一个足够大的正数 b，使得对于所有的实数 $x \leqslant a$ 和 $x \geqslant b$，均有 $f(x) < f(0)$.

又由于函数 $f(x)$ 在闭区间 $[a,b]$ 上连续，则存在 $c \in [a,b]$，使得 $f(x)$ 在区间 $[a,b]$ 上取得最大值 $M = f(c)$，且 $M \geqslant f(0)$.

因为对于所有的实数 $x \in (-\infty, a) \cup (b, +\infty)$，均有

$$f(x) < f(0) \leqslant M$$

所以，对于所有的 $x \in \mathbf{R}$，均有

$$f(x) \leqslant M$$

因此，在 $\mathbf{R}$ 上，函数 $f(x)$ 有最大值 M.

（2）由介值定理知，若 $f(x)$ 能取到两个值 $A,B(A<B)$，则 $f(x)$ 能取到区间 (A,B) 内的所有值.

下面分两种情况讨论.

①存在区间 (a,b)，使得 $c \in (a,b)$，$f(c) = M$，且对于所有的 $x \in (a,b) \setminus \{c\}$，均有 $f(x) < M$.

取 $A \neq c$，$B \neq c$，使得 $A,B \in (a,b)$ 且 $c \in [A,B]$. 下面只考虑这个区间.

因为 $f(x)$ 在区间 $[A,c]$，$[c,B]$ 上均是连续的，所以，在区间 $[A,c]$，$[c,B]$ 上均有最小值，分别记为 m_1,m_2.

设 $m = \max\{m_1,m_2\}$.

由介值定理知，存在 $x_1 \in [A,c]$，$y_1 \in [c,B]$，使得

$$f(x_1) = f(y_1) = m, x_1 < c < y_1$$

在区间 $[x_1,c]$，$[c,y_1]$ 上，由介值定理知，存在 $x_2 \in [x_1,c]$，$y_2 \in [c,y_1]$，使得

$$f(x_2) = f(y_2) = \frac{m+M}{2}, x_2 < c < y_2$$

类似地，定义数列 $\{u_n\}$，使得

$$u_1 = m, u_{n+1} = \frac{u_n + M}{2}, n \in \mathbf{Z}_+$$

则 $\{u_n\}$ 收敛于 M，对于每个正整数 $n \geqslant 2$，均存在 $x_n \in [x_{n-1},c]$，$y_n \in [c, y_{n-1}]$，使得

$$f(x_n) = f(y_n) = u_n, x_n < c < y_n$$

因为数列 $\{x_n\}$ 单调递增，且有上界 c，所以 $\{x_n\}$ 有极限.

设 $\lim\limits_{n \to +\infty} x_n = l$，则 $l \leqslant c$. 若 $l < c$，则由函数 $f(x)$ 的连续性得

$$M = \lim_{n \to +\infty} u_n = \lim_{n \to +\infty} f(x_n) = f(l) < M$$

矛盾. 因此，$\lim\limits_{n \to +\infty} x_n = c$. 类似地，$\lim\limits_{n \to +\infty} y_n = c$.

②若①中的区间不存在，则存在区间 $[a,b]$，使得对于所有的 $x \in [a,b]$，均有

$$f(x) = M$$

考虑数列

$$x_n = \frac{a+b}{2} + \frac{a-b}{2^n}, y_n = \frac{a+b}{2} + \frac{b-a}{2^n}$$

其中,$n=1,2,\cdots$. 则对于所有正整数 n,均有 $x_n, y_n \in [a,b]$. 于是,对于所有的正整数 n,均有

$$f(x_n) = f(y_n) = M, x_n < \frac{a+b}{2} < y_n$$

且

$$\lim_{n \to +\infty} x_n = \lim_{n \to +\infty} y_n = \frac{a+b}{2}$$

综上,结论成立.

当然,有人会说越南的数学教育是继承了法国的精英教育理念.《三联生活周刊》的主笔苗千曾采访过菲尔兹奖得主法国数学家 Cédric Villani.

"数学的本质是什么?"采访一开始,我就问了 Villani 一个有些大而不当的问题."抽象."他回答我.法国作为一个盛产数学家的国度,我想或许与这个民族善于处理抽象概念有关.但 Villani 并不认为数学研究与天分有太大的关系.他认为自己虽然在幼年时就展现出了数学天分,但也是后来通过艰苦的训练才取得了杰出的成就.

一些数据可以支持 Villani 的观点.法国 15 岁左右的中学生在国际学生评估项目(Programme for International Student Assessment)中的排名并不算高,在全世界仅排在第 25 位,法国学生 495 分的得分也仅仅略高于平均分 494 分.法国数学家能够不断涌现,更重要的原因很可能在于精英学生在预科学校(Prépa)和大学里接受的高强度数学训练.

所以,只要数学上不能搞平均主义,就必须用精英教育理念来制定政策.越南已经尝到了甜头,吴宝珠成为首位大学期间在越南接受教育的菲尔兹奖得主.而我们则望眼欲穿,遥遥无期.出版业也要贡献一份力量.出版业是一个小行业,但这个行业对社会的撬动作用却是巨大的.

刘培杰
2023 年 5 月 29 日
于哈工大

刘培杰数学工作室
已出版(即将出版)图书目录——原版影印

书 名	出版时间	定 价	编号
数学物理大百科全书.第1卷(英文)	2016—01	418.00	508
数学物理大百科全书.第2卷(英文)	2016—01	408.00	509
数学物理大百科全书.第3卷(英文)	2016—01	396.00	510
数学物理大百科全书.第4卷(英文)	2016—01	408.00	511
数学物理大百科全书.第5卷(英文)	2016—01	368.00	512
zeta函数,q-zeta函数,相伴级数与积分(英文)	2015—08	88.00	513
微分形式:理论与练习(英文)	2015—08	58.00	514
离散与微分包含的逼近和优化(英文)	2015—08	58.00	515
艾伦·图灵:他的工作与影响(英文)	2016—01	98.00	560
测度理论概率导论,第2版(英文)	2016—01	88.00	561
带有潜在故障恢复系统的半马尔柯夫模型控制(英文)	2016—01	98.00	562
数学分析原理(英文)	2016—01	88.00	563
随机偏微分方程的有效动力学(英文)	2016—01	88.00	564
图的谱半径(英文)	2016—01	58.00	565
量子机器学习中数据挖掘的量子计算方法(英文)	2016—01	98.00	566
量子物理的非常规方法(英文)	2016—01	118.00	567
运输过程的统一非局部理论:广义波尔兹曼物理动力学,第2版(英文)	2016—01	198.00	568
量子力学与经典力学之间的联系在原子、分子及电动力学系统建模中的应用(英文)	2016—01	58.00	569
算术域(英文)	2018—01	158.00	821
高等数学竞赛:1962—1991年的米洛克斯·史怀哲竞赛(英文)	2018—01	128.00	822
用数学奥林匹克精神解决数论问题(英文)	2018—01	108.00	823
代数几何(德文)	2018—04	68.00	824
丢番图逼近论(英文)	2018—01	78.00	825
代数几何学基础教程(英文)	2018—01	98.00	826
解析数论入门课程(英文)	2018—01	78.00	827
数论中的丢番图问题(英文)	2018—01	78.00	829
数论(梦幻之旅):第五届中日数论研讨会演讲集(英文)	2018—01	68.00	830
数论新应用(英文)	2018—01	68.00	831
数论(英文)	2018—01	78.00	832

刘培杰数学工作室
已出版(即将出版)图书目录——原版影印

书　名	出版时间	定　价	编号
湍流十讲(英文)	2018—04	108.00	886
无穷维李代数:第3版(英文)	2018—04	98.00	887
等值、不变量和对称性(英文)	2018—04	78.00	888
解析数论(英文)	2018—09	78.00	889
《数学原理》的演化:伯特兰·罗素撰写第二版时的手稿与笔记(英文)	2018—04	108.00	890
哈密尔顿数学论文集(第4卷):几何学、分析学、天文学、概率和有限差分等(英文)	2019—05	108.00	891
偏微分方程全局吸引子的特性(英文)	2018—09	108.00	979
整函数与下调和函数(英文)	2018—09	118.00	980
幂等分析(英文)	2018—09	118.00	981
李群,离散子群与不变量理论(英文)	2018—09	108.00	982
动力系统与统计力学(英文)	2018—09	118.00	983
表示论与动力系统(英文)	2018—09	118.00	984
分析学练习.第1部分(英文)	2021—01	88.00	1247
分析学练习.第2部分,非线性分析(英文)	2021—01	88.00	1248
初级统计学:循序渐进的方法:第10版(英文)	2019—05	68.00	1067
工程师与科学家微分方程用书:第4版(英文)	2019—07	58.00	1068
大学代数与三角学(英文)	2019—06	78.00	1069
培养数学能力的途径(英文)	2019—07	38.00	1070
工程师与科学家统计学:第4版(英文)	2019—06	58.00	1071
贸易与经济中的应用统计学:第6版(英文)	2019—06	58.00	1072
傅立叶级数和边值问题:第8版(英文)	2019—05	48.00	1073
通往天文学的途径:第5版(英文)	2019—05	58.00	1074
拉马努金笔记.第1卷(英文)	2019—06	165.00	1078
拉马努金笔记.第2卷(英文)	2019—06	165.00	1079
拉马努金笔记.第3卷(英文)	2019—06	165.00	1080
拉马努金笔记.第4卷(英文)	2019—06	165.00	1081
拉马努金笔记.第5卷(英文)	2019—06	165.00	1082
拉马努金遗失笔记.第1卷(英文)	2019—06	109.00	1083
拉马努金遗失笔记.第2卷(英文)	2019—06	109.00	1084
拉马努金遗失笔记.第3卷(英文)	2019—06	109.00	1085
拉马努金遗失笔记.第4卷(英文)	2019—06	109.00	1086
数论:1976年纽约洛克菲勒大学数论会议记录(英文)	2020—06	68.00	1145
数论:卡本代尔1979:1979年在南伊利诺伊卡本代尔大学举行的数论会议记录(英文)	2020—06	78.00	1146
数论:诺德韦克豪特1983:1983年在诺德韦克豪特举行的Journees Arithmetiques数论大会会议记录(英文)	2020—06	68.00	1147
数论:1985—1988年在纽约城市大学研究生院和大学中心举办的研讨会(英文)	2020—06	68.00	1148

刘培杰数学工作室
已出版(即将出版)图书目录——原版影印

书　名	出 版 时 间	定　价	编号
数论:1987年在乌尔姆举行的Journees Arithmetiques数论大会会议记录(英文)	2020-06	68.00	1149
数论:马德拉斯1987:1987年在马德拉斯安娜大学举行的国际拉马努金百年纪念大会会议记录(英文)	2020-06	68.00	1150
解析数论:1988年在东京举行的日法研讨会会议记录(英文)	2020-06	68.00	1151
解析数论:2002年在意大利切特拉罗举行的C.I.M.E.暑期班演讲集(英文)	2020-06	68.00	1152
量子世界中的蝴蝶:最迷人的量子分形故事(英文)	2020-06	118.00	1157
走进量子力学(英文)	2020-06	118.00	1158
计算物理学概论(英文)	2020-06	48.00	1159
物质,空间和时间的理论:量子理论(英文)	2020-10	48.00	1160
物质,空间和时间的理论:经典理论(英文)	2020-10	48.00	1161
量子场理论:解释世界的神秘背景(英文)	2020-07	38.00	1162
计算物理学概论(英文)	2020-06	48.00	1163
行星状星云(英文)	2020-10	38.00	1164
基本宇宙学:从亚里士多德的宇宙到大爆炸(英文)	2020-08	58.00	1165
数学磁流体力学(英文)	2020-07	58.00	1166
计算科学:第1卷,计算的科学(日文)	2020-07	88.00	1167
计算科学:第2卷,计算与宇宙(日文)	2020-07	88.00	1168
计算科学:第3卷,计算与物质(日文)	2020-07	88.00	1169
计算科学:第4卷,计算与生命(日文)	2020-07	88.00	1170
计算科学:第5卷,计算与地球环境(日文)	2020-07	88.00	1171
计算科学:第6卷,计算与社会(日文)	2020-07	88.00	1172
计算科学.别卷,超级计算机(日文)	2020-07	88.00	1173
多复变函数论(日文)	2022-06	78.00	1518
复变函数入门(日文)	2022-06	78.00	1523
代数与数论:综合方法(英文)	2020-10	78.00	1185
复分析:现代函数理论第一课(英文)	2020-07	58.00	1186
斐波那契数列和卡特兰数:导论(英文)	2020-10	68.00	1187
组合推理:计数艺术介绍(英文)	2020-07	88.00	1188
二次互反律的傅里叶分析证明(英文)	2020-07	48.00	1189
旋瓦兹分布的希尔伯特变换与应用(英文)	2020-07	58.00	1190
泛函分析:巴拿赫空间理论入门(英文)	2020-07	48.00	1191
卡塔兰数入门(英文)	2019-05	68.00	1060
测度与积分(英文)	2019-04	68.00	1059
组合学手册.第一卷(英文)	2020-06	128.00	1153
-代数、局部紧群和巴拿赫-代数丛的表示.第一卷,群和代数的基本表示理论(英文)	2020-05	148.00	1154
电磁理论(英文)	2020-08	48.00	1193
连续介质力学中的非线性问题(英文)	2020-09	78.00	1195
多变量数学入门(英文)	2021-05	68.00	1317
偏微分方程入门(英文)	2021-05	88.00	1318
若尔当典范性:理论与实践(英文)	2021-07	68.00	1366
伽罗瓦理论.第4版(英文)	2021-08	88.00	1408
R统计学概论	2023-03	88.00	1614
基于不确定静态和动态问题解的仿射算术(英文)	2023-03	38.00	1618

刘培杰数学工作室
已出版(即将出版)图书目录——原版影印

书　名	出版时间	定　价	编号
典型群,错排与素数(英文)	2020—11	58.00	1204
李代数的表示:通过 gln 进行介绍(英文)	2020—10	38.00	1205
实分析演讲集(英文)	2020—10	38.00	1206
现代分析及其应用的课程(英文)	2020—10	58.00	1207
运动中的抛射物数学(英文)	2020—10	38.00	1208
2—纽结与它们的群(英文)	2020—10	38.00	1209
概率,策略和选择:博弈与选举中的数学(英文)	2020—11	58.00	1210
分析学引论(英文)	2020—11	58.00	1211
量子群:通往流代数的路径(英文)	2020—11	38.00	1212
集合论入门(英文)	2020—10	48.00	1213
酉反射群(英文)	2020—11	58.00	1214
探索数学:吸引人的证明方式(英文)	2020—11	58.00	1215
微分拓扑短期课程(英文)	2020—10	48.00	1216
抽象凸分析(英文)	2020—11	68.00	1222
费马大定理笔记(英文)	2021—03	48.00	1223
高斯与雅可比和(英文)	2021—03	78.00	1224
π 与算术几何平均:关于解析数论和计算复杂性的研究(英文)	2021—01	58.00	1225
复分析入门(英文)	2021—03	48.00	1226
爱德华·卢卡斯与素性测定(英文)	2021—03	78.00	1227
通往凸分析及其应用的简单路径(英文)	2021—01	68.00	1229
微分几何的各个方面.第一卷(英文)	2021—01	58.00	1230
微分几何的各个方面.第二卷(英文)	2020—12	58.00	1231
微分几何的各个方面.第三卷(英文)	2020—12	58.00	1232
沃克流形几何学(英文)	2020—11	58.00	1233
彷射和韦尔几何应用(英文)	2020—12	58.00	1234
双曲几何学的旋转向量空间方法(英文)	2021—02	58.00	1235
积分:分析学的关键(英文)	2020—12	48.00	1236
为有天分的新生准备的分析学基础教材(英文)	2020—11	48.00	1237
数学不等式.第一卷.对称多项式不等式(英文)	2021—03	108.00	1273
数学不等式.第二卷.对称有理不等式与对称无理不等式(英文)	2021—03	108.00	1274
数学不等式.第三卷.循环不等式与非循环不等式(英文)	2021—03	108.00	1275
数学不等式.第四卷.Jensen 不等式的扩展与加细(英文)	2021—03	108.00	1276
数学不等式.第五卷.创建不等式与解不等式的其他方法(英文)	2021—04	108.00	1277

刘培杰数学工作室
已出版(即将出版)图书目录——原版影印

书　名	出版时间	定　价	编号
冯·诺依曼代数中的谱位移函数:半有限冯·诺依曼代数中的谱位移函数与谱流(英文)	2021—06	98.00	1308
链接结构:关于嵌入完全图的直线中链接单形的组合结构(英文)	2021—05	58.00	1309
代数几何方法.第1卷(英文)	2021—06	68.00	1310
代数几何方法.第2卷(英文)	2021—06	68.00	1311
代数几何方法.第3卷(英文)	2021—06	58.00	1312
代数、生物信息和机器人技术的算法问题.第四卷,独立恒等式系统(俄文)	2020—08	118.00	1199
代数、生物信息和机器人技术的算法问题.第五卷,相对覆盖性和独立可拆分恒等式系统(俄文)	2020—08	118.00	1200
代数、生物信息和机器人技术的算法问题.第六卷,恒等式和准恒等式的相等 问题、可推导性和可实现性(俄文)	2020—08	128.00	1201
分数阶微积分的应用:非局部动态过程,分数阶导热系数(俄文)	2021—01	68.00	1241
泛函分析问题与练习:第2版(俄文)	2021—01	98.00	1242
集合论、数学逻辑和算法论问题:第5版(俄文)	2021—01	98.00	1243
微分几何和拓扑短期课程(俄文)	2021—01	98.00	1244
素数规律(俄文)	2021—01	88.00	1245
无穷边值问题解的递减:无界域中的拟线性椭圆和抛物方程(俄文)	2021—01	48.00	1246
微分几何讲义(俄文)	2020—12	98.00	1253
二次型和矩阵(俄文)	2021—01	98.00	1255
积分和级数.第2卷,特殊函数(俄文)	2021—01	168.00	1258
积分和级数.第3卷,特殊函数补充:第2版(俄文)	2021—01	178.00	1264
几何图上的微分方程(俄文)	2021—01	138.00	1259
数论教程:第2版(俄文)	2021—01	98.00	1260
非阿基米德分析及其应用(俄文)	2021—03	98.00	1261
古典群和量子群的压缩(俄文)	2021—03	98.00	1263
数学分析习题集.第3卷,多元函数:第3版(俄文)	2021—03	98.00	1266
数学习题:乌拉尔国立大学数学力学系大学生奥林匹克(俄文)	2021—03	98.00	1267
柯西定理和微分方程的特解(俄文)	2021—03	98.00	1268
组合极值问题及其应用:第3版(俄文)	2021—03	98.00	1269
数学词典(俄文)	2021—01	98.00	1271
确定性混沌分析模型(俄文)	2021—06	168.00	1307
精选初等数学习题和定理.立体几何.第3版(俄文)	2021—03	68.00	1316
微分几何习题:第3版(俄文)	2021—05	98.00	1336
精选初等数学习题和定理.平面几何.第4版(俄文)	2021—05	68.00	1335
曲面理论在欧氏空间 E_n 中的直接表示(俄文)	2022—01	68.00	1444
维纳—霍普夫离散算子和托普利兹算子:某些可数赋范空间中的诺特性和可逆性(俄文)	2022—03	108.00	1496
Maple 中的数论:数论中的计算机计算(俄文)	2022—03	88.00	1497
贝尔曼和克努特问题及其概括:加法运算的复杂性(俄文)	2022—03	138.00	1498

刘培杰数学工作室
已出版（即将出版）图书目录——原版影印

书 名	出版时间	定 价	编号
复分析:共形映射(俄文)	2022—07	48.00	1542
微积分代数样条和多项式及其在数值方法中的应用(俄文)	2022—08	128.00	1543
蒙特卡罗方法中的随机过程和场模型:算法和应用(俄文)	2022—08	88.00	1544
线性椭圆型方程组:论二阶椭圆型方程的迪利克雷问题(俄文)	2022—08	98.00	1561
动态系统解的增长特性:估值、稳定性、应用(俄文)	2022—08	118.00	1565
群的自由积分解:建立和应用(俄文)	2022—08	78.00	1570
混合方程和偏差自变数方程问题:解的存在和唯一性(俄文)	2023—01	78.00	1582
拟度量空间分析:存在和逼近定理(俄文)	2023—01	108.00	1583
二维和三维流形上函数的拓扑性质:函数的拓扑分类(俄文)	2023—03	68.00	1584
齐次马尔科夫过程建模的矩阵方法:此类方法能够用于不同目的的的复杂系统研究、设计和完善(俄文)	2023—03	68.00	1594
周期函数的近似方法和特性:特殊课程(俄文)	2023—04	158.00	1622
扩散方程解的矩函数:变分法(俄文)	2023—03	58.00	1623
多赋范空间和广义函数:理论及应用(俄文)	2023—03	98.00	1632
分析中的多值映射:部分应用(俄文)	2023—06	98.00	1634
数学物理问题(俄文)	2023—03	78.00	1636
函数的幂级数与三角级数分解(俄文)	2024—01	58.00	1695
星体理论的数学基础:原子三元组(俄文)	2024—01	98.00	1696
素数规律:专著(俄文)	2024—01	118.00	1697
狭义相对论与广义相对论:时空与引力导论(英文)	2021—07	88.00	1319
束流物理学和粒子加速器的实践介绍:第2版(英文)	2021—07	88.00	1320
凝聚态物理中的拓扑和微分几何简介(英文)	2021—05	88.00	1321
混沌映射:动力学,分形学和快速涨落(英文)	2021—05	128.00	1322
广义相对论:黑洞、引力波和宇宙学介绍(英文)	2021—06	68.00	1323
现代分析电磁均质化(英文)	2021—06	68.00	1324
为科学家提供的基本流体动力学(英文)	2021—06	88.00	1325
视觉天文学:理解夜空的指南(英文)	2021—06	68.00	1326
物理学中的计算方法(英文)	2021—06	68.00	1327
单星的结构与演化:导论(英文)	2021—06	108.00	1328
超越居里:1903年至1963年物理界四位女性及其著名发现(英文)	2021—06	68.00	1329
范德瓦尔斯流体热力学的进展(英文)	2021—06	68.00	1330
先进的托卡马克稳定性理论(英文)	2021—06	88.00	1331
经典场论导论:基本相互作用的过程(英文)	2021—07	88.00	1332
光致电离量子动力学方法原理(英文)	2021—07	108.00	1333
经典域论和应力:能量张量(英文)	2021—05	88.00	1334
非线性太赫兹光谱的概念与应用(英文)	2021—06	68.00	1337
电磁学中的无穷空间并矢格林函数(英文)	2021—06	88.00	1338
物理科学基础数学.第1卷,齐次边值问题、傅里叶方法和特殊函数(英文)	2021—07	108.00	1339
离散量子力学(英文)	2021—07	68.00	1340
核磁共振的物理学和数学(英文)	2021—07	108.00	1341
分子水平的静电学(英文)	2021—08	68.00	1342
非线性波:理论、计算机模拟、实验(英文)	2021—06	108.00	1343
石墨烯光学:经典问题的电解解决方案(英文)	2021—06	68.00	1344
超材料多元宇宙(英文)	2021—07	68.00	1345
银河系外的天体物理学(英文)	2021—07	68.00	1346
原子物理学(英文)	2021—07	68.00	1347
将光打结:将拓扑学应用于光学(英文)	2021—07	68.00	1348
电磁学:问题与解法(英文)	2021—07	88.00	1364
海浪的原理:介绍量子力学的技巧与应用(英文)	2021—07	108.00	1365

刘培杰数学工作室
已出版(即将出版)图书目录——原版影印

书　名	出版时间	定　价	编号
多孔介质中的流体:输运与相变(英文)	2021—07	68.00	1372
洛伦兹群的物理学(英文)	2021—08	68.00	1373
物理导论的数学方法和解决方法手册(英文)	2021—08	68.00	1374
非线性波数学物理学入门(英文)	2021—08	88.00	1376
波:基本原理和动力学(英文)	2021—07	68.00	1377
光电子量子计量学.第1卷,基础(英文)	2021—07	88.00	1383
光电子量子计量学.第2卷,应用与进展(英文)	2021—07	68.00	1384
复杂流的格子玻尔兹曼建模的工程应用(英文)	2021—08	68.00	1393
电偶极矩挑战(英文)	2021—08	108.00	1394
电动力学:问题与解法(英文)	2021—09	68.00	1395
自由电子激光的经典理论(英文)	2021—08	68.00	1397
曼哈顿计划——核武器物理学简介(英文)	2021—09	68.00	1401
粒子物理学(英文)	2021—09	68.00	1402
引力场中的量子信息(英文)	2021—09	128.00	1403
器件物理学的基本经典力学(英文)	2021—09	68.00	1404
等离子体物理及其空间应用导论.第1卷,基本原理和初步过程(英文)	2021—09	68.00	1405
磁约束聚变等离子体物理:理想MHD理论(英文)	2023—03	68.00	1613
相对论量子场论.第1卷,典范形式体系(英文)	2023—03	38.00	1615
相对论量子场论.第2卷,路径积分形式(英文)	2023—06	38.00	1616
相对论量子场论.第3卷,量子场论的应用(英文)	2023—06	38.00	1617
涌现的物理学(英文)	2023—05	58.00	1619
量子化旋涡:一本拓扑激发手册(英文)	2023—04	68.00	1620
非线性动力学:实践的介绍性调查(英文)	2023—05	68.00	1621
静电加速器:一个多功能工具(英文)	2023—06	58.00	1625
相对论多体理论与统计力学(英文)	2023—06	58.00	1626
经典力学.第1卷,工具与向量(英文)	2023—04	38.00	1627
经典力学.第2卷,运动学和匀加速运动(英文)	2023—04	58.00	1628
经典力学.第3卷,牛顿定律和匀速圆周运动(英文)	2023—04	58.00	1629
经典力学.第4卷,万有引力定律(英文)	2023—04	38.00	1630
经典力学.第5卷,守恒定律与旋转运动(英文)	2023—04	38.00	1631
对称问题:纳维尔—斯托克斯问题(英文)	2023—04	38.00	1638
摄影的物理和艺术.第1卷,几何与光的本质(英文)	2023—04	78.00	1639
摄影的物理和艺术.第2卷,能量与色彩(英文)	2023—04	78.00	1640
摄影的物理和艺术.第3卷,探测器与数码的意义(英文)	2023—04	78.00	1641
拓扑与超弦理论焦点问题(英文)	2021—07	58.00	1349
应用数学:理论、方法与实践(英文)	2021—07	78.00	1350
非线性特征值问题:牛顿型方法与非线性瑞利函数(英文)	2021—07	58.00	1351
广义膨胀和齐性:利用齐性构造齐次系统的李雅普诺夫函数和控制律(英文)	2021—06	48.00	1352
解析数论焦点问题(英文)	2021—07	58.00	1353
随机微分方程:动态系统方法(英文)	2021—07	58.00	1354
经典力学与微分几何(英文)	2021—07	58.00	1355
负定相交形式流形上的瞬子模空间几何(英文)	2021—07	68.00	1356
广义卡塔兰轨道分析:广义卡塔兰轨道计算数字的方法(英文)	2021—07	48.00	1367
洛伦兹方法的变分:二维与三维洛伦兹方法(英文)	2021—08	38.00	1378
几何、分析和数论精编(英文)	2021—08	68.00	1380
从一个新角度看数论:通过遗传方法引入现实的概念(英文)	2021—07	58.00	1387
动力系统:短期课程(英文)	2021—08	68.00	1382
几何路径:理论与实践(英文)	2021—08	48.00	1385

刘培杰数学工作室
已出版(即将出版)图书目录——原版影印

书　名	出版时间	定　价	编号
论天体力学中某些问题的不可积性(英文)	2021−07	88.00	1396
广义斐波那契数列及其性质(英文)	2021−08	38.00	1386
对称函数和麦克唐纳多项式:余代数结构与 Kawanaka 恒等式(英文)	2021−09	38.00	1400
杰弗里·英格拉姆·泰勒科学论文集:第 1 卷.固体力学(英文)	2021−05	78.00	1360
杰弗里·英格拉姆·泰勒科学论文集:第 2 卷.气象学、海洋学和湍流(英文)	2021−05	68.00	1361
杰弗里·英格拉姆·泰勒科学论文集:第 3 卷.空气动力学以及落弹数和爆炸的力学(英文)	2021−05	68.00	1362
杰弗里·英格拉姆·泰勒科学论文集:第 4 卷.有关流体力学(英文)	2021−05	58.00	1363
非局域泛函演化方程:积分与分数阶(英文)	2021−08	48.00	1390
理论工作者的高等微分几何:纤维丛、射流流形和拉格朗日理论(英文)	2021−08	68.00	1391
半线性退化椭圆微分方程:局部定理与整体定理(英文)	2021−07	48.00	1392
非交换几何、规范理论和重整化:一般简介与非交换量子场论的重整化(英文)	2021−09	78.00	1406
数论论文集:拉普拉斯变换和带有数论系数的幂级数(俄文)	2021−09	48.00	1407
挠理论专题:相对极大值,单射与扩充模(英文)	2021−09	88.00	1410
强正则图与欧几里得若尔当代数:非通常关系中的启示(英文)	2021−10	48.00	1411
拉格朗日几何和哈密顿几何:力学的应用(英文)	2021−10	48.00	1412
时滞微分方程与差分方程的振动理论:二阶与三阶(英文)	2021−10	98.00	1417
卷积结构与几何函数理论:用以研究特定几何函数理论方向的分数阶微积分算子与卷积结构(英文)	2021−10	48.00	1418
经典数学物理的历史发展(英文)	2021−10	78.00	1419
扩展线性丢番图问题(英文)	2021−10	38.00	1420
一类混沌动力系统的分歧分析与控制:分歧分析与控制(英文)	2021−11	38.00	1421
伽利略空间和伪伽利略空间中一些特殊曲线的几何性质(英文)	2022−01	68.00	1422
一阶偏微分方程:哈密尔顿—雅可比理论(英文)	2021−11	48.00	1424
各向异性黎曼多面体的反问题:分段光滑的各向异性黎曼多面体反边界谱问题:唯一性(英文)	2021−11	38.00	1425
项目反应理论手册.第一卷,模型(英文)	2021−11	138.00	1431
项目反应理论手册.第二卷,统计工具(英文)	2021−11	118.00	1432
项目反应理论手册.第三卷,应用(英文)	2021−11	138.00	1433
二次无理数:经典数论入门(英文)	2022−05	138.00	1434

刘培杰数学工作室
已出版(即将出版)图书目录——原版影印

书　名	出版时间	定　价	编号
数,形与对称性:数论,几何和群论导论(英文)	2022—05	128.00	1435
有限域手册(英文)	2021—11	178.00	1436
计算数论(英文)	2021—11	148.00	1437
拟群与其表示简介(英文)	2021—11	88.00	1438
数论与密码学导论:第二版(英文)	2022—01	148.00	1423

书　名	出版时间	定　价	编号
几何分析中的柯西变换与黎兹变换:解析调和容量和李普希兹调和容量、变化和振荡以及一致可求长性(英文)	2021—12	38.00	1465
近似不动点定理及其应用(英文)	2022—05	28.00	1466
局部域的相关内容解析:对局部域的扩展及其伽罗瓦群的研究(英文)	2022—01	38.00	1467
反问题的二进制恢复方法(英文)	2022—03	28.00	1468
对几何函数中某些类的各个方面的研究:复变量理论(英文)	2022—01	38.00	1469
覆盖、对应和非交换几何(英文)	2022—01	28.00	1470
最优控制理论中的随机线性调节器问题:随机最优线性调节器问题(英文)	2022—01	38.00	1473
正交分解法:涡流流体动力学应用的正交分解法(英文)	2022—01	38.00	1475

书　名	出版时间	定　价	编号
芬斯勒几何的某些问题(英文)	2022—03	38.00	1476
受限三体问题(英文)	2022—05	38.00	1477
利用马利亚万微积分进行 Greeks 的计算:连续过程、跳跃过程中的马利亚万微积分和金融领域中的 Greeks(英文)	2022—05	48.00	1478
经典分析和泛函分析的应用:分析学的应用(英文)	2022—03	38.00	1479
特殊芬斯勒空间的探究(英文)	2022—03	48.00	1480
某些图形的施泰纳距离的细谷多项式:细谷多项式与图的维纳指数(英文)	2022—05	38.00	1481
图论问题的遗传算法:在新鲜与模糊的环境中(英文)	2022—05	48.00	1482
多项式映射的渐近簇(英文)	2022—05	38.00	1483

书　名	出版时间	定　价	编号
一维系统中的混沌:符号动力学,映射序列,一致收敛和沙可夫斯基定理(英文)	2022—05	38.00	1509
多维边界层流动与传热分析:粘性流体流动的数学建模与分析(英文)	2022—05	38.00	1510
演绎理论物理学的原理:一种基于量子力学波函数的逐次置信估计的一般理论的提议(英文)	2022—05	38.00	1511
R^2 和 R^3 中的仿射弹性曲线:概念和方法(英文)	2022—08	38.00	1512
算术数列中除数函数的分布:基本内容、调查、方法、第二矩、新结果(英文)	2022—05	28.00	1513
抛物型狄拉克算子和薛定谔方程:不定常薛定谔方程的抛物型狄拉克算子及其应用(英文)	2022—07	28.00	1514
黎曼-希尔伯特问题与量子场论:可积重正化、戴森-施温格方程(英文)	2022—08	38.00	1515
代数结构和几何结构的形变理论(英文)	2022—08	48.00	1516
概率结构和模糊结构上的不动点:概率结构和直觉模糊度量空间的不动点定理(英文)	2022—08	38.00	1517

刘培杰数学工作室
已出版(即将出版)图书目录——原版影印

书 名	出 版 时 间	定 价	编号
反若尔当对:简单反若尔当对的自同构(英文)	2022—07	28.00	1533
对某些黎曼—芬斯勒空间变换的研究:芬斯勒几何中的某些变换(英文)	2022—07	38.00	1534
内诣零流形映射的尼尔森数的阿诺索夫关系(英文)	2023—01	38.00	1535
与广义积分变换有关的分数次演算:对分数次演算的研究(英文)	2023—01	48.00	1536
强子的芬斯勒几何和吕拉几何(宇宙学方面):强子结构的芬斯勒几何和吕拉几何(拓扑缺陷)(英文)	2022—08	38.00	1537
一种基于混沌的非线性最优化问题:作业调度问题(英文)	2023—03	38.00	1538
广义概率论发展前景:关于趣味数学与置信函数实际应用的一些原创观点(英文)	2023—03	48.00	1539
纽结与物理学:第二版(英文)	2022—09	118.00	1547
正交多项式和q—级数的前沿(英文)	2022—09	98.00	1548
算子理论问题集(英文)	2022—09	108.00	1549
抽象代数:群、环与域的应用导论:第二版(英文)	2023—01	98.00	1550
菲尔兹奖得主演讲集:第三版(英文)	2023—01	138.00	1551
多元实函数教程(英文)	2022—09	118.00	1552
球面空间形式群的几何学:第二版(英文)	2022—09	98.00	1566
对称群的表示论(英文)	2023—01	98.00	1585
纽结理论:第二版(英文)	2023—01	88.00	1586
拟群理论的基础与应用(英文)	2023—01	88.00	1587
组合学:第二版(英文)	2023—01	98.00	1588
加性组合学:研究问题手册(英文)	2023—01	68.00	1589
扭曲、平铺与镶嵌:几何折纸中的数学方法(英文)	2023—01	98.00	1590
离散与计算几何手册:第三版(英文)	2023—01	248.00	1591
离散与组合数学手册:第二版(英文)	2023—01	248.00	1592
分析学教程.第1卷,一元实变量函数的微积分分析学介绍(英文)	2023—01	118.00	1595
分析学教程.第2卷,多元函数的微分和积分,向量微积分(英文)	2023—01	118.00	1596
分析学教程.第3卷,测度与积分理论,复变量的复值函数(英文)	2023—01	118.00	1597
分析学教程.第4卷,傅里叶分析,常微分方程,变分法(英文)	2023—01	118.00	1598